Applied Hydrogeology of Fractured Rocks

Verily, all this is water. All the created beings are water. The vital breaths in the body are water. Quadrupeds are water. Edible crops are water. Ambrosia is water. The luminaries are water. Vedic formulae are water. Truth is water. All deities are water.

from **Mahanarayanopanishad, Section 29(1)** (an ancient Indian scripture originally in Sanskrit)

Contents

Preface

In recent years, particular attention has been focused on the hydrogeology of low permeability rocks as is evidenced by a large number of seminars, symposia and workshops held on the above theme, throughout the world. The hydrogeological aspects of such rocks have attracted greater attention of scientists and engineers, both as a source of water supply, mainly in developing countries, and also as potential repositories for the safe disposal of high-level radioactive waste, particularly in Europe and North America. Development of geothermal resources is another area of interest.

White teaching graduate and post-graduate students of Earth Sciences and Hydrology at the University of Roorkee, Roorkee, and in several other specialist courses, we realized the need of a textbook devoted to fractured rock hydrogeology – to emphasize various aspects of exploration, development, water quality, contamination and assessment, including the application of newer tools of remote sensing and geographical information systems etc. to the problem. With this in view, we have endeavoured to write this book, which we hope will be useful to all earth scientists and engineers engaged in the field of fractured rock hydrogeology.

Scientific tools and methods of study in fractured rock hydrogeology include a number of aspects, viz. structural mapping, remote sensing, geophysical exploration, geographical information systems, field and laboratory hydraulic testing, including drilling, pumping tests, modelling, and assessment etc. Each of these is a topic in itself, such that separate books are available on individual topics. We have, however, endeavoured to strike a balance so that the reading material is suitable for a graduate/post-graduate level study.

Parts of the manuscript were reviewed by a number of fellow colleagues – A.K. Bhar, S. Balakrishnan, D. Kashyap, G.C. Mishra, B. Prakash, A. Prakash, G. Ramaswamy, R.G.S. Sastry, D.C. Singhal and B.S. Sukhija. We are greatly obliged to them for their help and comments in arriving at the final presentation. We are specially indebted to C.P. Gupta who has contributed section 17.7 of the book. The financial support to one of us (BBSS) received from Council of Scientific and Industrial Research, Government of India, during 1993–94 and from Association of Geoscientists for International Development (1994–97) is gratefully acknowledged. We are grateful to the University of Roorkee, for extending the facilities. We also appreciate the assistance provided by Yash Pal and N.K. Varshney in typing and drafting of the work.

Finally, we are also indebted to our families for enduring four years of our pre-occupation with this book.

B.B.S. Singhal
R.P. Gupta
Roorkee

Chapter one

Introduction and basic concepts

1.1 Need for and scope of the book

More than half of the surface area of the continents is covered with hard rocks of low permeability. These rocks may acquire moderate to good permeability on account of fracturing and hence are broadly grouped under the term fractured rocks, in the context of hydrogeology.

The importance of systematic hydrogeological studies in fractured rocks was realized by the international community about thirty years ago when an international symposium was held at Dubrovnik in the former Yugoslavia (UNESCO/IAHS, 1967). Such studies have gained greater importance in recent years as is evidenced by several special publications and international seminars and workshops sponsored by IAH, IAHS, IGC and other organizations worldwide (e.g. Wright and Burgess, 1992; Sheila and Banks, 1993; Anon, 1996).

In several developing countries in Asia, Africa and Latin America, greater emphasis is being given these days to supplying safe drinking water to the vast population living in hard rock terrains. This has necessitated evolving efficient and economic methods of groundwater exploration, assessment and development. In this context, integrated geological, remote sensing and geophysical methods have proved very successful. Further, disposal of solid and liquid wastes is an important problem in both developing and developed countries which has threatened water quality. In this context, low permeability rocks hold greater promise. Therefore, special attention is required to locate suitable sites for the safe disposal of such wastes, especially high level waste (HLW).

Thus, studies on the hydrogeology of fractured rocks are being pursued for different purposes and with widely differing objectives, such as:

1. Development of safe groundwater supplies for domestic and irrigation purposes.
2. Contaminant migration studies, in order to estimate the movement of pollutants through fractures etc.
3. Tapping of geothermal resources involving estimation of extractable amount of hot fluids from the natural geothermal gradients.
4. Development of petroleum and gas reservoirs.
5. Construction of underground rock cavities for storing water, oil and gas etc. and underground passages such as tunnels.
6. In several other geotechnical problems, e.g. stability of rock slopes and seepage from dams.

All the above studies require a clear understanding and proper description of the hydrogeological characteristics of fracture systems. Without such data, it is difficult to determine the velocity distributions in fractured rocks and hence the main pathways for the movement of fluids and contaminants including attenuation of solutes by dispersion and diffusion etc. In this work, although the main emphasis is on fractured rocks as a source of water supply, the other aspects such as water quality,

TABLE 1.1 Comparison of granular and fractured-rock aquifers (modified after Anon, 1995)

Aquifer characteristics	*Aquifer type*	
	Granular rock	*Fractured rock*
Effective porosity	Mostly primary	Mostly secondary, through joints, fractures etc.
Isotropy	More isotropic	Mostly anisotropic
Homogeneity	More homogeneous	Less homogeneous
Flow	Laminar	Possibly rapid and turbulent
Flow predictions	Darcy's law usually applies	Darcy's law may not apply; cubic law applicable
Recharge	Dispersed	Primarily dispersed, with some point recharge
Temporal head variation	Minimal variation	Moderate variation
Temporal water chemistry variation	Minimal variation	Greater variation

contaminant transport and development of geothermal resources are also included.

Earlier, the principles of groundwater occurrence and movement were studied mainly for the homogeneous porous medium, due to simplicity and widespread groundwater development in such formations. The characteristics of fluid flow in fractured rocks differ from those of the homogeneous medium, primarily due to their heterogeneity (Table 1.1). Therefore, the principle of flow mechanism and solute transport as developed for the homogeneous medium is applied to the fractured medium with appropriate modifications. These are highlighted in this text.

1.2 Hydrological cycle

The hydrological cycle can be simply defined as the circulation of water between ocean, atmosphere and land (Figure. 1.1). It is best to describe the start of the cycle from the oceans as they represent vast reservoirs of water although there is no definite start or end-point

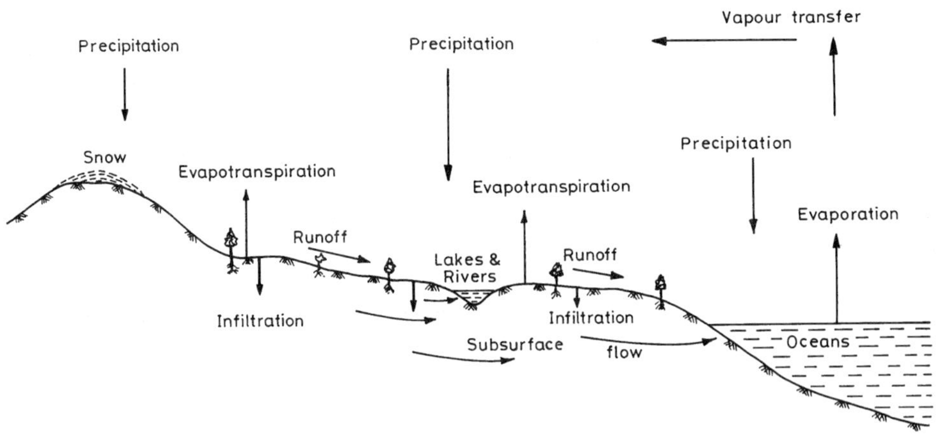

FIGURE 1.1 Schematic diagram of the hydrological cycle.

of the cycle. Water evaporates from the ocean surface and goes to the atmosphere where in vapour form it undergoes circulation depending upon the distribution of temperature and wind velocity. Under suitable atmospheric conditions, the water vapour condenses resulting in precipitation in the form of rainfall or snow.

The main interest of hydrologists is on the distribution of precipitation in time and space and its subsequent disposal after reaching the land surface. Some of the water which reaches the land surface drains as surface flow or run-off, some seeps down into the ground by infiltration. Part of this infiltration flows down to be added to a groundwater reservoir while part of the moisture from shallow depths returns to the atmosphere by the process of evapotranspiration. A part of the water, which is added to the groundwater reservoir, flows laterally under its hydraulic gradient and is discharged to the rivers, lakes or oceans as base flow from where part of it may return to the atmosphere by evaporation. This will form the longest cycle and it is of interest to a hydrogeologist. A part of the precipitation, before falling on the land surface, may return to the atmosphere by evaporation or interception by vegetation, thereby representing the shortest duration of the hydrological cycle.

An estimate of the world's water balance in terms of various components of the hydrological cycle is attempted by several workers, e.g. Berner and Berner (1987). They have computed that for the earth as a whole, precipitation is of the order of 0.496×10^6 km^3/year, out of which 0.110×10^6 km^3 falls on the continents and the remaining 0.386×10^6 km^3 on the oceans. The subsurface and surface run-off to the oceans is about 37 000 km^3 and the remaining 73 000 km^3 is lost by evapotranspiration from the continents. The evaporation exceeds precipitation over the oceans, the difference is made up by run-off from the continents.

The distribution of water between land and sea has varied with time. For example, during the last Pleistocene glaciation, about 18 000 years ago, sea level was lowered by around 130 m due to a transfer of about 47×10^6 km^3 of water (equal to about 3.5% of the oceanic volume) from the oceans to the land (Berner and Berner, 1987).

Recent changes in atmospheric temperature due to various human activities, particularly the burning of fossil fuels and deforestation, are also bound to affect the hydrological cycle. Records show that due to continuous increase in the atmospheric concentration of CO_2 and other greenhouse gases worldwide, the Earth's surface temperature may rise by about 2 °C by 2050 and more than 3 °C by 2100 (Dingman, 1994) which is likely to disturb the hydrological balance. This will require major adjustments in the demand and supply of water at different locations.

The various components of the hydrological cycle are described briefly in the following paragraphs.

Precipitation

The major types of precipitation are drizzle, rain, snow and hail. Atmospheric precipitation is a result of condensation of water vapour around hygroscopic nuclei in the atmosphere. Rainfall is measured with the help of rain gauges and is expressed in terms of depth of water in millimetres or centimetres. In the case of snow, generally the water equivalent is used as a measure of the precipitation; roughly 10 to 12 mm depth of snow equals 1 mm depth of water. Precipitation intensity, i.e. precipitation per unit of time, influences the recharge–run-off relation as rainfall of moderate intensity will be effective for groundwater recharge rather than short spells of high intensity. The latter will cause greater run-off leading to floods.

The areal distribution of rainfall in a basin for a given storm or period is expressed in terms of isohyets which are drawn by joining points of equal precipitation in an area. The average depth of precipitation over a basin can be computed by (a) arithmetic mean method, (b) isohyetal method, and (c) Thiesen polygon method. The reader may refer to any textbook on hydrology for the details of these methods, (e.g. Dingman, 1994). The isohyetal method is more accurate as it takes into consideration the influence of topography on precipitation.

Evapotranspiration (ET)

Evapotranspiration consists of two components – evaporation and transpiration. Evaporation is the loss of water to the atmosphere from water and land surfaces due to the vapour pressure gradient between the evaporating surface and the air; solar radiation is the principal energy source for evaporation. Evaporation rate is also affected by wind velocity as it brings unsaturated air to the evaporation surface. Evaporation from surface water bodies in arid and semi-arid regions may exceed the rainfall. Evaporation from the soil surface depends on depth to the water-table and type of soil (Figure 1.2); it will be equal to evaporation from a surface water body if the soil surface remains saturated with water.

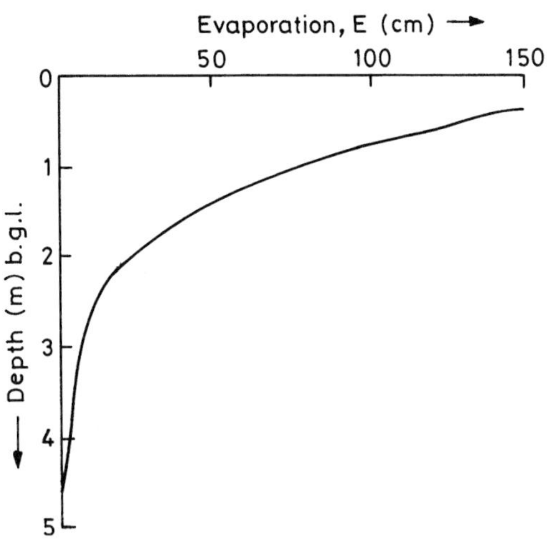

FIGURE 1.2 Relation between depth of water-table and rate of evaporation based on data from an experimental site in Shangqiu, PR China. The aeration zone is composed of clayey sand (after Chen and Cai, 1995).

Transpiration is the loss of water from vegetation when the vapour pressure in the air is less than that in the leaf cells. The amount of transpiration depends on the type, density and size of plants. Phreatophytes cause heavy loss of water due to transpiration while xerophytes, which have a shallow root system, cause minimal transpiration loss.

In field conditions, it is difficult to separate evaporation from transpiration. The total loss of water from the soil surface and plants is termed evapotranspiration (ET). It is also known as 'water loss' as this element of water is not available for the water supply. The terms 'evapotranspiration' and 'consumptive use' are frequently used interchangeably by agricultural scientists as they indicate the amount of water required to raise plants. Potential evapotranspiration (PET), a term introduced by Thornthwaite (1948), is the maximum amount of water that will be removed by evapotranspiration if the soil has sufficient water to meet the demand. The actual evapotranspiration (AET) is the amount of water evapotranspired under the existing field conditions. AET is usually less than PET being about 50 to 90% of the potential value. However, when soil is saturated with water as in waterlogged or swampy areas, the AET may equal the PET value. Losses due to evapotranspiration depend on the depth of the water-table and type of vegetation (Figure 1.3).

Evaporation from surface water bodies is measured by evaporation pans. Lysimeters are used for estimating ET losses by performing experiments under various soil moisture and vegetative cover conditions. Empirical formulae are also developed for estimating ET losses from some readily available meteorological data. The commonly used formulae are summarized by Marsily (1986) and Dingman (1994).

Under shallow water-table conditions, the following simple relation (White's formula) can be used to estimate ET losses:

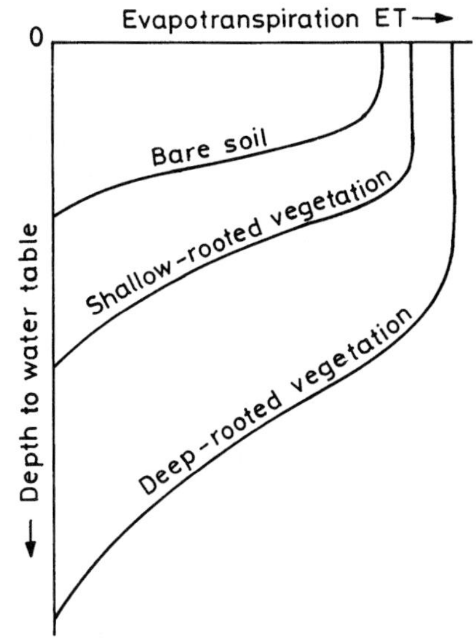

FIGURE 1.3 Relation between evapotranspiration and water-table depth for different terrain conditions (after Bouwer, 1975).

$$E = S_y (24h \pm s) \quad (1.1)$$

where E = daily rate of evapotranspiration, S_y = specific yield, h = water-table rise between midnight and 4 a.m., and s = rise or fall of the water-table during the 24 h period.

Infiltration

Infiltration is the process of absorption of water from rainfall or other surface water bodies into the soil. Infiltration is an important component of the hydrological cycle. It supplies water for the growth of plants, contributes to the base flow of streams and recharges the groundwater reservoirs. **Infiltration capacity**, f_p, is defined as the maximum rate at which water can penetrate into the soil in a given condition. **Infiltration rate**, f, is the actual rate at which water enters into the soil. The relation between f and f_p is $0 \le f \le f_p$. The actual rate of infiltration will be less than infiltration capacity unless the rainfall intensity is equal to or greater than the infiltration capacity. The rate of infiltration decreases exponentially with time due to swelling of clay particles and growth of algae. The maximum amount of water that soil can hold is called **field capacity**. Infiltration can be estimated either by using infiltrometers or by hydrograph analysis, and mainly depends on the properties of the soil/rock and vegetative cover.

Run-off

A part of the precipitation undergoes run-off which reaches the stream by several routes. Run-off from a basin consists mainly of four components, namely surface run-off, interflow, channel precipitation and groundwater (base) flow (Figure 1.4). The part of precipitation that filterates downward is added to the groundwater reservoir. A part of the infiltration moves laterally under the existing hydraulic gradient and is discharged into the stream with a greater time-lag as effluent seepage.

After a storm, on account of direct run-off, water-level in the stream rises as compared with the groundwater level, thereby causing a reverse flow from the stream into the groundwater

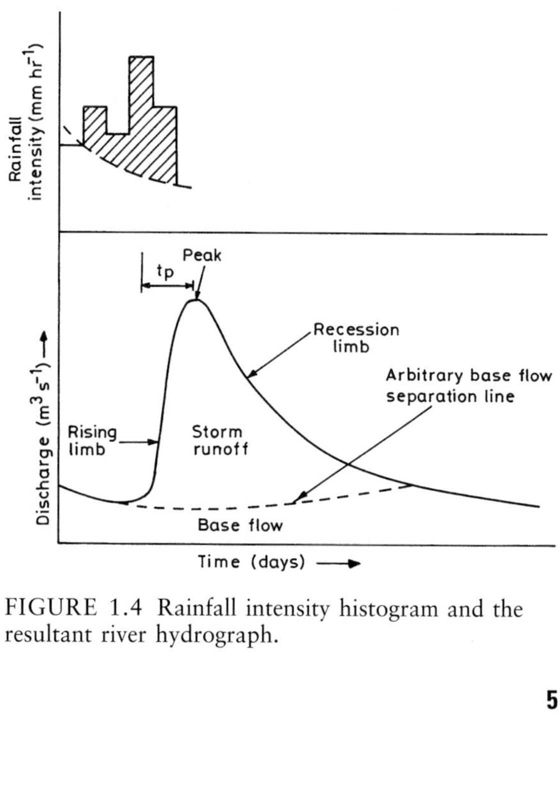

FIGURE 1.4 Rainfall intensity histogram and the resultant river hydrograph.

reservoir resulting in bank storage (Figure 1.5). As the stream stage falls, the water from the bank storage flows back to the stream causing a temporary rise in stream flow.

In actual practice, the total stream flow is divided into two parts – direct run-off and base flow. Direct run-off consists of overland flow (surface run-off), interflow and channel precipitation, whereas base flow is mainly from a groundwater source. In water management studies, it is important to estimate the groundwater component of a stream discharge, which can be achieved by the technique of hydrograph separation. In some areas the contribution from base flow is quite significant, for example in Britain, streams flowing over Cretaceous chalk and Tertiary sandstones receive about 75% of their total discharge from base flow annually (Ineson and Downing, 1964).

A typical stream hydrograph shows the variation in stream discharge with time as a result of a storm. A concentrated storm rainfall will result in a single-peaked hydrograph while multiple peaks are formed where there are a number of successive storms or a variation in rainfall intensity. A single-peaked hydrograph consists of three parts – rising limb, crest (peak) segment and falling or recession limb (Figure 1.4). The shape of the rising limb depends upon the character of the storm, i.e. duration and intensity distribution of the rainfall. The crest or peak of the hydrograph represents the maximum concentration of the run-off. It occurs after a certain time-lag with respect to rainfall intensity. The recession limb represents withdrawal of water from storage within a basin after all the inflow to the channel has stopped. It is mainly a function of geomorphological characteristics, i.e. topography, drainage pattern, soil and geology of the drainage basin.

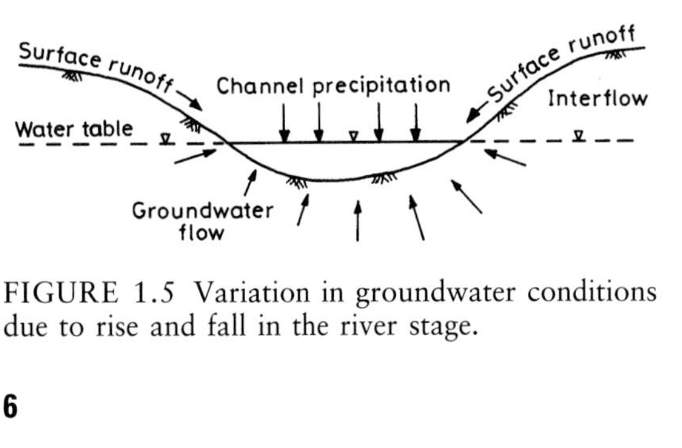

FIGURE 1.5 Variation in groundwater conditions due to rise and fall in the river stage.

The recession curve, also called a depletion curve, is described by a recession equation (1.2).

$$Q_t = Q_0 e^{-\alpha t} \tag{1.2}$$

where, Q_t is discharge at time, t, after the recession started, Q_0 is initial discharge at the start of the recession, α is recession constant dependent on the basin characteristics, and t is time since the recession began.

The procedure of separation of base flow (groundwater flow) is given in several books on hydrology (e.g. Linsley *et al.*, 1982; Dingman, 1994). There are several modifications of this approach; all these methods are of an arbitrary nature. The influence of hydrogeology in selecting a suitable method of hydrograph separation is discussed in UNESCO (1972). Groundwater flow component (Q_{GW}) can also be estimated from the chemical composition of stream water, groundwater and surface water using equation (1.3) (Dingman, 1994).

$$Q_{GW} = Q\left(\frac{C - C_{SW}}{C_{GW} - C_{SW}}\right) \tag{1.3}$$

where Q is stream flow, C is the concentration of chemical species in the stream water, and C_{SW} and C_{GW} are its concentration in surface water and groundwater respectively. For this purpose, a variety of dissolved constituents (Cl^-, SO_4^{2-}, HCO_3^-), environmental isotopes (^2H, ^3H and ^{18}O) and electrical conductivity have been used (Freeze and Cherry, 1979; Dingman, 1994).

In areas where field data are scarce or unavailable, remote sensing–GIS technology can be used to generate surrogate hydrogeological data (e.g. Meijerink *et al.*, 1994) – also see Chapter Four.

1.3 Classification of subsurface water

Water which occurs below the ground surface is termed as subsurface water, to distinguish it from surface water. A simple classification of subsurface water with respect to its depth of occurrence, and the extent to which it saturates the soil, is given in Figure 1.6. Depending on the degree of saturation, two depth zones can be broadly identified, i.e. the zone of aeration (vadose zone) and the zone of saturation (phreatic or groundwater zone). In the vadose zone, the intergranular space is only partly filled with water, the remaining space being occupied by air. Therefore in this zone, processes of oxidation and leaching are more prominent. The zone of saturation, as it is named, is saturated with water with the exclusion of air. This forms the zone of reduction and deposition of minerals. In an unconfined aquifer, the water-table represents the upper surface of the zone of saturation. As the water-table fluctuates depending on recharge and discharge conditions, the thickness of the two zones also changes seasonally. Being the main source of water supply to wells and springs, phreatic water or groundwater is of major interest to hydrogeologists.

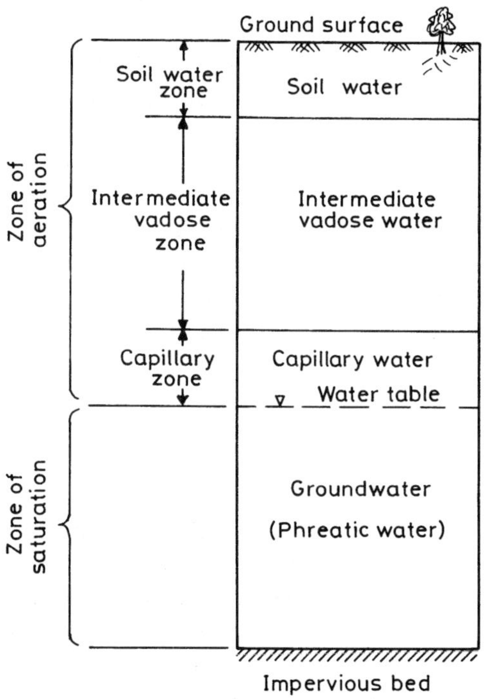

FIGURE 1.6 Classification of subsurface water.

The zone of aeration is divided into three zones from top to bottom – soil water zone, intermediate vadose zone and capillary zone; there are no sharp boundaries between the various zones. The soil water zone is of interest to agricultural scientists as it provides water for the growth of vegetation. Water in this zone occurs either as hygroscopic water which remains adsorbed or as thin film by surface tension. The moisture content in the soil water zone changes as a result of loss of water due to evapotranspiration, and therefore it shows diurnal variations.

Water in the intermediate zone is termed intermediate vadose water or suspended subsurface water as it is held due to intermolecular forces against the pull of gravity. The thickness of this zone may be zero when the water-table is close to the ground surface, or it may be even greater than 100 m under deep water-table conditions as in the arid regions.

The **capillary zone,** also termed **capillary fringe,** extends above the water-table up to the height of the capillary rise which depends mainly on the size of the intergranular openings. The height of the capillary rise, h_c, for water of specific weight (γ) 1 g cm^{-3} and at 20 °C is given by the relation

$$h_c = 0.15/r \qquad (1.4)$$

where r is the radius of the capillary opening which can be taken to be equivalent to pore size. The height of the capillary rise varies from about 2.5 cm in gravel to more than 100 cm in silt. As the pore sizes in natural material will vary within short distances, microscopically the upper surface of the capillary zone will be uneven.

The zone of aeration and zone of saturation are separated by a water-table or phreatic surface, which is under atmospheric pressure. The water-table may be either very close to the ground surface in areas of intensive recharge or may be several hundred metres deep in arid regions. Fluctuations in the water-table indicate changes in groundwater storage, either due to natural reasons or by man's activity. Therefore, the monitoring of water-table fluctuation is of importance for the management of water resources. We return to this subject in section 17.4.

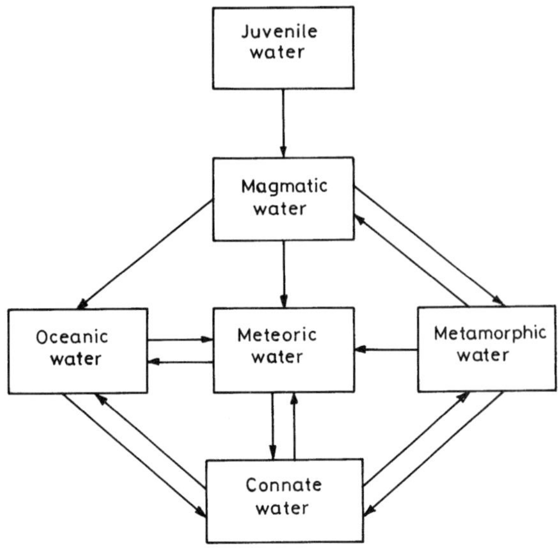

FIGURE 1.7 Relationships between different genetic types of water (after White, 1957, reproduced with permission of the Geological Society of America, Boulder, Colarado, USA).

1.4 Classification of water with respect to origin

The relationship between various genetic types of water is shown in Figure 1.7. Most of the water is of meteoric type as it is a result of atmospheric precipitation being a part of the present-day hydrologic cycle. It is the main source of water to wells and springs. The other types of water which are more of academic interest are connate water, juvenile or magmatic water and metamorphic water.

Juvenile water is also known as **new** water, as it is introduced into the hydrosphere for the first time. Magmatic water is mainly of juvenile origin derived from either deep-seated magma or may be of shallow volcanic origin.

Connate water is the remnant of ancient water retained in the aquifers and is not in hydraulic continuity with the present-day hydrological cycle. It is therefore also known as fossil water, although this term for water is a misnomer. It may be either of marine or freshwater origin. Connate water is commonly associated with oil and gas where it is usually of marine origin. In arid regions it represents the past pluvial climate, as in the Sahara desert.

Metamorphic water or rejuvenated water is the term used for water derived from hydrous minerals like clays, micas, etc., due to the process of metamorphism. It is more of academic interest rather than a source of water supply. The various genetic types of water can be distinguished, to some extent, on the basis of hydrochemical and isotopic data (White, 1957; Matthess, 1982).

1.5 Hydrological classification of geological formations

The occurrence and movement of groundwater depends on the geohydrological characteristics of the subsurface formations. These natural formations vary greatly in their lithology, texture and structure which influence their hydrological characteristics. The geological formations are accordingly classified into the following three types depending on their relative permeabilities:

1. Aquifer: Aquifer is a natural formation or a geological structure saturated with water which has good hydraulic conductivity to supply a reasonable quantity of water to a well or spring. Unconsolidated sedimentary formations like gravel and sand form excellent aquifers. Fractured igneous and metamorphic rocks and carbonate rocks with solution cavities also form good aquifers. Generally the hydraulic conductivity of an aquifer should be more than 10^{-6} ms^{-1}.
2. Aquitard: Aquitard is a formation having insufficient permeability to make it a source of water supply but allows interchange of groundwater in between adjacent aquifers due to vertical leakage. Therefore, aquitards serve as semi-confining layers. Examples are those of silt, shale and ***kankar*** (calcrete).
3. Aquiclude: This is a confining formation which is impermeable like unfractured crystalline rocks, clays and shales. In nature, truly impermeable formations are rare as every geological unit has some hydraulic conductivity.

1.5.1 Types of aquifers

The lateral continuity and vertical boundaries of aquifers are often not well defined. The aquifers may be of localized occurrence or may extend over distances of several hundred kilometres. Aquifers in the great Australian basin and in the Sahara desert (Nubian sandstones) have been traced over lateral distances of several hundred kilometres. Based on hydraulic characteristics of confining layers, aquifers are classified into the following types (Figure 1.8)

1. Confined aquifer

A confined aquifer, also known as an artesian aquifer, is overlain and underlain by a confining

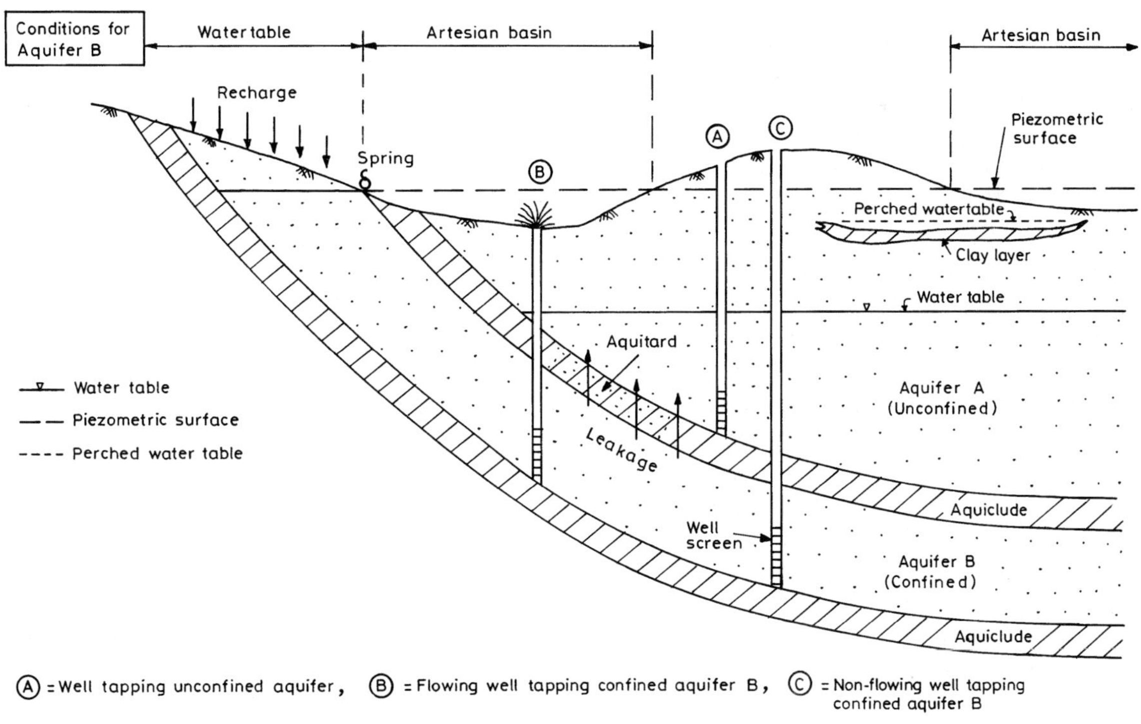

FIGURE 1.8 Types of aquifers.

layer. Water in a confined aquifer occurs under pressure which is greater than the atmospheric pressure. The piezometric (potentiometric) surface, which is an imaginary surface to which water will rise in wells tapping the confined aquifer, should be above the upper surface of the aquifer, i.e. above the base of the overlying confining layer. Confined aquifers are mainly recharged at the outcrops which form the intake areas. Groundwater movement in a confined aquifer is similar to a conduit flow. A confined aquifer may change to being an unconfined aquifer either with the time or space, depending on the position of the potentiometric surface which in turn depends on the recharge and discharge from the aquifer. Confined aquifers, with the potentiometric surface above the land surface, support flowing wells. In India, for example, flowing well conditions exist in alluvial aquifers of the ***Terai*** belt in the sub-Himalayan region and in Tertiary sandstone aquifers in South India (section 14.1).

Confined aquifers may be formed under the following three types of geological conditions:

(a) Stratiform multilayered formations. They occur either as gently dipping beds, monoclines or synclines. They usually form high pressure systems, viz. Dakota sandstones of the USA and Cuddalore sandstone in South India.
(b) Fractures and joints. In igneous and metamorphic rocks groundwater may occur under confined conditions in joints and fractures. In volcanic rocks, like Deccan basalts of India, vesicular horizons and interflow spaces may form confined aquifers in some places.
(c) Solution cavities. Groundwater in soluble rocks, like limestones, may also occur under confined conditions, viz. the Roswell basin in New Mexico, USA.

2. Leaky or semi-confined aquifer

In nature, truly confined aquifers are rare because the confining layers are not completely impervious. In leaky aquifers, aquitards form the semi-confining layers, through which vertical leakage takes place due to head differences across it (Figure 1.8).

3. Unconfined aquifer

An unconfined or phreatic aquifer is exposed to the surface without any intervening confining layer, but it is underlain by a confining layer. It is partially saturated with water; the upper surface of saturation is termed water-table which is under atmospheric pressure. It is recharged directly over the entire exposed surface of the aquifer. In unconfined aquifers, gravity drainage due to pumping is not instantaneous; they show delayed drainage which is more in clay and silty formations as compared to coarse-grained material.

4. Perched aquifer

The perched aquifer is a type of unconfined aquifer separated from the main regional aquifer by localized clay lens or any other impervious material in the zone of aeration. The thickness and lateral extent of the perched aquifer is controlled by the shape and size of the clay layer. Being of limited extent, perched aquifers are only a source of limited water supply. A perched aquifer can be distinguished from the main unconfined aquifer, by a sudden fall of water-level in the borehole during drilling, as it cuts across the underlying clay layer. Therefore the borehole will become dry until the regional water-table is reached.

5. Double porosity aquifer

A double or dual porosity aquifer, viz. fractured rocks consists of two parts – the matrix blocks and the fractures. The blocks have low permeability but high storativity while fractures have high permeability but lower storativity. The hydaulic behaviour of a double porosity aquifer is described in section 7.2.2.

The distinction between different types of aquifers is often difficult. It is necessary to have data about subsurface lithology, water-levels and hydraulic parameters of aquifer and confining layers to identify a particular type of aquifer. The response of different aquifer types to pumping is discussed in Chapter Nine.

Further, the distinction between unconfined and confined aquifers is also a matter of scale. An aquifer may locally appear to be of the confined type but on a regional scale the different aquifers may be interconnected forming an unconfined aquifer system.

1.6 Methods and stages of investigations

The exploration of groundwater resources should be done in stages as depicted schematically in Figure 1.9. In the first instance a search should be made for existing data available in published or unpublished reports of the federal and state departments of Geology and Water Resources, academic theses and dissertations. Additional information can be obtained from local landowners, construction firms and highway departments. The following information is of particular relevance in this context:

(a) Surface and subsurface data about the rock types and their attitude. In the case of hard rocks, data about depth to bedrock and structural aspects like folds, faults and other rock discontinuities is of more relevance;
(b) Geomorphological and drainage characteristics: satellite images, aerial photographs and topographic maps should be used for this purpose;
(c) Borehole logs about subsurface geology;
(d) Hydrometerological data on rainfall, snowfall, evapotranspiration, run-off etc.;
(e) Groundwater levels and their seasonal variation;
(f) Hydraulic characteristics, viz. hydraulic conductivity, transmissivity, storativity and well yield, etc.;
(g) Water quality;
(h) Groundwater withdrawals.

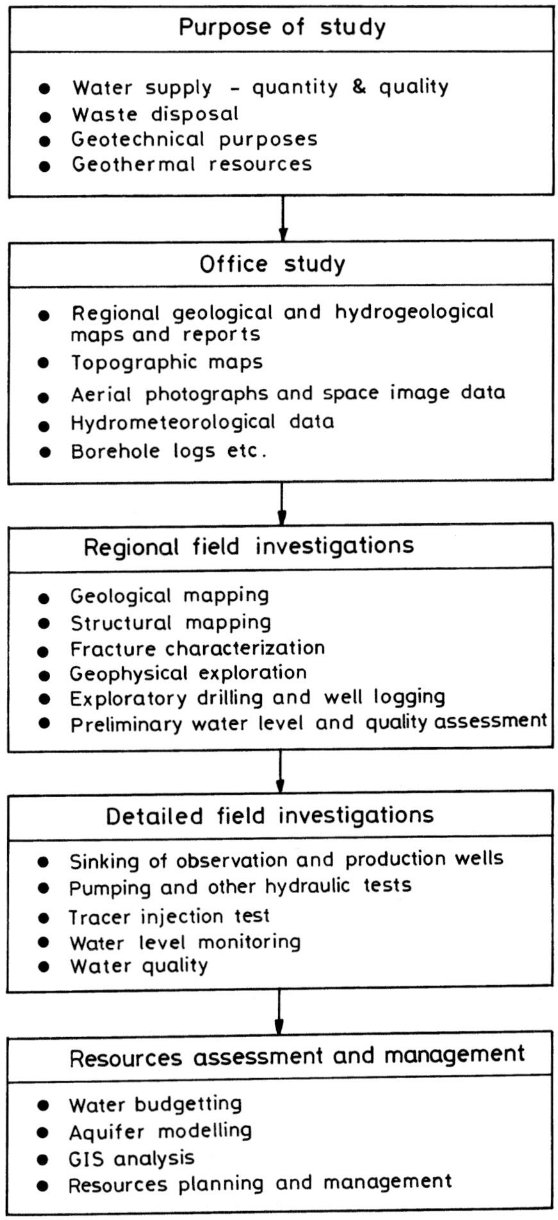

FIGURE 1.9 Steps in exploration, assessment and management of groundwater resources.

The existing data can be compiled and reinterpreted wherever necessary. In the absence of required information, dedicated investigations, depending on the nature of the problem, ought to be planned.

Chapter two

Fractures and discontinuities

2.1 Introduction

From the hydrogeological point of view, fractures and discontinuities are among the most important of geological structures. Most rocks possess fractures and other discontinuities (Figure 2.1) which facilitate storage and movement of fluids through them. On the other hand, some discontinuities, e.g. faults and dykes may also act as barriers to water flow. Porosity, permeability and groundwater flow characteristics of fractured rocks, particularly their quantitative aspects, are rather poorly understood. Main flow paths in fractured rocks are along joints, fractures, shear zones, faults and other discontinuities. There is a great need to understand hydraulic characteristics of such rocks, in view of (a) groundwater development to meet local needs, and (b) as depositories for nuclear and other toxic wastes.

There could be multiple discontinuities in fractured rocks along which groundwater flow takes place. A number of factors including stress,

FIGURE 2.1 Examples of fractured rocks: (a) Metamorphic rocks (meta-argillites) in Khetri Copper Belt, India. Several sets of fractures including shear planes are developed; some of the fractures possess infillings. (b) Sandstones of Vindhyan Group, India; bedding planes constitute the dominant discontinuity surfaces (Photograph (b) courtesy of A.K. Jindal).

temperature, roughness, fracture geometry and intersection etc. control the groundwater flow through fractures. For example, fracture aperture and flow rate are directly interrelated; non-parallelism of walls and wall roughness lead to friction losses; hydraulic conductivity through fractures is inversely related to normal stresses and depth, as normal stress tends to close the fractures and reduce the hydraulic conductivity.

It has also been noted that fracture permeability reduces with increasing temperature. As temperature increases with depth, thermal expansion in rocks takes place which leads to reduction in fracture aperture and corresponding decrease in permeability. Further the permeability is also affected by cementation, filling, age and weathering (see Chapter eight).

Parallel fractures impart a strong anisotropy to the rock mass; on the other hand, a greater number of more interconnected fractures tends to reduce anisotropy. Further, larger fracture lengths, greater fracture density and larger apertures increase hydraulic conductivity.

Therefore, summarily, for hydrogeological studies, it is extremely important to understand and describe the structure of the rock mass and quantify the pattern and nature of its discontinuities (van Golf-Racht, 1982; Sharp, 1993; Lee and Farmer, 1993; Marsily, 1986).

2.2 Discontinuities: types, genetic relations and significance

Discontinuity is a collective term used here to include joints, fractures, bedding planes, rock cleavage, foliation, shear zones, faults and other contacts. In this discussion using a genetic approach, discontinuities are grouped into the following categories:

(1) bedding plane
(2) foliation including cleavage
(3) fractures (joints)
(4) faults and shear zones, and
(5) other geological discontinuties.

2.2.1 Bedding plane

Primary bedding and compositional layers in sedimentary rocks form the bedding plane. Usually it is the most significant discontinuity surface in all sedimentary rocks such as sandstones (Figure 2.1b), siltstones, shales etc., except in some massive sandstones or limestones. The bedding plane can be readily identified in the field owing to mineralogical – compositional – textural layering. The bedding plane, being the most important discontinuity, imparts anisotropy and has a profound influence on groundwater flow in the vadose zone. The groundwater flow is by-and-large downdip (Figure 2.2).

Folds are flexures in rocks formed due to warping of rocks. Although a wide variety of folds are distinguishable, the two basic types are anticlines (limbs dipping away from each other) and synclines (limbs dipping towards each other). Folding leads to change or reversal in dip directions of beds, and this affects groundwater flow. Further, folding is accompanied by fracturing of rocks. In an anticline, the crest undergoes higher tensional stresses and hence develops open tensile fractures, which may constitute better sites for groundwater development.

2.2.2 Foliation

Foliation is a property of rocks, whereby they break along approximately parallel surfaces. The term is restricted to the planes of secondary origin occurring in metamorphic rocks. Foliation develops due to parallel-planar alignment of platy mineral grains at right angles to the direction of stress, which imparts fissility. The parallel alignment takes place as a result of recrystallization during regional dynamothermal metamorphism, a widespread and common phenomenon in crystalline rocks. **Rock cleavage** is almost a synonymous term. It is also used for planes of secondary origin along which the rock

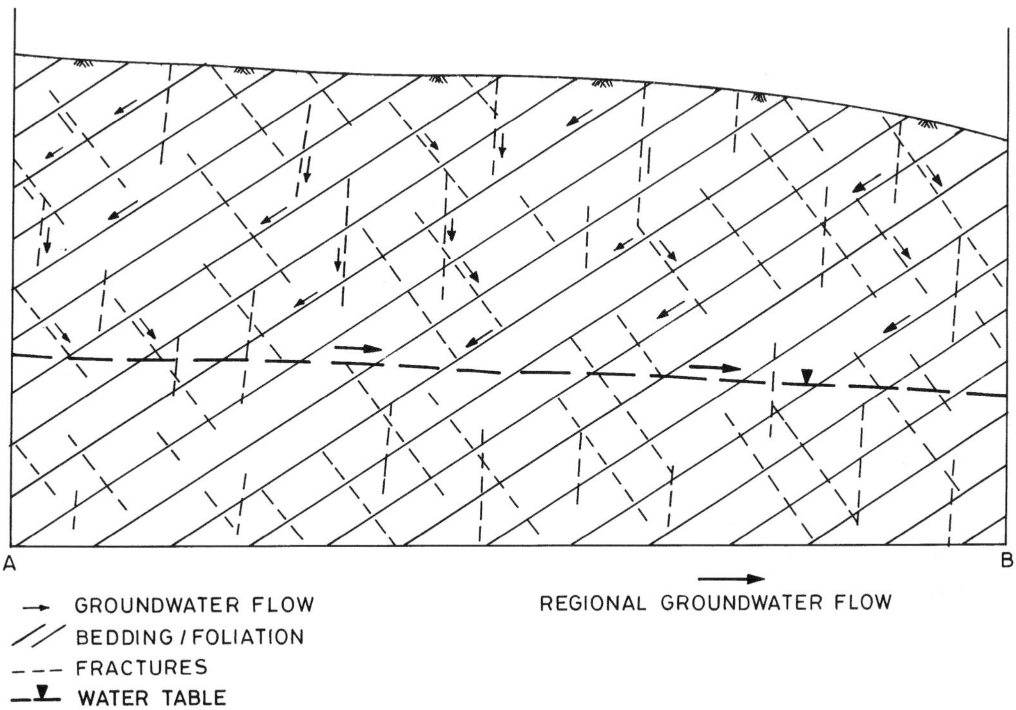

FIGURE 2.2 Schematic diagram showing the role of bedding planes and fractures on groundwater movement in the vadose zone.

has a tendency to break in near-parallel surfaces. Some terms are used for specific metamorphic rocks; thus the term **slaty cleavage** is used for rock cleavage in slates, **schistosity** is used for schists and **gneissosity** for gneisses. Foliation planes may or may not be parallel to bedding. Foliation that is parallel to the bedding is often referred to as bedding foliation. Fracture cleavage is produced by closely spaced jointing. In many schistose rocks, shear cleavages are developed due to closely spaced shear-slip planes, known as slip-cleavage. In a folded region, the foliation often developed parallel to the axial plane of folds is called axial plane foliation.

Foliation in metamorphic rocks has a profound influence on groundwater movement, possessing almost the same role as bedding in the sedimentary rocks, both being the most significant discontinuities in the respective rock categories (e.g. Figure 2.2).

2.2.3 Fractures and joints

1. Introduction and terminology

Fractures, also called joints, are planes along which stress has caused partial loss of cohesion in the rock. It is a relatively smooth planar surface representing a plane of weakness (discontinuity) in the rock. Conventionally, a fracture or joint is defined as a plane where there is hardly any visible movement parallel to the surface of the fracture; otherwise, it is classified as a fault. In practice, however, a precise distinction may be difficult, as at times within one set of fractures, some planes may show a little displacement whereas others may not exhibit any movement. Slight movement at right angles to the fracture surface will produce an open fracture, which may remain unfilled or subsequently may get filled by secondary minerals or rock fragments.

Fracture-discontinuities are classified and described in several ways using a variety of nomenclature, such as: joints, fracture, fault, shear, gash, fissure, vein. Generally, the term fracture is used synonymously with joint, implying a planar crack or break in the rock without any displacement. The terms fault and shear are used for failure planes exhibiting displacement, parallel to the fracture surfaces. Gash is a small-scale open tension fracture that occurs at an angle to a fault. Fissure is a more extensive open tensile fracture. A filled fissure is called a vein. 'Fracture zones' are zones of closely spaced and highly interconnected discrete fractures. They may be quite extensive (length > several kilometres) and may even vary laterally in hydraulic properties.

An individual fracture has limited spatial extent and is discontinuous in its own plane (Figure 2.3). On any outcrop, fractures have certain trace lengths and fracture spacings. By mutual intersection, the various fracture sets may form an interconnected continuous network, provided that the lengths of the joints in the different sets are much greater than the spacings between them (Figure 2.18). The interconnectivity of fractures leads to greater hydraulic conductivity.

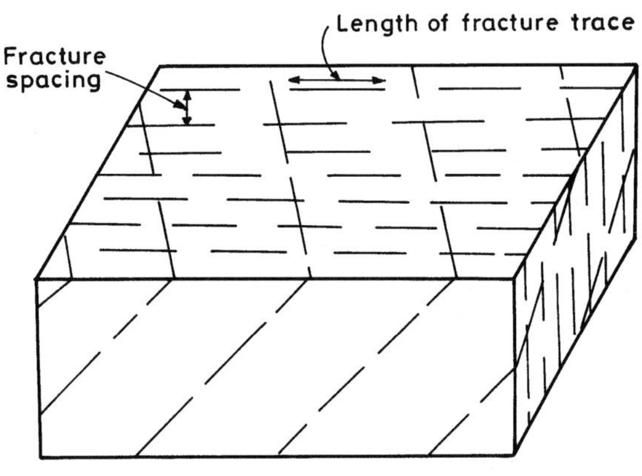

FIGURE 2.3 Two sets of fractures are schematically shown in the block. An individual fracture has limited spatial extent and is discontinuous in its own plane. Fracture spacing and fracture trace length are indicated for one set.

2. Causes of fracturing

Although fractures are extremely common and widespread in rocks, geologically they are still not studied well enough (Price and Cosgrove, 1990). Complex processes are believed to be involved in the origin of fractures, which are related to the geological history of the area. Fractures are created by stresses which may have diverse origin, such as: (a) tectonic stresses related with deformation of the rocks, (b) residual stresses due to events that happened long before the fracturing, (c) contraction due to shrinkage because of cooling of magma or dessication of sediments, (d) surficial movements such as landslides or movement of glaciers, (e) erosional unloading of deep-seated rocks, and (f) weathering, in which dilation may lead to irregular extension cracks and dissolution may cause widening of cavities, cracks etc.

3. Types of fractures

Firstly, fractures may be identified into two broad types (a) systematic, which are planar, and more regular in distribution, and (b) non-systematic, which are irregular and curved (Figure 2.4). The non-systematic fractures meet but do not cross other fractures and joints, are curved in plan and terminate at the bedding surface. They are minor features of dilational type and develop in the weathering zone. Curvilinear pattern is their genetic characteristic. Systematic fractures that are developed in parallel are treated as a set of fractures.

Geometric classification – Considering geometric relationship with bedding/foliation, the systematic fractures or joints are classified into several types. Strike joints are those that strike parallel to the strike of the bedding/foliation of the rock. In dip joints, the strike direction of joints runs parallel to the dip direction of the rock. Oblique or diagonal joints strike at an angle to the strike of the rocks. Bedding joints

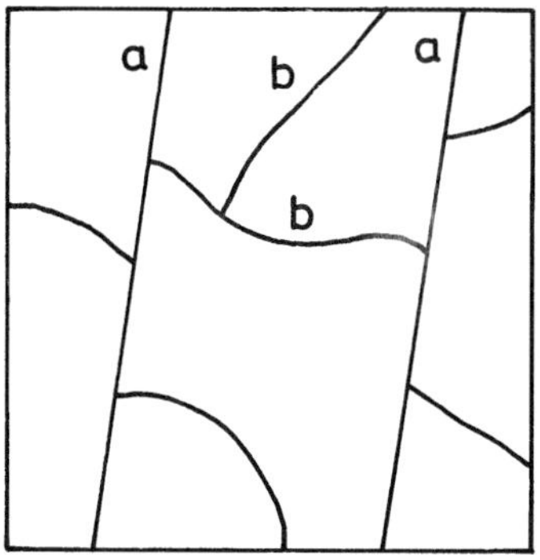

FIGURE 2.4 Systematic and non-systematic types of fractures.

are essentially parallel to the bedding plane of the associated sedimentary rock.

Depending upon their extent of development, fractures may be classified into two types: first-order and second-order. First-order fractures cut through several layers of rocks; second-order fractures are limited to a single rock layer. Further, depending upon the strike trend of fractures with respect to the regional fold axis, fractures are designated as longitudinal (parallel), transverse (perpendicular) or oblique (Figure 2.8). **Genetic classification** – Genetically, systematic fractures can be classified into three types:

a) Shear fractures, which may (or may not) exhibit shear displacement and are cogenetically developed in conjugate sets with a dihedral angle $2\theta > 45°$.
b) Dilational fractures, which are of tensile origin, commonly developed perpendicular to the bedding plane, and are open fractures with no evidence of shear movement.
c) Hybrid fractures, which exhibit features of both shear and dilational origin. They may occur in conjugate sets with a dihedral angle $2\theta < 45°$. They are open (extension!), may be partly filled with veins, and may also exhibit some shear displacement.

The physical stress conditions under which these three types of fractures develop are illustrated by the Mohr diagram in Figure 2.5. The curve ABC is a Mohr envelope; the stress circles touching the Mohr envelope at A, B and C points indicate different failure conditions of the rock. In condition 'A', the principal maximum compressive stress is negative, i.e. extensional, and therefore it leads to a dilational failure. In condition 'C', a typical conjugate shear failure takes place, such that the dihedral angle $2\theta > 45°$. 'B' represents a condition where is a positive maximum principal compressive stress and a negative minimum principal compressive stress, i.e. the effective normal stress perpendicular to the fracture plane is negative (extensional). This can be attributed to high fluid pressure conditions at depth. Hence, there is a tendency for such shear fractures to open and also get filled with minerals. Typically, in such hybrid shear–extension fractures, the dihedral angle is $2\theta < 45°$.

Conjugate shear fractures developing at greater depths are of ductile nature and possess a large 2θ ($\approx 90°$). On the other hand, conjugate brittle shears develop at a shallower depth and possess a smaller 2θ ($\approx 60°$). Further, brittle deformation causes derivative shears of several orders to form successively deviating trends, which cause a spread in the trends of conjugate shears (Ruhland, 1973). The process of shearing is also accompanied by tensile deformation. Thus, brittle deformation may produce fractures of different magnitude and direction in successive orders. In a rock mass fractured by three orders of brittle deformation, tensile fracturing may spread over a range of about 75° and shear fracturing over a range of nearly 135° (Figure 2.6).

4. Discrimination between shear and extension fractures

The rheological principles indicate that there is no sharp categorization between extension and shear fractures. In fact, all gradations from one

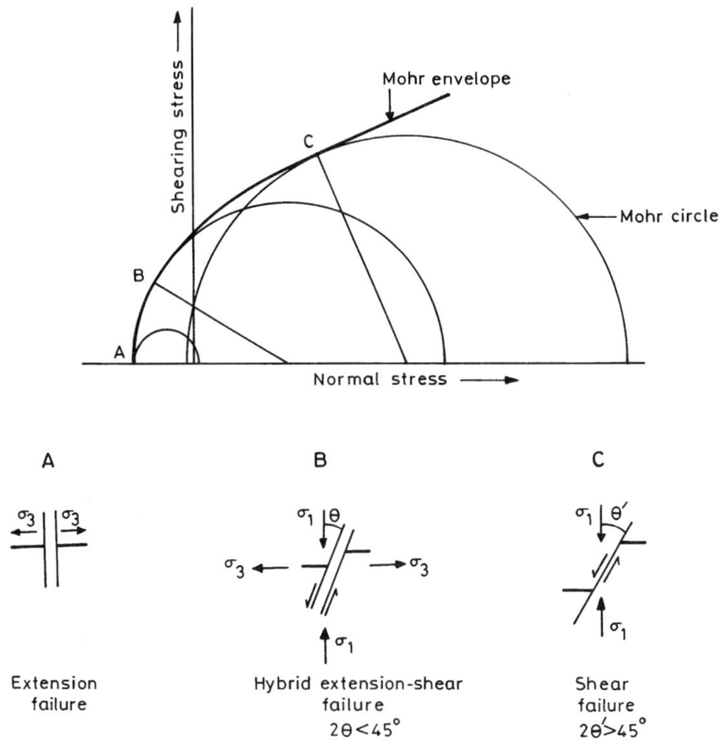

FIGURE 2.5 Basic genetic types of fractures: A = extension fracture; B = hybrid extension-shear fracture; C = shear fracture. The Mohr diagram indicates the stress conditions for these failures. σ_1 and σ_3 are the maximum and minimum principal compressive stresses respectively.

category to the other take place. However, from a hydrogeological point of view, it is important to distinguish between shear and extension joints as dilational joints are more open and possess greater hydraulic conductivity than shear joints. Discrimination between shear and tension joints may be difficult, particularly in complexly deformed areas. However, the following features may help in their discrimination:

(a) Shear joints may exhibit displacement parallel to the plane of the joints, which is absent in the case of extension joints.
(b) Shear joints commonly occur in conjugate sets which may be indicated by a statistical analysis.
(c) In the field, slickensides and other criteria of relative movement may be observed in the case of shear joints.
(d) Generally, extension joints are open and shear joints are tight.
(e) The orientation of the joints with respect to the bedding foliation and/or fold axis can provide information on shear versus tensile origin of fractures, as shear joints occur in oblique conjugate sets whereas extension joints occur as longitudinal and transverse joints forming an orthogonal pair (Figure 2.7).
(f) The cumulative trend diagram of fractures may also provide information on the related stress field, and therefore the likely trends of shear and extension fractures; the maximum principal compressive stress bisects the dihedral angle of conjugate shear fractures and is parallel to the tensile fracture.

5. Orientation of fractures *vis-à-vis* regional structure

Ideally, in the case of a simple-dipping strata, four sets of fractures (systematic) develop (Figure 2.7). S_1 and S_2 form a conjugate set of

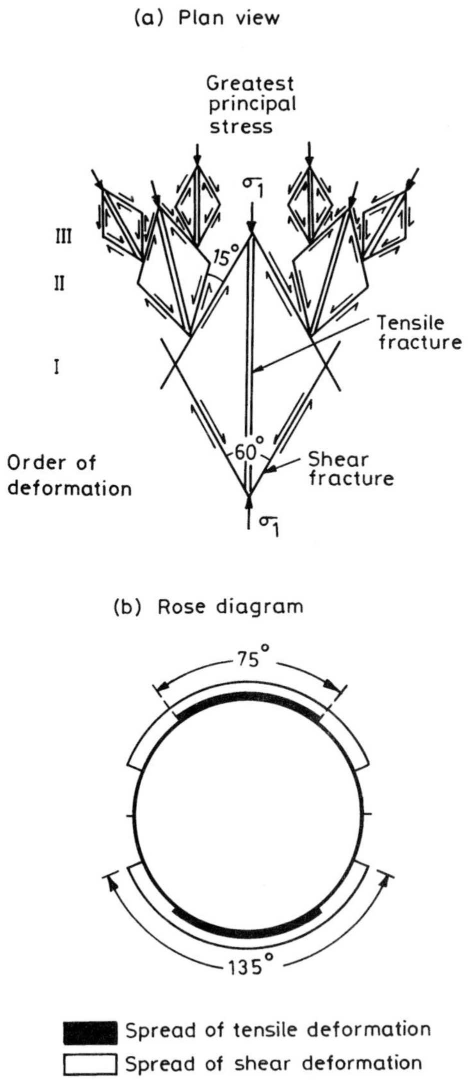

FIGURE 2.6 Scheme of brittle deformation of a homogeneous rock mass. The rose diagram illustrates the ranges of orientations of tensile and shear fractures due to three orders of deformation (after Ruhland, 1973).

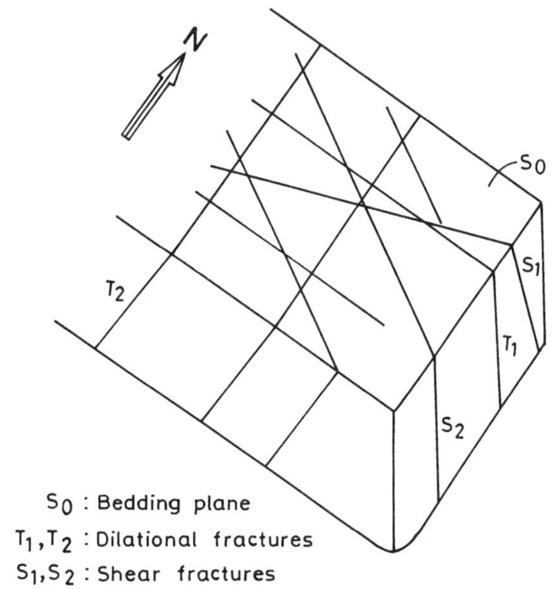

FIGURE 2.7 Development of four sets of fractures in the case of a simple dipping strata (see also Figure 2.15).

shear fractures and T_1 and T_2 are extension fractures. All these fractures are perpendicular to the bedding plane and contain the intermediate principal compressive stress σ_2.

Figure 2.8 shows a simplified ideal relationship between fractures and folds. σ_1 is the maximum principal compressive stress perpendicular to the fold axis (b). A conjugate set of oblique trending right-lateral and left-lateral shear fractures develops. There are two sets of extension fractures, one longitudinal and the other transverse to the fold axis, both being mutually orthogonal.

During folding, bending of a bed causes extension on the convex side and compression on the concave side (Figure 2.9). This results in extension fractures and normal faults on the crests of anticlines. Less commonly, thrust faults also develop in the inner areas of compression.

6. Other types of fractures

Sheeting joints – These joints are generally flat, somewhat curved and nearly parallel to the topographic surface, often developed in granitoid rocks. They are closely developed near to the surface, and their spacing increases with depth. They are generated due to release of overburden stress.

Columnar joints – Joints of this type are tension fractures generated due to shrinkage in

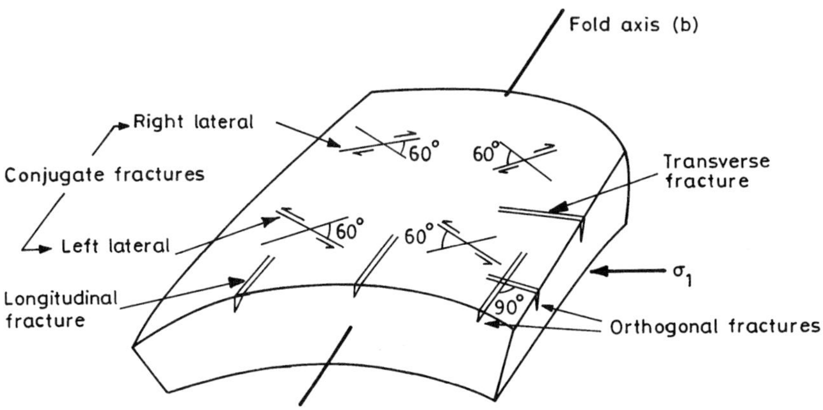

FIGURE 2.8 Ideal relationship between major joint sets in a folded bed. There are two sets of conjugate shear fractures and two sets of mutually orthogonal dilational fractures. All the fractures are shown vertical.

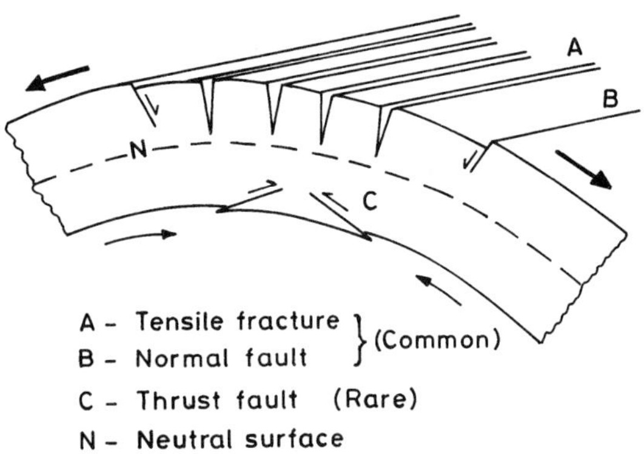

FIGURE 2.9 Development of extensional fractures and normal faults on the crest and the upper axial zone of an anticline. Thrust faults may also develop, occasionally, in the inner area of compression.

rocks. Shrinkage may occur due to cooling or dessication. Igneous rocks contract on cooling; mud and silt shrink because of dessication. As a result, polygonal and columnar joints develop. The columns are generally a few centimetres to a metre in diameter, and several metres long (high). Frequently, the columns are four-, five- or six-sided in shape (see Figure 12.5).

2.2.4 Faults and shear zones

Rupture and shear movement due to stresses leads to faulting. The stress in rocks is mostly a result of mountain building tectonic activity. From a hydrogeological point of view, faults and shear zones constitute very important types of discontinuities in rocks. Faults are planes and zones of rupture along which the opposite walls have moved past each other, parallel to the surface of rupture. The orientation of a fault plane is defined in terms of strike and dip, as that of any other plane in structural geology. Faults vary in dimension from a few millimetres long with minor displacement to several hundred kilometres in strike lengths with movement of several tens of kilometres.

1. Terminology

In describing faults, a range of terminology is used; only some of the more important terms are introduced here (e.g. Billings, 1972; Price and Cosgrove, 1990). The fault block above the dipping fault plane is called **hanging wall**; the block below the fault plane is called **footwall** (Figure 2.10). The angle which a fault plane makes with the vertical plane parallel to the strike of the fault is called **hade**; it is a complement of the dip. In many instances the displacement is distributed through a zone, called the fault zone, which may be a few centimetres to hundreds of metres wide. Faults may exhibit simple translational or rotational movements. Slip is the relative displacement as measured on the fault surface; strike-slip and dip-slip are the displacements along the strike direction and dip direction respectively on the fault plane. **Throw** is the vertical displacement caused by the fault and the blocks which have moved up and down are called **upthrow** and **downthrow blocks**, respectively.

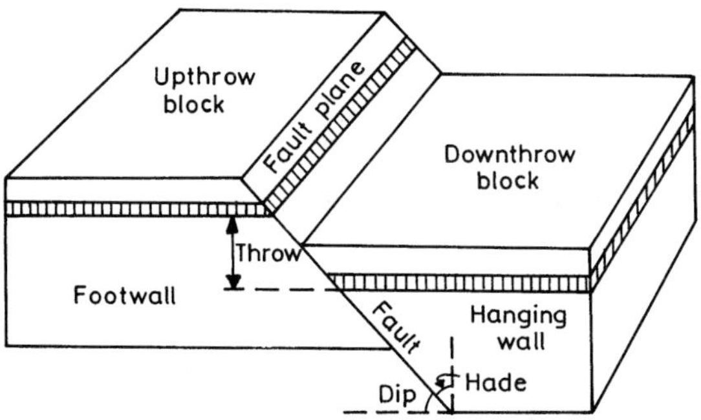

FIGURE 2.10 Common terms used in describing a fault.

Shear zones are generally filled with broken and crushed rocks, which may be embedded in clay matrix. Shear zones tend to be more extensive and continuous than joints.

2. Classification

Faults are classified in various ways in the literature. Some classifications are based on geometrical relations between the fault plane and country rocks. Genetically, faults are classified into three types: (a) normal, (b) reverse and (c) strike-slip. They are related to the stress conditions (Figure 2.11).

In the case of a normal fault, the hanging wall moves relatively downward; normal faults are generally high-angle faults caused when σ_1 is vertical. Reverse faults are generally low-angle (gently dipping) faults caused when σ_3 is vertical. It is characterized by the relative upward movement of the hanging wall. Strike faults are vertical faults marked by movement only in the strike direction of the fault. These are caused when σ_2 is vertical.

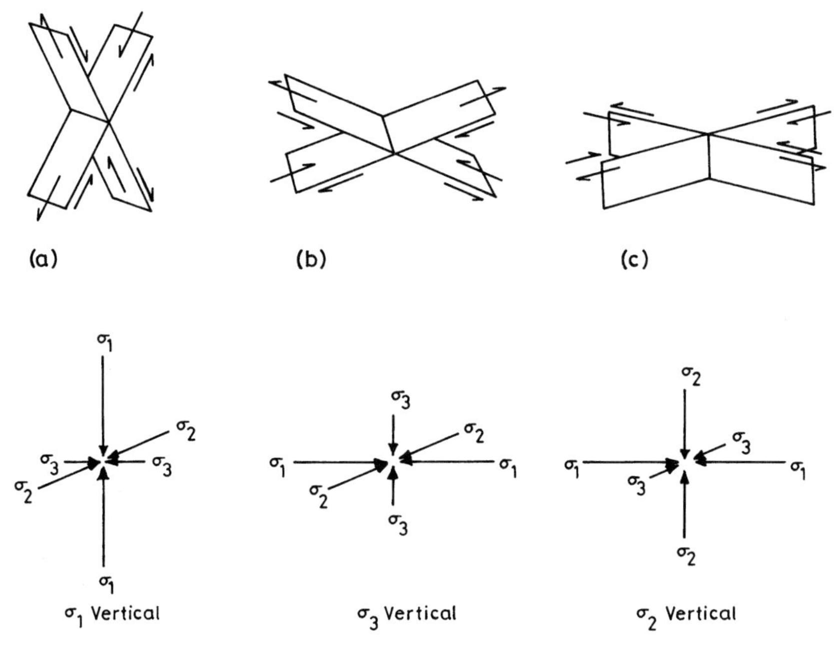

FIGURE 2.11 The basic genetic types of faults: (a) normal, (b) reverse and (c) strike-slip. Orientations of principal compressive stresses are also shown.

3. Recognition of faults in the field

A number of criteria are used to decipher the presence of faults, though in a specific case only some of the features may be present. Some of the more important criteria include: (a) Displacement of key beds. (b) Truncation of beds and structures. (c) Repetition and omission of strata. (d) Presence of features indicating movement on a fault surface such as slickensides, mylonite, breccia, gouge, grooving. (e) Evidence of mineralization or silicification along fault zones. (f) Physiographic features such as fault scarps, offset ridges, etc. (g) Alignment such as springs alignment, pond alignment, vegetation alignment, rectilinearity of a stream. (h) Indication of sudden anomalous changes in a river course, such as knick points, offset of streams, anomolous or closed meanders. (i) Erosional features such as triangular facets, unpaired terraces etc.

Investigations for faults may be made in outcrops, road cuttings, mines or other excavations, where smaller faults can often be directly observed. A larger fault, on the other hand, may be identified on stratigraphic and physiographic evidence, and particularly on remote sensing images, as only small segments of the fault may be exposed in the field, and the feature may be largely covered under soil, debris or vegetation (see section 4.8.11).

4. Effect of faults on groundwater regime

Faults may affect groundwater regime in numerous ways, some of the more important being the following:

1. It is well known that faults may have such effects as truncation, displacement, repetition or omission of beds. In this light, the distribution and occurrence of aquifers may be affected by faults as locally an aquifer unit may get displaced/truncated/omitted (Figure 2.12a,b).
2. A fault may bring impervious rock against an aquifer, which would affect groundwater flow and distribution (Figure 2.12a).
3. Truncation of an aquifer by a fault may lead to seepage and formation of a spring line along the fault (Figure 2.12c).
4. A fault may lead to a scarp; intensive erosion of the upthrow block and deposition of extensive piedmonts on the downthrow block may follow; the piedmont deposits may serve as good aquifers (Figure 2.12d).
5. An aquifer may get repeated in a borehole due to thrust faulting; further it may also get re-exposed on the surface for recharge (Figure 2.12e).
6. Vertical dykes, veins etc. which generally act as barriers to groundwater flow, may be breached by faults and this may produce local channelways across the barrier (Figure 2.12f).
7. A fault may lead to a groundwater cascade (Figure 2.12g).
8. Faults create linear zones of higher secondary porosity; these zones may act as preferred channels of groundwater flow, leading to recharge/discharge.
9. A fault may lead to inter-basinal subsurface flow.
10. A fault zone, when silicified, may act as a barrier for groundwater flow.

2.2.5 Other geological discontinuities

In addition to the above structural features, there could be other geological boundaries such as unconformities and intrusive contacts which may act as discontinuities.

Unconformity is a surface of erosion and nondeposition separating overlying younger strata from the underlying older rocks. Conglomerate beds and palaeosols usually occur along the unconformity surface which often form good aquifers. An unconformity implies that a hydrologeological unit may get laterally pinched out and spatially replaced by another unit (Figure 2.13).

Intrusive contacts are other geological bound-

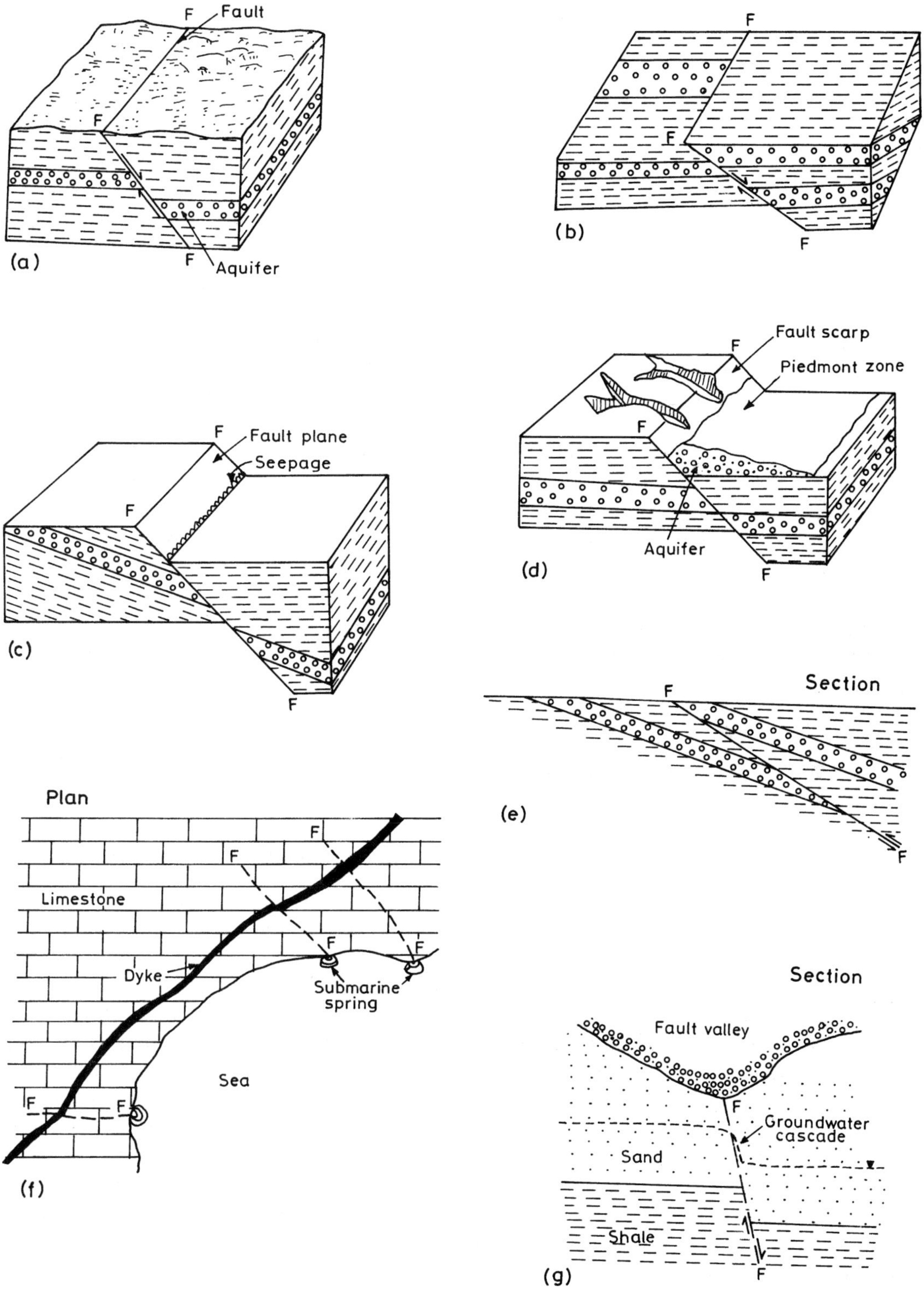

FIGURE 2.12 Effects of faults on aquifers (for details see text).

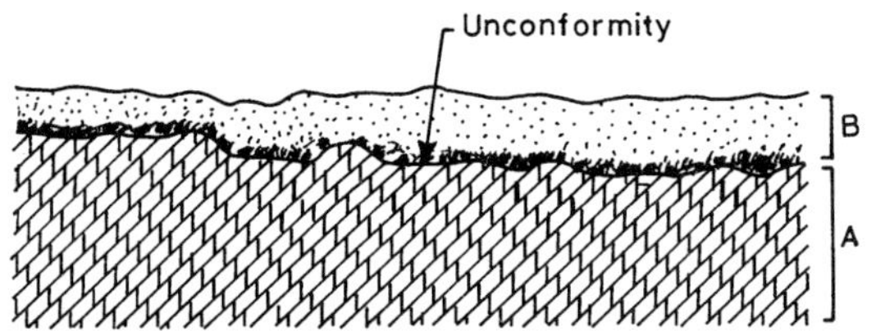

FIGURE 2.13 Development of an aquifer along an unconformity between two impervious beds A and B.

aries of significance in the context of hydrogeology. Intrusive bodies occur in a variety of shapes and sizes, such as batholiths, dykes, sills etc. Their relation with the host rocks could be concordant, transgressive, or discordant. The igneous plutonic bodies crystallize under high pressure and temperature; they are devoid of primary porosity. Therefore, hydrogeological characters of the host rocks and intrusive rocks may be vastly different from each other. Hence igneous contacts act as regional boundaries from a hydrogeological point of view.

2.3 Fracture characterization and measurements

A fractured rock mass can be considered to be made up of three basic components: a) fracture network, b) matrix block, and c) infillings along the fractures, if present (Figure 2.1). A single fracture or discontinuity plane is characterized by its orientation, nature (shear/tensile), persistence and aperture etc. Several fracture planes of the same type create a fracture set. They have a certain spacing (frequency). Several intersecting intercommunicating fracture sets create a fracture network which facilitates fluid flow. Thus, it is extremely important to characterize discontinuities and take their measurements, for a meaningful application. The various impor-

TABLE 2.1 Parameters for discontinuity characterization

S. No.	*Parameter*	*Description*
1.	Number of sets	Number of sets of discontinuities present in the network
2.	Orientation	Attitude of discontinuity in space defined in terms of dip amount and direction
3.	Spacing	Perpendicular distance between adjacent discontinuities of the same set
4.	Persistence	Trace length of the discontinuity seen in exposure
5.	Density	
	– linear	Number of fractures per unit length
	– areal	Cumulative length of fractures per unit area of exposure
	– volumetric	Cumulative fractured surface area per unit bulk rock volume
6.	Fracture area and shape	Area of fractured surface and its shape
7.	Volumetric fracture count	Number of fractures per cubic metre of rock volume
8.	Matrix block unit	Block size and shape resulting from the fracture network
9.	Connectivity	Intersection and termination characteristics of fractures
10.	Aperture	Perpendicular distance between the adjacent rock-walls of a discontinuity, the space being air or water-filled
11.	Asperity	Projections of the wall-rock along the discontinuity surface
12.	Wall coatings and infillings	Solid materials occurring as wall coatings and filling along the discontinuity surface

tant parameters are summarized in Table 2.1.

The concept of representative elementary volume (REV) is very important and may be introduced here. The REV is the minimum rock mass volume which has the hydraulic and/or mechanical properties similar to those of the rock mass. For mechanical properties, a sample size of a few cubic metres may be sufficient for approaching a REV; however, in the case of hydraulic flow, REV may be substantially larger, and in some cases, it may not exist due to strong anisotropy and variability of rock characters (section 8.4).

1. Number of sets

Several sets of discontinuities are often developed in a rock mass, three to four sets being most common. The number of sets of discontinuities in an exposure can be statistically determined by contouring the pole-plots (see Figure 2.15). Relevant data such as orientation, spacing, length, aperture etc. has to be collected for each set of discontinuity.

2. Orientation

Orientation is the parameter to define a single fracture plane in space, using angular relationships, as for any geological planar surface. It is defined in terms of dip direction (angle with respect to north) and dip amount (angle with horizontal). The orientation is expressed in terms of a pair of numbers, such as 25°/N 330°, implying a plane, dipping at 25° in the direction 330° measured clockwise from the north. In the field, inaccuracies often creep into the measurements, and therefore statistical analysis is desirable.

Rose diagram. This is a method of displaying the relative statistical prevalence of directional trends, e.g. strike direction of fractures, lineaments etc. It can be prepared for parameters like length or number, viz. number of joints direction-wise, or length of joints direction-wise. Frequently, the directions are grouped at 10° intervals. Frequency in a group interval is represented along the radial axis. The length of petals becomes a measure of relative dominance of the trend and the resulting strike petals have a mirror image about the centre of the rosette (Figure 2.14). Data on the amount of dip cannot be incorporated in the rosette, and may, however, be shown outside the circumference. **Histogram plot** is another way to represent the relative prevalence of trends.

Spherical projection. This is for representing the orientation of geological planar surfaces; the method of equal-area projection is frequently employed, as it accurately denotes the spatial distribution of data. Basic concepts on great-circle plots and π-pole plots to represent planes can be found in any standard text on structural geology (e.g. Billings, 1972; Price and Cosgrove, 1990). The pole plotting method has a relative advantage over the great-circle

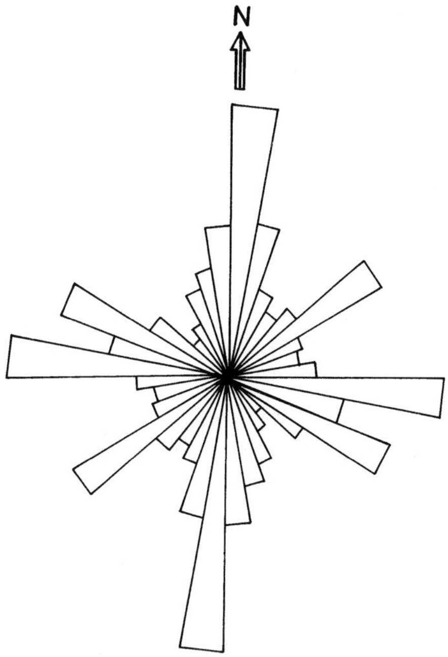

FIGURE 2.14 Rose diagram of strike trends of discontinuities showing their relative prevalence.

Sr. No.	Surface	Strike	Dip direction	Dip amount
1	S_0	N - S	E	45°
2	T_1	E - W	Vertical	—
3	T_2	N - S	W	45°
4	S_1	N 68°- N 248°	N 338°	80°
5	S_2	N 112°- N 292°	N 202°	80°

(a) Fracture orientation data

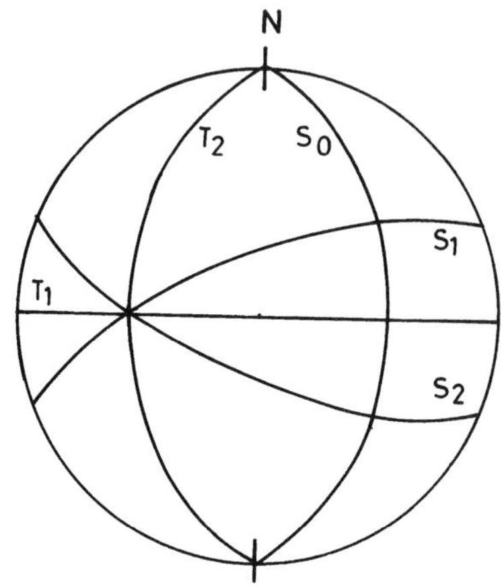

(b) Great circle diagram

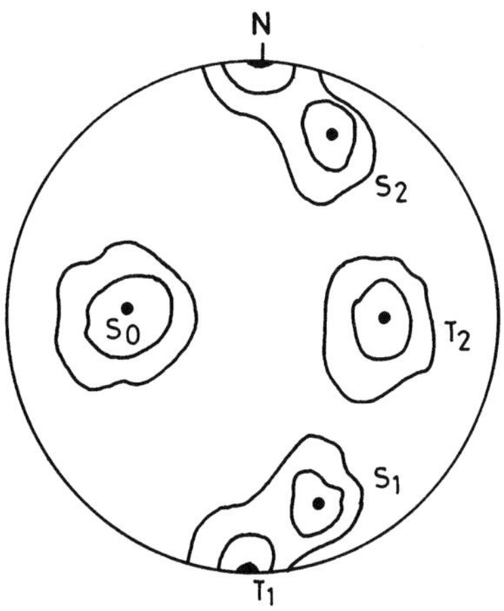

(c) π-Pole diagram

FIGURE 2.15 (a) Orientation data of discontinuities. S_0 is bedding plane, T_1 and T_2 are tensile fractures and S_1 & S_2 are shear fractures. Their great circle and π-pole diagrams are shown in Figures (b) and (c) respectively.

method in that clusters of poles and their relative concentrations can be readily ascertained on such plots by contouring. A Schmidt net is often used for density contouring to provide information on the highest concentration, i.e. most prominent fracture plane. Figure 2.15 gives an example.

It may be important to find the overall effect of various discontinuities. The mean direction of a group of poles can be represented by a simple vector-sum of all the constituting poles, following the Fisher distribution (Figure 2.16) (e.g. Lee and Farmer, 1993). Similarly, a resultant vector can be calculated by summing all the clusters, to give a net directional effect of all the sets of discontinuities.

It should be mentioned here that only selected and not all of the discontinuities present may play a significant role in fluid movement in the rock mass. Therefore, selection and data integration ought to be done judiciously. Further, Sharp (1993) gave a more useful concept for integration of discontinuity trend, frequency and aperture data to make hydraulic zonation maps (see section 7.2.5).

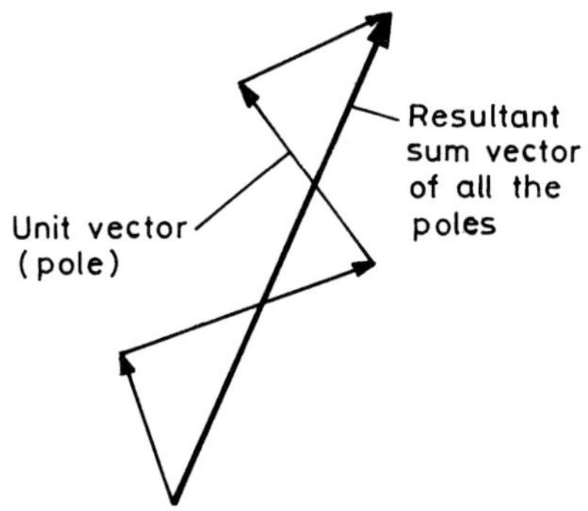

FIGURE 2.16 Principle of determining the mean pole direction as the resultant vector sum of all the vectors (poles), following the Fisher method.

3. Spacing (interval)

Systematic joints are roughly equidistant and possess parallelism, and therefore the parameter statistical spacing has significance. It describes the average (or modal) perpendicular distance between two adjacent discontinuities of the same set. It has a profound influence on rock mass permeability and groundwater flow. Fracture spacing is reciprocal of the fracture frequency or linear fracture density. It also controls fracture intensity and matrix block size.

Fracture separation, f_s, is related to lithology and thickness of beds, b, and is given as (Price and Cosgrove, 1990):

$$f_s = Y \cdot b \tag{2.1}$$

where Y is a constant related to lithology. Modelling and theoretical approaches also show that fracture spacing and bed thickness should have a linear relationship, for a given lithologic material.

By spreading a tape in any convenient direction on an outcrop face, average apparent spacing (d_m) between fractures of a set can be measured. This measurement has to be corrected for angular distortion (θ) to give the value of true fracture interval, perpendicular to the fracture orientation. The correction angle (θ) equals the angle between the direction of tape alignment and the pole to the fracture plane, and can be easily computed using a stereographic net (Figure 2.17). The true fracture spacing (f_s) can be obtained from the measured fracture spacing (d_m) as:

$$f_s = f_{sa} \cdot \cos\theta \tag{2.2}$$

Further, it has been reported that fairly reliable estimates of fracture spacing can also be

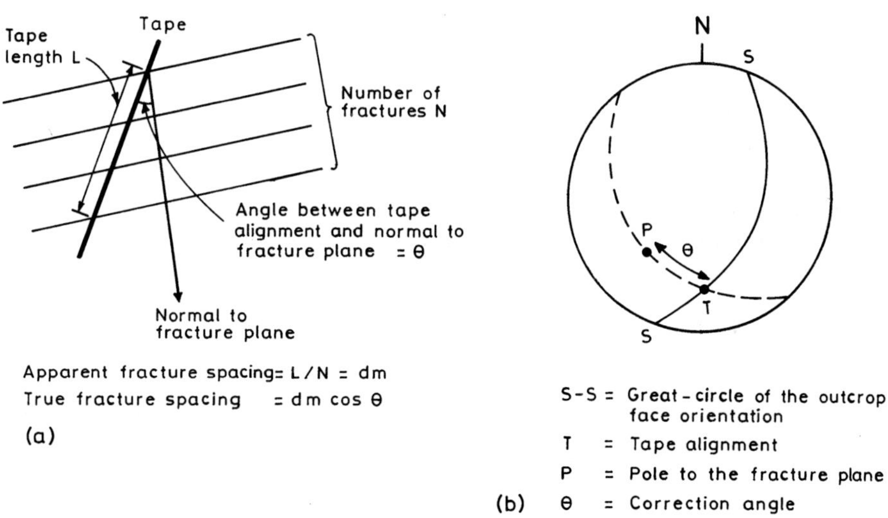

FIGURE 2.17 Spacing of fractures and computation of true fracture spacing. (a) Measurements in field for apparent fracture spacing. (b) Angle of correction, i.e. the angle between the line of tape alignment and pole to the fracture plane, computed by stereographic method.

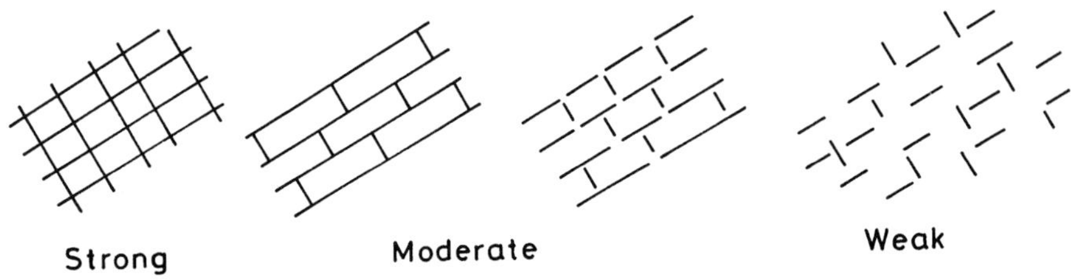

FIGURE 2.18 The influence of persistence of discontinuity on degree of fracturing and interconnectivity.

given by the P-wave velocity using seismic refraction techniques (see section 5.4).

4. Persistence (fracture length)

Fracture persistence or length is a measure of the extent of development of discontinuity surface (Figure 2.18). This carries the notion of size and controls the degree of fracturing. It is a crude measure of the penetration length of a fracture in a rock mass. Fracture trace length is also related to fractured surface area; as some of the discontinuities are more persistent and continuous than others, it becomes a very important parameter in controlling groundwater flow.

Persistence is rather difficult to quantify, as it will differ in the dip and strike directions. It can be measured by observing the discontinuity trace length in an exposure, in both dip and strike directions. The observed trace length may be only an apparent value of the true trace length due to various types of bias creeping into the data during measurements in exposures, drifts, excavations, benches etc. For example, the biases could be like: (a) inability to recognize fracture traces shorter than a certain threshold length will lead to a bias (truncation of the histogram); (b) inability to measure full length of the traces owing to incomplete exposures in drift walls, excavations etc., will lead to recording of censored length data; (c) the observed length of fracture trace depends on the relative orientation between the fracture plane and the exposure face; (d) in the sampling area or scanline, a stronger discontinuity is more likely to appear than a weaker one. Considering such aspects, methods for true trace length estimation have been discussed by a few workers (e.g. Pahl, 1981; Laslett, 1982; Chiles and Marsily, 1993).

5. Fracture density

Fracture density is measured for each set of fracture sets and corresponds to the degree of rock fracturing. It can be described in three ways – linear, areal and volumetric, depending upon whether the measurement/computation corresponds to length (1D), area (2D) or volume (3D), respectively.

(a) Linear fracture density (1D fracture density, d_1) is the average number of fractures of a particular set, per unit length measured in a direction perpendicular to the fracture plane. It equals fracture frequency (F_f) and is the reciprocal of fracture spacing.
(b) Areal fracture density (2D fracture density, d_2) is a way to quantify persistence of the discontinuity. It is the average fractured length (of traces) per unit area, on a planar surface.
(c) Volumetric fracture density (3D fracture density, d_3) is the average fractured surface area per unit rock volume, created by all the fractures of a given set.

All types of fracture densities, d_1, d_2 and d_3 have the same dimension (L^{-1}). The volumetric density is independent of direction and is a static parameter, like porosity. On the other hand, areal and linear densities are directional parameters and have bearing on fluid flow.

Both d_1 and d_2 depend on the orientation of the fractures *vis-à-vis* that of the scanline/exposure face. However, d_3 is independent of direction and can be estimated from a survey with boreholes or scanlines, with the help of proper weighting of the observed fractures (Chiles and Marsily, 1993). For computing the correct weighting factors, consider first the case of a borehole or a scanline survey (Figure 2.19). The

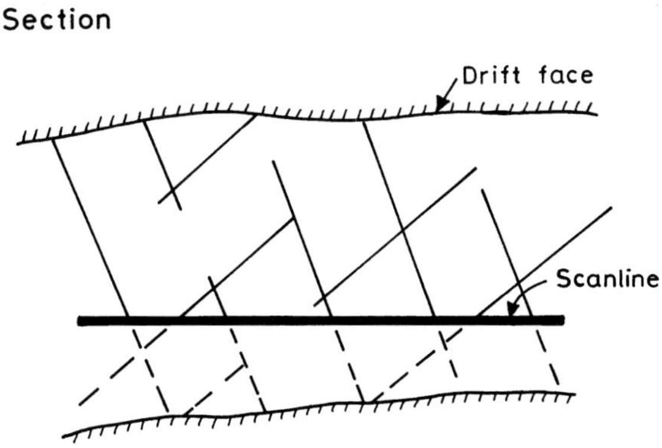

FIGURE 2.19 Scanline method of discontinuity survey.

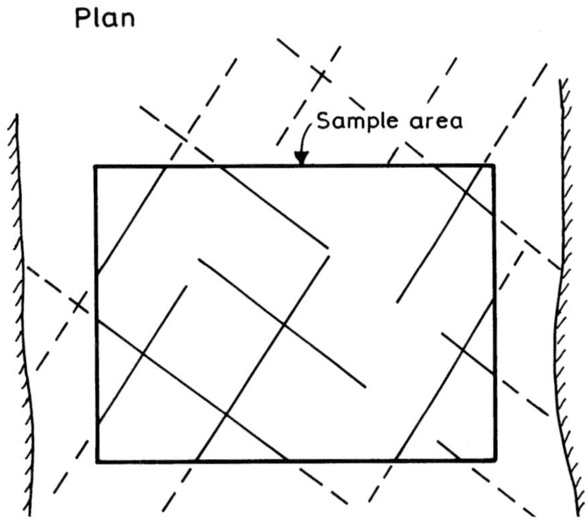

FIGURE 2.20 Areal method of discontinuity survey.

surveyed straight line can be considered as a cylinder of length L with a small section p (as in the case of a borehole). If n fractures intersect the survey line and θ_i is the acute angle made by the ith fracture plane with the borehole, then the fracture surface within the cylinder is $p/\sin \theta_i$, for the ith fracture. Hence, 3D fracture density d_3 is:

$$d_3 = \frac{1}{L \cdot p} \sum_{i=1}^{n} \frac{p}{\sin \theta_i} = \frac{1}{L} \sum_{i=1}^{n} \frac{1}{\sin \theta_i} \quad (2.3)$$

Thus, the weighting factor is related to the acute angle between the fracture plane and the scanline.

Similarly, considering the case of an areal survey, the exposure can be considered as a layer of area S and a small thickness e. Within the surveyed rectangle, if n fractures are traced on the exposure (Figure 2.20), and the ith fracture has a trace length l_i and makes an angle θ_i with the exposure plane, then the fractured surface area for the ith fracture is $e \cdot l_i \sin \theta_i$. Hence, 3D fracture density is:

$$d_3 = \frac{1}{S \cdot e} \left(\sum_{i=1}^{n} \frac{e \cdot l_i}{\sin \theta_i} \right) = \frac{1}{S} \left(\sum_{i=1}^{n} \frac{l_i}{\sin \theta_i} \right) \quad (2.4)$$

Thus, the weighting factor is related to acute angle and fracture trace length. If all the fractures have the same trace length, then l is constant. For parallel fractures, θ_i can be replaced by θ. For example, in an area in Bundelkhand granites (India), the 3D fracture density was computed using the scanline method at 64 observation sites. It is observed that d_3 in the area varies generally from about 6 m^{-1} to 21 m^{-1}, whereas there are smaller pockets of higher values of d_3, of the order of 31 m^{-1}. The variation in d_3 across the study area is shown in Figure 2.21, where the magnitude of d_3 is plotted as a circle of appropriate radius.

With simplifications and assumptions, d_1, d_2 and d_3 can be interrelated; if fracture orientations are purely random, then (Chiles and Marsily, 1993):

$$d^1 = l / 2 \cdot d^3 \quad (2.5)$$

$$d^2 = \pi / 4 \cdot d^3 \quad (2.6)$$

6. Fracture area and shape

Fracture area can be estimated from the strike trace length and dip trace length, assuming that the fractured surface has a certain regular

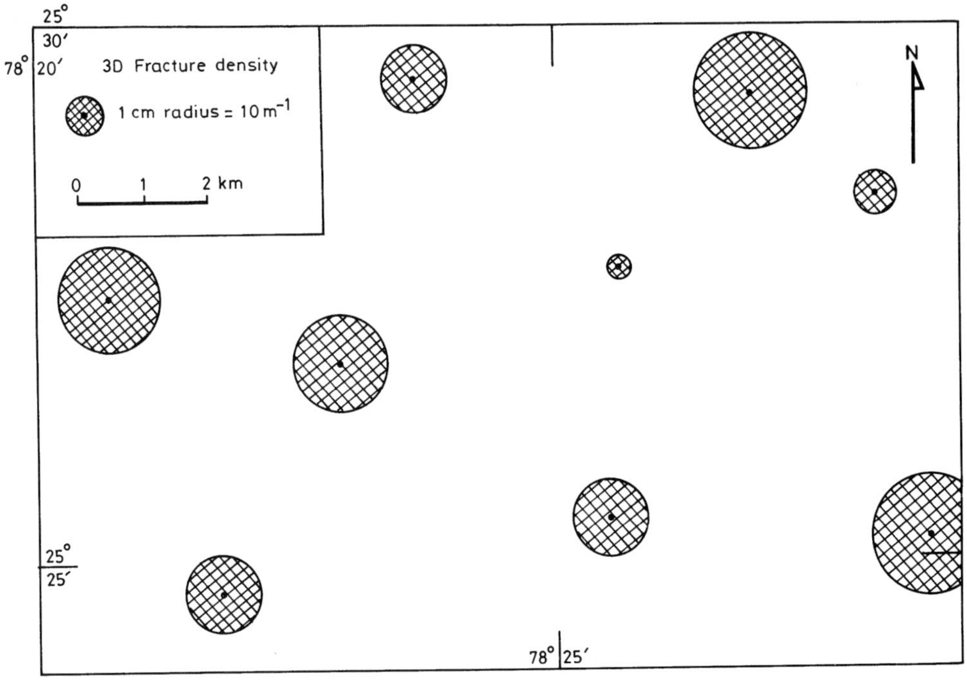

FIGURE 2.21 Map showing variation in 3D fracture density in a part of Bundelkhand granites, Central India.

shape, e.g. circular, square, elliptical, rectangular or polygonal. Out of these the case of circular discs is the simplest. Disc diameter D can be related to fracture surface area A as:

$$A = (\pi / 4) \cdot \left(D^2 + S_D^2\right) \tag{2.7}$$

where S_D is the standard deviation of disc diameter distribution. Statistical aspects on the bearing of fracture shape on area estimations are discussed by a few workers (Lee and Farmer, 1993).

The 3D density of disc centres τ, average disc surface area A and the 3D fracturation density d_3 are interrelated as:

$$d_3 = \tau \times A \tag{2.8}$$

7. Volumetric fracture count

Volumetric fracture count (V_f) is the total number of fractures per cubic metre of rock volume and is determined from the mean fracture spacing as:

$$V_f = 1/f_{s1} + 1/f_{s2} + 1/f_{s3} \cdots + 1/f_{si} \tag{2.9}$$

where f_{si} is the mean fracture spacing of the *i*th fracture set in metres. This also carries the notion of fracture intensity which is defined as the number of discontinuities per unit length, measured along a line, area or volume. Volumetric fracture count has a direct bearing on the size of matrix blocks and representative elementary volume (REV).

8. Matrix block unit

The rock block bounded by a fracture network is called the matrix block unit. Each matrix block unit can be considered to be hydrogeologically separated from the adjacent block. The shape of the matrix block unit could be prismatic, cubical or tabular, as governed by the

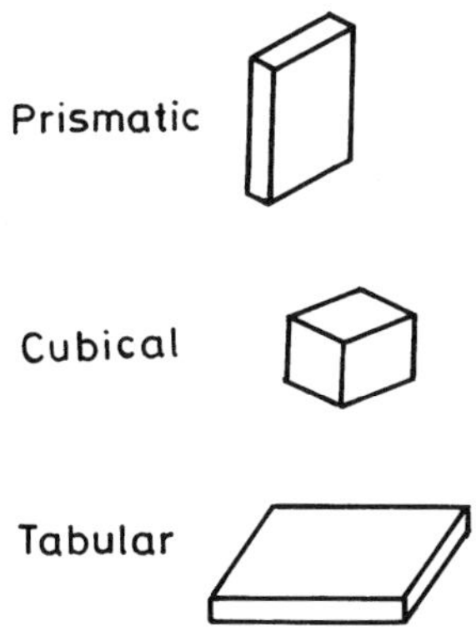

FIGURE 2.22 Shape of matrix block units: prismatic, cubical and tabular.

orientation of fractures and their distribution (Figure 2.22). For example, predominantly vertical fractures produce columnar and parellelopiped blocks (such as columnar joints in basalts); dominantly horizontal joints lead to plates and sheets (e.g. sheeting joints in granitoid rocks). These features impart hydraulic anisotropy to the geologic unit.

Consider an ideal case where beds are horizontal and fractures only vertical. It is known that fracture spacing and bed thickness are directly related (equation 2.1). It follows that a particular lithology has a tendency to develop block units of a certain shape, the block unit volume being dependent upon the bed thickness.

Block size is also related to the volumetric fracture count V_f. The maximum number of matrix blocks N_{bmax} can be expressed as (Kazi and Sen, 1985):

$$N_{b\,max} = \left(\frac{V_f}{3} + 1\right)^3 \qquad (2.11)$$

Fractal concepts are also used to define fragmented rocks. It is found that for fragmented materials including rocks, there is a size-frequency relationship of the form:

$$N(r) \propto (r^{-D}) \qquad (2.11)$$

where N(r) is the number of fragments with a characteristic linear dimension greater than (r) and D is the fractal dimension. It is believed that in future, fractal dimensions could be very useful in defining rock mass characteristics (e.g. Mojtabai *et. al.*, 1989; Ghosh, 1990).

9. Fracture connectivity

Discontinuities may exhibit differing termination and connectivity characteristics (Figure 2.23a). Intersection of discontinuities is important as groundwater flow takes place through multiple fractures. Greater continuous intercommunication among the fracture network is provided by a higher degree of fracturing. Fracture connectivity increases with increasing fracture length and fracture density, as the chance of fracture intersection increases.

For evaluating connectivity it is necessary to study how the fractures terminate. Barton *et al.* (1987) classified fractures into three categories: abutting, crossing and blind. The fractures of the blind type do not intersect other fractures and remain unconnected. Laubach (1992) suggested that in many cases fracture connectivity may be gradual and that many fractures earlier classified as abutting, were really diffuse (interfingering type). He grouped fracture terminations into blind, diffuse and connected (which includes abutting). The data can be plotted in a ternary diagram to represent the bulk condition of fracture intersection in the rock mass (Figure 2.23b). As an example, it is shown in the figure that most of the joints in Bundelkhand granites (BG) are of the connected type.

10. Rock quality designation (RQD)

RQD is a semi-quantative measure of fracture density which can be estimated from core recovery data. RQD is defined as the ratio of the recovered core more than 4 in. (about 10 cm) long and of good quality to the total drilled length and is expressed as a percentage. Although RQD is mainly used in assessing the

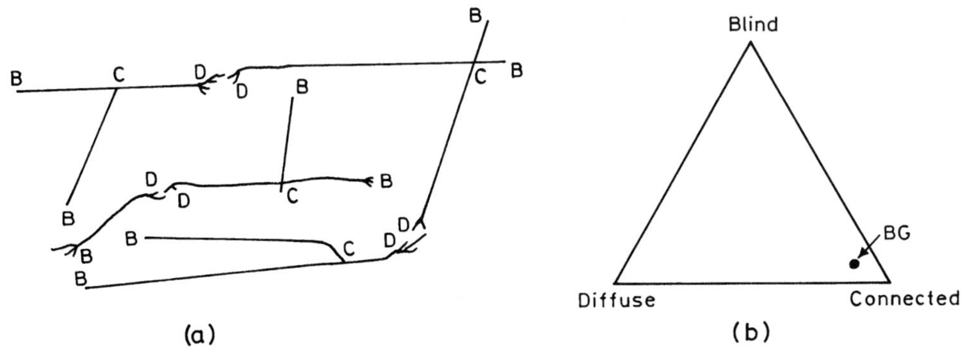

FIGURE 2.23 Fracture connectivity. (a) Different types of fracture terminations: B = blind; C = crossing; D = diffusely connected. (b) Ternary diagram of fracture terminations (after Laubach, 1992); the point B.G. corresponds to data from Bundelkhand granites indicating high degree of fracture interconnectivity.

geomechanical properties of rocks, it is also considered to be an important parameter in assessing relative permeability.

11. Aperture

Aperture is the perpendicular distance separating the adjacent rock walls of an open discontinuity, in which the intervening space is air- or water-filled. Apertures may vary from very tight to wide; and commonly, subsurface rock masses have small apertures. Tensile stresses may lead to larger apertures. Often shear fractures have much lower aperture values than dilational fractures.

An aperture may increase by dissolution, erosion etc. particularly in the weathered zone. It may decrease with depth due to lithostatic pressure, and the fracture wall compression strength is an important parameter governing an aperture as the lithostatic pressure tends to close the fracture opening.

Table 2.2 gives aperture ranges as usually classified in rock mechanics. Measurement of apertures in surface exposures is made with a vernier or gauge. The data on discontinuity sets with corresponding apertures is to be recorded. Asperities affect the aperture size and also render its measurement difficult in the field. Therefore, when considering fluid flow, apertures are defined in terms of flow properties, as volumetric flow rate is governed by the cube of the aperture. An aperture can be integrated with fracture density to give an integrated function representative of hydraulic conductivity (section 7.2.5).

TABLE 2.2 Aperture classification by size (after Barton, 1973)

Aperture (mm)	*Term*
< 0.1	very tight
0.1–0.25	tight
0.25–0.50	partly open
0.50–2.50	open
2.50–10.0	moderately wide
> 10.0	wide

12. Asperity

Fracture walls are not flat parallel smooth surfaces but contain irregularities, called asperities (Figure 2.24). The asperity reduces fluid flow and leads to a local channelling effect of preferential flow. This reduces the effective porosity and makes flow velocities irregular. Observations on asperities should be made for each type of fracture surface and measurement made. Mean height of asperities together with Reynolds number R_e has a direct influence on flow regime, i.e. laminar versus turbulent flow (section 7.1.1).

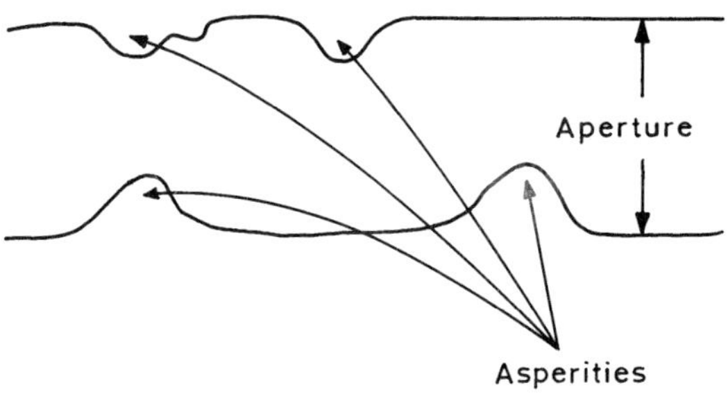

FIGURE 2.24 Asperities in fracture walls.

13. Wall coatings and infillings

This is the solid material occurring between the adjacent walls of a discontinuity, e.g. clay, fault gouge, breccia, chert, calcite. Filling material could be homogeneous or heterogeneous, and could partly or completely fill the discontinuity. The material may have variable permeability, depending upon mineralogy, grain size, width etc. The net effect of wall coatings and infillings is a reduced aperture.

2.4 Methods of field investigations

Methods of field investigations can be classified into two broad types (Jouanna, 1993) – 2D and 3D (Table 2.3). The 2D methods are based on observations made at the rock surface, at surface or subsurface levels. They include scanline surveys, borehole surveys, and different types of areal surveys (Figure 2.25). These methods give an idea of the hydrogeologic properties at and around the site of observation.

The 3D methods are aimed at gathering information on the bulk volumetric properties involving inner structure of the fractured rock mass. There can be direct or indirect 3D methods. Brief descriptions of the various 2D and 3D methods are given below.

1. Scanline surveys

Scanline surveys involve direct observation of rock features along a line on the rock surface, such as on an outcrop, drift face, excavation, or adit (Figure 2.19). Scanlines are usually horizontal; however, vertical scanlines are preferred where fractures are mostly horizontal.

A suitable rock exposure or face is selected. A sample scanline is marked on the face, and its orientation (rake on the face) is recorded. Fractures intersecting the line are collected, and each fracture is represented by its trace which can be measured. Observations are made for various parameters, such as: location of the fracture trace intersection with the scanline; orientation of the fracture and angle made with the scanline; termination type if seen and connectivity; alternatively, whether the fracture extends beyond the top of the face/batter; fracture type and other relevant fracture characteristics.

TABLE 2.3 Methods of field investigations

1.	2D Methods	– Based on rock surface observations on lithology, structure, fractures, and their characteristics; made at surface or subsurface levels.
	1.1	Scanline surveys
	1.2	Areal surveys – on outcrops, pits, trenches, adits, drift etc. including terrestrial geophotogrammetry and remote sensing
	1.3	Borehole surveys – including drilling, study of oriented cores, borehole logging, dipmeter, borehole cameras and formation microscanner methods
2.	3D Methods	– Investigations aimed at bulk volumetric properties of rock mass in 3D.
	2.1	Hydraulic well tests
	2.2	Hydrochemical methods
	2.3	Geophysical methods including seismic, electrical, EM, gravity, magnetic and georadar.

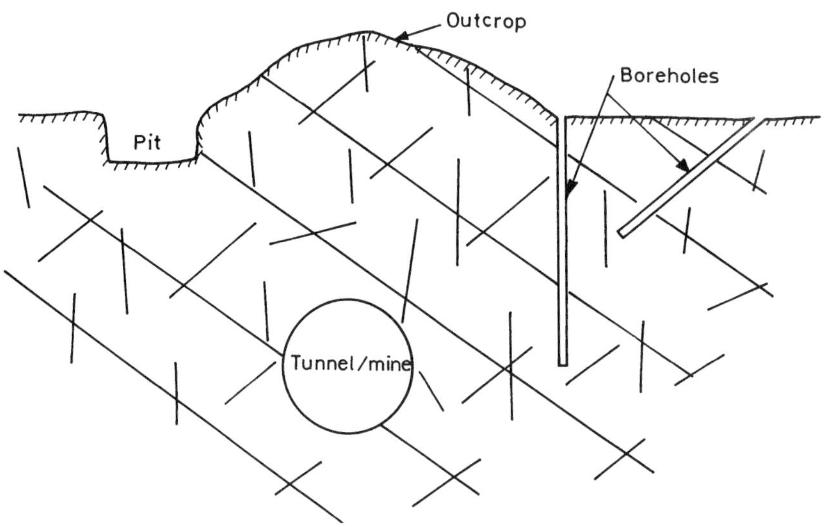

FIGURE 2.25 The various 2D methods of field investigations.

2. Areal surveys

Areal surveys can be treated as extensions of the scanline surveys. They are used for surveying fracture characteristics on a rock surface area, e.g. on an outcrop, drift face, adit or tunnel (Figure 2.20). In the field, an area is first demarcated on a rock surface for observation and statistical sampling. Detailed observations on fracture characteristics are made within the marked area where all the fracture data are collected.

Direct observations and field mapping at natural rock outcrops is an old established technique. Weathering, surficial cover, soil, vegetation etc., influence the accessibility and visibility of good outcrops. Excavations, pits and trenches are made to expose the fresh rocks for visual inspection at a shallow depth. Subsurface direct observation can be made in adits and tunnels; geological maps of rock faces exposed can be prepared and fracture characteristics measured.

Remote sensing includes the study of photographs and images acquired from aerial and space platforms. This technique can give valuable information on geology, structure, fractures, lineaments etc. and forms an important mapping tool (Chapter Four). Further, stereo-photographs of rocks exposed in outcrops, scarps, excavations etc. can be taken from a ground-based (terrestrial) platform. These stereo pairs can be studied and measurements of fracture characteristics can be done in the laboratory (see section 4.6.2).

3. Borehole surveys

These are the only tools for direct observations of rock surface and features occurring at depth (Figure 2.25). A number of methods are available. As drilling is expensive, an optimum combination of methods is employed for getting maximum information from drilling. Study of drill cores, particularly orientated cores, provides data on orientation of structures, fractures etc. Further, borehole walls can be studied in several ways – geophysical well logging is a standard technique, including electrical, caliper, radioactivity and magnetic logging. These give information on lithology and structure. The borehole televiewer provides images of the borehole walls with joints and fractures. The dipmeter and formation microscanner help to measure orientation of structural features at depth *in situ*. Details on drilling and well observation techniques are given in Chapter Five.

4. 3D methods

As mentioned above, 3D methods are aimed to provide information on bulk volumetric properties of the fractured rock mass. These methods include hydraulic well tests, hydrochemical methods and geophysical techniques. The hydraulic well tests comprise pumping tests of various configurations and types, and give bulk volumetric assessment (Chapter Nine). In hydrochemical methods various types of geochemical tracer studies and solute transport studies are carried out for bulk volumetric hydrogeological characterization (see section 9.3). Further, a number of geophysical methods are used such as seismic, electrical, EM, gravity, magnetic and georadar. They are briefly described in Chapter Five from a hydrogeological investigation point of view.

Chapter three

Hydrogeological investigations

The purpose of most hydrogeological investigations is to locate potential sites for development of an adequate quantity of reasonably good quality groundwater for a particular use, domestic, irrigation or industrial. The quantity and quality criteria will depend on the local needs and the socio-economic conditions of the people. Alternative possible sources of water supply, e.g. import of surface water or groundwater from adjoining areas may also be considered. Further, hydrogeological investigations are also taken up for the selection of suitable sites for waste disposal, tapping of geothermal power and in several other geotechnical problems, viz. tunnelling, mining, hill slope stability etc. These include geological, geomorphological, and hydrogeological studies described in this chapter. Besides, remote sensing and geophysical studies are discussed in Chapter Four and Five.

3.1 Geological investigations

Geological investigations include surface mapping of different lithological units and their structural and geomorphological features on a scale of about 1:20 000 to 1:200 000. Aerial photographs and satellite images are preferred as base maps for geological and geomorphological mapping.

In this text, from a hydrogeological point of view, the various rock types are broadly classified into four groups (Table 3.1). Their detailed hydrogeological characters are described in Chapters Eleven–Fourteen.

In fractured rocks, special emphasis has to be provided for mapping of lineaments, fractures and other rock discontinuities (Chapter Two). Field mapping should also include information about the orientation and density of fractures, although their subsurface distribution may be different and which can be deciphered from subsurface investigations (Chapter Five).

Data about the thickness and composition of the weathered zone (regolith) is particularly important in crystalline rocks (Chapter Eleven). In volcanic rocks, in addition to lineaments, attention should also be paid to the characters of individual flow units and interflow formations (Chapter Twelve). In carbonate rocks, mapping of various solution (karst) features and springs are of special importance (Chapter Thirteen).

Based on exploratory drilling and well log data, subsurface maps and sections, viz. fence diagrams, isopach maps, structural contour maps etc. are prepared to project the subsurface distribution and configuration of aquifers, aquitards and aquicludes (e.g. Walton, 1970; Erdelyi and Galfi, 1988).

3.2 Geomorphological investigations

Geomorphological investigations include delineation and mapping of various landforms and drainage characteristics. These contribute significantly in deciphering areas of groundwater

TABLE 3.1 Hydrogeological classification of rocks

Rock group	*Rock types*	*Mode of occurrence*	*Main features important for groundwater occurrence*
Crystalline rocks	Non-volcanic igneous and metamorphic rocks, viz. granites, gneisses, schists, slates and phyllites, etc.	Large size massifs and plutons; regional metamorphic belts	Weathered horizon, fractures and lineaments
Volcanic rocks	Basalts, andesites and rhyolites	Lava flows at places interbedded with sedimentary beds	Fractures, vesicles and interflow sediments
Carbonate rocks	Limestones and dolomites	Mostly as chemical precipitates with varying admixtures of clastics in a layered sedimentary sequence	Fractures and solution cavities
Clastic rocks	Consolidated: sandstones and shales; unconsolidated: gravel, sand, clay etc.	Interbedded sedimentary sequence	Intergranular pore spaces and fractures

recharge and their potential for groundwater development. Geomorphic mapping can best be done from remote sensing data using satellite images and aerial photographs on suitable scales (say 1:20 000 to 1:50 000) (Chapter Four). Preferably, a river basin should be taken as an unit for geomorphological data analysis, though also it may be realized that the surface water basin may not always coincide with the groundwater basin.

3.2.1 Landforms

Landform is an end product of natural weathering. It depends on three main factors: (a) present and past climatic conditions, (b) rock types and their structural features, and (c) the time span involved in weathering.

Genetically, the landforms are divided into two groups: erosional landforms, and depositional landforms. Erosional landforms are typically associated with the resistant hard rock terrains. They comprise: (a) residual hills, (b) inselbergs, (c) pediments, (d) buried pediments with weathered basements, and (e) valley fills. These are described in section 11.2. Depositional landforms are developed by depositional processes of various natural agencies, e.g. river, glacier and wind. The depositional landforms are typically made up of unconsolidated sediments and may occur in the regional setting of hard rock terrains. Therefore, they may play an important role in the groundwater development for local needs (Chapter Fourteen).

3.2.2 Drainage characteristics

Qualitative and quantitative drainage characteristics of a basin provide an indirect clue to the hydrogeological characteristics of the area and therefore are useful in the assessment of groundwater resources. The important characteristics are drainage pattern and drainage density. These are related with the lithology, structure and permeability of the bedrock.

Drainage is said to be internal when few

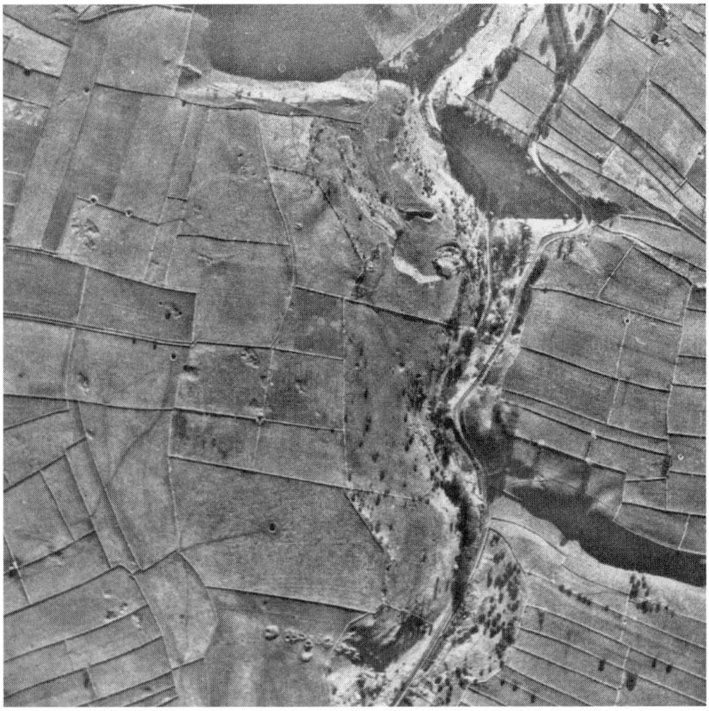

FIGURE 3.1 Internal drainage in limestone. (Photograph courtesy of Aerofilms Ltd.)

drainage lines are seen on the surface and drainage appears to be mostly subsurface, commonly seen in limestone and gravel deposits (Figure 3.1). External drainage is where the surface drainage network is seen to be well developed. Drainage channelways are characterized by the typical serpentine-sinuous shape and pattern. Their manifestation depends upon dimensions of the channel, scale of study and also on physical conditions of the channel, which may change from season to season, viz. due to flooding, drought, change in sediment load and vegetation cover.

Drainage pattern is the spatial arrangement of streams and is, in general, very characteristic of rock structure and lithology. Drainage patterns can be studied and mapped on topographic maps and aerial photographs as well as satellite images, on scales as per requirements. Common types of drainage patterns, characteristic of fractured and other hard rock terrains are listed in Table 3.2 and illustrated in Figure 3.2.

Among various other drainage characteristics of a basin, drainage density, D_d, is an important parameter. Drainage density (D_d) is the ratio of total channel lengths of streams within a basin to the area of the basin. It has the dimensions of L^{-1} and can be expressed as

$$D_d = \sum L / A \tag{3.1}$$

where D_d is drainage density in km^{-1}, $\sum L$ is the total channel length in km, and A is the basin area, in km^2. The accuracy of measurement of the total lengths of all streams would depend on the scale of the map, as all the streams may not be shown on a small-scale map. Large-scale aerial photographs therefore provide a better idea of drainage network for computing D_d.

Drainage density exhibits a very wide range of values in nature depending upon the relief, climate, resistance to erosion and permeability of rock material (Table 3.3). In general, low drainage density is characteristic of regions of

TABLE 3.2 Common drainage patterns in fractured and other hard rock formations and their geological significance

Type	*Description*	*Geological control*
Dendritic	Irregular, branching of streams, resembling a tree	Homogeneous materials and crystalline rocks; horizontal beds; gentle regional slope
Rectangular	Streams having right-angled bands	Jointed/faulted rocks, e.g. sandstones, quartzites etc.
Rectilinear	Straight line pattern	Strong linear structural control
Angulate	Streams joining at acute angles	Joints/fractures at acute angles to each other
Parallel	Channels running nearly parallel to each other	Steep slopes; also in areas of parallel elongate landforms
Trellis	Main streams running parallel, and minor tributaries joining the main streams nearly at right angles	Dipping or folded sedimentary or low-grade meta-sedimentary rocks; areas of parallel fractures
Radial	Streams originating from a central point or region	Volcanic cones, domes and residual erosional features
Annular	Ring-like pattern	Structural domes

highly resistant or highly permeable surface and low relief. High drainage density is found in regions of weak or impermeable subsurface materials, sparse vegetation and mountainous relief (Strahler, 1964). Although low drainage density is considered to indicate regions of high permeability, this criterion should be used with care as other factors such as relief, climate and resistance to erosion may outweigh the influence of permeability of surficial material in a certain area. In areas of low relief, D_d may be more indicative of permeability of surface material and therefore could be used as a criterion for the selection of suitable sites for shallow wells. Drainage density also influences run-off pattern and thereby infiltration capacity of the rock material. For example, high drainage density removes surface run-off rapidly, decreasing lag time and increasing the peak of the hydrograph. Among other drainage network parameters, bifurcation ratio, stream frequency and basin shape are also of hydrological interest. Readers may refer to Strahler (1964), Gregory and Walling (1973) and other texts.

3.3 Hydrogeological investigations

These include groundwater level measurements, estimation of hydraulic properties of aquifers and assessment of groundwater quality. The study of springs including their location, variability of discharge, chemical composition and temperature also provide useful hydrogeological information.

An observation well network is established for monitoring groundwater levels and groundwater quality. For estimating hydraulic characteristics of aquifers, pumping tests and tracer injection tests are carried out. Tracer tests help in estimating not only hydraulic conductivity but also flow mechanism which is important from the point of view of evaluating contami-

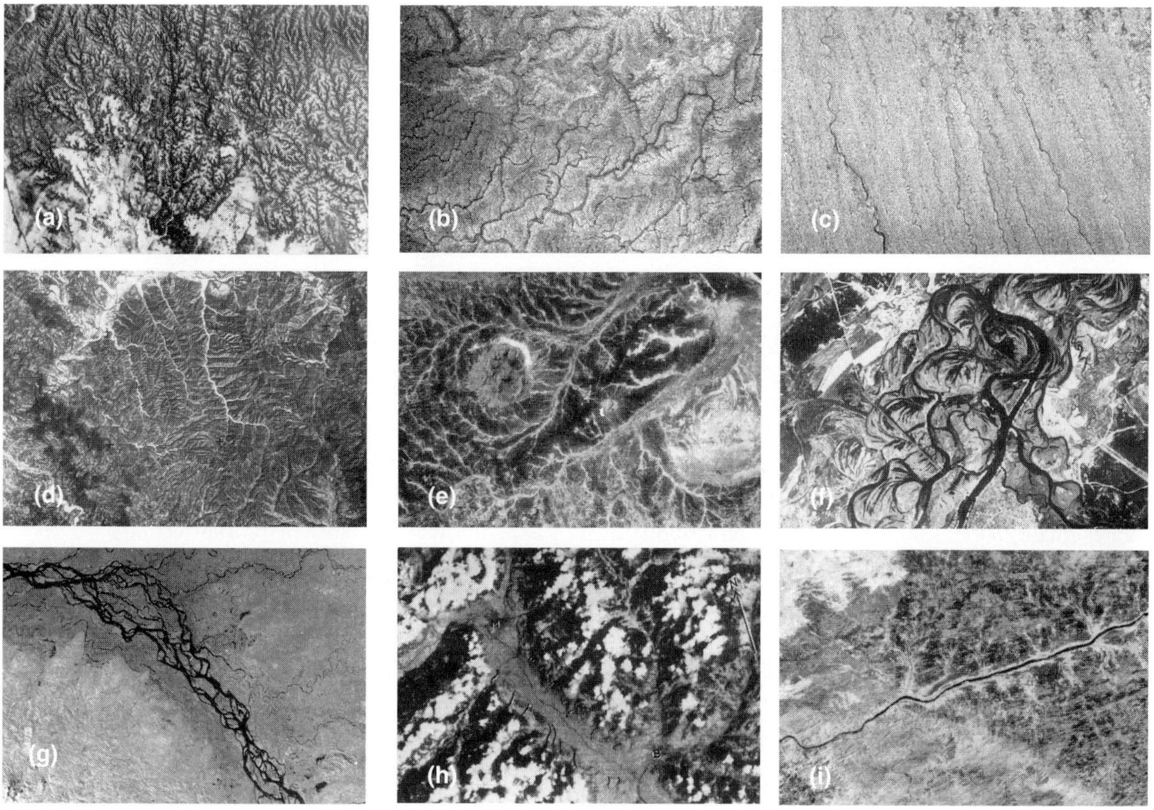

FIGURE 3.2 Drainage patterns: (a) dendritic, (b) rectangular and angulate, (c) parallel, (d) trellis, (e) annular and subradial, (f) meandering, (g) braided, (h) barbed, (i) rectilinear (modified after Gupta, 1991)

TABLE 3.3 Drainage density (D_d) values for some terrains in USA (after Strahler, 1964)

D_d (km^{-1})	*Geological terrain*
1.9–2.5	Resistant sandstone strata of the Appalachian Plateau
5.0–10.0	Rocks of moderate resistance in humid central and eastern USA
12.5–19.0	Strongly fractured and deeply weathered igneous and metamorphic rocks in southern California under dry summer and sub-tropical conditions
125–250 and at places 688–812	In bad land developed on weak clays, barren of vegetation

nant transport. In fractured rocks, packer (Lugeon) tests and slug tests are preferred, in the first instance, as these do not require observation wells. Cross-hole hydraulic tests and pumping tests are necessary for detailed estimation of hydraulic parameters (Chapter Nine). As pumping tests are expensive and require time and manpower, one has to be very selective in implementing such a programme. From the groundwater development point of view, groundwater resources, both static and dynamic, are to be computed, and a management strategy for groundwater development has to be worked out (Chapter Seventeen).

TABLE 3.4 Types of observation well networks (modified after Heath, 1976; UNESCO, 1977)

Types of networks	*Objectives*	*Products*
Hydrogeologial	(a) Status of storage (b) Areal extent and interconnection of aquifers (c) Demarcation of recharge and discharge areas and groundwater flow regime	Regional water-table and/or potentiometric surface map showing water-level fluctuations over a selected period
Water management	(a) Effect of stresses on recharge and discharge conditions (b) Degree of confinement (c) Hydraulic characteristics of aquifers	(a) Local water-level maps (b) Hydrographs showing changes in water-level with time (c) Time-drawdown curves
Baseline	To determine effect of climate, topography and geology on groundwater storage	Hydrographs showing effect of climate, topography and geology
Water quality	(a) Water rock interaction (b) Contaminant transport	Water quality maps and sections
Special purpose	Effect of mining, irrigation and drainage projects	Water-level and quality maps

3.3.1 Water level measurements

Observation well networks

The purpose of a groundwater observation well network is to install observation wells or piezometers to monitor the quantity and quality of a groundwater reservoir. Keeping in view the regional and local requirements, the planning and design of such a network should depend on hydrogeological situations, purpose of investigations, stage of development, as well as political and social demands (UNESCO, 1972, 1977). The various types of observation networks are listed in Table 3.4. Placement and depth of observation wells depend on fracture characteristics. In the case of uniformly and densely fractured rocks, the placement and construction of such wells can be similar to granular aquifers.

Observation wells can be drilled by using any conventional method of water-well drilling (Chapter Fifteen); in unconsolidated formations, wells are provided with screens tapping the zone of interest – in consolidated rocks, open-end wells can serve this purpose. Observation wells are usually of small diameter to accommodate the water-level measuring device.

The purpose of the hydrogeological network is to allow preparation of water-table or piezometric surface maps. Such maps provide information on the direction of groundwater movement, effect of stresses and changes in groundwater storage as well as areal extent of aquifers and nature of boundaries. In heterogeneous formations, it is necessary to construct a group of observation wells (well nests) tapping different aquifer zones, in order to determine the possible aquifer interconnections, and comparison of water-levels in various aquifers (Figure 3.3). In fractured rocks, the observation wells should be put in relation to rock discontinuities tapping different fracture sets for assessing the extent of interconnection and the role of different fractures as conduits for groundwater flow. Lineaments and intersection

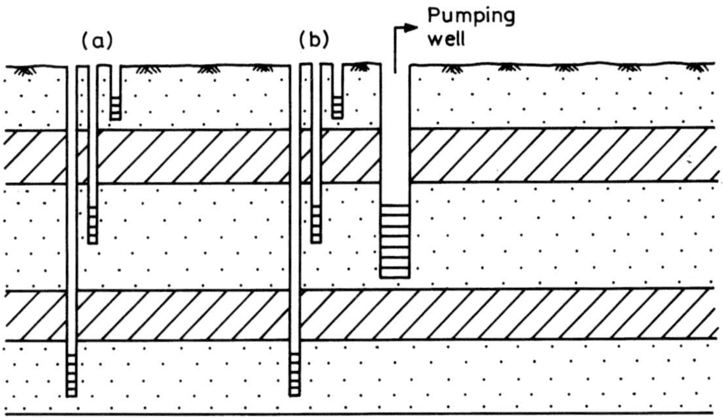

FIGURE 3.3 Network of observation wells tapping different aquifers in the vicinity of a pumping well.

of vertical fractures are potential sites for monitoring wells. In situations where the matrix blocks have appreciable porosity, it may be necessary to monitor the blocks as well as the high permeability zones (Anon, 1995).

The objective of a water management network is to obtain information about the effect of recharge and withdrawal on the groundwater system. Such wells should be located in proximity to pumping well fields in clusters tapping the pumped aquifer as well as the overlying and underlying aquifers. In order to determine the extent of interconnection, separate observation wells should be installed in unconfined and confined aquifers (Figure 3.3).

The baseline network consists of observation wells located in areas which are not significantly affected by withdrawals or other man-made stresses. The purpose of this network is to determine the effect of climate, geology and topography on groundwater levels.

The water quality network is used to determine natural changes in water quality due to rock-aquifer interaction in the direction of groundwater flow and also with depth, and the effect of man-induced contaminants on water quality. Water quality characteristics can be extremely variable in fractured rocks and karst aquifers. This is particularly true during and after recharge. In order to study the effect of contamination from a point source, at least one observation well should be located upstream from the possible source of pollution, in order to know the background quality. The other observation wells are installed in the downgradient side from the source of contamination at suitable distances, depending on the rock permeability. In fractured rocks, the monitoring well network should be designed considering the fracture network to enable initial detection of the contaminant (Meyer *et al.*, 1994). In addition to observation wells, data from springs and caves etc. can also provide useful information.

The special purpose network is designed to determine the effect of mining and other engineering structures like dams, canals and drainage channels on groundwater levels; in the case of opencast mines, the observation well network is put around the periphery of the excavation. In order to determine the effect of rivers and canals on groundwater regime, the observation wells should be put in line across the surface water bodies. If an area is bounded on both sides by rivers or canals, the observation wells should be located not only near these hydraulic boundaries, but also in the interfluve.

Timing and frequency of measurements

The frequency of water-level measurements in wells and discharge from springs depends on the nature of the aquifer and the purpose of study, the broad guidelines being as follows:

	Frequency of measurement
confined aquifer	: one per month
unconfined aquifer	: three per month
wells adjacent to rivers	: daily during flood times
snowmelt areas	: daily
irrigated areas	: related to time of irrigation

In drainage studies and also in the study of the effect of external forces, viz. changes in barometric pressure, ocean tides, earthquakes, etc., continuous recording of water-level using automatic water-level recorders is necessary.

Spring discharge measurements are of importance for water resources assessment, especially in fractured and karstic rocks (section 13.6). Plots of spring discharge (or stage) and water quality in relation to time, known as spring hydrograph and chemograph respectively, have been used extensively in karst aquifer studies. Spring discharge in the range of 0.005–0.01 m^3s^{-1} is measured by volumetric methods while higher discharges can be measured using weirs; sampling frequency will depend upon the trend of variability of the parameter. Samples should be collected through at least one major recharge event, as such an event is likely to cause rapid changes in the spring discharge, hydraulic head in wells and also in water quality. Samples must also be collected during base flow conditions.

3.4 Hydrogeological maps

Data obtained from field and laboratory studies are plotted on hydrogeological maps and sections. It is necessary to know the purpose of preparing such maps in order to decide the type of information to be provided and the manner of its presentation. Unlike geological maps, hydrogeological/hydrological maps deal with essentially transient phenomena, viz. groundwater resources, groundwater levels, water quality, etc. Therefore, hydrogeological maps should also indicate the dates of surveys, reliability of data and methods of survey.

3.4.1 Regional maps

Hydrogeological maps depict a number of parameters including the geology of the area, major rock discontinuities, extent of aquifers, hydraulic characteristics, contours of water-table and piezometric surface, water quality, surface water bodies and important meteorological characteristics. Hydrogeological maps may be international, national, regional or local. They usually vary from 1:1 000 000 to 1:250 000 in scale. There may be considerable variation in the amount of information included in different maps depending on the chosen scale, size of the area and the purpose of the map. For example, a hydrogeological map of the Czech Republic was prepared on the scale 1:1 000 000, showing variations in transmissivities and yield characteristics (Krasny, 1996). Further, Russian hydrogeologists (Zektser and Dzhamalov, 1988) compiled a World Groundwater Flow map on a 1:10 000 0000 scale depicting data of specific groundwater discharge.

For geothermal studies, geothermal resource maps are prepared showing temperature and pressure distribution of a hydrogeological unit which are of practical importance for development of geothermal power.

In earlier years, different workers and organizations used differing legends and colour codes to depict various groundwater features. This created difficulties in proper interpretation, comparison and compilation of data. To overcome this problem, a common legend to include a colour code for hydrogeological maps has been suggested for international adoption (UNESCO, 1983; Castany *et al.*, 1985). Hydrogeological maps of several countries have been published using the above code (Castany *et al.*, 1985; Verma and Jolly, 1992; Struckmeier, 1993).

One of the earliest attempts to prepare a hydrogeological map of India was made by Taylor (1959). Subsequently, the Geological Survey of India published a hydrogeological map of the country on the scale of 1:2 000 000 (GSI, 1969). In 1976, the Central Ground Water Board compiled and published a hydrogeological map of India on a 1:5 000 000 scale which was updated in 1989. The revised map (CGWB, 1989) depicts major lithological units, their groundwater potential in terms of well yields and groundwater quality. A map of India showing major groundwater provinces is given in Figure 17.2. Recently, hydrogeological atlases of some States (Provinces) have also been published (e.g. CGWB, 1995b) which provide detailed information on groundwater quality and yield characteristics of aquifers.

3.4.2 Map media – hard copy and electronic

Hard copy maps use paper, photographic film or other similar materials for map record and display in colour or black-and-white. Most of the hydrogeological maps are available as hard copy prints throughout the world. However, with the transformation in information technology recently, computer coded maps in electronic media are now becoming available in some countries. These maps may use raster or vector format in GIS, and may be restructured and reformatted as required. Further, such maps can depict different hydrogeological parameters and also use the standard internationally accepted legend or colour scheme.

The computer coded maps have the advantage of digital compilation and processing in the GIS environment (Chapter Six). Further, the original resolution of data may be retained in such maps in comparison to the hard copy medium where the resolution generally gets reduced. These maps and data may be available in electronic media such as cartridges, tapes and diskettes for transfer to other systems where they could be transferred to soft copy for display, processing, editing etc. With the developments in information superhighways, it is possible that soon hydrogeological data in the public domain may be available on networks like the internet/world wide web etc. facilitating world- wide super-fast accessibility.

3.4.3 Special purpose maps

Special thematic maps can also be prepared giving certain specific information, e.g. water-table maps, isopiestic maps, water quality maps and other such parameters as spatial distribution of permeability, transmissivity, specific yield, well yields, and specific capacity.

Groundwater level maps

Based on water-level measurements, various types of contour maps, such as water-table maps, piezometric surface maps, depth to water-level and water-level fluctuation maps can be prepared. These maps depict the spatial and temporal variations of the groundwater resources of an area. Water-table maps represent the position of water-level in unconfined aquifers while a piezometric surface map represents the pressure conditions in a single or a group of confined aquifers. The construction of such maps assumes that the vertical hydraulic gradients are not significant and that the well intersects enough fractures. Measurements of water-levels over an area should be completed within a short time interval. The contours represent lines of equal potential and therefore flow lines can be drawn at right angles to these contours. The spacing of the contours depends upon flow rate, aquifer thickness and hydraulic conductivity (Figure 3.4). The shape of contours and their values indicate recharge and discharge areas. The upward and downward swing of water-table contours along a stream would indicate effluent and influent condition respectively. Groundwater mounds represent areas of recharge while groundwater depressions are developed due to excessive withdrawal (Figure 3.5).

Care should be taken in the interpolation of water-table contours across the hydrogeological boundaries as illustrated in Figures 3.6 and 3.7.

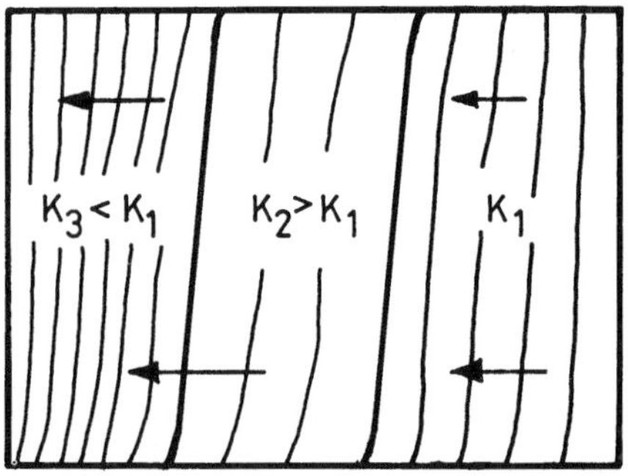

FIGURE 3.4 Effect of hydraulic conductivity (K) on the spacing of water table contours. K_1, K_2 and K_3 are the hydraulic conductivities of the three layers from right to left.

This is of even greater importance when the observation points are limited and are not uniformly distributed. In sparsely fractured rocks, the water-table will have a 'stair-step' appearance due to large contrasts in hydraulic conductivity between blocks and fractures. In these settings, water-levels measured in the wells may not be in continuity and the concept of 'water-table' becomes less clearly defined. Therefore maps prepared on the basis of such data may not give correct data on the groundwater flow. Figure 3.8 indicates that the direction of groundwater flow estimated from water-table contours is different to the actual direction of flow determined from tracer studies.

Depth of water-level maps

These provide information only about the spatial variation of depth to water-level, and cannot be used for determining direction of groundwater flow.

Water-level fluctuation maps

These indicate the change of water-level in wells during a given period of time. Such maps indicate rates of recharge and discharge to aquifers and changes in storage of water within the aquifer and aquitards. The reasons for such fluctuations are described in section 17.4. Groundwater level trends can also be depicted by hydrographs where variation in groundwater level is plotted against time of observation (Figure 17.3). These are of importance in groundwater management studies.

Granular aquifers show minor fluctuations due to recharge, but in sparsely fractured media, the fractures and other conduits respond quickly to recharge events. In such settings, during recharge, higher-lying fractures may get saturated although the low-permeability blocks may remain unsaturated. Therefore, moderately

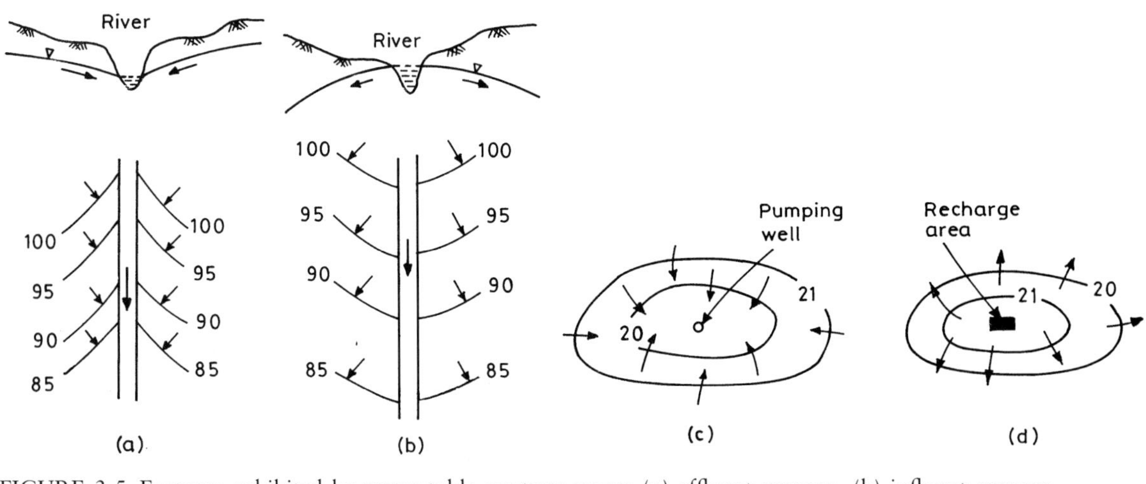

FIGURE 3.5 Features exhibited by water-table contour maps: (a) effluent seepage, (b) influent seepage, (c) groundwater depression, (d) groundwater mound.

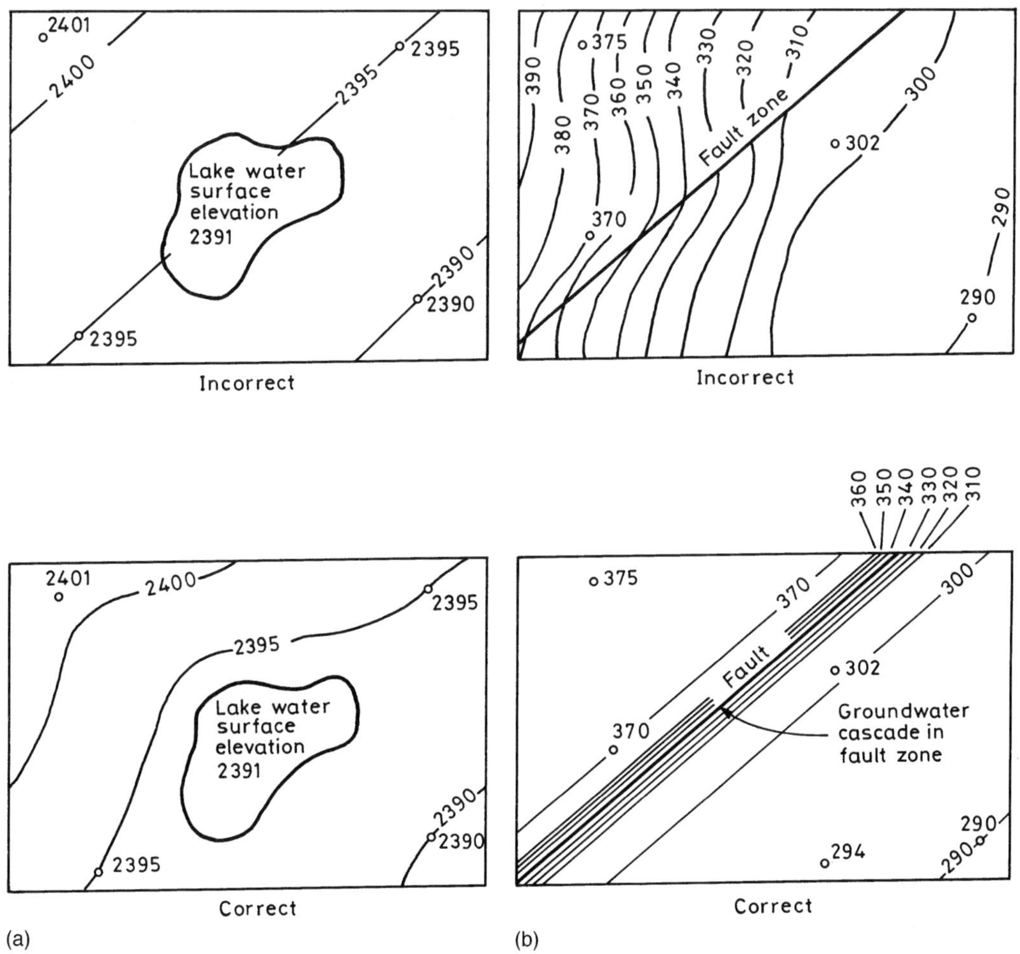

FIGURE 3.6 Contouring water-table maps in areas of (a) topographic depressions occupied by lakes, and (b) fault zones. Incorrect and correct contour maps are shown (after Davis and DeWiest, 1966).

fractured and karst aquifers may show complex response to recharge events.

3.5. Tracer studies

The study of groundwater movement using a conventional approach of analysis of water-level gradients and transmissivities generally envisages a continuous through-flow. However, in actual groundwater systems because of certain geological controls, especially in the fractured rocks, there are likely to be hydrogeological discontinuities. Thus in such rocks, groundwater flow – direction and velocity – can be best studied using tracers because the interpretation of these tests does not require the assumptions of a porous medium (continuum approach).

Tracers are defined as chemical substances (inorganic or organic molecules, including isotopes), being present naturally or introduced into the environment or added artificially, which follow the movement of water as faithfully as possible. Though there exists no single

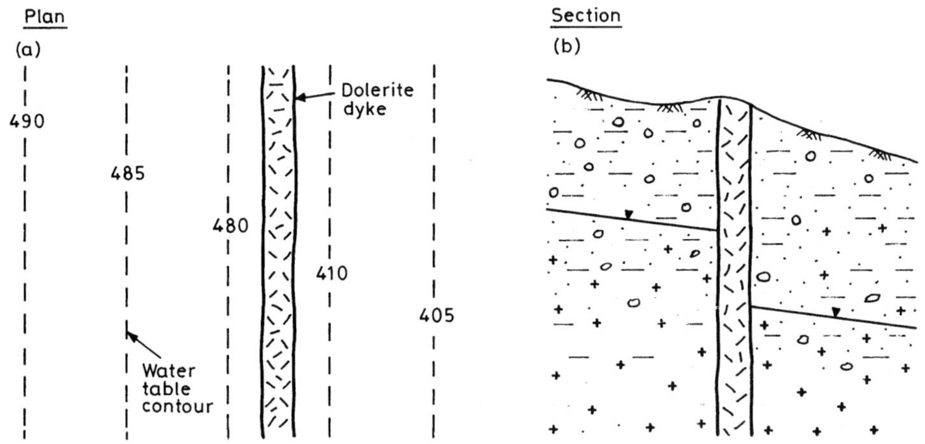

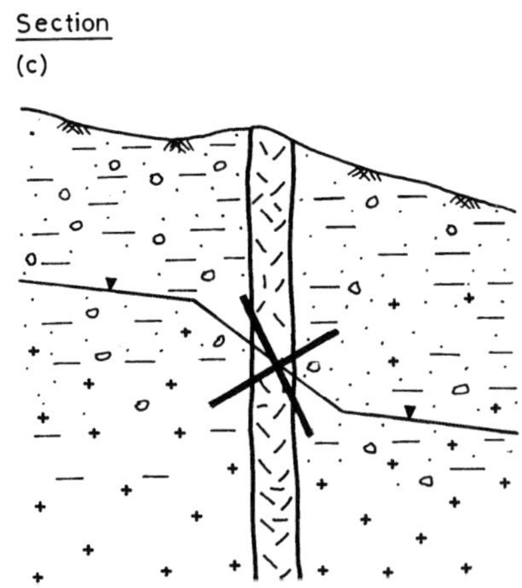

FIGURE 3.7 Effect of dyke leading to groundwater compartments, inhibiting linear interpolation of water-table contours. (a) plan; (b) and (c) are correct and incorrect interpolations respectively.

tracer which may meet all the requirements, an ideal tracer should be (Todd, 1980):

1. quantitatively detectable in very small concentrations;
2. more or less absent from the natural water to be traced;
3. neither adsorbed by the aquifer material nor chemically reacting with natural water causing its precipitation or biological degradation;
4. non-toxic to the people and the ecosystem;
5. inexpensive and readily available;
6. in the case of chemical tracers, the resulting solution should have approximately the same density as water.

Non-toxicity is the most important tracer characteristic. A variety of tracers is used in hydrogeological investigations for estimating the rate and direction of groundwater movement, groundwater recharge and its residence time. Tracers are also used for the study of origin of groundwaters including saline and geothermal waters, contaminant transport and interconnection between surface water and groundwater, stream discharge measurement and sediment transport.

3.5.1 Types of tracers

A tracer can be natural or artificial, inert (stable) or radioactive. Natural tracers are those

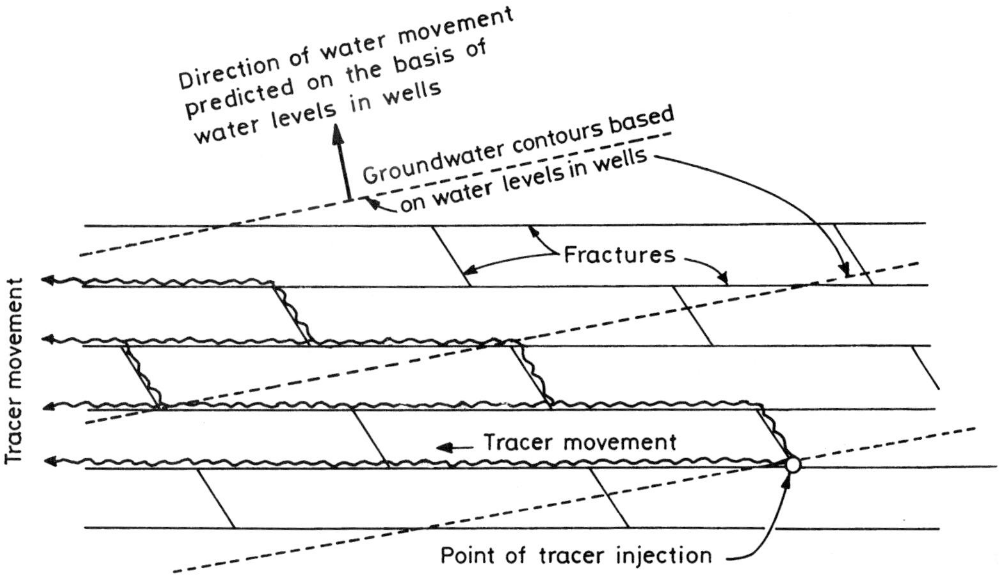

FIGURE 3.8 Comparison of groundwater flow direction inferred from water-table contours with the actual flow direction determined from tracer studies in a fractured rock (after Davis and DeWiest, 1966).

which are present in the natural waters. For example, surface water with a comparatively high concentration of nitrate can be used as a natural inert tracer. In other situations, injection of fresh water into a deep seated formation having highly saline water may act as a tracer and can be useful in studying the mixing phenomenon and thereby estimating hydraulic conductivity of formations.

Silica is a good example of a natural reactive tracer to trace the movement of water in hot fractured formations because its solubility increases with a rise in temperature. Increase of silica caused by rock dissolution during an injection backflow experiment in the hot dry rock (HDR) experiment in the Rhine graben gave a rough idea of the quantity of rock dissolved during the interaction (Jouanna, 1993).

The selection of a tracer is site-specific, depending on both the formation fluid and rock characteristics and also their temperature. A list of tracers used in groundwater investigations is given in Table 3.5.

Sometimes it is also necessary to use multiple tracers to have a better understanding of the aquifer characteristics. One of the limitations of the artificially injected tracers is that they can, at best, give the hydraulic characteristics of only a small segment of the aquifer investigated. Secondly, they provide information about the system valid only for the duration of the experiment. A good knowledge of the mineralogical characteristics of the rock helps to obtain information about the adsorption behaviour of various tracers. Matrix diffusion, especially in

TABLE 3.5 List of commonly used tracers in groundwater studies

Dyes	Sodium fluorescein, methylene blue, congo red
Salts	Sodium chloride, calcium chloride, ammonium chloride, lithium chloride
Organic compounds	Chlorofluorocarbons
Radioactive isotopes	^{3}H, ^{14}C, ^{32}P, ^{36}Cl, ^{51}Cr-EDTA, ^{60}Co ^{82}Br, ^{86}Rb, ^{129}I
Stable isotopes	^{2}H, ^{4}He, ^{15}N, ^{18}O

fractured rocks, is also important as it will considerably influence the flow pattern of tracers. This subject is considered further in section 7.4.

Dyes and salts

Dyes, like fluorescein, congo red and methylene blue are commonly used as groundwater tracers. These can be detected visually or by fluorometers. The amount of dye required for determining the groundwater velocity depends on the permeability and thickness of the aquifer as well as the distance between the injection and observation wells (Table 3.6).

Salt solutions are also commonly used as tracers; the water soluble salts like chlorides and sulphates can be detected chemically or by measuring electrical conductivity of groundwater. For *in situ* measurement of electrical conductivity in a well, special electrodes insulated from the well casing are lowered at a position opposite to the screen. Approximate doses of the salt tracers used for injection and recommended distances between injection and observation wells are given in Table 3.7.

TABLE 3.6 Dye quantities used in measuring groundwater velocity (in grammes per 10 m of flow path) (after UNESCO, 1972)

Dye	*Type of water-bearing fomation*			
	Clayey	*Sandy*	*Fractured*	*Karstic*
Fluorescein	5–20	2–10	2–20	2–10
Congo red	20–80	20–60	20–80	20–80
Methylene blue	20–80	20–60	20–80	20–80

Table 3.7 Recommended doses of salt tracers for injection into wells (after UNESCO, 1972)

Salt	*Dose (kg)*	*Distance between injection and observation wells (m)*
Sodium chloride	10–15	5–7
Calcium chloride	5–10	3–5
Ammonium chloride	3–5	2–5
Lithium chloride	0.010–0.015	2–5

Organic compounds

Chlorofluorocarbons (CFCs) are very stable synthetic organic compounds containing chlorine and fluorine. They are used in refrigeration, air conditioning and as cleaning agents and solvents. Their concentration in the air has been well known since 1950. Recently the use of CFCs has been restricted as they deplete the ozone layer and contribute significantly to the greenhouse warming effect.

Although CFCs are not environmental isotopes, they are of great value in groundwater studies as tracers and for estimating rates of recharge to shallow aquifers. They are also useful for dating groundwater of recent origin (younger than 45 yr) except in urban and industrialized areas, due to the release of CFCs from man-made sources locally (Busenberg and Plummer, 1992). It is better to use CFCs with other environmental isotopes like ^{3}H and ^{15}N to study a particular hydrogeological problem (Long, 1995).

3.5.2 Isotopes

Isotopes of a particular element have the same atomic number but different atomic weights, e.g. ^{1}H, ^{2}H, ^{3}H. Isotopes are used for a variety of purposes in hydrogeological investigations, viz. as tracers, for determination of age and origin of groundwater, and estimation of recharge and hydraulic characteristics of aquifers (IAEA, 1983). The isotopes may be radioactive or stable. A number of radioisotopes produced naturally by the cosmic ray interaction with atmosphere get incorporated into the hydrological cycle. Added to the cosmogenically produced isotopes, there are isotopes introduced into the environment through the testing of nuclear weapons and atomic installations. Together, these are called environmental isotopes. The isotopes of hydrogen and oxygen, viz. ^{2}H, ^{3}H and ^{18}O are natural tracers that tag water directly and therefore are of particular interest in hydrological investigations; the commonly used radioactive environmental isotopes of interest are ^3H and ^{14}C. Stable isotopes are useful in determining the source of water and its genesis. A convenient way of expressing the ratio of two isotopes of an element is by the delta relation. For example, $\delta^{18}0$ is defined as

$$\delta\,^{18}O‰ = \frac{(^{18}O/^{16}O)_{standard} - (^{18}O/^{16}O)_{standard}}{(^{18}O/^{16}O)_{standard}} \times 1000 \qquad (3.2)$$

For deuterium ^2H and ^{18}O, the normal standard reference is SMOW, an acronym for Standard Mean Ocean Water.

The advantage of radioactive tracers is that they can be detected even in low concentrations in water. Although tritium, as HTO, has been used widely in a variety of hydrogeological problems, it can not be detected *in situ* in the field; it is also preferentially adsorbed by montmorillonite (Hamill and Bell, 1986). ^{14}C also cannot be measured *in situ* in the field. Radioactive tracers are commonly used for dating groundwater, which is of importance from the point of view of management of groundwater resources. For example, ^3H (half-life 12.3 yr) and ^{14}C (half-life 5730 yr) are widely used for this purpose. Because of the short half-life of ^3H, it is used to estimate the residence time of comparatively younger water which was recharged during the last three to four decades. Estimation of recharge using tritium is discussed in Chapter Seventeen. ^{14}C can be used for the age determination of much older (up to 50 000 yr) waters. Lloyd and Heathcote (1985) have given several interesting case histories of the use of ^{14}C for dating of groundwaters. As the origin of groundwater is also related to climatic cycles, the presence of isotopically different water can be attributed to recharge in pluvial intervals during glacial epochs (Payne, 1967). For example, ^{14}C dating of deep well waters from some arid zones in the United Arab Republic and Saudi Arabia, gave ages of 20 000 to 30 000 yr which shows that the waters were recharged during the Wisconsin ice age when the area had a wet climate (Todd, 1980). Studies in the arid tracts of western Rajasthan, India, also show that the shallow groundwater is recharged from river channels and flash floods but the groundwater in deeper aquifers has negligible tritium and depleted $\delta^{18}O$ indicating that it was recharged during pluvial times in the past (Navada *et al.*, 1993). Radioisotope (^3H and ^{14}C) measurements in groundwaters from the fractured crystalline rocks of the Vedavati basin in south India indicate that shallow groundwater is quite young but the deeper groundwater is old (200–4000 yr) indicating lack of interconnection between fractures (Sukhija and Rao, 1983). Similar results have been obtained from studies in the crystalline rocks of the Bohemian Massif, Czech Republic (Silar, 1996).

^{36}Cl (half-life, 301 000 yr) is suitable for dating much older groundwaters within a range of 5×10^4 to 1×10^6 yr, which is beyond the normal range of dating. It is chemically stable and is not adsorbed on soil particle surfaces. In addition to dating of groundwater, ^{36}Cl has also been used for recharge studies in the vadose zone especially in arid and semi-arid regions (Long, 1995). However, as in deep groundwaters, there may also be other sources of Cl, due to either dissolution of evaporite deposits or membrane filtration; therefore necessary correction in the observed concentration of ^{36}Cl is required for dating (Long, 1995).

The ratio of stable isotopes δ^2H and $\delta^{18}O$ is used for a variety of hydrogeological problems based on fractionation depending on altitude, distance from the ocean and evaporation. It is therefore possible to identify the source of groundwater through stable isotope composition. For example, groundwater which has the source of recharge at high altitudes will be richer in lighter isotopes, and may be distinguished from the groundwater recharged at lower altitudes. On the other hand the contribution from surface water bodies like lakes, ponds and reservoirs, subjected to the evaporation process, can be identified by enrichment in heavier stable isotopes. Further, isotope data can also be used to determine the possible mixing of different types of groundwaters, salinization processes and water–rock interaction.

Deuterium (^{2}H) and ^{18}O data from hot and cold springs help in determining the source of the recharge and flow pattern. A similarity of δ^2H values between spring water and precipitation would indicate local recharge; an increase in δ^{18}O values in hot spring waters is attributed to water–rock interaction (Chapter Sixteen).

An excellent example of the importance of ^{3}H and stable isotopes for studying flow mechanism in the crystalline rocks of the Mont Blanc tunnel in the Central Austrian Alps is given by Fontes *et al.* (Long, 1995). This study demonstrated that the groundwater infiltration is vertical and is controlled by the structure of the steeply dipping rock formations. A similar study in the crystalline rocks of Seoul, Korea, indicated that the main recharge to crystalline rock aquifers is by downward percolation from the overlying alluvial and unconsolidated deposits (Ahn, 1988).

^{15}N has been used for identifying sources of nitrate pollution in groundwater from agricultural fields and household septic systems. Its use so far has been limited because of analytical problems and poor understanding of ways to distinguish various sources of nitrates in natural waters. Preferably, in addition to ^{15}N, other tracers may also be used to give more dependable results. As an example, Barraclough *et al.* (1994) used NO_3 and ^{2}H as tracers for studying fracture flow mechanism in the British Chalk which helped in obtaining a better understanding of pollutant migration.

Chapter four

Remote sensing

4.1 Introduction

Remote sensing, encompassing the study of satellite data and aerial photographs, is an extremely powerful technique for earth resources exploration, mapping and management. It involves measurements of the electromagnetic (EM) radiation in the wavelength range of about 0.4 μm-1 m, from sensors flying on aerial or space platforms to characterize and infer properties of the terrain. Remote sensing has evolved primarily from the methods of aerial photography and photo-interpretation used extensively in the 1950s and 1960s. The technique has grown rapidly during the last 3–4 decades. In the context of groundwater studies, remote sensing is of great value as a very first reconnaissance tool, the usual sequence of investigations being: satellite images, aerial photographs, geophysical data, drilling. The remote sensing data are used for studying various groundwater indicators and mapping their distribution, as discussed later.

Fundamental principle

The basic principle involved in remote sensing is that each object, depending upon its physical characteristics, reflects, emits and absorbs varying intensities of radiation at different EM wavelength ranges. The curves depicting relative intensity of light reflected/absorbed/emitted by the objects at different wavelengths are called spectral response curves (Figure 4.1). Using information from one or more wavelength ranges, it is possible to discriminate between different types of ground objects (e.g. water, dry soil, wet soil, vegetation) and map their distri-

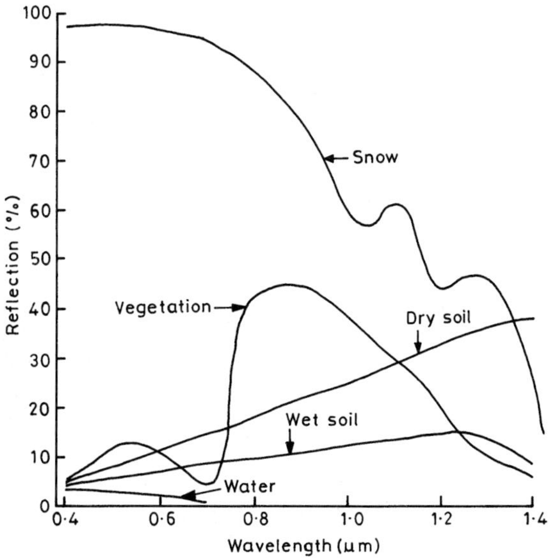

FIGURE 4.1 Spectral response curves of selected objects.

bution on the ground. A generalized schematic of energy data/flow in a typical remote sensing programme is shown in Figure 4.2.

Advantages and limitations

The chief advantages of remote sensing techniques over other methods of data collection are due to the following:

(1) Synoptic overview. Remote sensing permits delineation of regional features and trends.
(2) Feasibility. In some inaccessible areas, remote sensing may be the only way to get the information.
(3) Time saving. The technique saves time and manpower as information about a large

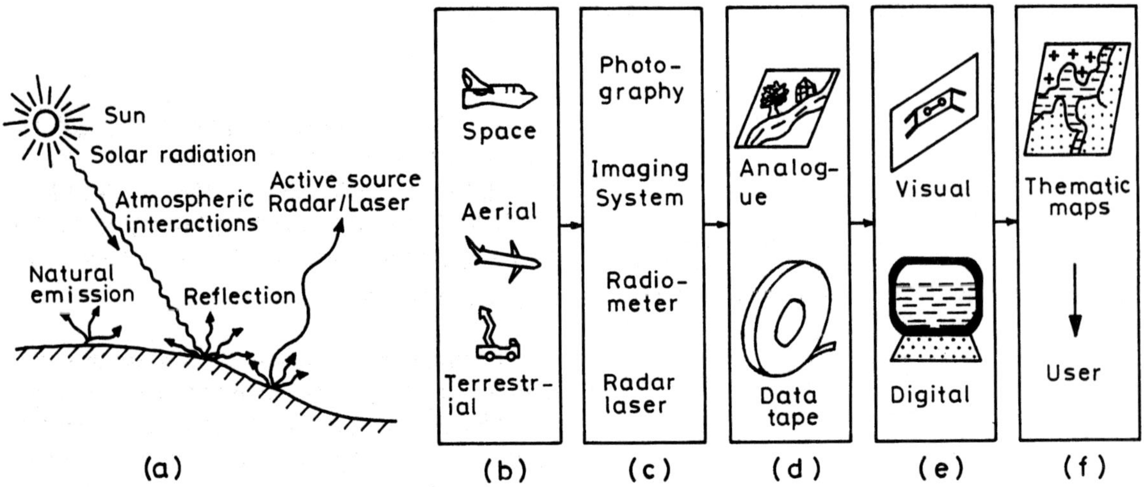

FIGURE 4.2 Schematic data flow in a typical remote sensing programme.

area is quickly gathered.

(4) Multidisciplinary applications. The same remote sensing data can be used by workers in different disciplines of natural sciences.

The main limitation of the remote sensing technique for geological–hydrogeological applications arises from the fact that the EM radiations have a limited depth of penetration – say a fraction of a millimetre in the visible-near IR range to a couple of metres (in dry desert conditions) at the most in the microwave region. Therefore, subsurface hydrogeological interpretations have to be based, by and large, on indirect surface evidence and features such as landform, lithology, structure, vegetation, soil, drainage, land use, surface anomalies and other guides (see section 4.8).

4.2 Physical principles

EM spectrum and sources of radiations

Electromagnetic spectrum is the ordering of EM radiations according to wavelength, frequency, or energy. The nomenclature used for different parts of the EM spectrum is shown in Figure 4.3c.

For remote sensing purposes, the sensors may utilize either naturally available radiations from the Sun or the Earth or artificial radiations. The technique involving artificial illumination is called active, in contrast to the one utilizing naturally available radiations, which is termed passive. Solar reflected radiations dominate in the ultraviolet-visible-near IR parts of the spectrum (Figure 4.3a), which, therefore, is called the solar reflection region. The Earth-emitted radiations dominate in the 3–20 μm wavelength region, and this spectral region is therefore called thermal IR. Besides, artificial illumination using radar is frequently used in the microwave region.

Atmospheric interactions

The EM radiations while passing through the Earth's atmosphere interact with atmospheric constituents and are selectively scattered, absorbed and transmitted. Raleigh scattering is the most important type of scattering. It leads to haze and low-contrast pictures in the UV-blue parts of the EM spectrum. Further, selective absorption of the EM radiations takes place by atmospheric gases such as H_2O-vapour, CO_2 and O_3 etc. The spectral regions of least atmospheric absorption are called **atmospheric win-**

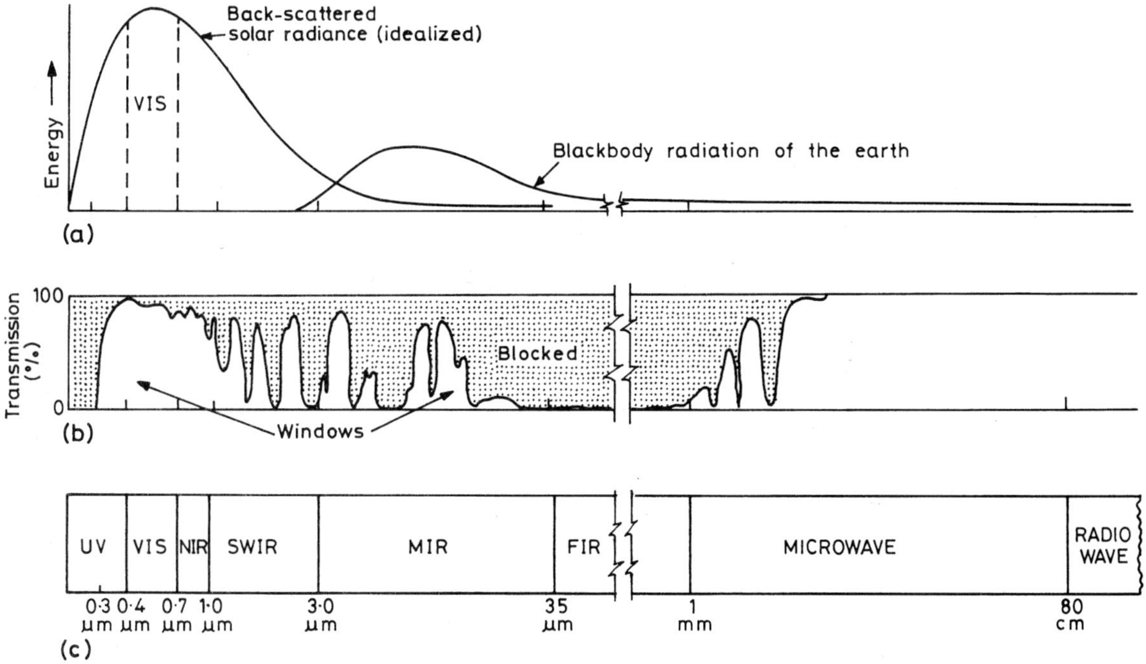

FIGURE 4.3 (a) Backscattered solar radiation (idealized) and black body radiation of the earth; (b) atmospheric windows; (c) EM spectrum and nomenclature.

dows. They can be used advantageously for looking at the Earth's surface from aerial/space platforms. Major atmospheric windows occur in the visible (VIS), near-IR (NIR), shortwave-IR (SWIR), thermal IR (TIR), and microwave regions (Figure 4.3b).

Platforms

The sensors for EM radiations can be operated from airborne, spaceborne or ground-based platforms. Some of the typical airborne platforms are helicopters, aircraft and balloons, whereas satellites, rockets and space shuttles constitute the space platforms. The hydraulic platforms and hand-held field instruments are used for collecting ground truth, to help better interpret aerial and space remote sensing data. Often satellite multispectral image data and aerial stereo photography are employed to provide information on different scales in a complementary manner.

4.3 Remote sensors

A sensor is a device to measure radiation intensity; it comprises optical components and detectors. A wide variety of sensors operate in various parts of the EM spectrum. For a comprehensive discussion on remote sensors, the reader is referred to other texts (Colwell, 1983; Sabins, 1987; Lillesand and Kiefer, 1987; Gupta, 1991). Keeping in mind operational aspects for groundwater investigations in fractured rocks, three main types of sensors are considered here: (a) photographic systems, (b) scanning systems, and (c) imaging radar systems.

4.3.1 Photographic systems

Perhaps the most familiar remote sensing systems are photographic. Their main advantages are high resolution and good geometric accuracy, whereas their chief limitation includes logistics related to retrieval of films from spacecraft. The working principle of an aerial/space camera system is simple. Ground objects falling into the field-of-view of the lens are focused and imaged on the photographic film. Most commonly, vertical aerial photography is carried out with an overlap of 70–75% to enable stereoviewing.

A standard film is panchromatic black-and-white and gives output in shades of grey. The other films are colour, of which colour infrared film is most important for hydrogeological investigations. Filters permit transmission of selected wavelengths of light and form an integral part of the modern photographic system. Multiband cameras with different film–filter combinations are used for spectrozonal photography, i.e. photographing the same ground area in different spectral ranges, such as blue, green, red, near-IR bands etc. In comparison to the photographic techniques, the spaceborne scanners provide better quality multispectral image data at lower costs, and therefore the use of multiband cameras is restricted to specific cases.

The photographic systems can be deployed from space, aerial or terrestrial platforms. From aerial platforms, the vertical stereophotography is a standard technique (Miller and Miller, 1961; Wolf, 1983). Its unique advantages of 3D perspective viewing with good geometric accuracy have not been taken over by any spaceborne remote sensing technique to date. Aerial photographs normally of 1:25 000 to 1:80 000 scale are used, the larger-scale photography being ideally employed for detailed site selections. From groundbased platforms, a relatively new method called terrestrial geophotogrammetry has been advantageously developed during the last two to three decades, and is briefly mentioned in section 4.6.2.

4.3.2 Scanning systems

Introduction and working principle

Multispectral scanners constitute the most important device for remote sensing from satellite platforms. They are extensively used in the visible, near-infrared and thermal infrared parts of the spectrum. A scanner is a non-photographic device and generates digital data on intensity of ground radiance. The entire scene is considered to be comprised of a large number of smaller cells. The radiations from each unit cell are collected and integrated by the scanner, to yield a brightness value, or digital number (DN), which is ascribed to the unit cell (Figure 4.4). This process is called scanning.

Broadly, a scanner consists of two main parts: an optical part and a detector. The optical part collects the radiations and divides them into different wavelength ranges. The detector

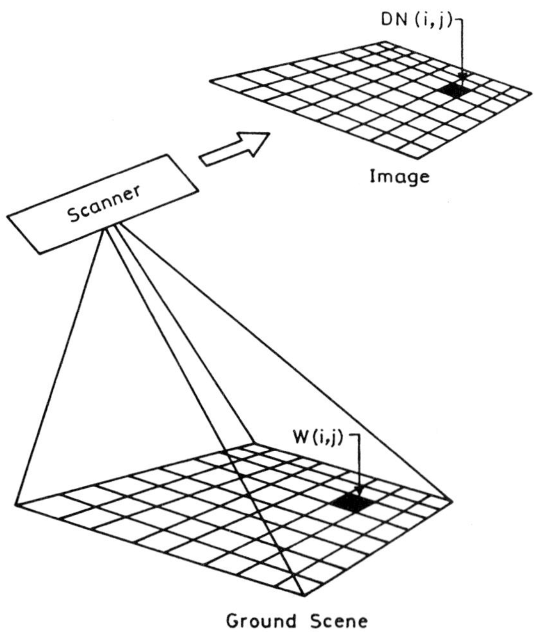

FIGURE 4.4 Working principle of a scanner. The ground radiance $W_{(i,j)}$ from a ground cell is converted into a digital number $DN_{(i,j)}$ at the corresponding position in the image.

part transforms the optical energy into an electrical signal. The amplified signal is classified, depending upon its intensity, into one of the various levels, called quantization levels (or digital numbers, DNs). These data are relayed from satellite sensors to ground receiving stations on microwave links. Thereafter, the scanner data are rearranged as a two-dimensional data array, to form an image. The ground resolution cell corresponds to a picture element (pixel) on the image/screen. There are several advantages of scanners over photographic techniques related to retrieval of data, sensing in extended wavelength range with high spectral, spatial and radiometric resolutions. Further, the digital information is amenable to computer processing enhancement etc.

Resolution of a sensor is an important parameter and is given in terms of spatial, spectral, radiometric and temporal aspects. Spatial resolution implies the ground resolution and corresponds to a pixel on the image; spectral resolution means the span of wavelength range over which a spectral channel operates. Radiometric resolution generally implies the total number of quantization levels used in the sensor. Temporal resolution refers to the repetition of observations over an area.

Types of scanners

There are two basic types of line scanners: (1) opto-mechanical (OM)-line scanners and (2) charge-coupled device (CCD)-line scanners. The OM-line scanners have been extensively used from aerial and space platforms in multispectral mode in a wide range of wavelengths from visible, near-IR, and SWIR to thermal-IR. In this, the collector optics include a moving plane mirror to reflect radiations from the ground which are collected and directed onto the filter and detector assembly. The MSS and TM on Landsats are typical OM-scanners. The CCD-line scanners are the modern solid-state scanners, used in many recent earth-observation satellites, such as SPOT-HRV, IRS-LISS, JERS-OPS, and MOMS. The CCD sensors utilize photoconductors as the detector material. A CCD-array comprising several thousand detector elements is placed at the focal plane of a camera lens system which senses the radiations and converts the same into electrical signals. Typically, for each spectral band there is one CCD-array viewing the entire swath simultaneously. Thus, there is no moving part in the sensor.

4.3.3 Imaging radar system

Introduction and working principle

The electromagnetic (EM) spectrum range 1 mm to 80 cm is designated as microwave. The atmosphere is nearly transparent at these wavelengths. Basically, the microwave technique is of two types: (a) passive, and (b) active. The passive microwave radiometry has a major limitation of coarse spatial resolution (about 20–30 sq km). We therefore, in the context of groundwater applications, restrict this discussion to active microwave remote sensing, also called imaging radars.

Radar (Radio Detection and Ranging) basically operates on the principle that artificially generated microwaves sent in a particular direction collide with the objects and are scattered. The back-scattered radiations are received, amplified and analysed to determine location, electrical properties and surface configuration of the objects. The imaging radar technique has all-time and all-weather capability, independent of solar illumination and atmospheric–meteorologic factors.

For imaging purposes, the radar is mounted in the configuration of side-looking airborne radar (SLAR), at the base of the sensor platform (Figure 4.5). The radar transmits short pulses of EM energy, illuminating narrow strips of ground, perpendicular to the flight direction. Radiations back-scattered from the ground are recorded to give information about the strip. As

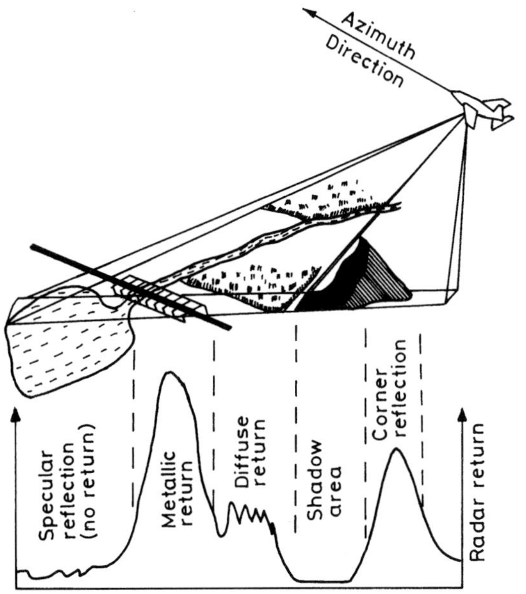

FIGURE 4.5 Principle of side-looking airborne radar and typical radar returns (modified after Sabins, 1987).

the sensorcraft moves forward, a two-dimensional image is generated.

SLAR system characteristics

The SLAR sensors are specified in terms of a number of parameters, viz. radar wavelength, EM beam polarization, depression angle, swath width, aperture type and resolution etc. (for details, see Elachi, 1983; Trevett, 1986). Commonly, SLAR systems have a ground resolution of the order of 5–30 m, depending upon the altitude and sophistication in data processing. A number of SLAR experiments have been conducted from aircraft and spacecraft. Important spaceborne SLAR missions have been Seasat, SIR-A, SIR-C, ERS-1, ERS-2, JERS-1 and Radarsat.

The geometry of SLAR imagery is fundamentally different from that of photographs and scanner imagery and therefore deserves specific mention. Many geometric distortions arise due to image displacement, particularly in mountainous areas. Further, shadows and look-direction effects are also prominently manifested on SLAR images.

The back-scattered signal received at the SLAR antenna is called 'radar return' and carries information about the properties of the object. It depends upon numerous factors, such as radar wavelength, EM beam polarization, local incidence angle, target surface roughness and complex dielectric constant. Thus, it has to be kept in mind during interpretation that the radar response is governed by a complex interplay of radar system and ground terrain factors.

4.4 Important space sensors

Landsat programme

To date, the Landsat programme has provided the most extensively used remote sensing data the world over. Its chief advantage has been in delivering unrestricted global data of good geometric accuracy. Under the Landsat programme until now six satellites (Landsat 1, 2, 3, 4, 5 and 6) have been launched (Landsat-6 having failed). These satellites have been placed in near-polar, near-circular, sun-synchronous orbit. In this configuration, as the satellite orbits in the north–south plane, the Earth below spins around its axis, from west to east (Figure 4.6). Thus, different parts of the globe are 'seen' by the satellite during different north–south passes. The remote sensing data are acquired in the descending node, i.e. as the satellite moves from the north pole to south pole. The L – 1, 2, 3 were placed at an altitude of 918 km with a repeat cycle of 18 days, and L – 4, 5 at an altitude of 705 km and a repeat cycle of 16 days.

The Landsats have carried on-board two main sensors, Multispectral Scanning System (MSS) and Thematic Mapper (TM), both being OM-line scanners and producing ground scenes of nominally 185 × 185 km size. The MSS sensor has been a regular payload of the Landsats and has made this programme a tremendous success.

FIGURE 4.6 Typical near-polar orbit of an earth observation satellite such as Landsat, IRS etc. As the satellite revolves in N–S orbit, the earth spins from W to E; therefore different parts of the earth are observed by the satellite in different N–S passes.

The TM is an advanced multispectral scanner used in Landsat 4 and 5 missions. It operates in seven wavelength bands, out of which six are in the solar reflection region and one in the thermal-IR region. Salient specifications of MSS and TM sensors are given in Table 4.1.

SPOT programme

The French satellite system SPOT is a typical second-generation earth resources satellite. These satellites have been placed in near-polar sun-synchronous 830 km high orbit with a repeat cycle of 26 days. The sensor here is called HRV (High Resolution Visible) which is a CCD-line scanner. The HRVs acquire data in two interesting modes: (a) panchromatic mode in a swath of 60 km with ground resolution of 10 × 10 m and (b) multispectral mode, in three channels (green, red and infrared) with a ground resolution of 20 × 20 m in a swath width of 60 km. Further, the HRVs can be tilted to acquire data in off-nadir viewing mode, for more frequent repetitive coverages, and for stereoscopy.

IRS programme

Under the Indian remote sensing satellite (IRS) programme, the Indian Space Research Organisation (ISRO) has launched a series of land observation satellites, named as IRS-1A (1988), IRS-1B (1991), IRS-1C (1995) and IRS-1D (1997). The

TABLE 4.1 Salient specifications of Landsat MSS and TM bands

Band no.	*MSS Spectral range) (μm)*	*Ground resolution (m)*	*Band no.*	*TM Spectral range (μm)*	*Ground resolution (m)*	*Band name*
			1	0.45–0.52	30	Blue
1	0.5–0.6	79	2	0.52–0.60	30	Green
2	0.6–0.7	79	3	0.63–0.69	30	Red
3	0.7–0.8	79 }	4	0.76–0.90	30	Near–IR
4	0.8–1.1	79 }				
			5	1.55–1.75	30	SWIR–I
			7	2.08–2.35	30	SWIR–II
			6	10.4–12.5	120	Thermal–IR

TABLE 4.2 Salient specifications of IRS sensors

Sensor	*LISS-I*		*LISS-II*		*LISS-III*	*Pan*
Spacecraft	IRS–1A, 1B		IRS–1A, 1B		IRS–1C	IRS–1C
Spectral bands	4*		4*		4	1
Name	Code	Spectral range (μm)	Code	Spectral range (μm)		Broad spectral range
Blue	B1	0.45–0.57	(B1	absent)		0.50–0.75 μm, with off-nadir viewing capability
Green	B2	0.52–0.59	B2	0.52–0.59		
Red	B3	0.62–0.68	B3	0.62–0.68		
Near-IR	B4	0.77–0.86	B4	0.77–0.86		
SWIR-I	–	–	B5	1.55–1.70		
Ground resolution	72.5 m	36.25 m	23.5 m for B2–4, 70.5 m for B5			5.8 m
Swath (km)	148.5	74.2 × 2	141 (B2–4) & 148 (B5)			70

* LISS-I and LISS-II have identical spectral bands and differ only in ground resolution

first three of these satellites have been placed in near-polar, sun-synchronous orbit, at nominal altitudes of 904 km (IRS-1A and -1B) and 817 km (IRS-1C) whereas there have been some problems with the orbit of IRS-1D. The sensors utilize CCD linear arrays and are called linear imaging self-scanning (LISS) cameras. Both IRS-1A and -1B have carried on board identical LISS-I and LISS-II sets of cameras, whereas IRS-1C and IRS-1D use Pan and LISS-III cameras (Table 4.2).

JERS-1 (Fuyo-1) programme

Japan's Earth Resources Satellite-1 (renamed Fuyo-1) (1992) employs both an optical sensor (OPS) and a radar sensor (SAR) on the same spacecraft. Fuyo-1 has been placed in a near-circular, sun-synchronous, near-polar orbit at a nominal altitude of 568 km with a 44-day repeat cycle. The OPS sensor system uses CCD linear arrays, and can be broadly divided into two sub-systems: VNIR and SWIR. Its salient specifications are summarized in Table 4.3. The Fuyo-1 SAR operates at 23.5 cm wavelength and provides data in a swath of 75 km with a ground resolution of 18 m (Nishidai, 1993).

TABLE 4.3 Salient specification of JERS-1 OPS sensor system (after Nishidai, 1993)

Band		*Wavelength (μm)*	
VNIR	1	Green	0.52–0.60
	2	Red	0.63–0.69
	3	NIR	0.76 –0.86
	4*	NIR	0.76–0.86
	*	Forward viewing for stereo images	
SWIR	5	1.60	1.71
	6	2.01	2.12
	7	2.13	2.25
	8	2.27	2.40
Detector		: CCD linear array	
Ground resolution		: 18.3 × 24.2 m	
Swath width		: 75 km	

Other programmes

In addition to the above, a number of other space sensors have also been flown, e.g. shuttle-borne experiments like the German MOMS and Stereo-MOMS, Indian-IPS 1, 2, 3, Seasat, SIR-A, SIR-B, SIR-C, ALMAZ, European ERS-1, 2 etc. Most of their specifications are of the same order as the sensors described above (for details see e.g. Colwell, 1983; Ryerson *et al.*, 1997).

4.5 Interpretation principles

Procedures for study

To start a study, first, the remote sensing photographs and images are indexed to provide information on location, scale, orientation and areal coverage. Mosaicing provides a bird's-eye view of the entire area. Overlapping photographs and images may be studied stereoscopically. Products of a multispectral set can be combined in colour mode. Digital data can be processed for enhancement, classification and interpretation and it can also be combined with ancillary geodata in a GIS approach, for a comprehensive interpretation.

Interpretation parameters

Several photo-characteristics are used to study and interpret features on photographs and images. These are called elements of interpretation and geotechnical elements (Table 4.4) (for details see Colwell, 1960; Mekel, 1970, 1978). Convergence of evidence is a general underlying principle in photo-interpretation. It implies integrating all the evidence and analysing where all the photo-parameters are leading to, collectively.

Ground truth

Ground truth is necessary for reliable interpretation of remote sensing data. It aims to provide a reference base for interpretation, and also to verify anomalous responses, if any. Parameters for field survey may include: ground surface character, geological structure, rock/soil type, soil moisture, vegetation type and density, groundwater level. Often the ground truth observations are made by spot field-checking.

TABLE 4.4 Elements of photo-interpretation and geotechnical elements

Sl.		*A. Elements of photo-interpretation*
1.	Tone	It is a measure of the relative brightness of an object in shades of grey
2.	Colour	Appropriate terms are used to describe the colour; use of colour dramatically increases the capability of subtler distinctions
3.	Texture	It means tonal arrangements and changes in a photographic image; it is the composite effect of unit features too small to be discerned individually
4.	Pattern	It refers to spatial arrangement of features, like drainage, vegetation, etc.
5.	Shadow	Shadow helps in studying the profile view of objects, and is particularly useful for man-made objects
6.	Shape	Outline of an object in plan
7.	Size	Size of an object is related to scale
8.	Site/association	Mutual association of objects is one of the most important guides in photo/image interpretation
		B. Geotechnical elements
1.	Landform	Constructional and depositional landforms are associated with various natural agencies of weathering and erosion, and are extremely helpful in identifying hydrogeomorphological characteristics of the site
2.	Drainage	It is one of the most important geotechnical element and includes the study of drainage density, valley shape and drainage pattern
3.	Soil	Soil characteristics depend upon the bedrock material and agencies of weathering. They have characteristic hydrological properties which govern surface run-off vis-a-vis infiltration
4.	Vegetation	Vegetation distribution and type may be related to lithology and subsurface seepage zone

At times, adequately large-scale aerial photographs may also serve as ground truth for smaller-scale satellite image data.

4.6 Interpretation of remote sensing data

This discussion is divided into the following sub-headings:

1. Panchromatic photographs and images
2. Terrestrial geophotogrammetry
3. Multispectral solar reflection data
4. Colour infrared photographs and standard FCCs
5. Thermal IR data
6. SLAR imagery.

4.6.1 Panchromatic photographs and image data

Panchromatic sensors are broadband sensors operating in the visible range (0.4-0.7 μm wavelength). One major advantage of panchromatic band data is the higher spatial resolution (such as SPOT-Pan = 10 m; IRS-IC Pan = 5.8 m), in comparison to that of multispectral sensors. Panchromatic photographs and images may be obtained from a variety of platforms and sensors, e.g. stereo aerial photography, Metric Camera and Large Format Camera photography from a space shuttle, Stereo-MOMS from a space shuttle, SPOT-HRV-Pan sensor, and IRS-1C Pan sensor. These data from various sensors may differ in their geometry and scale; however, they are quite similar to one another as far as the spectral characters are concerned, and therefore, their interpretation must follow the same line of logic. The technique of air photo-interpretation has been described in numerous standard publications (Colwell, 1960; Miller and Miller, 1961; Ray, 1965). A number of factors such as atmospheric condition, topography and slope aspect affect the image data (Table 4.5). Target reflectance, i.e. albedo and its variation across the scene, is very important and deciphering this parameter holds the key to remote sensing applications in the solar reflection region. Spectral characters and identification features of selected features are described below.

On panchromatic products, a deep and clear water body appears dark; on the other hand, a turbid and shallow water body appears in shades of light grey to grey, and vegetation appears dark grey to light grey, depending upon the density and type of foliage. Soils appear in various shades of grey, the variation being related to moisture content, organic matter and grain size of the soil. Discrimination between various rock groups is based on convergence of

TABLE 4.5 Salient factors affecting image data in the solar reflection region

	Primary variables	*Secondary variables*	*Comments*
1	Atmospheric factors	Composition of the atmosphere leading to absorption; particulates and aerosols, leading to scattering; relative humidity, cloud cover, rain etc.	May vary within a scene from place to place
2	Topography and slope aspect	Goniometric aspects, landscape position and slope direction; sun-sensor look angle	Vary from place to place within a scene, depending upon sun-local topography relation
3	Target reflectance	Albedo of the object; surface texture and coating	Deciphering this attribute holds the clue in remote sensing

evidence derived from a number of parameters including landform, drainage and structure (section 4.8.8). Cultural features like cities, settlements, roads and railway tracks can often be recognized by their shapes and outlines.

Stereo aerial black-and-white photographs have been routinely used for the study of landforms, lithology, structure, lineaments, soil, vegetation etc. Interpretation of panchromatic data can provide valuable input to geohydrological studies, particularly in the context of recharge, run-off, evaporation and groundwater – all of which are inter-related.

4.6.2 Terrestrial geophotogrammetry

Terrestrial geophotogrammetry is an interesting tool for the study of fractured rocks. Stereo photographs of rocks exposed in outcrops, scarps, excavations etc. can be taken from a ground based (terrestrial) platform (Figure 4.7). The stereopair is studied under a photogrammetric optical projection plotting system. Measurements on fracture characteristics are done, as planes can be defined by a set of co-ordinates of three (or more) points. Data on all the fracture planes can be generated from stereopairs and the orientation data are handled by a computer to provide pole-plots; thus, extensive statistical data on fracture planes in rocks can be generated. This technique reduces field time and allows acquisition of data from inaccessible outcrops.

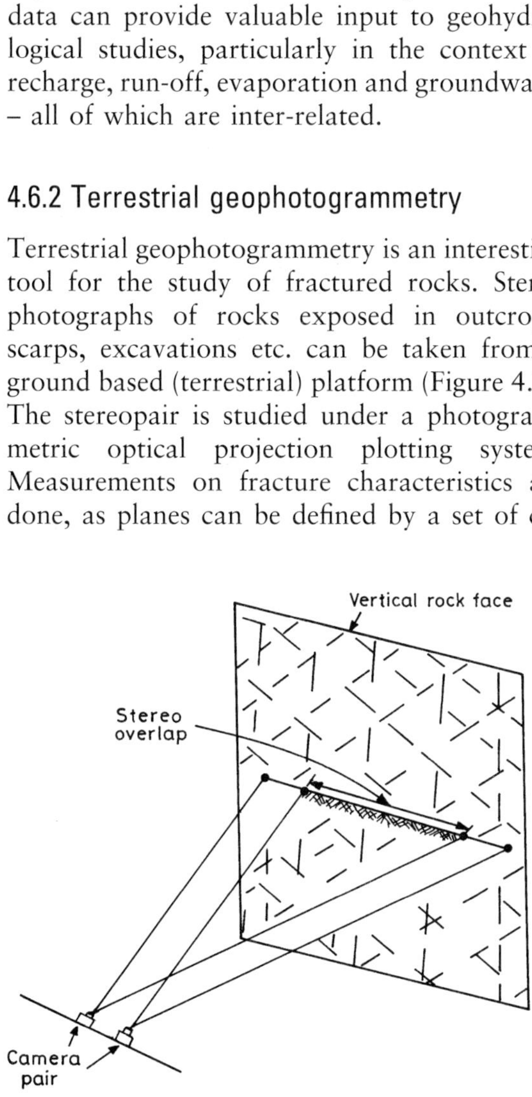

FIGURE 4.7 Schematic representation of terrestrial stereophotography. The camera axes are horizontal facing the vertical rock face.

4.6.3 Multispectral solar reflection data

Multispectral data in the solar reflection region are available, on a routine basis, from a number of space sensors such as Landsat MSS, TM, SPOT-HRV-Multi, IRS-LISS-I, LISS-II, LISS-III sensors, and JERS-OPS. Spectral distribution of the sensors is comparable (Table 4.6). Therefore, interpretation of the image products from these sensors follows a common line of argument. Multiband photographic products, although not available on a regular basis, are also included in this discussion, since the spectral considerations are the same as for the above sensors. Spectral characters of common objects are briefly mentioned below.

Water exhibits different characteristics, depending upon its suspended load and the depth of the water body. Clear deep water bodies are generally dark and turbid and shallow water bodies reflect in the shorter wavelengths. In the NIR and SWIR, all water bodies appear black. Vegetation in general appears medium-dark in the VIS and bright in the NIR. Cropland is medium grey and is marked by the characteristic field pattern. Soils and fallow fields (dry) are light in the VIS and medium grey in the NIR–SWIR bands. Moist ground is medium grey in the VIS but very dark in the NIR and SWIR. Rocky terrain (bare) is usually brighter in the VIS than in the NIR and SWIR, and may be characterized by landforms and structures. Clayey soils are medium to light toned in the VIS, NIR and SWIR-I, but are very dark in the SWIR-II, due to strong absorption by the hydroxyl anions. The multispectral data from

TABLE 4.6 Important multispectral sensors and their characteristics

	Ground resolution	*Blue*	*Green*	*Red*	*NIR*	*SWIR-I*	*SWIR-II*
Multiband photography	Variable	✔	✔	✔	✔		
Landsat-							
MSS	79 m		✔	✔	✔		
TM	30 m	✔	✔	✔	✔	✔	✔
IRS-							
LISSI	72.5 m	✔	✔	✔	✔		
LISSII	36.25 m	✔	✔	✔	✔		
LISSIII	23.5 m (G, R, NIR) & 70.5 m (SWIR-I)		✔	✔	✔	✔	
SPOT-HRV	20 m		✔	✔	✔		
JERS-OPS	18.3 × 24.2 m		✔	✔	✔	✔	✔ (3 bands)

Characteristics					
Blue	*Green*	*Red*	*NIR*	*SWIR-I*	*SWIR-II*
Very strong absorption by vegetation and Fe-O; good water penetration; high scattering by suspended particles	Some vegetation reflectance; good water penetration; scattering by suspended particles	Very strong absorption by vegetation; some water penetration; scattering by suspended particles	High reflectance by limonite and vegetation; total absorption by water	General higher reflectance; insensitive to moisture contained in vegetation or hydroxyl bearing minerals; absorption by water, snow	High absorption by hydroxyl-bearing minerals, carbonates hydrous minerals vegetation, water and snow

these sensors have opened up vast opportunities for mapping and monitoring of surface features, including hydrogeological investigations together with a multitude of other applications.

4.6.4 Colour infrared photographs and standard FCCs

Colour infrared (CIR) photography has been carried out from aerial platforms and also from selected space missions like Skylab, Metric Camera and Large Format Camera etc. Further, false colour composite (FCC) is a widely used technique for combining multispectral images of the same area through colour coding. In an FCC, three photographs or images of a multispectral set are projected concurrently, each one in a different primary colour, and geometrically superimposed to yield a colour composite. This optical combination carries information from all three input images in terms of colour variation across the scene. As these colours are not shown by the objects in nature, these are called false colour composites (FCCs). Though any spectral band may be coded in any primary colour, a standard FCC uses a definite colour coding scheme as follows.

- response in blue wavelength – cut-off;
- response in green wavelength – shown in blue colour;
- response in red wavelength – shown in green colour;
- response in NIR wavelength-shown in red colour.

This is analogous to a colour infrared (CIR)

photograph in terms of colour variation. On these images, vegetation appears in shades of red; deep and clear water bodies are dark, whereas shallow, silted water bodies are cyan; dry sands are white-yellow while wet fields are cyan-light greyish. Bare rocky slopes appear in shades of grey. Examples of FCCs are given in Figures 4.10 and 4.17.

4.6.5 Thermal IR data

The EM wavelength region of 3–25 μm is popularly called thermal infrared (TIR) region. Any sensor operating in this wavelength region would primarily detect the thermal radiative properties of the ground materials. Out of this the 8–14 μm region has been of greatest interest, as this is characterized by the peak of the Earth's blackbody radiation and is an excellent atmospheric window (Figure 4.3). The radiated energy is a function of two parameters: surface temperature and emissivity. The surface temperature is governed by a number of physical properties, among which thermal inertia is most important in governing diurnal temperature variations in nature. If thermal inertia is less, diurnal temperature variation is more and vice versa.

A typical thermal IR remote sensing experiment uses a set of pre-dawn (night) and day (noon) passes. Commonly, TIR image is a simple (radiant) temperature image. Other processed images, like a diurnal temperature difference image or thermal inertia image, can also be generated.

Topographical features are enhanced on daytime thermal images due to differential heating and shadowing; however, on the night-time images, the topography becomes subdued. All materials undergo cyclic heating and cooling in the day–night cycle, thus showing systematic temperature variations in a diurnal cycle (Figure 4.8). Standing water appears brighter (relatively warmer) on the night-time image and darker (cooler) on the daytime image. Soils with higher moisture are cooler than those with a lower moisture content. Temperature contrasts of between 5–10 °C could be produced by moisture variations in the soil alone. Effluent seepage would also lead to a cooler ground surface; soil moisture estimation has practical utility in agriculture and water budgetting. Drainages are also well shown on thermal data. Vegetation in general is warmer on the night-time and cooler on the daytime image, in comparison to the adjoining unvegetated land. The daytime lower temperature is related to the transpiration process in the plants and the night-time higher temperature to the higher moisture content in leaves.

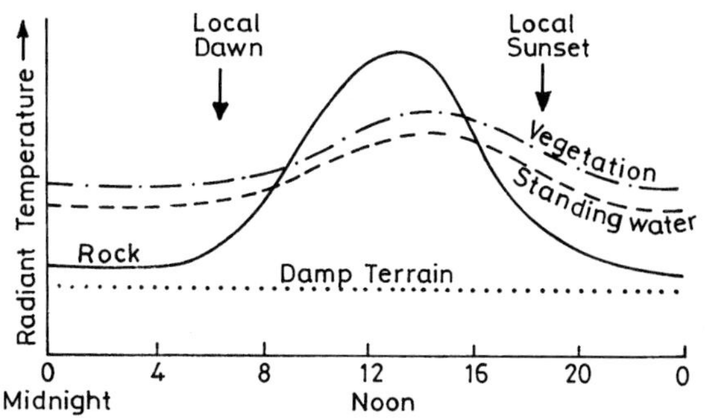

FIGURE 4.8 Thermal response of rock, vegetation, standing water and damp terrain in a diurnal cycle.

Structural features like folds, faults etc. may be manifested due to spatial differences in thermal characters of the rocks. Bedding and foliation planes appear as subparallel linear features due to thermal contrasts of compositional layering. Faults and lineaments may be conduits for groundwater flow and may be associated with springs. This would lead to evaporative cooling along a line or zone, producing linear features (Figure 4.18), therefore, thermal data can be used for exploration of shallow aquifers and water-bearing fractures. In general, the spatial resolution of sensors is an important aspect in applications. The aerial thermal scanners have spatial resolution of the order of 2-6 m, quite suitable for such applications. In contrast, the available imagery from spaceborne TIR sensors is generally of coarser resolution (120 m for Landsat TM6, 600 m for HCMR) and this has limited the application of space-acquired TIR imagery to mapping of only regional structural features.

4.6.6 SLAR imagery

As mentioned earlier, radar return carries valuable information about the physical, geometric and electrical properties of the ground objects. On a SLAR image, intensity of radar return is shown in shades of grey, such that areas of higher backscatter appear brighter (Figure 4.5). For common visual image interpretation, the simple radar images are widely used, though they can also be processed and combined with other data sets for enhancing hydrogeologic information.

Commonly on a SLAR image, most of the area is dominated by diffuse scattering caused by rough ground surface and vegetation; a very strong return is caused by metallic objects and corner reflectors (hills and buildings). Specular reflection is produced by smooth surfaces, like quiet water bodies, playa-lakes etc. and there is little return at the SLAR antenna (Figure 4.5).

Study of a drainage network is often the first step in radar image interpretation. The waterways are generally black to dark on the radar images, as the water surface leads to specular reflection. However, in semi-arid to arid conditions, the valley bases are often vegetated, and appear bright on the radar images (Figure 4.24). General terrain ruggedness is another important geotechnical element. Consideration of individual landform facet is very important for proper SLAR image interpretation and soil moisture is another important parameter. The microwave response is governed by complex dielectric constant (k) of the objects. Water has very high k, as compared to dry soil and rock. When moisture content in soil or rock is increased, a regular increase of backscattering coefficient of the mixture takes place (Figure 4.9). This variation can be used to map soil moisture on the ground.

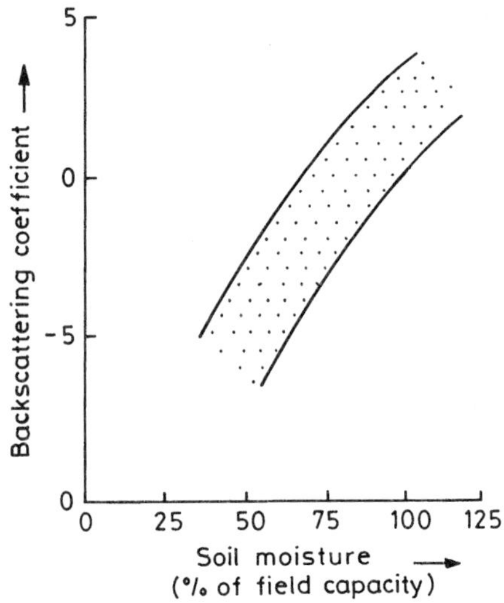

FIGURE 4.9 Increase in radar backscattering coefficient with rise in soil moisture (redrawn after Bernard *et al.*, 1984).

Owing to the fact that minor details are suppressed, the SLAR images have value for regional landform studies. Radar images are found to be immensely useful for structural delineation. Lineaments are extremely well manifested on SLAR images. The lithologic interpretation on a radar image has to be based on indirect criteria like surface roughness, vegetation, soil moisture, drainage, relief, and special features like sinkholes. The radar image texture is an important parameter in deciphering lithological units.

Depth penetration is another important point in the context of geological-hydrogeological applications. Depth penetration increases with longer wavelengths as shorter wavelengths undergo a sort of 'skinning' effect. Further, the depth penetration is inversely related to the complex dielectric constant (k). (Deane and Domville, 1973). Higher moisture content in soils inhibits depth penetration, whereas some limited depth information can be obtained in hyper-arid regions.

4.7 Groundwater indicators on remote sensing data products

As mentioned elsewhere, remote sensing data provides surface information, whereas groundwater occurs at depth and may be a few metres or several tens of metres deep. The depth penetration of EM radiations is barely of the order of fractions of a millimetre in the visible region, to hardly a few metres in the microwave region, at best. Therefore, remote sensing data are unable to provide any direct information on groundwater in most cases. However, the surface morphological-hydrological-geological regime, which primarily governs the subsurface water conditions, can be well studied and mapped on remote sensing data products. Therefore, remote sensing acts as a very useful guide and efficient tool for regional and local groundwater exploration, particularly as a forerunner in a cost-effective manner.

In the context of groundwater exploration, the various surface features or indicators can be grouped into two categories: (1) first-order or direct indicators, and (2) second-order or indirect indicators. The first-order indicators are directly related to the groundwater regime (viz. recharge zones, discharge zones, soil moisture and vegetation). The second-order indicators are those hydrogeological parameters which indicate the groundwater regime regionally, e.g. rock/soil types, structures, including rock fractures, landform, drainage characteristics etc. (Table 4.7).

TABLE 4.7 Important indicators of groundwater on remote sensing data (after Ellyett and Pratt 1975; Gupta, 1991)

(A) First-order or direct indicators
- (1) Features associated with recharge zones; rivers, canals, lakes, ponds etc.
- (2) Features associated with discharge zones: springs etc.
- (3) Soil moisture
- (4) Vegetation (anomalous)

(B) Second-order or indirect indicators
- (1) Topography
- (2) Landforms
- (3) Depth of weathering and regolith
- (4) Lithology
- (5) Geological structure
- (6) Lineaments, joints and fractures
- (7) Faults and shear zones
- (8) Soil types
- (9) Soil moisture
- (10) Vegetation
- (11) Drainage characteristics
- (12) Special geological features, like, karst, alluvial fans, dykes and reefs, unconformities, buried channels, salt encrustations etc. which may have a unique bearing on groundwater occurrence and movement.

Image data selection

Selection of remote sensing data for groundwater applications has to be done with great care as detection of features of interest is related to spatial resolution and spectral bandwidth of the sensor as well as seasonal conditions of data acquisition. For example, small-scale image data are good for evaluating regional setting of landforms, whereas large-scale photographs are required for locating actual borehole sites. Similarly, an understanding of spectral response of objects is crucial for selecting spectral bands of remote sensing data, as discussed earlier. Further, temporal conditions (rainfall, soil moisture, vegetation etc.) greatly affect the manifestation of features. Figure 4.10 gives an example of the same area (granitic terrain in Central India) imaged by the same sensor (IRS-LISS-II) in two different seasons: post-monsoon and pre-monsoon. The post-monsoon image shows a thin vegetation cover which is widespread, and therefore distribution of various landforms and lineaments is not clear on this image. On the other hand, it is the summer (pre-monsoon) image on which various landforms like buried pediments, valley fills and lineaments are clearly brought out.

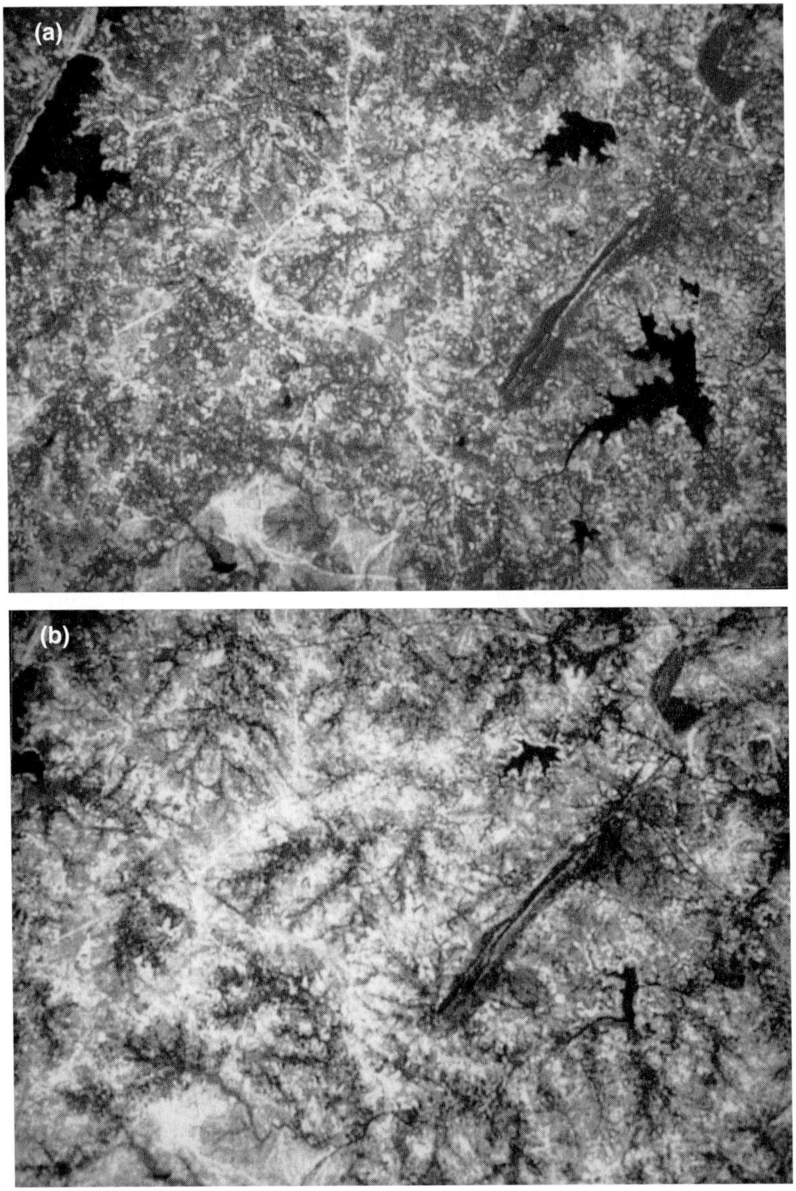

FIGURE 4.10 A set of (a) post-monsoon (dated 13 October 1996) and (b) pre-monsoon (7 April 1997) IRS LISS-II FCCs of a part of Bundelkhand granites in Central India. Note that the various landforms (buried pediments, valley fills etc.) and lineaments are better deciphered on the pre-monsoon (summer) image.

4.8 Thematic applications

Application-potential of remote sensing technique based on various groundwater indicators is briefly discussed in the following paragraphs, with examples.

4.8.1. Features associated with recharge zones

Surface water bodies constitute an important source of groundwater recharge and hence their identification is useful. Water bodies such as streams, canals, ponds and reservoirs are often very well marked on all types of optical and microwave image data. However, certain spectral bands such as near-infrared, thermal-infrared and microwave are highly sensitive to surface moisture. Figure 4.11 is a plot of NIR reflectance (MSS4) against depth of water-table. The groundwater discharge zones have shallow water-tables and lower reflectance than the recharge zones.

Multispectral image data depicting spatial characteristics, patterns and shapes of surface water bodies can indicate whether they are recharging the groundwater. For example Figure 4.12 is a Landsat image of a region marked by an overall low drainage density. The streams appear to get thinner and thinner when going downstream; evidently, they are losing water into the ground. Another example is furnished by the Landsat image (Figure 4.23) of the Himalayan foothill ranges. The streams are gradually lost into the alluvial sands, implying influent seepage.

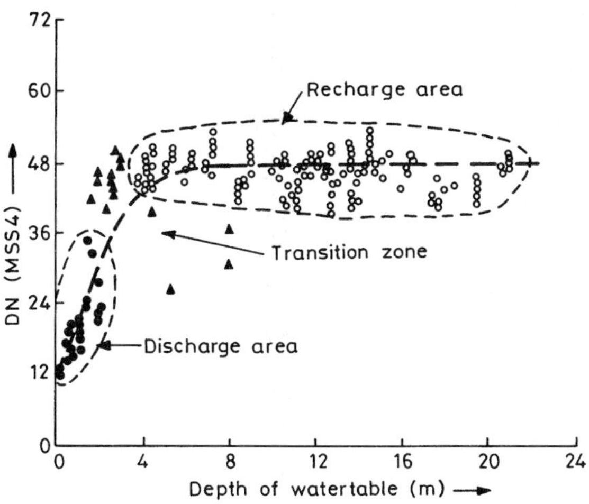

FIGURE 4.11 Relationship between NIR reflectance (MSS4 DN values) and depth of water-table; the distribution of groundwater discharge and recharge areas is indicated (modified after Bobba *et al.*, 1992).

FIGURE 4.12 Features associated with groundwater recharge in the sub-Himalayas; the thinning of streams towards downstream indicates influent seepage. (Landsat MSS 4 image).

4.8.2 Features associated with discharge sites

Springs (effluent seepage) may be detected on panchromatic aerial photographs, multispectral image data and thermal IR data, due to the fact that higher surface moisture may result in abnormal lower ground reflectance, anomalous local vegetation or temperature distribution. For example, in the area shown in Figure 4.13, a number of springs and gradual building-up of streams southwards is due to effluent groundwater seepage. Another example of springs in coastal areas is shown in Figure 13.11 detected by thermal-IR data. Effluent seepage may also be detected by SLAR, as the microwaves are sensitive to surface moisture and vegetation. Remote sensing has been utilized to estimate spring discharge from swamp areal extent in Kenya (Meijerink *et al.*, 1994) and from vegetation areal extent in Botswana (Hatton and Dincer, 1979).

4.8.3 Soil moisture

Soil moisture is considered as a direct guide, and also as an indirect guide, for groundwater studies. Moisture content influences the response of soil to EM radiations, throughout the EM spectrum. On the panchromatic band, a higher moisture in soils leads to darker photo-tones. As the infrared reflectance is highly sensitive to moisture content,

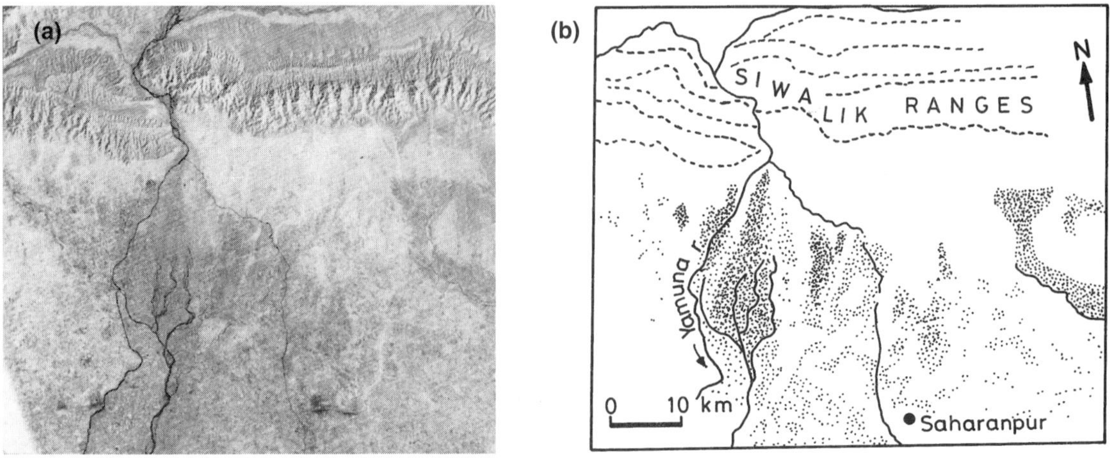

FIGURE 4.13 Features associated with groundwater discharge: (a) Landsat MSS4 image; (b) interpretation map. Note the building up of streams due to effluent seepage.

the soil moisture variation can be detected using the NIR and SWIR bands. For example, very high reflectance in the visible and NIR bands and low drainage density point towards near dry surface conditions, absence of vegetation and deep water-table conditions (e.g. Figure 4.13). The water-logged areas are also easily identified on NIR. Figure 4.14 is an IRS image showing quartzite ridges and intervening waterlogged phyllite valleys. However, reflectance data can hardly be used for quantitative estimation of moisture content, owing to a non-linear variation.

FIGURE 4.14 IRS LISS-II image of a part of northern Rajasthan (India). The region is underlain by quartzite (Q) (ridges), and phyllites (P) (valleys). The valleys are waterlogged (L); white patches are salt encrustations (S).

The thermal-IR region is also sensitive to surface moisture. The wet areas are sites of evaporative cooling and appear cooler (darker) than the adjacent dry areas. The radar response can also be used for quantitative scaling of soil moisture variation (Figure 4.9). The passive microwave radiometry has a rather limited application owing to coarse spatial resolution.

4.8.4 Vegetation

Vegetation forms an important general guide for groundwater investigations, and anomalous vegetation may serve as a direct guide for the same. Remote sensing data of the dry season may bring out vegetation anomalies as water-bearing fractures may sustain vegetation even under dry/drought conditions. Vegetation can be studied on panchromatic aerial photography, multispectral images, thermal-IR data and SLAR images. The spectral response curve of healthy vegetation is shown in Figure 4.15. In the visible band, the pigments in leaves govern the reflectance response; therefore, on B, G, R

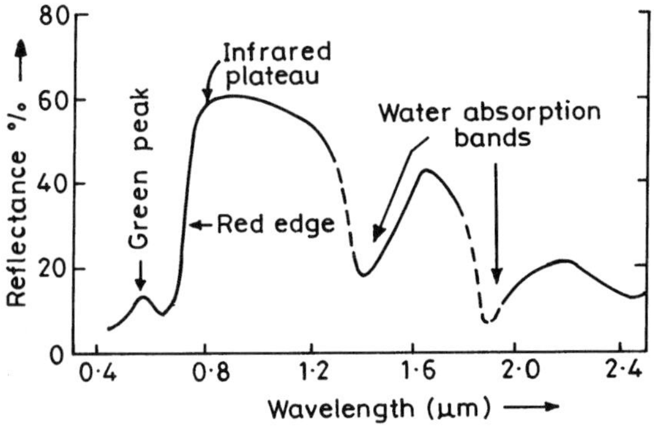

FIGURE 4.15 Spectral reflectance curve of healthy vegetation in the solar reflection region.

and panchromatic bands, the vegetation is generally darker than the adjacent rock or soil. Healthy vegetation is strongly reflective in the NIR and this characteristic helps in discriminating healthy *vis-à-vis* diseased/decaying vegetation. In the SWIR region, the reflectance is governed by leaf water content and this spectral band has importance in evaluating vegetation stress. In SLAR images, denser canopy leads to greater scattering and a corresponding higher radar return (Figure 4.24).

Suitably large-scale stereo-photo pairs may yield information on the type of plants that will indicate groundwater conditions in the area. For example, the presence of phreatophytic vegetation denotes a shallow water-table, whereas xerophytic plants would indicate dry arid conditions. Palm trees indicate a shallow water-table but highly mineralized water. Mangroves flourish along the seashore.

4.8.5 Topography

Surface topography has a pronounced influence on groundwater occurrence; topography also governs boundaries of the basin. The regional topography is best studied on stereo aerial photographs. Besides, topographic maps, multispectral images and radar images are also useful for this study.

4.8.6 Landforms

One of the widest applications of remote sensing data and aerial photography has been in the study of landforms, because: (a) remote sensing particularly with stereoscopic capability gives direct information on the landscapes, involving slopes, relief and forms, and (b) landform features can be better studied on a regional scale using synoptic coverages provided by remote sensing data, rather than in the field. Some special landforms such as karst, buried channels, dykes, alluvial fans etc. are of particular interest in groundwater studies, and they are discussed separately at relevant places in this book. Genetically, the landforms are divided into two broad groups: (1) erosional landforms, and (2) depositional landforms.

(a) **Erosional landforms:** Erosional landforms are typically associated with the resistant hard rock terrains and they are discussed in more detail in Chapter Eleven. From the point of view of groundwater potential, valley fill is the most important landform unit. At times, it may be difficult to place the boundary between buried pediments and valley fills, on panchromatic aerial photographs and other raw data. However, dedicated image processing of multispectral data could lead to better discrimination of such landform features as illustrated in section 4.9.

b) **Depositional landforms:** Depositional landforms are a result of depositional processes of various natural agencies, viz. river, wind, and glacier. They differ in form, shape, size, occurrence, composition, association, and hence also in their groundwater potential. However, as they may also occur in the regional setting of fractured hard rock terrains, they may play an important role in groundwater development for local needs.

4.8.7 Depth of weathering and regolith

In hard rock terrains, the depth of weathering and thickness of regolith are important in locating dug wells. Spatial distribution of rock outcrops can give a reasonable idea of the likely thickness of regolith/overburden, and depth to bedrock. Thickness of regolith is also related to landform and is likely to be more at lower slopes and in valleys. For this purpose, studies of stereo aerial photographs and FCCs of multispectral images are particularly useful (Figures 11.3 and 11.4).

TABLE 4.8 Photocharacters of broad lithologic groups

Crystalline rocks
Widely different weathering characters are exhibited, depending upon composition, structure, and climate; more prone to weathering in humid, warm climates than in cold, dry climates; commonly occur as large size bodies (plutons), forming low-lying or undulating terrains; in arid conditions, sometimes show woolsack weathering in which isolated outcrops protrude through the weathered mantle. Drainage is generally medium density; massive rocks have dendritic pattern while jointed and foliated rocks exhibit angular–rectangular patterns; plutonic igneous rocks are massive; metamorphic rocks show foliation; lithological layering in gneisses leads to bandings. Several sets of joints may be present; granites also exhibit sheeting joints. Vegetated cover is variable. Spectral characters depend upon weathering, soil and vegetation; generally fresh rock surfaces are lighter toned. Also generally, if acidic they are lighter toned than basic and ultrabasic rocks.

Carbonate rocks
These are highly susceptible to solution by water; in humid climates, carbonate rocks show karst topography. In arid regions, the carbonates form ridges and hills. Drainage density in both arid and humid terrains is low; internal drainage is high in humid areas. Bedding may be weakly developed; intercalations of other lithologies may enhance lithological layering. Generally 3 to 4 sets of well-developed joints are observed; joint surfaces may become curved due to solution activity. Light coloured, thin calcareous soil is often observed in carbonate terrain; removal of carbonates may result in an iron oxide layer called terra-rosa. Vegetation is variable, depending upon weathering and climate. Carbonate rocks usually appear light toned; very often exhibit mottling due to variation in surface moisture.

Volcanic rocks
These are highly susceptible to weathering, which obliterates characteristic landforms and features in older flows. Drainage is highly variable; absent or very coarse due to high porosity in newer flows; often older flows display high drainage density and dendritic drainage. Foliation or bedding is absent in volcanic rocks; successive flows interbedded with weathered material and non-volcanic sediments may resemble coarse bedding (pseudo-bedding) shown on the scrap faces. Joints may be well developed; columnar joints are typically present in basaltic flows. Acidic flows have light toned soil cover and basaltic flows have darker toned soil cover. Vegetation is sparse on younger flows; weathered areas may support vegetation and cultivation. Tonal characters are related to composition, age, weathering, soil moisture and vegetation.

Clastic rocks
Sandstones are resistant to weathering, whereas shales are incompetent and easily eroded. Sandstones form hills, ridges, scarps, and topographically prominent features; shales form valleys and lower hill slopes. Sandstones have low to medium drainage density, and rectangular–angular pattern; shales have high drainage density and dendritic drainage; intercalated sequences of sandstone-shale produce trellis pattern. Bedding is often very prominent. Sandstones often exhibit compositional layering; massive and pure varieties lack compositional layering. Shales are less commonly exposed, as they are usually covered under weathered debris. Several sets of joints may be developed. Soil cover is variable. Shales usually possess thick soil cover. Vegetation bandings may be present, sandstones being vegetated and shales used as agricultural grounds. Tone is highly variable, depending upon vegetation and soil moisture.

4.8.8 Lithology

Different lithologic units, i.e. rock types, can be inferred indirectly based on the rock structures and resultant landforms. The first task is to identify the terrain, that is fractured hard rocks versus unconsolidated sediments. On photographs and images, the hard rock areas are characterized by the presence of a number of features such as compositional bandings, bedding, foliation, fractures, folds etc. Further, they have characteristic erosional landforms as described earlier. The unconsolidated sediments are marked by the absence of the rock structures and presence of various depositional landforms like alluvial fans, point bars, moraines, dunes etc. From the point of view of hydrogeological investigations, the fractured hard rocks have been divided into four groups in this work: (a) crystalline rocks, (b) volcanic rocks, (c) carbonate rocks, and (d) clastic rocks. Their salient photocharacters are summarized in Table 4.8.

4.8.9 Geological structures

Remote sensing techniques have found extensive application in structural studies. They help decipher planar discontinuities in the terrain, and understand their disposition and mutual relations. The discontinuities may be visible on simple photographs and images, or the remote sensing data may be processed specifically to enhance certain structural features. In general, the remote sensors like aerial photography, Landsat MSS, TM, SPOT-HRV, IRS-LISS etc., provide plan-like information, the data being collected while the sensor views the earth vertically from above. Vertical and steeply dipping planar discontinuities are therefore prominently displayed on these images and photographs. On the other hand, the gently dipping or sub-horizontal structural discontinuities are comparatively subdued, and are difficult to delineate, especially in rugged mountainous terrain.

Geological structures include bedding, foliation, folds, joints, lineaments, faults and shear zones. In this section, the discussion is confined to bedding/foliation and folds. Bedding is the primary discontinuity in rocks, and is due to compositional layering; on remote sensing images, bedding appears as regular and often prominent linear features, marked by differential weathering, tone, texture, soil and vegetation; the linear features due to bedding appear longer, even-spaced and more regular, in contrast to those produced by foliation and joints.

Orientation of bedding or foliation is important. The strata may be sub-horizontal, inclined or vertical, and may form segments of larger fold structures. The outcrop pattern on photographs and images is influenced by structure and relief in the area. Folds can be delineated by tracing the bedding/marker horizon along the swinging strike and recognition of dips. Broad, open, longitudinal folds are relatively easier to locate on satellite images than tight overturned isoclinal folds, owing to the small areal extent of hinge areas, which may provide the only clues of their presence. Therefore, such folds need to be studied on appropriately larger scales, such as on aerial photographs.

4.8.10 Lineaments including joints and fractures

Lineament studies are most important for groundwater investigation in hard rock terrains. The technique of mapping fracture traces and local lineaments from aerial photographs, which has now been extended to satellite imagery for locating zones of higher permeability in hard rock terrain, was developed by Parizek (1976), also Lattman and Parizek, (1964). It has been shown that wells located on or close to fracture traces yield many times more water than the wells away from the fracture traces, and they are more consistent in their yield when located on lineaments rather than under other conditions.

What is a lineament?

The photolinears, i.e. linear alignment of features, are one of the most obvious features on aerial photos and satellite images. The term lineament is now used, by and large, in a geomorphological sense, as 'a mappable simple or composite linear feature of a surface whose parts are aligned in a rectilinear or slightly curvilinear relationship (Figure 11.3), and which differs distinctly from the pattern of the adjacent features, and presumably reflects a subsurface phenomenon' (O'Leary *et al.*, 1976). Hence, this category includes all structural, topographical, vegetational, soil and lithological alignments, which are very likely to be the surface expression of buried fractures and structures. This definition seems to be the most practical in the context of remote sensing image interpretation.

Manifestation of lineaments

Lineaments occur as straight, curvilinear, parallel or en-echelon features (Figure 4.16). Generally, the lineaments are related to fracture systems, discontinuity planes, faults and shear zones in rocks. The term also includes fracture traces described from aerial photographic interpretation by some workers. Dykes and veins also appear as lineaments; whereas dykes and veins can be identified in field or large-scale photographs, fractures are usually covered under regolith. The pattern of a lineament is important on the image. Lineaments with straighter alignments indicate steeply dipping surfaces; by implication, they are likely to extend deeper below the ground surface (section 11.5.1).

Ground element corresponding to a lineament would depend on scale of the remote sensing data. Both small scale (such as satellite data) and large scale (aerial photographs) are useful and ought to be used in a complementary manner for hydrogeological studies. On regional scales (say 1:250 000), lineament features may be more than *ca.* 5 km in length, representing long valleys and complex fractured zones; such data facilitate regional selection. On larger scales (e.g. aerial photographs or IRS 1-C Pan images, say 1:20 000 scale), shorter, local, minor drainage features, individual fracture traces and soil changes are represented as lineament features; such data are useful in detailed investigations including well siting.

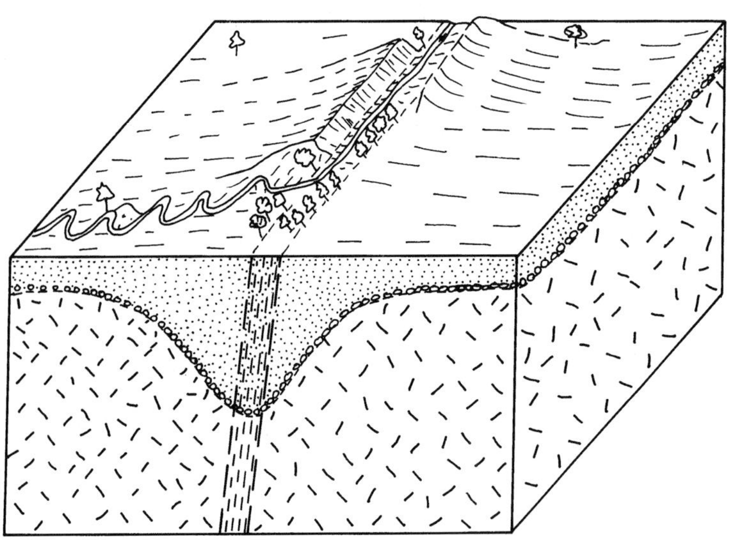

FIGURE 4.16 Schematic showing surface manifestation of lineament in terms of topographic/drainage/vegetation alignment. The lineament zone is possessing greater depth of weathering.

Different types of fractures and their characteristics were described earlier (Chapter Two). The distinction between dilational and shear fractures is very important for groundwater studies as the former are more productive. This distinction can sometimes be made on remote sensing data on the basis of two considerations: (a) relative movement along an individual discontinuity, and (b) orientation and statistical distribution of lineaments. It is always desirable for both these types of evidence to be mutually supportive. Sometimes, relative movements along lineaments can be seen, to indicate faults and shear fractures (Figure 4.17); in contrast, lineaments related to dilational fractures would not show any relative displacement. Further, in

FIGURE 4.17 Mapping of various types of fractures on IRS-1C LISS-III FCC of Cuddapah region, India (courtesy D.P. Rao, A. Bhattacharya and P.R. Reddy).

areas of less complex deformation, statistical analysis of lineament trends can distinguish between fractures of shear origin and tensile origin. It is better if such interpretations are supported by evidence of relative movement observed in the field or on remote sensing data. Hydrotectonic models have also been attempted by some workers (e.g. Buckley and Zeil, 1984) to interpret the erstwhile stress field and thence identify tensile fracture trends, as they are believed to be more productive.

On an image, the lineaments can be easily identified by visual interpretation using tone, colour, texture, pattern, and association. Alternatively, the automatic techniques of digital edge (or line) detection can also be applied for lineament detection. However, these techniques are found to have limited application due to the cropping up of many non-meaningful linears. Therefore, the visual interpretation technique is preferred and extensively applied, although it involves some degree of subjectivity, due to human perception.

Mapping of lineaments can be done on all types of remote sensing images: stereo panchromatic photographs, multispectral and thermal-IR images and SLAR images. The panchromatic, NIR, SWIR, and thermal-IR images contain information of the near-surface, whereas SLAR images may provide limited depth penetration (of the order of a few metres at best) in arid conditions. Manifestation of lineaments is related to ground conditions and the sensor spectral band. Figure 4.18 shows an example where an aerial thermal-IR image exhibits a number of structural features (bedding planes and fractures), which are not seen on the panchromatic band aerial photograph. The structural features appear dark due to evaporative cooling along them. Further, often lineaments are better manifested on SLAR images than on panchromatic photographs, owing to the fact that radar response is intensely affected by surface roughness, micro-relief, vegetation and surface moisture. Figure 4.19 shows a comparison of lineament features observed on air photos and SLAR images of the same area in Brazil.

Lineaments can be mapped on the original data products, as well as on enhanced images. Further, a variety of digital techniques to enhance linear features can also be applied, in the form of isotropic and anisotropic filters (section 4.9). An anistropic filter can be used to enhance linear information in a certain preferred direction; however, due to the presence of artifacts, the interpretation of such filtered products has to be done cautiously. This is illustrated by the example in Figure 4.20 where Sobel-filtered, Robert-filtered and hybrid FCC images of the same area (Athur Valley, South India) are shown. These products bring out dif-

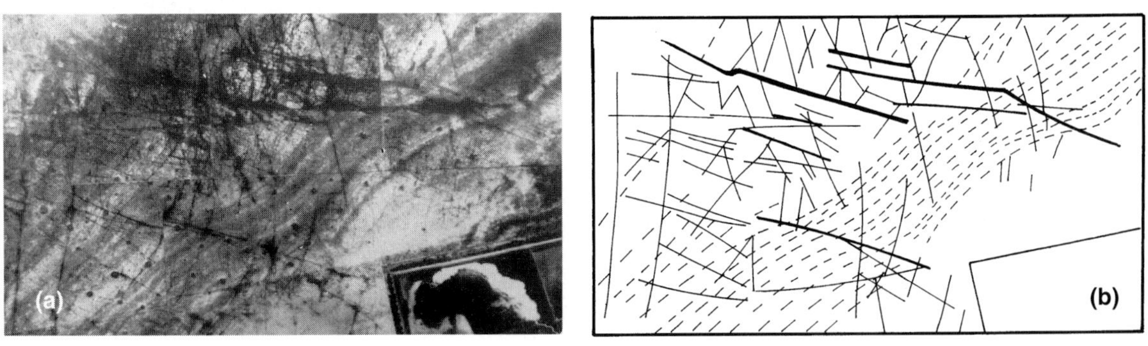

FIGURE 4.18 (a) Thermal IR image showing numerous structural features such as bedding and lineaments; (b) interpretation map (Stilfonstein area, Transvaal, South Africa) (after Warwick *et al.*, 1979).

(a) 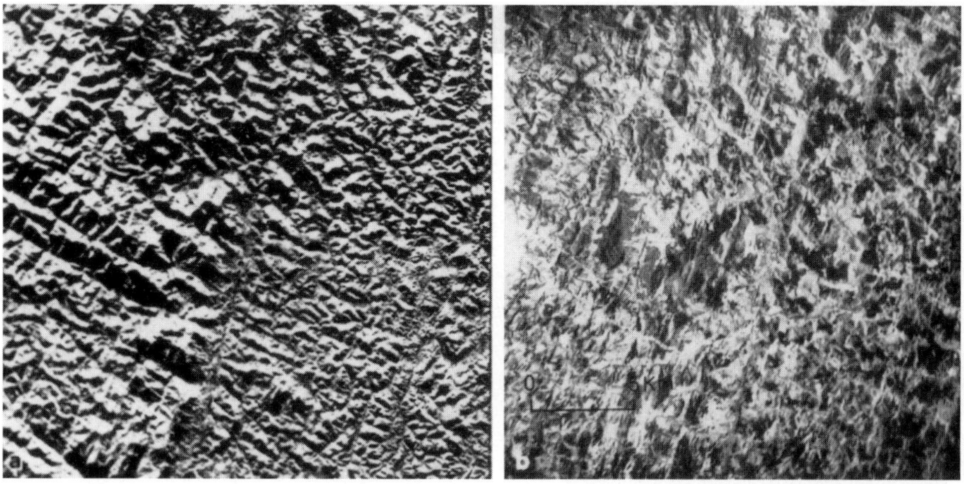 (b)

FIGURE 4.19 (a) X-band radar image of an area in Bahia, Brazil; (b) aerial photograph of the same area. Note that the structural features are better manifested on the radar image than on the aerial photograph (courtesy A.J. Pedreira).

ferent lineaments and the interpretation ought to be done on the basis of commonality and prominence of features.

Statistical treatment of lineament data

There are a number of ways to assess the statistical distribution of lineaments in an area. One is by considering the number of lineaments per unit area; the second, by measuring the total length of lineaments per unit area; and the third, by counting the number of lineament intersections per unit area. The method of lineament intersection per unit area (density) is generally faster and more convenient. The intersections of two (or more) lineaments are plotted as points, and the number of points falling within a specified grid area is counted. The data are contoured to give the lineament intersection density contour map (Figure 4.21). This corresponds to the zones of higher fracturing in the area. Further, the lineament intersection density maps can be rasterized and converted into lineament intersection density images. In the same way, it is possible to create lineament length images and lineament number images. Such processed data products provide statistical information on distribution of fractured zones.

Relation of lineaments with other features

As lineaments frequently represent fracture zones in the rock, they are associated with zones of greater weathering, thicker soil cover, denser vegetation, higher moisture and valley alignments (Figure 4.16). They are therefore also often characterized by lower electrical resistivity values. Correlation between lineament and vegetation anomalies has been shown in several areas, for example in Botswana based on SPOT-HRV data, by Gustafsson (1993). Figure 4.22 is an interesting example illustrating that such a strong correlation in fractured rocks between lineament and vegetation anomalies may exist even for long linear stretches of 5–10 km.

Lineament density and drainage density maps can also be combined for regional groundwater exploration; the target zones are marked by higher lineament density (representing greater fracturing) and lower drainage density (implying greater infiltration) (Figure 11.15).

Lineaments together with other features like landforms and drainage have to be considered

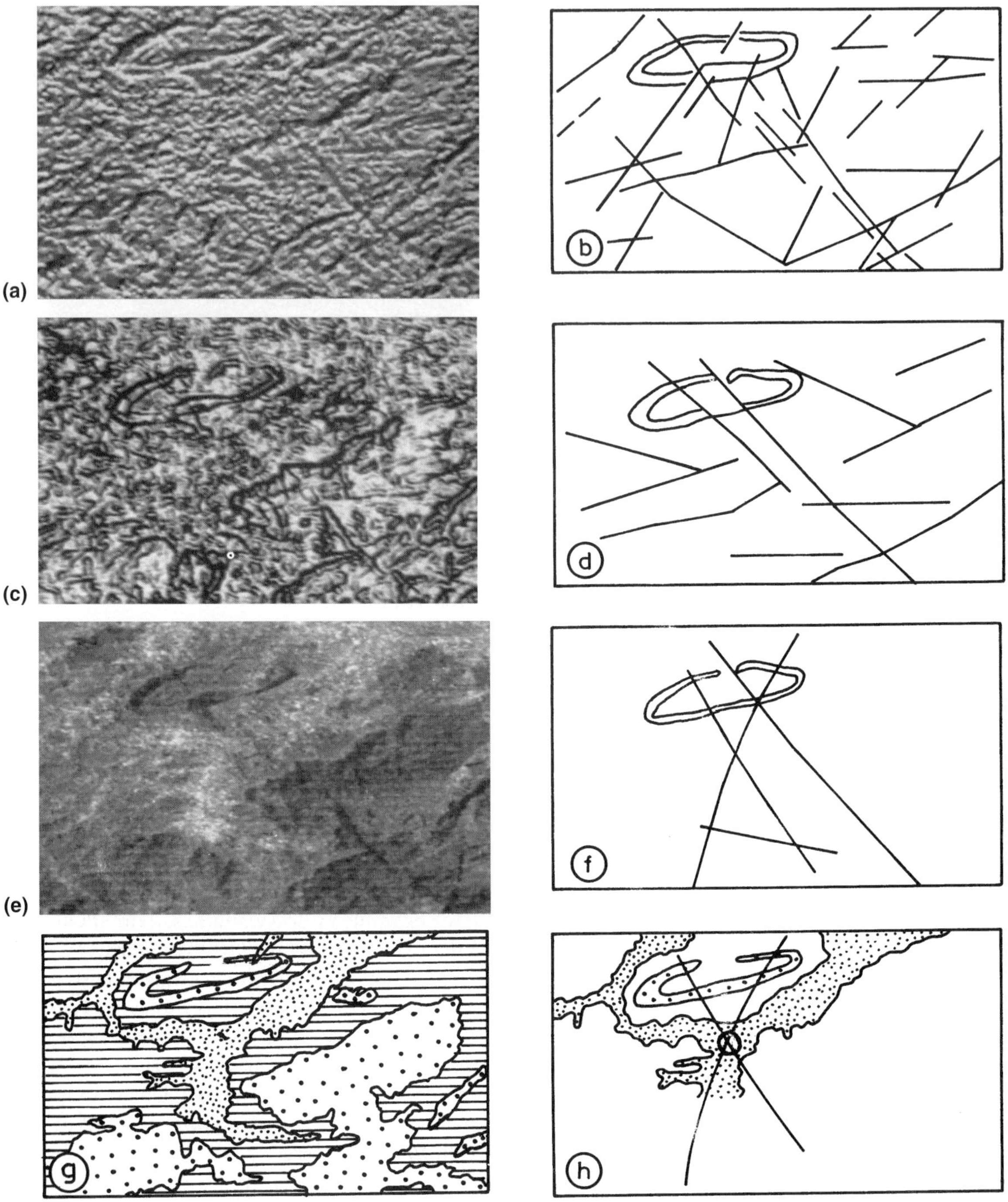

FIGURE 4.20 Image processing and dedicated interpretation of a region in the Athur valley (India) using MSS data. (**a**), (**c**) and (**e**) are Sobel-filtered, Roberts-filtered and hybrid FCCs respectively; (**b**), (**d**), and (**f**) are their respective lineament interpretation maps; (**g**) is a landform interpretation map; (**h**) shows a selection of groundwater potential targets based on lineament and landform interpretations. ((**a**),(**c**) and (**e**) courtesy of A. Perumal).

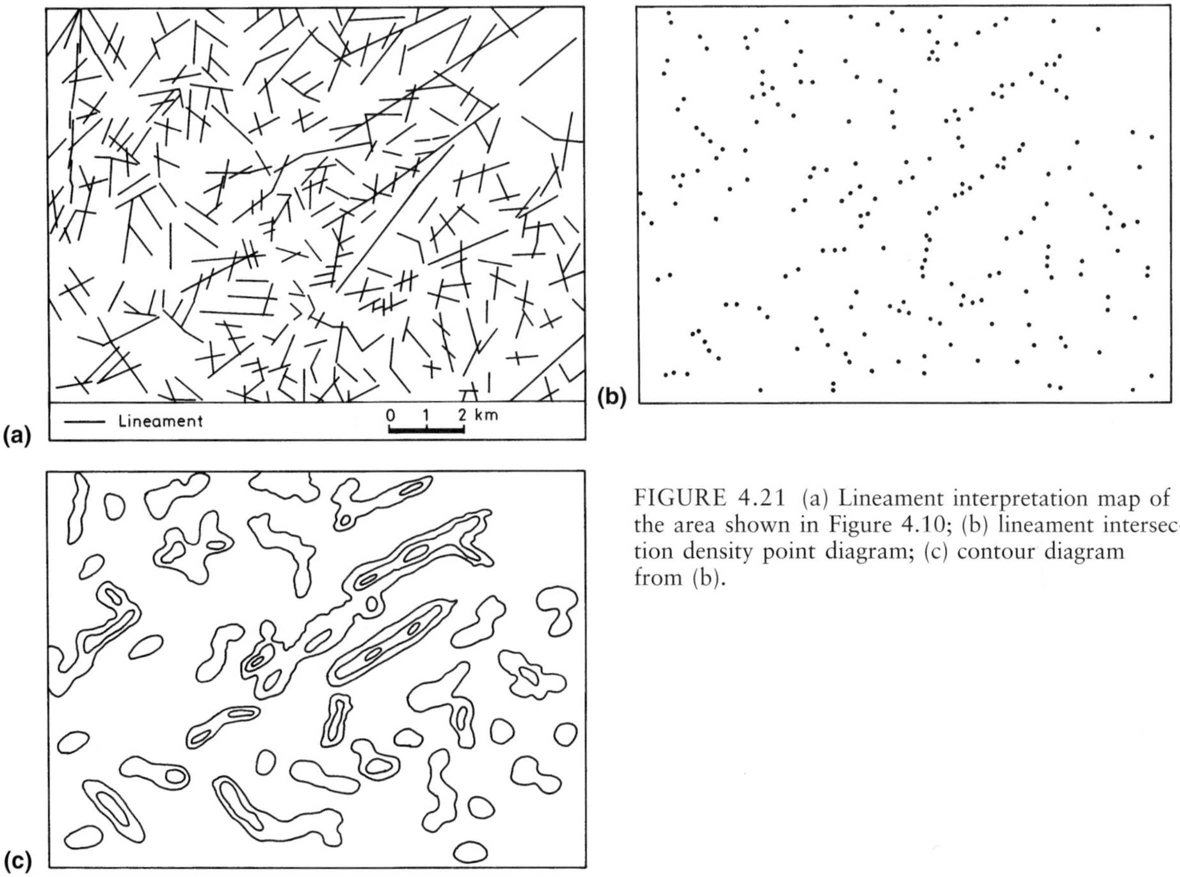

FIGURE 4.21 (a) Lineament interpretation map of the area shown in Figure 4.10; (b) lineament intersection density point diagram; (c) contour diagram from (b).

for delineating groundwater potential areas and siting of wells. For example, in an area the groundwater potential targets are marked by the intersection of lineament and valley fill (Figure 4.20h). Further, it may also be recalled that often lineament traces appear as surface drainage ways, and it is advisable to locate boreholes away from drainage lines to avoid chances of flooding. This consideration of sitting wells may be responsible for lower yields, in some cases.

Lineaments detected on satellite images often correspond to wide zones on the ground which may be debris/soil covered. Therefore, delineating the main zone of fracturing (lineament) is often necessary and can be best accomplished using geophysical tools. Zeil *et al.* (1991) describe a case study from Botswana in which they geometrically co-registered Landsat TM and geophysical (EM) data to locate lineament-fracture zones more accurately in the field. Further, intersection of tensile and shear fractures in association with geophysical anomalies and suitable landforms may produce exceptionally high yields (Boeckh, 1992, Figure 15.10).

Additional examples of the use of lineament in conjunction with other data in a GIS approach for groundwater study are discussed in section 6.4.3. Examples of well yield as related to lineaments are given in sections 11.5.1 for crystalline rocks and 13.5 for carbonate rocks.

(a)

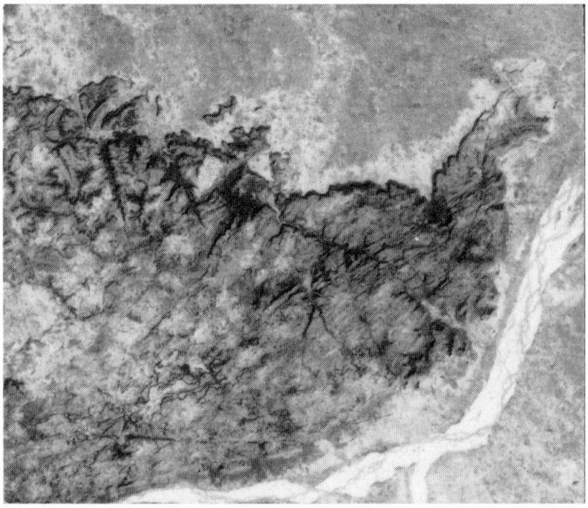

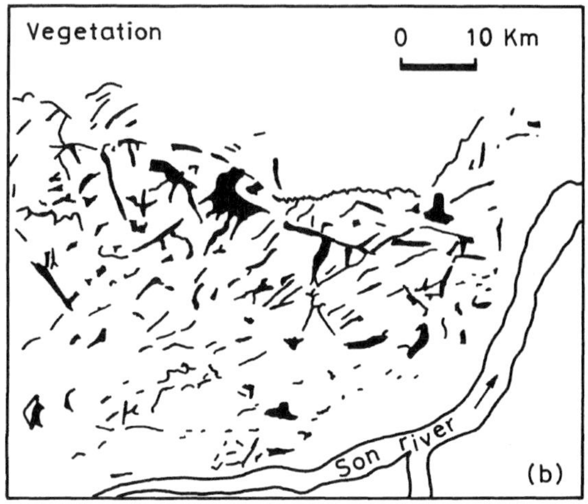

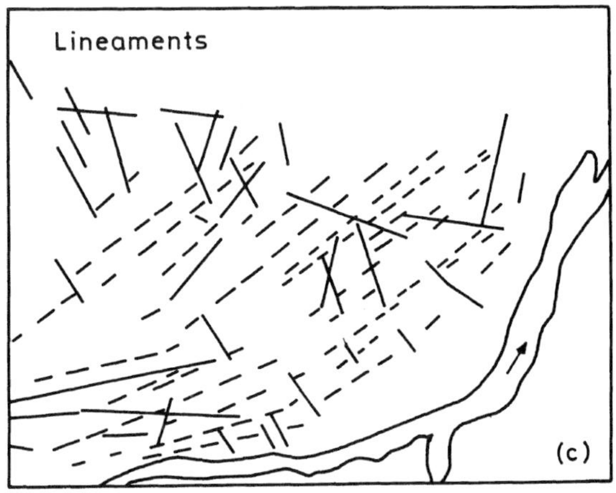

FIGURE 4.22 Preferential alignment of vegetation for tens of kilometres along lineaments in the Son river area in Central India. (a) Landsat MSS2 (red band) image; (b) and (c) interpretation maps showing vegetation and lineaments respectively.

Below left and right

FIGURE 4.23 (a) Landsat TM 4 image of sub-Himalayan region, India–Nepal border; (b) interpretation map. Note the presence of several N–S running parallel dry channels which become water bearing after the spring line. The abrupt eastward shift of the river course and higher surface moisture in the eastern block are due to a Recent N–S trending normal fault with downthrow towards east.

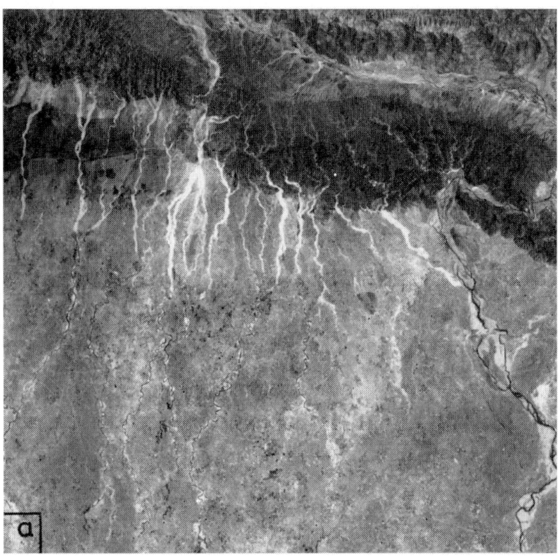

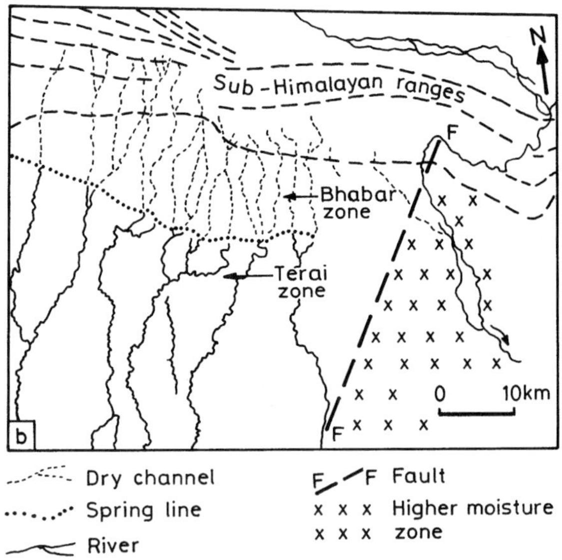

4.8.11 Faults and shear zones

Characteristics of faults and shear zones have been discussed in Chapter Two. The chief criteria for delineating faults on remote sensing data are as follows: (1) displacement of beds or key horizons, (2) truncation of beds, (3) drag effects, (4) presence of scarps, (5) triangular facets, (6) alignment of topographic features, (7) offsetting of streams, (8) alignment of ponds or closed depressions, (9) spring alignment, (10) alignment of vegetation, (11) straight segment of streams, (12) waterfalls across stream courses, (13) knick points (local steepening of stream gradient), (14) disruption of channels and valleys.

As such, a distinction between faults and lineaments is difficult to make, the main difference being that faults show evidence of movement. On remote sensing images, faults with strike-slip movement are more readily discerned than those with only dip-slip movement. Like lineaments, faults can be delineated on all types of remote sensing data – panchromatic, multispectral, thermal-IR, and SLAR data. Figure 4.23 shows the presence of a Recent fault affecting Siwalik mountain ranges and Gangetic alluvial deposits. The downthrow character of the eastern block is indicated by sudden eastward flow of the river and higher surface moisture on the eastern block.

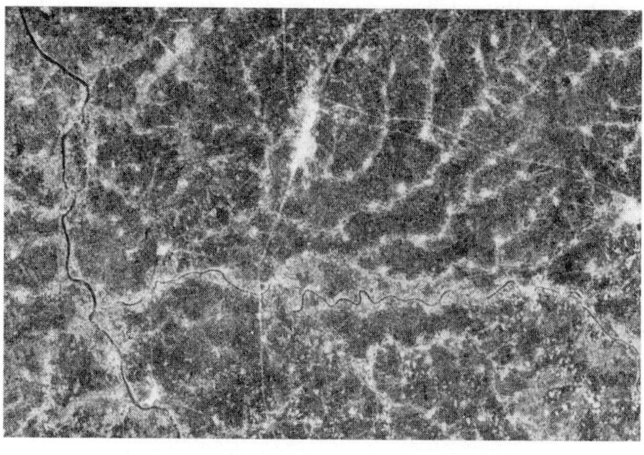

FIGURE 4.24 SIR-A radar image of a part of Central India. On the SLAR image vegetated channels are bright due to higher radar returns whereas water-bearing channels are dark due to specular reflection.

4.8.12 Soil types

Soil types and humus content can be estimated from remote sensing data, particularly panchromatic photographs. Soil types are intimately related to bedrock – for example, sandy soils originate from arenaceous and siliceous rocks; clayey soils develop from argillaceous rocks; basalts typically produce black cotton soils. Coarse-grained soils and sandy soils are generally light toned. Saline soils are light coloured due to salt encrustation. Fine-grained soils and clayey soils are moisture rich and generally dark toned. Clayey soils also have absorption bands in the SWIR region. Humus content in soils is related to vegetation and moisture which leads to darker tones.

4.8.13 Drainage characteristics

For hydrogeological investigations, drainage is a very important parameter of study. The drainage channels are characterized by the typical serpentine-sinuous shape and pattern; they may appear differently on different spectral bands depending upon depth of water, turbidity, vegetation cover etc. The condition of the channel including its pattern and density may give a good idea of the hydrogeological characteristics of the terrain. Figure 4.10 shows an IRS-LISS-II image of the granitic area (Bundelkhand, India) where the drainage is controlled by fractures and the channel is marked by good vegetation; there is little surface water. Figure 4.24 is a SIR-A image of basaltic terrain from Central India showing a dendtritic drainage pattern; the drainage-ways are mostly vegetated (bright) with little surface water. Both Figure 4.10 and Figure 4.24 are typical of semi-arid conditions.

4.8.14 Special geological features

A number of geological features may be locally important, such as karst landforms (in carbonate rocks), alluvial fans and buried channels (in clastic formations), dykes and reefs (in volcanic and crystaline rocks), and unconformities (in almost all geological units). These are discussed in the respective lithology chapters.

4.9 Digital image processing

4.9.1 Introduction and methodology

Digital image processing is carried out for image data correction, superimposing digital image data, enhancement and classification. A digital image is an array of numbers, in the form of rows and columns, depicting spatial distribution of a certain parameter, usually reflectivity of EM radiations in the case of remote sensing images.

Image processing can be carried out for various purposes; for detailed information, the reader is referred to standard texts (e.g. Moik, 1980; Hord, 1982, 1986; Mather 1987; Jensen, 1986). Digital image processing for geological applications has been discussed by Drury (1987), Sabins (1987), and Gupta (1991) among others. Based on the objective, digital image processing may be grouped into four types: (a) correction (b) registration, (c) enhancement, and (d) classification. The general sequence of working with remote sensing data is: preprocessing (geometric and radiometric corrections) followed by image enhancement, georeferencing, visual interactive interpretation, and finally output.

Image correction is also called image restoration. Several types of radiometric errors and anomalies, including striping, bad data lines and atmospheric scattering effects and geometric distortions such as those related to the Earth's rotation may be present in the image data. These are rectified before further processing. **Registration** is the process of superimposing images, maps, or data-layers over one another with geometric congruence.

Image enhancement is the modification of an image to alter its impact on the viewer and render it easier to interpret. A wide variety of enhancement techniques are available; salient features of a few commonly used techniques are summarized in Table 4.9. Contrast enhancement deals with rescaling of grey tones to improve the image contrast and is almost invariably carried out. Edge enhancement is basically a process whereby borders of objects get enhanced. Main applications of edge-enhancement filtering are: (a) to obtain a sharper image with more information on local characters, and (b) to enhance fractures, structural features and lineaments, overall, or alternatively in a certain preferred direction. For example, Figure 4.20a and 4.20c show FCCs of Sobel-filtered and Robert-filtered images of the same area, which bring out different lineaments. Therefore interpretation of such data needs a careful evaluation.

Addition and subtraction is the simplest method to digitally combine multi-image data. After digitally registering image sets, pixel-wise addition of two or more bands produces a new addition image. In a similar manner, pixel-wise subtraction of one image data from another generates a new subtraction image. In general, addition produces a high contrast image and is good for general study. On the other hand, a difference image has a reduced contrast. A difference image is particularly useful for change detection; for example, change in vegetated area, flood area, wet ground area etc. between two dates of satellite passes may be detected by this method. Another technique of combining multispectral data, is by computing the brightness index (BI). BI for any two spectral bands is equal to the square root of the sum of the squares of the two input bands, i.e. $BI\ 1/3 = [(Band\ 1)^2 + (Band\ 3)^2]^{1/2}$. In essence, it produces a type of addition image, and has been useful for discriminating landforms (Figure 4.20e).

TABLE 4.9 Common types of digital image enhancements

Name	*Description*
1. Contrast enhancement	Rescaling of grey levels to improve the image contrast
1.1 Linear stretch	Range of image values is uniformly linearly expanded to fill the range of display device
1.2 Piece-wise linear stretch	Linear stretches carried out in discrete parts
1.3 Histogram equalized stretch	Assigning new image values on the basis of their frequency of occurrence, resulting in a uniform population density of pixels in the new image, also called ramp stretch; this results in a very high image contrast
1.4 Logarithmic stretch	Image data stretched using a logarithmic function; useful for enhancing features in the lower DN-range
1.5 Exponential stretch	Image data are stretched using an exponential function; useful for enhancing features in the upper DN-range
1.6 Density slicing	The old image data range is subdivided into a number of zones and each zone is assigned one particular level in the new image data range; the steps could be of equal or unequal width
2. Edge enhancement	Application of a filtering technique to obtain a sharper image; information on local variations, which may be related to local morphology, soil, vegetation, moisture, fractures etc. are enhanced
3. Addition and subtraction	Simplest method to combine multi-image data; pixel-wise addition or subtraction is carried out to generate a new image
4. Principal component transformation	PC-Analysis builds up a set of new axes which are mutually orthogonal, i.e. non-corrected; images are built according to the new axes, called PC images
5. Ratio images	Generated by dividing the pixel value in one band by the corresponding pixel value in another band for each pixel. On ratio images, effects of illumination/topography are reduced
6. Colour enhancement	Uses colour display, density slicing colour coding is for single images; the grey levels are sliced into a series of density intervals and each range is displayed in separate colour, RGB

Principal component transformation is a standard statistical technique for reducing data redundancy. The PCT builds up a new set of axes which are mutually orthogonal.The entire data set can be represented in terms of these new axes, which are much less in number. New images are built corresponding to the new axes. These images may be suitably contrast-stretched and used for interpretation individually or combined into colour composites.

Ratio images are generated by dividing the pixel value in one band by the corresponding pixel value in another band, for each pixel. Ratio images are made to reduce the effects of variation in illumination due to topography and enhance spectral information. Ratioing is particularly useful in estimating vegetation density (Figure 4.25).

Colour enhancement leads to feature enhancement and facilitates better interpretation. Density-slicing colour coding and RGB (red-green-blue) coding are the basic techniques

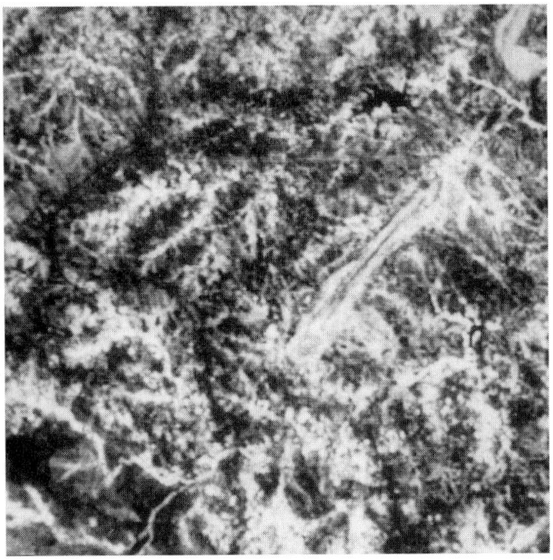

FIGURE 4.25 Ratio NDVI image of a part of the area shown in Figure 4.10. Brighter tones correspond to greater density of vegetation.

of colour enhancement. Density-slicing colour coding is used to enhance differences within a single image. In colour composites, a set of three co-registered images is co-projected for concurrent display, using blue, green and red primary colours. Such colour displays have been found to be extremely useful for geoscientific investigations and terrain evaluation. For example, a hybrid colour composite of Athur Valley, India, using MSS data (such that PC3 = red, BI MSS 1/3 = green and BI 2/4 = blue) is shown in Figure 4.20e. Its landform interpretation map is given in Figure 4.20f. It is obvious that valley fills, which form potential targets for graoundwater (and are otherwise difficult to differentiate from buried pediments), can be readily delineated from the rest of the geomorphic features.

Classification is carried out to construct thematic maps, where each pixel is assigned to a particular class by name. The methods of image classification are based on digital image pattern recognition. For subsurface geological exploration, the classification approaches have rather limited applications, and enhancement has been a more rewarding approach for subsurface geological-hydrogeological investigations (cf. Siegal and Abrams, 1976; Gupta, 1991).

4.9.2 Digital image enhancement of groundwater indicators

It must be stated at the outset that there is no universal best digital image processing technique for enhancing all the groundwater indicators, in all conditions. Many combinations have to be made by trial and error, to suit particular settings, and in general, the convergence of evidence principle is very helpful. In these paragraphs some important guidelines for enhancing groundwater indicators are outlined. The discussion here is based on the Landsat-MSS, TM, SPOT-HRV, IRS-LISS and ERS-SAR types of data, in view of their general worldwide availability.

(1) Structures, faults and lineaments

For mapping of various structures, viz. faults and lineaments, on remote sensing images, the following types of processings are useful.

(a) NIR band image, suitably edge-enhanced and contrast-stretched.
(b) FCCs utilizing edge-enhanced and contrast-manipulated images of Green-Red-NIR bands; FCCs of Red–SWIR-I–SWIR-II bands.
(c) Difference image of NIR-Red bands, particularly when vegetation is associated with lineaments.
(d) Laplacian, Sobel and Robert filters for edge enhancement.
(e) Directional filtering using anisotropic filters in horizontal, vertical and diagonal directions; however, careful interpretation is necessary owing to artifacts in such products.
(f) FCCs of the above filtered images.
(g) Hybrid FCCs of SLAR Image-Red band-NIR band images.

(2) Lithology

Lithologic discrimination is usually based on indirect evidence like landform, drainage, soil and vegetation. Therefore, the enhancement procedures for all the above geotechnical elements may be useful. Besides, the following type of products may be particularly interesting: colour composites of edge-enhanced contrast-manipulated bands, involving green-red-NIR bands, or Red–SWIR-I–SWIR-II bands, or NIR–SWIR-I–SWIR-II bands; colour composites of PC images.

(3) Vegetation

For vegetation studies, the following procedures of digital image enhancement may be useful.

(a) Red and NIR images, suitably edge-enhanced and constrast manipulated.
(b) FCCs: vegetation can be most easily and clearly detected on false colour composites, incorporating Red and NIR bands. FCCs of other colour schemes would display vegetation showing different corresponding colours on the images.
(c) A difference image of NIR Band–Red band.
(d) The health and vigour of vegetation is indicated by a spectral ratio, involving Red to NIR bands, giving a vegetation index (VI). For example, one may use spectral ratios like Red/NIR, VIS/NIR or [(NIR–Red)/(NIR+Red)], the last one being called normalized difference vegetation index (NDVI). NDVI has higher values for vegetated areas and lower values for barren areas. Typically, the vegetation index requires a ratio image computation which is invariably more noisy than a single-band image, and demands more careful computation and interpretation.

(4) Landform

Landforms are best studied by stereoscopic techniques. Further, the general image quality is vital for landform evaluation. The following techniques of image processing may be particularly helpful.

(a) Standard FCC of edge-enhanced contrast-manipulated Green-Red-NIR bands.
(b) FCC of NIR–SWIR-I–SWIR-II bands, suitably enhanced.
(c) Colour composites of PC-1-2-3 images.
(d) Hybrid colour composites involving PC3 (or PC2) and computed brightness parameter.

(5) Drainage

(a) In general, the edge-enhanced contrast-manipulated NIR band is most useful in drainage studies.
(b) Red-band image suitably sharpened is useful when drainage channels are vegetated.
(c) Standard FCC (Green-Red-NIR bands) and the FCC of Red–SWIR-I–SWIR-II bands will also bring out the drainage network.

(6) Soil moisture and springs

Usually, an edge-enhanced contrast-manipulated NIR band image is good for delineating surface moisture zones. The SLAR image is also suitable for such studies.

4.10 Application of remote sensing in estimating other hydrological parameters

The importance of remote sensing in estimating surrogate hydrological data has been emphasized by a few workers and well reviewed by Meijerink *et al.* (1994). Main applications could be of the following types.

(1) Topography and DTM: In areas where accurate base maps are not available, topographic data and Digital Terrain Model (DTM) can be generated from stereo aerial photos and stereo pushbroom sensors. DTM may be subjected to filtering to produce a slope map, flow-path map, etc., particularly with GIS functions.
(2) Rainfall: In areas of rainfall < 1500 mm

yr^{-1}, regional vegetation patterns on the Landsat-type data are related to spatial rainfall patterns. Therefore, the NDVI image can be related to rainfall distribution. As the rainfall pattern is affected by topography, local DTM can also provide some information on spatial variation in rainfall. Thus, field data from rain gauge stations, local DTM, and vegetation patterns can be merged in GIS to yield a more realistic isohyetal map.

(3) Snow cover extent: Owing to synoptic overviews and repetitive coverages, the satellite data are extremely useful for mapping snow cover areas and delineating snow cover depletion patterns.

(4) Evaporation and soil moisture: These are important hydrological parameters, governing soil moisture retention and infiltration. Remote sensing data can be used to generate maps showing variation in soil moisture.

(5) Interception and infiltration: Interception of surface water is affected by vegetation while rooting depth of plants governs infiltration. Remote sensing data can be used to generate maps of various land-cover categories for estimating coefficients of infiltration.

(6) Water bodies: Repetitive remote sensing data is highly useful in mapping distribution and temporal variations in water bodies, zones of influent/effluent seepage and flooding.

Chapter five

Geophysical exploration

5.1 Introduction

Geophysical survey is necessary to ascertain the subsurface geological and hydrogeological conditions. Various physical properties utilized in geophysical exploration include electrical conductivity, magnetic susceptibility, density, elasticity, and radioactivity. Depending upon the scale of operations, geophysical surveys can help delineate regional hydrogeological features or even pinpoint locations for drilling of boreholes. Geophysical surveys, though more expensive than hydrogeological investigations, can appreciably reduce much more costly drilling operations. Details of geophysical methods can be found in other standard texts (e.g. Dobrin, 1976; Parasnis, 1986; Telford *et al.*, 1988). Geophysical applications specifically for groundwater are reviewed by a few workers (Zohdy *et al.*, 1974; de Stadelhofen, 1994; Beeson and Jones, 1988; Palacky *et al.*, 1981).

The application of geophysical methods in groundwater exploration is mainly for estimating weathered zone thickness and for fracture delineation. For estimating the weathered zone thickness, the methods of electrical resistivity, EM, and seismic refraction are most appropriate. For fracture characterization, the methods of azimuthal resistivity surveys, EM surveys, magnetic surveys, cross-hole seismic tomography, and borehole geophysical methods are found to be useful. It is essential that geophysical surveys are not undertaken in isolation but are fully integrated with remote sensing, geological, hydrogeological and drilling work.

A reliable interpretation of geophysical survey data should take into account prior knowledge of subsurface geology in the area. At the outset, these surveys should be carried out at locations where the subsurface geology is better known. This will provide useful controls for interpretation of geophysical data in terms of subsurface geology, which can be extrapolated to other areas.

5.2 Electromagnetic methods

In areas where surface layers are highly resistive, electromagnetic methods may be used with advantage for groundwater exploration. The inductive coupling avoids the need for direct electrical contact, thus eliminating problems associated with resistive dry or rocky surface conditions. Electromagnetic data can be collected from aeroplanes or helicopters, allowing survey of large areas at relatively low cost. As EM profiling is also faster and efficient, lately it has superseded the earlier popular resistivity profiling as a reconnaissance tool.

In case of deeper regolith (say > 20 m), the variation in thickness of the weathered zone in an area is important, and this can be estimated from EM profile data. On the other hand, in areas of shallow bedrock (regolith < 10 m), fractures in rocks are the target. The lineaments inferred from aerial photography and remote sensing data need to be surveyed by profiling for precise location and potential. In this context, the EM method has assumed the highest utility as an inventory tool. The recommendations from an EM survey can then be checked and confirmed by resistivity techniques in more detail, and interpreted with other exploration data.

The EM survey can be carried out in various ways, such as horizontal loop EM (HLEM) (vertical dipole) and vertical loop EM (VLEM) (horizontal dipole). The HLEM is found to be particularly useful for detecting vertical and subvertical fractures (Boeckh, 1992). Further, the VLEM system is extremely sensitive to changes in azimuth relative to conductor (fracture) strike. Thus semi-quantitative evaluation of width, depth and strike of fractures zones can be made from the anomalous shapes of both HLEM and VLEM data collectively (Hazell *et al.*, 1992).

The coil spacing are suitably selected, usually 20 m/40 m or 50 m/100 m being most common. The survey is made in a grid such that one axis of the grid is parallel to the main hydraulic conductivity orientation (i.e. rock discontinuity trend). Shorter coil separation and closely spaced EM stations can give better estimates of regolith thickness and dip of the fracture zone.

A water-bearing fracture acts as a conductor. The vertical dipole EM profile across a vertical conductor is marked by two apparent conductivity maxima symmetrically placed on either side of a minimum, centred over the conductor (Figure 5.1a). In the case of a dipping conductor, the EM response curve is asymmetrical; the conductor dips beneath the greater of the two flanking maxima (Figure 5.1b). This anomalous pattern is extremely useful in deciphering dipping fracture zones. It can help to avoid faulty borehole siting (Wright, 1992; Boeckh, 1992) (section 15.2).

Another important EM method is the one utilizing the very low frequency (VLF) band (15–25 kHz) produced by distant powerful radio stations. The EM–VLF field is modified in the presence of an electrically conducting body at depth and this change in the field is measured. From this data, apparent resistivity

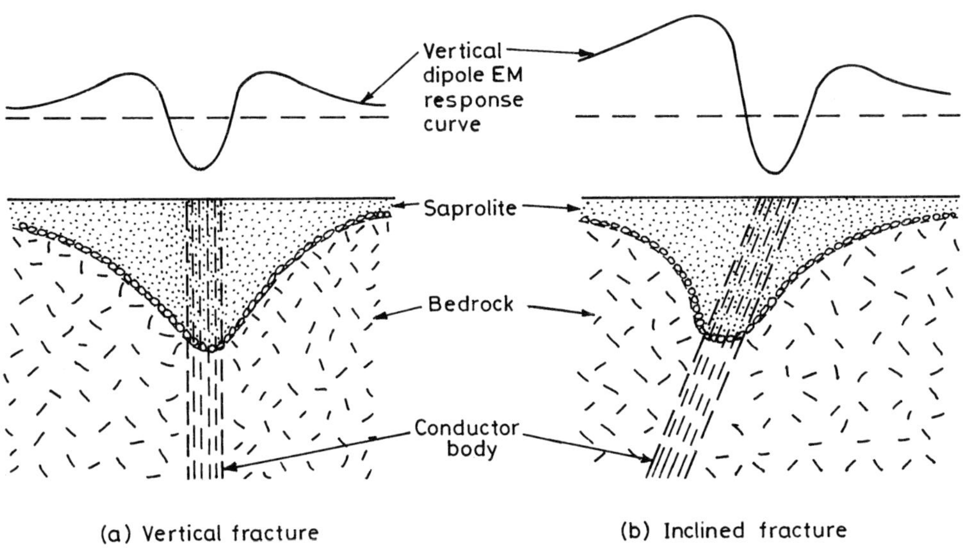

FIGURE 5.1 Schematic representation of horizontal loop EM (HLEM) (vertical dipole) response of a profile across a fracture (conductor body). (a) The fracture is vertical, and (b) the fracture is inclined.

of the buried horizon is computed. Although the interpretation of EM–VLF data based on forward modelling is mostly not unique, still it does provide useful information on subsurface conducting bodies. It is particularly useful in areas covered with resistive layers, such as dry desert sands or resistive hard rocks like basalts overlying fractured weathered water-bearing horizons. For example, Bromley *et al.* (1994) successfully used the EM–VLF technique for detecting fractured water-bearing horizons covered under basaltic rocks in Botswana.

5.3 Electrical resistivity methods

5.3.1 Basic concepts and procedures

This is the most widely used geophysical technique which gives a good indication of the occurrence of fractured horizons, groundwater and its quality. It is based on the principle that electrical resistivity of a geological formation is dependent upon the material as well as upon the bulk porosity and water content in the formation. The electrical resistivity of a dry formation is much higher than that of the same formation when it is saturated with water (Table 5.1).

In the electrical resistivity methods, electric current is introduced into the earth through two current electrodes. The resulting electric potential drop depends upon the resistance of the earth material and the amount of current flow. Two potential electrodes, positioned within the field of the current electrodes, give potential data. The data on current flow and potential drop are converted into resistivity values; the field resistivity values are apparent rather than true.

TABLE 5.1 Ranges of resistivity values of common rocks (ohm metre)

Material	*Almost dry*	*When saturated with water**
Quartzite	$4.4 \times 10^3 - 2 \times 10^8$	50 – 500
Granite	$10^3 - 10^6$	50 – 300
Limestone	$600 - 10^7$	50 – 1000
Basalt	$4 \times 10^4 - 1.3 \times 10^8$	10 – 50
Gneiss	$6.8 \times 10^4 - 3 \times 10^6$	50 – 350
Sand	$150 - 2 \times 10^3$	10 – 100

*Saturation of a hard rock implies that the formation has a secondary porosity due to weathering, fracturing etc.

The task of defining layers in a field survey is quite intricate as every measurement corresponds to a composite effect of all the layers in the rock volume penetrated. Several interpretation methods, many involving curve-matching with standard curves, have been developed to provide better delineation of resistivity layers in various conditions. Broadly, the depth of investigation of a resistivity survey is directly proportional to the electrode separation, and increases with increasing electrode spacing (for details refer to Koefoed, 1979; Bhattacharya and Patra, 1968; Zohdy *et al.*, 1974).

The electrical resistivity surveys for groundwater may comprise of sounding and/or profiling. The sounding technique is used to determine variations in electrical resistivity with depth, whereas profiling is used for delineating lateral variations of resistivity across an area. In vertical electrical sounding (VES), also called electrical drilling, the distances between electrodes are increased so that the electric current penetrates deeper and deeper levels, which allows resistivity measurement of a deeper and larger volume of the earth. The electrodes are normally linear, aligned parallel to the trend of rock discontinuity, which may be deciphered from other surveys or rock outcrops. The potential electrodes are placed in between the current electrodes and are symmetric with respect to the centre of the configuration. Most commonly used electrode configurations are Wenner and Schlumberger arrays. For these configurations the apparent resistivity *Pa* which is a measure of the effects of all the layers between the maximum depth of penetration

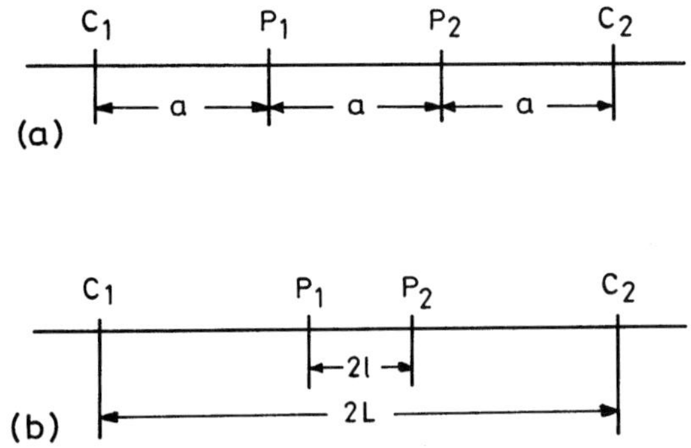

FIGURE 5.2 Electrode configurations for colinear resistivity survey. (a) Wenner configuration and (b) Schlumberger configuration.

and the surface, is calculated as follows.

$$\ell a = 2\pi\, aR \qquad \text{(Wenner arry)} \qquad (5.1)$$

$$\ell a = \pi\, (L^2 / 2l)\, R \qquad \text{(Schlumberger array)} \qquad (5.2)$$

where a, L and l are distances as shown in Figure 5.2, and R is the measured resistance (voltage/current) in each case. In the Wenner configuration, the electrode spacing a is kept equal. In the Schlumberger configuration, the distance L is varied, while keeping l constant. The distances a, L and l control the depth of investigation.

The apparent resistivity is plotted against the electrode separation (Figure 5.3). This resistivity curve is then interpreted in terms of horizontally stratified materials, thereby obtaining the number of layers, their respective thicknesses and resistivities, with the help of available standard curves. The interpretation can be further refined through forward or inverse modelling. Standard interactive computer programs carried out on PCs are now available for this purpose. Normally, three or four distinct layers are about the maximum number for a reasonably accurate interpretation of a resistivity sounding curve. The VES remains an extremely powerful technique for delineation of regolith thickness which is vital when the saprolite is potentially thick (> 20 m). Table 5.2 gives aquifer prospect as related to resistivity of layered regolith. However, VES is an inappropriate tool for detecting localized fracture systems.

TABLE 5.2 Aquifer prospect as related to resistivity (ohm metre) of layered regolith (after Bernardi *et al.*, in Wright, 1992)

0–20	Clays with limited prospect (or saline water)
20–100	Optimum weathering and groundwater prospect
100–150	Medium conditions and prospect
150–200	Little weathering and poor prospect
>250	Negligible prospect

In horizontal profiling or electrical mapping, resistivity surveys with constant electrode spacings are carried out at different locations. This enables gathering resistivity data over an area, with constant electrode separation or for a chosen depth of investigation. Contouring of resistivity data values and interpretation provides information on variation in bedrock/soil type, spatial variation in depth of weathering and moisture content etc. With this method, one can detect whether any anomalous zone exists, and whether it is resistive or conductive. This method is commonly used in reconnaissance, after which detailed study in the selected

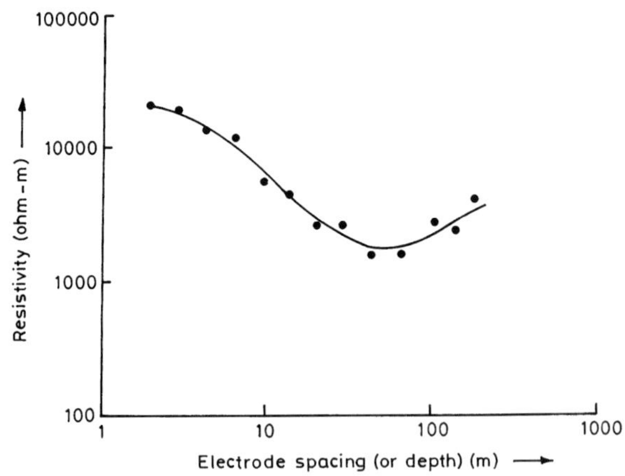

FIGURE 5.3 Electrical resistivity curve for depth sounding as an example.

target area is made through geoelectrical sounding. A microprocessor-based resistivity traversing (MRT) system has been developed by Griffiths and Turnbill (1985) to provide coverage which is a compromise between profiling and full VES. The resistivity data can be interpreted in terms of depth. Figure 5.4 shows resistivity traversing across a lineament in Zimbabwe.

The resistivity value in a specific case is influenced mainly by porosity (primary and secondary), degree of saturation and type of fluid. The resistivity varies with the degree of weathering and seasonal fluctuations in water salinity. In an area, low resistivity values may correspond to clays, highly fractured rocks, or saline sand. On the other hand, high resistivity values may correspond to tight (low porosity) rocks, freshwater-bearing sands or a relatively clean (clay-free) zone.

Normally, it is difficult to distinguish between permeable and impermeable fractures through electrical methods because their electrical properties are similar, especially when the impermeable fractures are filled with gouge, clay minerals or other alteration products. Such ambiguities which may occur are natural to geophysical methods. Therefore, it is necessary that geophysical data are interpreted with adequate control on surface and subsurface geology, which may be available from exposures or boreholes. Electrical methods are also useful in detecting changes brought about in the conductivities of pore fluids by the contamination of groundwater by the leachates from landfills and acid mine drainage.

5.3.2 Azimuthal resistivity method

Azimuthal resistivity survey (ARS) is a very powerful method of measuring *in situ* characteristics of fractured rocks. It reveals magnitude and azimuthal variation in the permeability of the bedrock. It has many advantages, for example in a poorly exposed terrain, field mapping of a fracture system may not be possible. Even

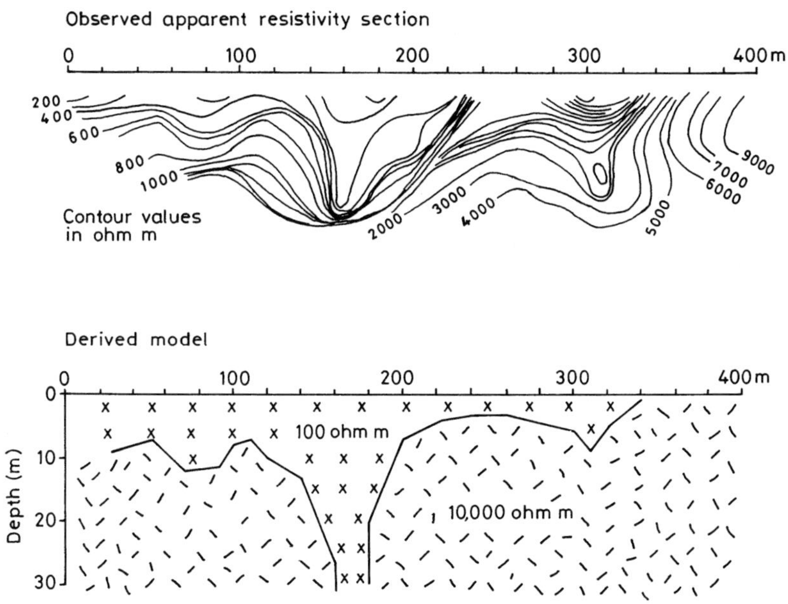

FIGURE 5.4 Resistivity traverse data across a lineament in Zimbabwe, with a two-layer interpretation (after Griffiths in Carruthers and Smith, 1992).

in areas of good exposures, extensive field data on fracture characteristics are required for evaluating the anisotropic character of the bedrock. The ARS measures bulk properties of fractured rocks; in such rocks, the parallel geometric arrangement of water-bearing fractures makes the resistivity anisotropic. In this regard, fracture connectivity plays a key role in controlling maximum permeability direction as well as groundwater velocity.

The technique utilizes conventional resistivity equipment and is performed by rotating a Wenner array (or Schlumberger array) about a fixed centre point. The apparent resistivity is measured as a function of azimuth, say at a 10° or 15° interval. The electrode spacing of about 5–25 m is used. When the apparent resistivities in different directions are plotted as radii, an anisotropy figure is generated, called the apparent resistivity figure (ARF) (Figure 5.5). In the case of one set of parallel fractures, this is an ellipse.

For a single set of steeply dipping saturated fractures, the true resistivity minimum would be oriented parallel to the fracture strike. However, in the azimuthal resistivity survey, the apparent resistivity maximum gets oriented parallel to the fracture strike (Figures 5.5; 5.6a,b). This is known as the paradox of anisotropy. It is a result of using current magnitude in the calculation of apparent resistivity, whereas the current density determines the actual differences in potential.

The coefficient of anisotropy of apparent resistivity ellipse is λ ($= \sqrt{\rho_y/\rho_x}$). It has been shown that joint porosity ø can be approximated under ideal non-shale conditions as $ø = \rho_o (\lambda^2 - 1)/\rho_y$, where ρ_o is equivalent to groundwater resistivity.

Extending the case of single-set fractures to multiple-set fractures, it is found that the effect of multiple set fractures is additive in nature. If there are two sets of fractures of unequal development, the azimuthal resistivity plot exhibits peaks of unequal magnitude (Figure 5.6c,d). Further, if in an area joint lengths are less than the electrode spacing and joints are poorly developed, the orientation of the ellipse will be intermediate to the trends of joints and will represent the direction of greatest connectivity. It becomes a function of both average fracture length and fracture frequency of both the joint sets (Figure 5.6e,f)

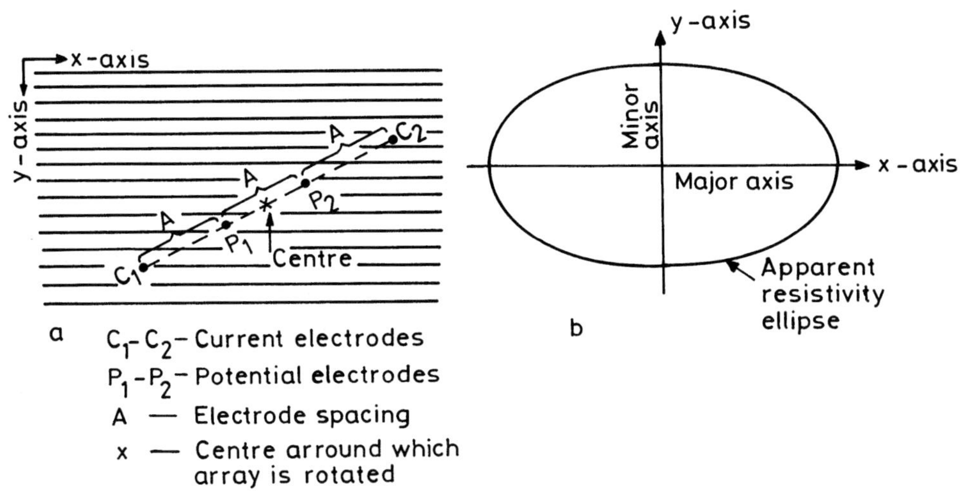

FIGURE 5.5 (a) Scheme of a colinear azimuthal resistivity survey. (b) The resulting azimuthal (apparent) resistivity figure (ARF) which is an ellipse in a simple case.

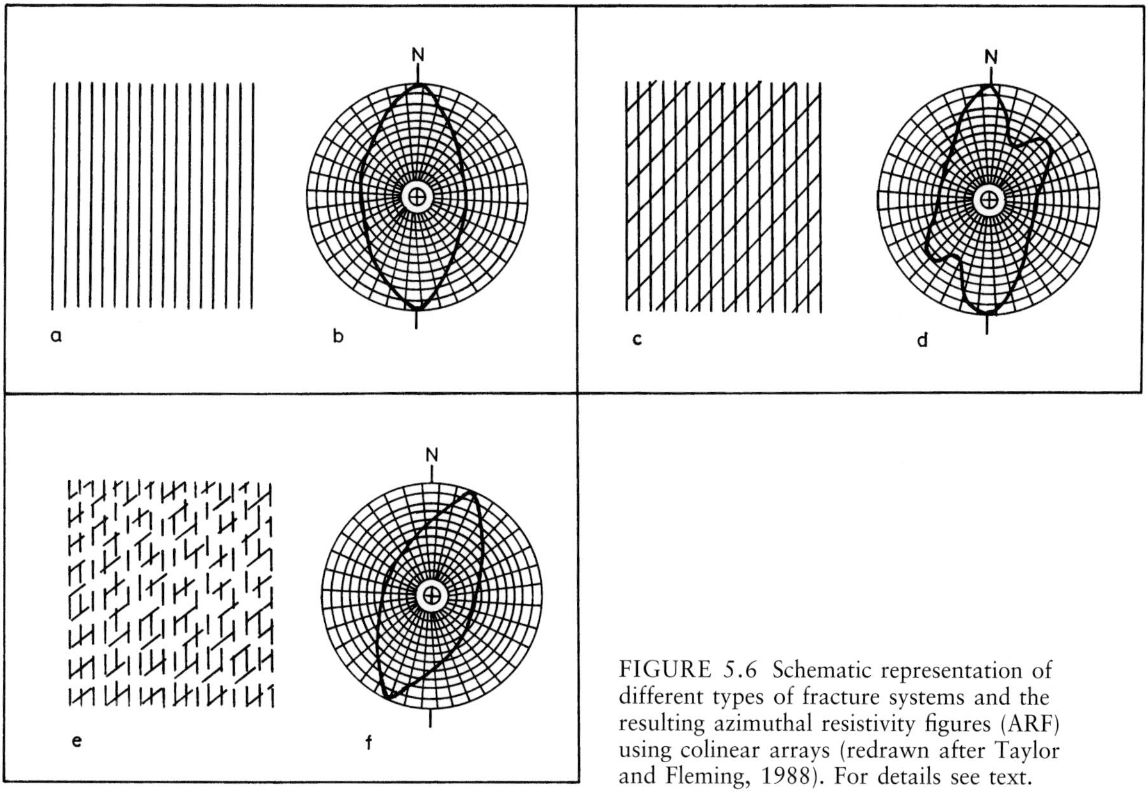

FIGURE 5.6 Schematic representation of different types of fracture systems and the resulting azimuthal resistivity figures (ARF) using colinear arrays (redrawn after Taylor and Fleming, 1988). For details see text.

Application of azimuthal resistivity surveys for the detection of fractures has been done by a number of workers, for example McDowell (1979) and Palacky *et al.* (1981) in granites; Mallik *et al.* (1983), Taylor and Fleming (1988), Haeni *et al.* (1993) and Skjernaa and Jorgensen (1993) in a variety of igneous and metamorphic rocks.

Ideally, the method is valid for homogeneous anisotropic rocks with near-vertical fractures and large fracture length and high fracture frequency, in comparison to electrode spacing. In such cases, the idealization for one set of fractures is shown in Figure 5.5 for a linear array method. If the fracture set is not vertical but dips at an angle, the apparent resistivity ellipse has a relatively increased minor axis. For a horizontal set of fractures, it would take the shape of a circle (both axes equal), as horizontal fractures contribute equally to the horizontal permeability in all azimuthal directions.

A close study of ARFs greatly helps in understanding the natural anisotropies in the bedrock. It provides a good representation of permeability anisotropies, which may be difficult to obtain even from field fracture measurements. A narrow ellipse (large-coefficient of anisotropy or large long to short axes ratio) indicates near-vertical continuous parallel fractures, with a large aperture. On the other hand, a broad ellipse suggests dipping or less continuous fractures with a low aperture. A single-peaked ARF indicates one set of fractures. A double-peaked ARF is formed by two sets of fractures, each peak corresponding to one set of fractures (Figure 5.6). In some cases, the direction of the long axis of the ARF lies in between the strike directions of two major fracture sets, and probably indicates the most conductive path through the fractured rock (Taylor and Fleming, 1988).

The square-array DC resistivity method, developed for areas with inclined strata, can also be applied in azimuthal mode. It is conducted in a manner similar to the traditional linear surveys. Major advantages of the square array method lie in its higher sensitivity to anisotropy as compared to Schlumberger linear arrays, and its requirement of less surface area for a given depth of penetration. For a single-set saturated steeply dipping fractures, the square-array method gives apparent resistivity minimum oriented in the same direction as the fracture strike, i.e. reverse of the linear array (Figure 5.7) (Haeni *et al.*, 1993). Thus, the fracture strike is perpendicular to the direction of maximum resistivity. After a comparison, Haeni *et al.* (op. cit.) consider azimuthal square array to be superior to azimuthal Schlumberger array in the method of implementation and results.

Before concluding, it may also be mentioned that the azimuthal resistivity method may not be able to distinguish between clay-filled and water-filled fractures, which have similar electrical properties but greatly different hydraulic conductivities. Such ambiguities are common in geophysical methods.

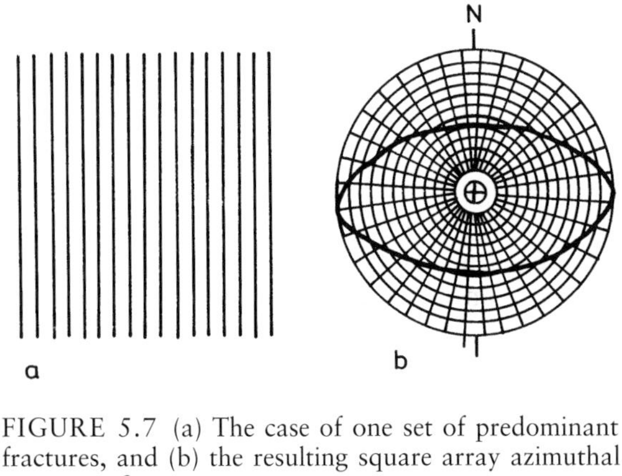

FIGURE 5.7 (a) The case of one set of predominant fractures, and (b) the resulting square array azimuthal resistivity figure.

5.3.3 Combined EM-resistivity method

Many of the geophysical surveys are run in combination which helps in resolving ambiguities and confirming interpretation from various angles. For example Randall-Roberts (1993) used EM–VLF, VES and SP techniques for hydrogeological exploration in fractured Precambrian gneiss in Mexico. EM–VLF was used to locate and define fractures in plan; VES soundings brought out horizontal sheeting as zones of low resistivity; SP permitted an interpretation of permeable intersections between vertical and horizontal fracturing. Thus, these data sets enabled a three-dimensional analysis from surface geophysical measurements, which was subsequently confirmed by a drilling and pumping test.

Bromley *et al.* (1994) describe a combination of aero-magnetic, VLF and coaxial EM surveys for groundwater studies in Botswana. In this region, the main aquifer is of the Karoo Formation, broken into a series of graben and horst structures by several faults and is completely masked under basalts and Kalahari beds. They used low-altitude (20 m height) airborne geophysical surveys to cover a 3300 km^2 area. The magnetic and VLF data were used to penetrate the masking cover. The drilling programme was guided by the geophysical data. Highest yields were obtained from fracture zones associated with VLF anomalies and an NW–SE set of lineaments.

5.4 Seismic methods

5.4.1 Basic concepts and procedures

The technique is based on the principle that the elastic properties of materials govern wave velocities. In general, a higher elastic modulus implies higher wave velocity in the material. In seismic surveys, waves are artificially generated by an explosion or impact of a sledgehammer, on the ground surface or at a certain depth. The resulting elastic waves are recorded in order of arrival at a series of vibration detectors (geophones), and the data is interpreted to give wave velocities. Seismic waves follow multiple paths from source to receiver; in the near-surface zone, the waves may take a direct path from source to receiver. Further, the waves moving downward into the earth may be reflected and refracted at velocity interfaces. Figure 5.8 shows wave travel by direct, refracted and reflected paths. For depth calculations involving two or more layers, algorithms have been developed and the field procedures for seismic investigations have been made greatly efficient in recent years, with the aid of compact, portable computer controlled instruments.

Important rock characteristics influencing wave velocity are: crystallinity, porosity, cementation, weathering, and discontinuities such as bedding and joints. Massive, compact, crystalline, low-porosity rocks possess higher seismic wave velocities, and unconsolidated formations possess least velocities (Table 5.3). The presence of fractures in a rock mass causes a reduction in seismic velocity and an increase in attenuation. These two effects form the basis for the characterization of fractures by seismic

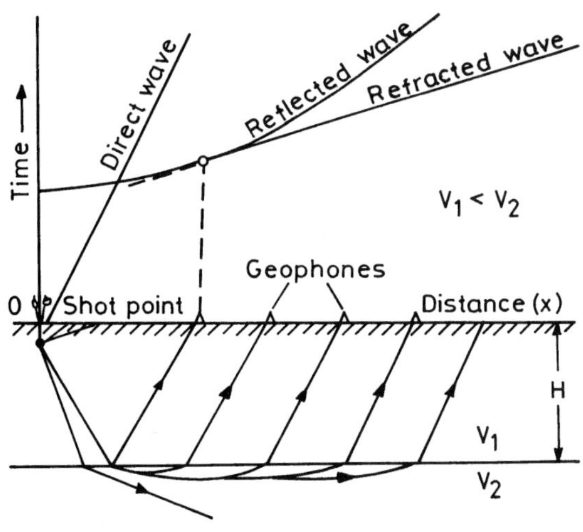

FIGURE 5.8 Direct, reflected and refracted waves in a seismic survey.

TABLE 5.3 Common ranges of P–wave velocities

Material	*Velocity* *(m s^{-1})*
Weathered soil horizon	100–500
Alluvium	500–2500
Sandstone	1500–4000
Limestone	2500–6500
Granite	4000–7000

methods (Anon., 1996). Seismic velocity is greatly influenced by fracturing in rocks; shock velocities measured in the field are much lower than those measured on intact (core) samples in the laboratory, for the same rock. Degree of fracturing can be estimated to some extent from a parameter called **velocity ratio**, computed as the ratio of the field (*in situ*) velocity (V_F) to the laboratory velocity (V_L), in a rock. As the number of fractures decreases, V_F tends to approach V_L. It has been suggested that in general, a velocity ratio (V_F/V_L) of less than 0.5 indicates a significantly fractured rock condition. Therefore, the velocity ratio (V_F/V_L) is also sometimes called 'fracture index'.

Seismic reflection methods are more suited for exploration of deeper structures whereas refraction techniques are more extensively used for investigation of shallower contacts. For groundwater studies, seismic refraction methods are more frequently used, the main application being deciphering the thickness of the weathered zone. However, in some cases, optimum use of seismic methods may involve a combination of refraction and reflection principles.

In the case of the velocity interface being inclined (e.g. dipping strata), there is an additional variable. In such cases, recording seismic data in up- and downdip directions, or reverse profiling is required to obtain true estimates of velocities and depth. The dip of the discontinuity may be calculated by comparing the reverse profile data.

5.4.2 Azimuthal seismic refraction method

The azimuthal seismic refraction method can detect strike direction of major fractures in the bedrock. However, small isolated fractures or fracture zones may not be detected by refraction surveys. A fractured rock mass exhibits anisotropy in wave velocity. For a single set of steeply dipping saturated fractures, a seismic velocity maximum occurs in the direction of the fracture strike and the velocity minimum occurs orthogonally to it. Azimuthal seismic refraction data (using the P-wave) can be collected by rotating the survey line at a constant angular increment, about a common centre point. For each survey incidence, a sledgehammer impact may serve as an energy source and geophones are spaced at equal intervals. Data analysis has to be carried out to obtain P-wave velocity for each direction. Figure 5.9 gives an example of the azimuthal plot of seismic velocity data and its interpretation.

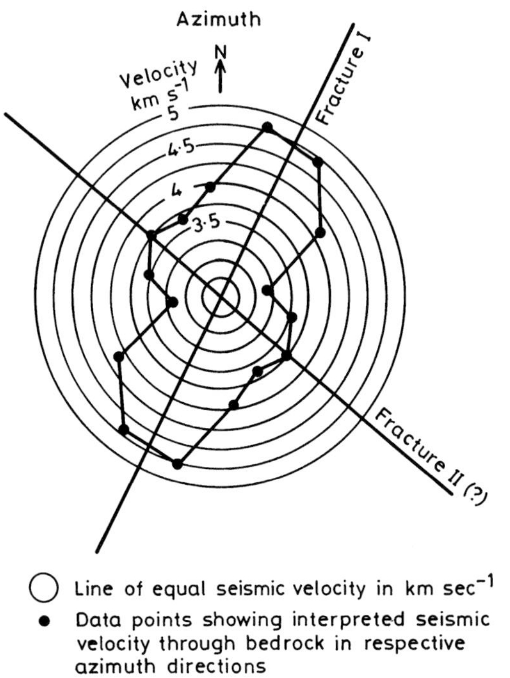

FIGURE 5.9 Azimuthal seismic survey indicating the presence of Fracture I and Fracture II (?) at a site in Mirror Lake area, New Hampshire. (after Haeni *et al*., 1993).

5.5 Radon survey

Radon (Rn) is an odourless, colourless gas produced by radioactive decay of uranium and thorium in nature. It is the only gas to be radioactive, emitting alpha particles, and is therefore hazardous for health. Its presence and concentration can be detected in water and solid material (soil/rocks) (Ball *et al*., 1991). A part of the radon generated in nature may escape in carrier fluids like CO_2 or H_2O through voids and fractures; this property is useful in geothermal and groundwater investigations. Other applications of radon may include delineation of faults, basement structure and possible prediction of earthquake and volcanic activity. There is often an increased content of radon in soil-gases over faults and fractured zones, due to the increased flow of water along these discontinuities. Radon survey has been successfully used in some areas for groundwater exploration in basement fractured rocks (Pointet, 1989; Wright, 1992).

5.6 Radar methods

The radar methods can be used from space, aircraft, the ground and boreholes. The radar techniques from aircraft and space platforms are discussed in Chapter Four. Here we discuss the radar technique from the ground (called ground penetrating radar) and boreholes (called borehole radar).

The **ground penetrating radar (GPR)** is a promising surface geophysical method for hydrogeological studies (Beres and Haeni, 1991) (Figure 5.10). The GPR system emits short pulses of radio frequency EM radiation into the subsurface from a transmitting antenna; the backscattered radiations are sensed by another receiving antenna. The reflected signal is dependent upon the inhomogeneities in electrical

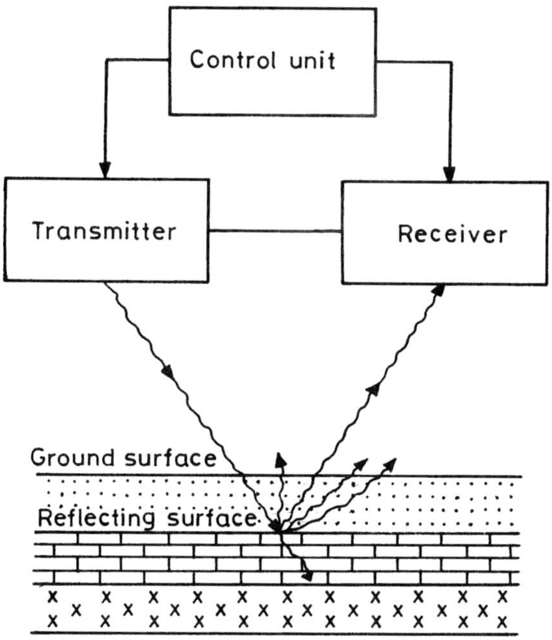

FIGURE 5.10 Working principle of a ground penetrating radar.

properties of the subsurface, such as water content, composition, structure and discontinuities in the rock. The received signal is amplified, processed and displayed. It shows the total two-way travel-time for a signal – to pass through the subsurface, reflect from an inhomogeneity and return to the surface. The GPR system is used in profiles across an area, moving at an average speed of 3–5 km per hour. The types of reflection configurations are interpreted in terms of physical characteristics of lithologic units. As far as depth penetration is concerned, the method is limited by attenuation of the EM waves in the subsurface material. The method produces profiles of the subsurface which are similar to seismic reflection products. GPR is an excellent tool for mapping fractures or fracture zones with high resolution in the absence of overburden, including in tunnels and boreholes (e.g. Cook, 1975; Rubin and Fowler, 1978).

Borehole radar method is similar to the surface radar method, except that the survey is carried out in a borehole, and therefore, the transmitter and receiver are both oriented vertically in a borehole. Radar readings are taken at constant intervals as a function of depth. Data are digitally recorded, processed and displayed; from this data, it is possible to interpret the presence of important fractures and master joints in the rock (e.g. Olsson *et al.*, 1992; Haeni *et al.*, 1993).

5.7 Gravity and magnetic methods

The gravity and magnetic methods indirectly yield some information about favourable structures for groundwater occurrence. These methods make use of natural fields of gravity and geomagnetism. Changes in gravitational and magnetic fields may be observed on the Earth's surface, which could be related to lateral changes in density and magnetic susceptibiity of the material at depth. The variation/contrast in density and susceptibility produce small but measurable changes of corresponding fields. The instruments used for these measurements are the gravimeter and the magnetometer. The practical units for measuring gravity and magnetic fields are milligal and gamma (nanotesla) respectively. In these surveys, the gravity and magnetic values are measured at pre-fixed observation points in profiles. Various corrections need to be done on the data (e.g. Dobrin, 1976). The corrected values of gravity and magnetic field are plotted at the stations/profiles and an iso-anomaly map is drawn. Quantitative interpretation is done by analysing the nature of contours, highs and lows.

Across a fault plane, a steep gradient of gravity is observed. A gravity high implies denser rocks closer to the ground surface, e.g. basic intrusions. Lower density materials and cavities produce gravity lows. Gravity methods have a rather coarse resolution, the method being suited for detecting large structures, such as regional folds, subsurface domes etc. For the specific purpose of aquifer location, for example, solution cavities in karst areas, microgravity measurements have an interesting potential. Both laterally and vertically extended low density features can also be mapped. The technique requires good knowledge of rock density and their variations, derived from measurements on rock samples.

Magnetic surveys (magnetometry) are among the most cost effective of geophysical techniques for geological mapping. These are quite effective in delineating subsurface mafic dykes, as well as quartz and pegmatite veins, the latter due to diamagnetic property (i.e. negative magnetic susceptibility of quartz). Magnetic surveys can be conducted from space, aerial- and ground-based platforms. In fact, low-altitude aerial magnetic survey is an extensively used technique in geoexploration.

Magnetic surveys can give an idea of the major geological-structural features. The anticlines would produce positive and synclines negative anomalies. A fault is indicated by a sharp gradient in the magnetic contour map. Since the basement rocks are more magnetic compared to overlying sediments, the trend of the magnetic contours is largely related to structural trends in the basement and distribution of basic intrusions. Modelling of magnetic data can bring out highly useful information on structure, dip of faults, contacts etc. This is particularly important as fractured zones along faults and contacts of dykes etc. also form potential aquifers. Orientation of these features can be delineated from magnetic surveys.

Both gravity and magnetic profile data may be portrayed in image mode. This facilitates spatial filtering and image enhancement, which renders image interpretation easy. Figure 5.11 is an example from Botswana, showing the image of aeromagnetic data together with an interpretation geologic map. Further discussion on image manipulation – GIS aspects of various geodata sets, to facilitate a coherent integrated interpretation – is discussed under GIS (Chapter Six).

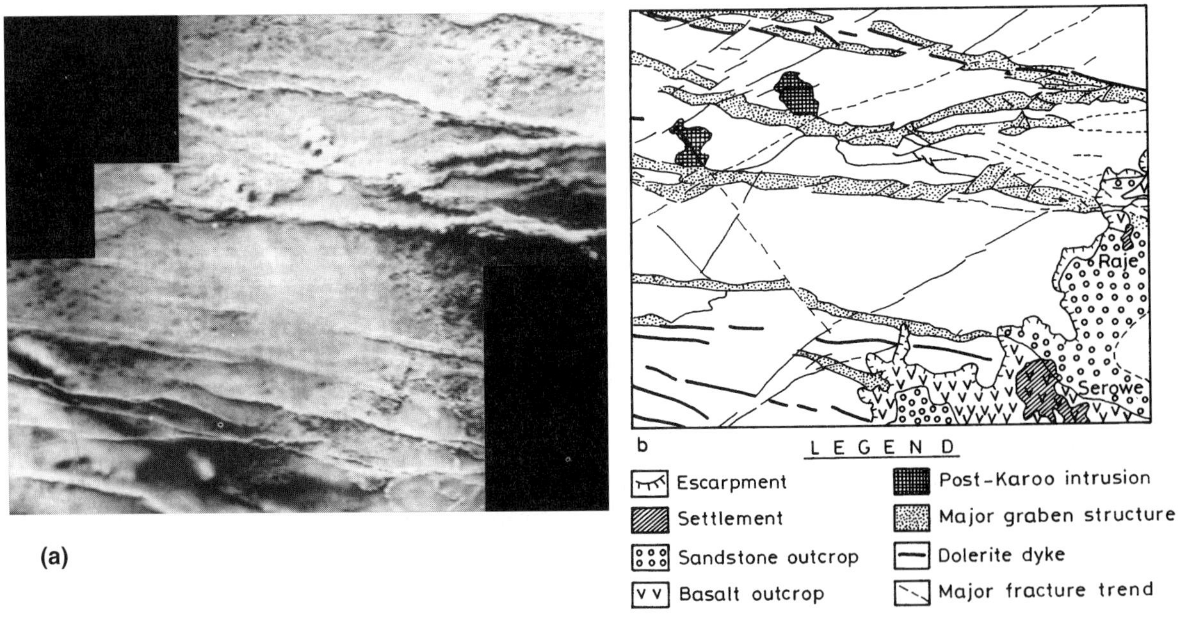

FIGURE 5.11 (a) Aeromagnetic residual field image of an area in Botswana. (b) Its structural geological interpretation map (after Bromley *et al.*, 1994, reprinted by permission of journal of *Ground Water*).

5.8 Geophysical tomography

The advent of computer aided tomography (CAT) has revolutionized medical sciences. Even though similar techniques have been traditionally utilized by seimologists as well as by exploration geophysicists in the field of seismics for quite some time, no such special term is used. But with the emergence of necessary mathematical tools viz., algebraic reconstruction technique (ART), simultaneous iterative reconstruction technique (SIRT), back-projection methods etc. in the field of medical imaging (including CAT and NMR tomography), a new discipline has eventually taken shape in the field of geophysics as well, viz. geophysical tomography. Under the umbrella of geophysical tomography, several methods can be included like seismic tomography, electromagnetic tomography, and resistivity tomography. 'Tomography' simply means a technique used to obtain an image of the selected plane of a solid object (Worthington, 1984). It comes from the Greek world *tome* meaning a slice.

Seismic travel time tomography

This basically entails imaging intervening media between the array of receivers and sources. As per geometric configuration of arrangement of sources and receivers, we can have cross-hole, surface-hole and surface-surface travel time tomography. A fracture zone either air- or water-filled in an otherwise massive rock constitutes a low velocity region. Depending upon the orientation of fracture, one of the above modes of seismic imaging can be effective. For example, surface-surface mode is good for horizontal or sub-horizontal fractures whereas near-vertical fractures can be better imaged by the cross-hole method, and so on.

Recently, the cross-hole method of seismic velocity measurement has been used more extensively. It helps in assessing the whole rock mass properties *in situ*, between the two boreholes. Cross-hole seismic survey is particularly useful in the location of cavities or old mine workings in urban areas. Cross-hole seismic measurement underneath the foundations of a building is probably the only effective method to assess the rock mass beneath a building. Identification of underground cavities is important to check damage to buildings due to ground subsidence on account of subsurface cavities. In hot dry rock (HDR) systems, the cross-hole seismic method is used to delineate the cavity created by explosive HDR stimulation (McCann *et al.*, 1986). This method is also used for groundwater investigations for mapping fractures (Carruthers *et al.*, 1993).

EM tomography

The interest in cross-hole EM tomography is mainly for imaging interwell electrical conductivity. The sensitivity of electrical conductivity to porosity, fluid type, saturation and temperature has led to the development of cross-well EM systems and imaging algorithms (Rector, 1995). Field examples and numerical simulations demonstrate remarkably good resolution of interwell features when compared to surface EM techniques (Spies and Habashy, 1995; Alumbaugh and Morrison, 1995; Lee and Xie, 1993; Wilt *et al.*, 1995). The cross-well EM tomography can be used to map fractures within highly resistive compact rock (Alumbaugh and Morrison, 1993). This method can also track an injected slug of water (Wilt *et al.*, 1995). The conductivity images of data collected before and after injection showed a clear anomaly as a result of saltwater plume and indicated the direction of plume migration.

Electrical resistivity tomography (ERT)

With the advent of multi-electrode, microprocessor-based resistivity measurements (Griffiths and Turnbill, 1985), it is now possible to carry out three-dimensional resistivity surveys in a variety of combinations like cross-hole, hole-surface, surface-hole and surface-surface. These are increasingly being applied to problems of groundwater flow, and pollutants movement studies. They can be used as well to delineate water-filled fractures within moderately resistive host rock.

5.9 Subsurface methods

Subsurface methods including exploratory drilling and well logging are essential for confirming results and interpretations made from surface geological and geophysical investigations. Although subsurface investigations are more expensive than surface methods, the precision and reliability of data which they provide more than offset this consideration.

5.9.1 Exploratory excavation and drilling

Exploratory excavation may be done by digging pits, trenches, adits and shafts, depending upon the type of problem, topography and needs. The technique provides a means for directly observing and mapping subsurface features, for example, observing how discontinuities continue and behave at depth, and also sampling subsurface rocks. In this stage, various faces and walls need to be very carefully mapped and logged and mutual relationships of various fractures and discontinuities recorded. The excavation methods should be such as to minimally disturb the rock conditions.

However, exploratory excavation being expensive it is limited to shallow reaches. For deeper exploration, exploratory drilling is carried out; it helps define the geometry and extent of the aquifer and assessing the groundwater potential. The test holes are preferentially located in such a way that in the case of prospects of a good aquifer, the same boreholes can be converted into production wells by redrilling, or reaming to a larger diameter. The test holes also serve as observation wells for monitoring groundwater levels.

The data obtained during drilling is recorded as a lithologic log. It is a description of geologic characters of various strata such as lithology, thickness, and core recovery encountered during drilling. The drilling time log, consisting of a record of time taken to drill every two metres of depth is helpful in indicating where there is a change of strata or intensity of fracturing and weathering. The rate of penetration of a stratum can be correlated with the formation characteristics and hence with its water-yielding capacity, in a relative manner.

5.9.2 Geophysical well logging

Geophysical logs are obtained by lowering a probing tool in a sonde down the borehole. They are used to study the variation of physical properties of subsurface rocks (including fractures etc.) and their fluids. Correlation of logs may reveal the nature of stratification and extension of structures and fractures. The boreholes are logged with a number of geophysical probing tools which provide direct and some quantitative data about the hydraulic characters of subsurface formations.

With the advancement in microelectronics, compact logging units are now available with software for interpretation. A number of properties of both formation and interstitial water, such as coefficient of diffusion, formation factor, hydraulic conductivity, specific yield, and concentration exponent can be estimated. Further, borehole geophysical data can be used to estimate water properties such as salinity, viscosity and density, and formation properties such as porosity and permeability in suitable conditions (Jorgensen, 1991).

As far as fracture evaluation in particular is concerned, a large number of logs is required to detect and interpret fracture characteristics properly. Fractured zones produce anomalies with respect to normal or constant hole size. Fractures when open, lead to highly permeable paths which can be detected in logging in terms of high drilling rate, loss of drilling fluid, poor core recovery and/or significant increase of hole size. The treatment here gives a brief resumé of the well logging methods with special reference to fractured rocks.

Spontaneous potential log

This gives a record of electric potential with depth in the borehole. It is useful for shale versus sandstone discrimination, but it has limited utility for fracture identification.

Gamma ray log

Gamma rays are emitted by all natural rock formations as a result of random disintegration of naturally present radioactive elements. The elements producing gamma rays are potassium, uranium and thorium (KUT), and the log records a total count of gamma rays. The KUT elements naturally concentrate in finer-grained materials (clays, silts) where they are adsorbed in minerals like clays. Therefore, the gamma ray log is regarded as a clay indicator. In fractured igneous and metamorphic rocks, the log response is relatively less consistent. For example, the log may show peak responses due to potassium-rich minerals (e.g. feldspars) and/or a clay-rich weathered zone, and/or a zone of leaching. Conversely, in some cases a lower activity against the weathered rock may also be observed. Fluid circulation or past fluid circulation in the rock mass can sometimes also be inferred from the gamma ray log. This is due to the presence of uranium oxide which is soluble and highly mobile and can be precipitated in joint and fracture surfaces which form fluid routes within the rock mass. Therefore, gamma logging can identify this local activity where the boreholes intersect such fractures.

Caliper log

This measures the diameter of an uncased drill hole as a function of depth. The measurement is obtained with a 3- or 4-arm probe which is electronically opened when the probe is at the bottom of the borehole, and the variation in the borehole diameter is recorded as the probe is winched to the surface. Permeable zones will usually show a reduction in hole size due to deposition of thicker mudcake on the borehole. Fractured horizons usually show an increase in the borehole size which may occur due to breaking of the formation wall during drilling. Fracture orientation is likely to affect the hole ellipticity.

Bore fluid logs

These include logging of bore fluid parameters such as electrical conductivity and fluid temperature. Of these, the temperature log has been the most common. Fractured permeable formations are characterized by low temperature anomalies, which occur due to locally increased mud circulation. Logs of fluid temperature and fluid electrical conductivity can be interpreted in terms of groundwater conditions. Fluid logs are run under different hydraulic conditions – usually at rest and also during pumping. A comparison of the two data sets reveals the vertical position where water enters the borehole. These logs are also helpful in investigating inter-aquifer migration of water, adequacy of grouting, quality of groundwater and other related aspects.

Resistivity logs

An electric resistivity log is a record of the apparent resistivity of the subsurface formation with depth. There are numerous variations in the resistivity logs. The electrical logs cannot be run in cased holes and may be operated in dry holes or preferably fluid-filled holes. The measured apparent resistivity depends upon the geometric fracture characteristics and the nature of fluid filling the fractures. It is influenced by fracture orientation, size, length, spacing etc.

The range of resistivity in hard rocks is quite large. Fractures filled with water tend to cause a decrease in apparent resistivity in hard rocks. A useful method is by using single point resistance (PR) technique. The PR log represents the varying electrical resistance between a single downhole electrode and a fixed surface electrode. It does not measure the true rock resistivity and is strongly influenced by borehole diameter change. However, unlike multi-electrode resistivity logging devices, its response is symmetrical with bed boundaries recorded in the correct position, and the relative response is useful for recording the junction between rock units and for correlation. Normal resistivity logs in adjacent vertical boreholes can help in mapping the lateral extension of subhorizontal fractures. Vertical or near vertical fractures may not be detected by induction log due to the fact that the induced current tends to flow in horizontal loops around the borehole and therefore a vertical fracture containing conductive fluid may go undetected in the log. On the other hand, horizontal fractures filled with conducting fluid appear as conductive anomalies.

Microresistivity logs are likely to miss fractures, as they measure only a small volume of rock around the wellbore. Fractures lead to increased conductivity due to higher local porosity and greater water saturation. Different combinations of laterologs and induction laterologs can be used to decipher the presence of fractures close to the wellbore and distant from the wellbore (van-Golf-Racht, 1982). However, these advanced techniques are used more in the petroleum industry.

Dipmeter logs

Basically they record the dip angle and dip direction of a bedding plane intersecting the borehole. The tool consists of four radial pads positioned at angular intervals of 90°. It is rotated in the borehole at a uniform speed, as it is winched up, yielding four microresistivity curves. The azimuth recording of electrode 1 is made continuously. The dipmeter response may describe all types of discontinuities from horizontal to vertical. The fracture identification log (FIL) is an improved tool for detecting fractures; higher efficiency in FIL is obtained by superimposing the response of a couple of electrodes, i.e. the combination of electrode responses in a specific manner.

Porosity log

Under porosity logs are also included density, neutron and sonic logs. These logs are capable of detecting fractures and evaluating secondary porosity. In principle, the secondary porosity must be evaluated as the difference between the bulk porosity and the matrix porosity, both of which are measured through logs. A double porosity model is used to link the bulk porosity and the matrix porosity (section 7.2.2). The density log is a gamma-gamma ray log. A gamma ray beam is emitted from an artificial source and a counting system detects the backscattered intensity, which is related to the density of the rock. A higher density causes a relatively lower level of gamma-gamma intensity. Fractures causing higher secondary porosity are indicated by a higher gamma-gamma ray count.

Neutron logs respond primarily to the amount of hydrogen present in the formations. In the case of open, water-filled fractures, neutron logs exhibit anomalies indicating higher porosity. The sonic log is very useful in fracture detection, particularly in dense rocks. It uses a transmitter and a wave receiver. As the transmitter makes an energy emission, different types of shear and compressional waves are generated and received at the receiver. The amplitude, velocity and attenuation of different wave types are influenced by the fracture characteristics. A study of amplitude, attenuation and arrival times of shear and compressional waves can provide an indication of fracture orientation and lithology.

Borehole televiewer (BHTV)

This is used to detect and evaluate fractures and formation boundaries by direct measurements. It may be treated as a partial substitute for continuum well coring. Combined with drill core data, it is a highly valuable tool. BHTV is carried out in boreholes filled with homogeneous, gas-free liquid such as drilling mud, fresh water etc. BHTV includes a source of acoustic energy and a magnetometer mounted in the tool. The tool is rotated at a uniform speed during logging; the changes in the uniformity of the borehole walls such as fractures, rugs, pits, traces etc. are reflected as changes in picture intensity. It produces a two-dimensional image of the borehole wall. The intersection of fractures with the borehole can also be observed in the BHTV image; the fractures perpendicular to the borehole appear as horizontal traces and those parallel to the borehole appear as vertical traces. Fractures intersecting at angles appear as sinusoids. From these data, the dip and strike of fractures can be calculated.

Flowmeter log

This is a type of borehole fluid log. It is basically a velocimeter and makes a continuous record of flow profile versus depth; it provides a confirmation of fracture location. Flowmeter logging during pumping can detect increased velocity of water flow moving to the pump, at various inlet points in the borehole, and is therefore highly useful in detecting water-bearing fracture zones. Fluid replacement logging is a recently developed technique for identifying the permeable fractures in a borehole. The technique involves replacement of borehole fluid by deionized water and subsequent measurement of variations in the electrical conductivity of the fluid in the borehole with time (Anon., 1996).

Figure 5.12 shows a suite of well logs from a high-yielding well in Zimbabwe as an example. The coincident local excursions in caliper, neutron and resistivity logs mark the individual fractures which provide major contributions to well yield.

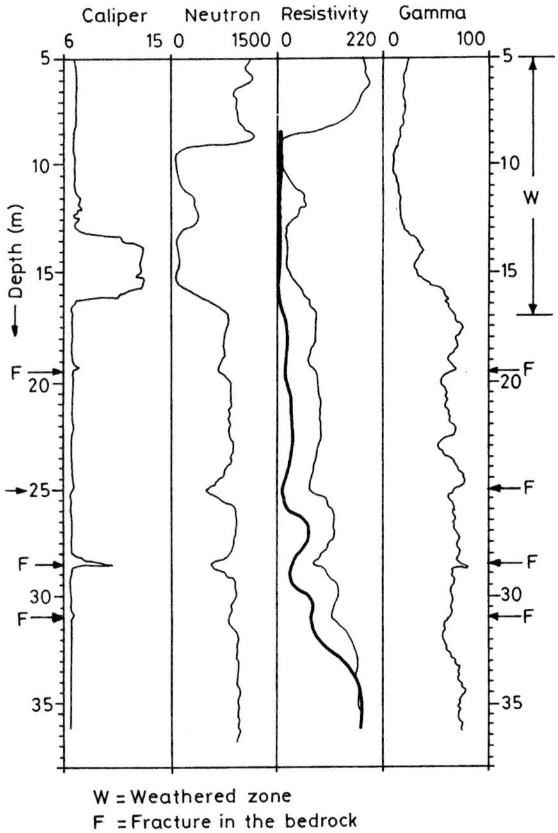

FIGURE 5.12 A suite of well logs including calliper log, neutron log and resistivity log from a site in Zimbabwe. The weathered zone extends up to about 17 m depth. Fractures are indicated at 19.5 m, 25 m, 27.5 m and 31 m levels (after Carruthers and Smith, 1992) (Scales: caliper–inches, neutron–CPS, resistivity–chm metre, gamma–cps).

Chapter six

Geographical information systems (GIS)

6.1 Introduction

Geographical information system, also called geobased information system (GIS), is a relatively new technology. It is a very powerful tool for processing, analysing and integrating spatial data sets (Aronoff, 1989; Star and Estes, 1990; Maguire *et al.*, 1991; Attenucci *et al.*, 1991; Heywood *et al.*, 1997). A GIS deals with information on locational patterns of features and their attributes (characteristics). It can be considered as a higher-order computer-coded map which permits storage, selective dedicated manipulation, display and output of spatial information. Figure 6.1 explains the working concept of a GIS; in a very comprehensive sense, GIS may mean identifying data needs, acquiring data, data management, processing and analysis of data and decision-making. In normal usage, however, GIS means spatial data processing, integration and analysis. GIS has become a standard, indispensable, tool for handling spatial information for the exploration, development and management of the Earth's resources.

6.1.1 Why GIS for groundwater studies?

For handling groundwater data, the GIS technology is aptly suited, for the following main reasons.

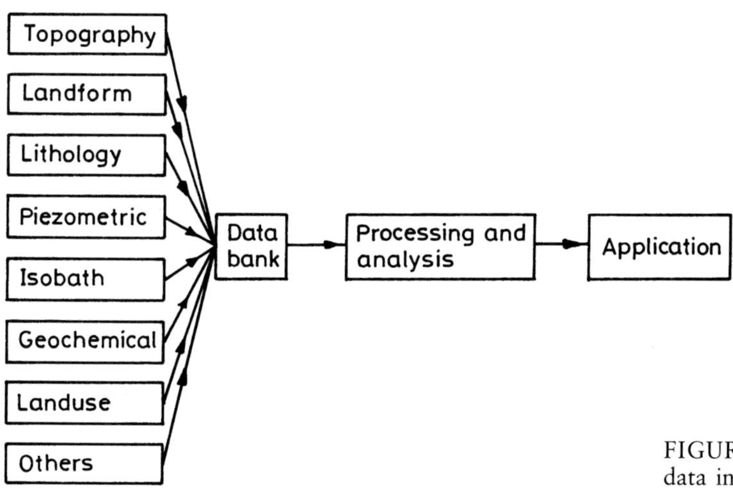

FIGURE 6.1 The concept of data integration in a GIS.

Concurrent handling of locational and attribute data

In groundwater studies, one has to deal with information comprising locational data (where is it?) and attribute data (what is it?). GIS packages have the unique capability of handling locational and attribute data; such capability is not available in other groups of packages (Figure 6.2).

Variety of data

Groundwater investigations often comprise diverse forms and types of data, such as the following: (a) topographic contour maps, (b) landform maps, (c) lithological maps, (d) structural geological maps, (e) isobath maps (contour maps of equal depth of water-table), (f) isogram (isocone) maps depicting groundwater characteristics by contours of equal concentration of

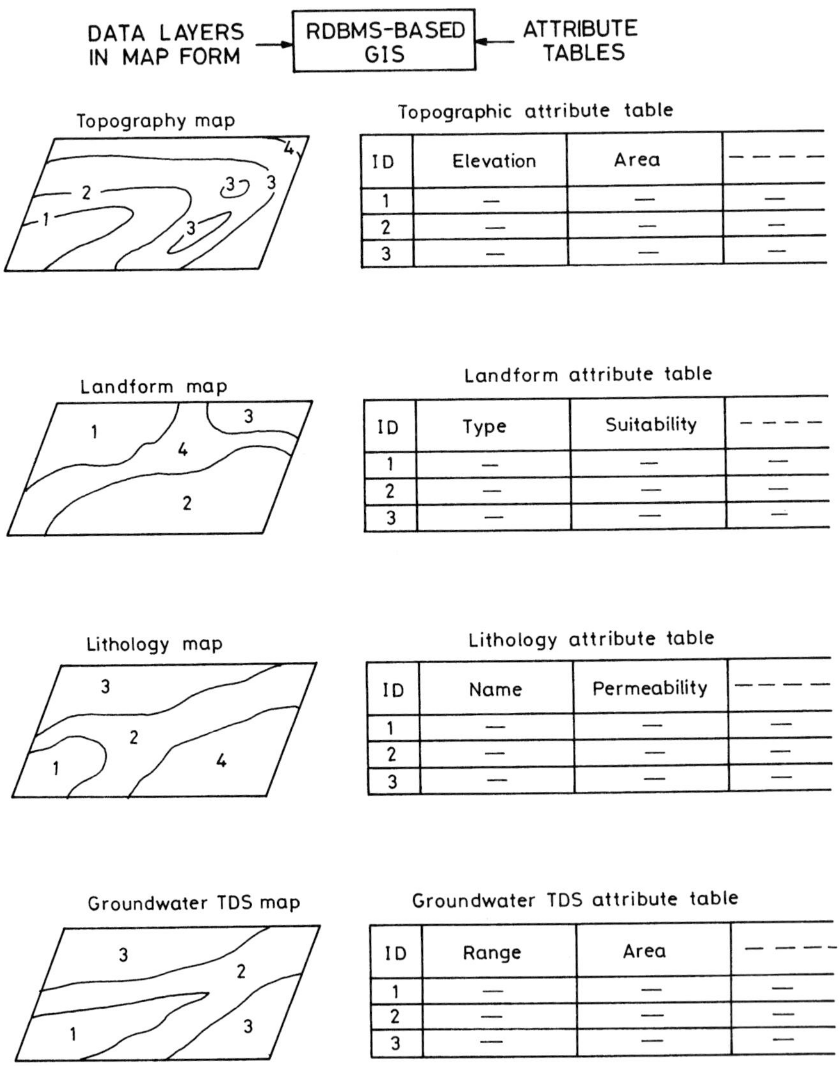

FIGURE 6.2 Schematic representation of GIS-working. The GIS maintains a link between the map feature and the corresponding tabular information.

dissolved solids (TDS) or ions, (g) drainage density and other geomorphic maps, (h) tables of various observations and data sets, and (i) point data, say locations and water-levels of observations wells or springs.

In these, some of the variables are of continuous type, such as TDS content, water-level data, and some others are of categorical type, such as low/medium/high drainage density, or gravel/marble/granite lithology (for continuous and categorical data types, see section 6.2). It is essential to integrate the spatial information for coherent and meaningful interpretation, and to avoid compartmentalization of data. GIS offers technological avenues for integrating the variety of data sets on both qualitative and quantitative lines, hitherto not available through any other route (e.g. Star and Estes, 1990; Gupta, 1991).

Flexibility of operations and concurrent display

Modern GIS are endowed with numerous functions for computing, searching for and classifying data, which allow processing and analysis of spatial information in a highly flexible manner and concurrent display, interactively.

Speed, time and costs of processing

Advances in microelectronics and computer technology have made it possible for modern GIS to store, process and analyse large volumes of data which otherwise would be too expensive, tedious and time-consuming to carry out by other methods.

Higher accuracy and repeatability of results

The technique, being digital computer-based, yields higher accuracy in comparison to manual cartographic products. The results are amenable to rechecking and confirmation.

6.2 Basics of GIS

We mention here some very basic concepts of GIS (for details, refer to Aronoff, 1989; Star and Estes, 1990). The hardware for a GIS comprises a basic computer (viz. CPU, storage devices, keyboard and monitor), digitizer and/or scanner for inputting spatial data, high resolution colour monitor for displaying spatial data in image mode, plotter for high quality production of maps, and printer for printing tables, data, raster maps etc.

In GIS, the database is created to collate and maintain information. It is a collection of information about things or objects. The geographical or spatial information has two fundamental components (Figure 6.2):

1. Location (position) of the feature (where is it?), e.g. location of a pumping well, spring, river or city.
2. Attribute character of the feature (what is it?), e.g. lithology type, or groundwater elevation, or TDS content in water.

Data pertaining to the above two aspects are explicitly or specifically recorded in GIS. There could be other critical characteristics such as time (when did the phenomenon occur?), or spatial-textural relationship (neighbourhood relations with other features), which are not explicitly recorded in GIS, but could possibly be searched for and deduced.

Location data

In GIS, the location or spatial position is given in terms of a set of latitude/longitude, or relative coordinates, or reference axes like Easting/Westing. From a geometrical point of view, all features on a map can be resolved into points, arcs and polygons (Figure 6.3). All features, whether points, lines or polygons can be described in terms of a pair of co-ordinates, viz: points – as a pair of x–y coordinates; lines – as a set of interconnected points; and polygons – as an area enclosed by a set of lines.

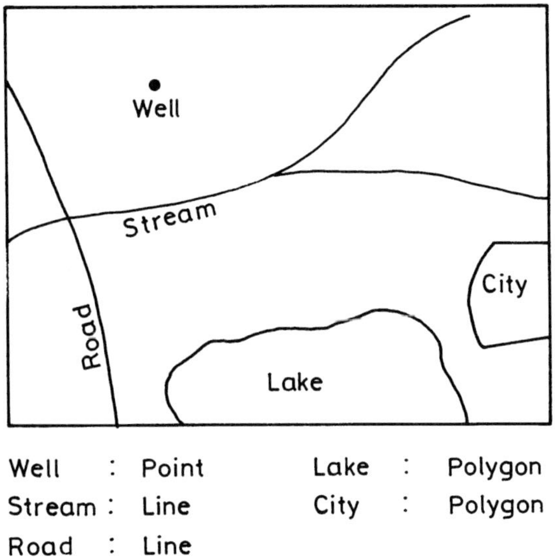

FIGURE 6.3 Types of location data: points, arcs and polygons.

Attribute data

Attribute data is the information pertaining to what the feature is, i.e. whether the point indicated is a pumping well or an observation well, or information at the specified location pertains to water quality, or depth of water-table, or topography, or rock type etc. The attribute data is considered to be of two basic types – categorical and continuous. Table 6.1 describes the possible varieties of attribute data with examples drawn from the field of groundwater. A major challenge or problem is to combine all this varied information, and for this image-based GIS is a powerful tool.

In GIS, all information is stored in digital form in a computer, in the form of maps (also called coverages) and tables. The link between the tables and the corresponding map feature is maintained, and schematically, the whole thing may appear as in Figure 6.2.

6.2.3 Basic data structures in GIS

Two basic types of data structures exist in GIS: vector and raster (Figure 6.4). Raster is a cellular organization; it comprises an array of regular cells, in rows and columns. Limitations in raster structure arise from degradation in information due to quantization and cell size. In vector structure, spatial information is digitized and stored as points, arcs (lines) and polygons. In such cases, topology (i.e. mutual relations between various spatial elements) is specifically defined. The vector structure is geometrically more precise and compact. The method of cartographic manual digitization, which is very widely used, utilizes vector mode. However, vector mode is relatively tedious for performing many GIS operations, such as overlay, neighbourhood functions etc. On the other

TABLE 6.1 Characteristics of continuous and categorical attribute data types

Types of attribute/property	*Type of scale*	*Remarks*	*Example*
Categorical	Nominal	Mutually exclusive categories of equal status	A, B, C; or quartzite, schist, marble etc.
	Ordinal	Hierarchy of states in which all intervening lengths are not equal	Drainage density; low, medium, high
Continuous	Interval	Possess lengths of equal increment, but no absolute zero	Linear contrast stretched image
	Ratio	Possess lengths of equal increment and also absolute zero	Rainfall (mm)

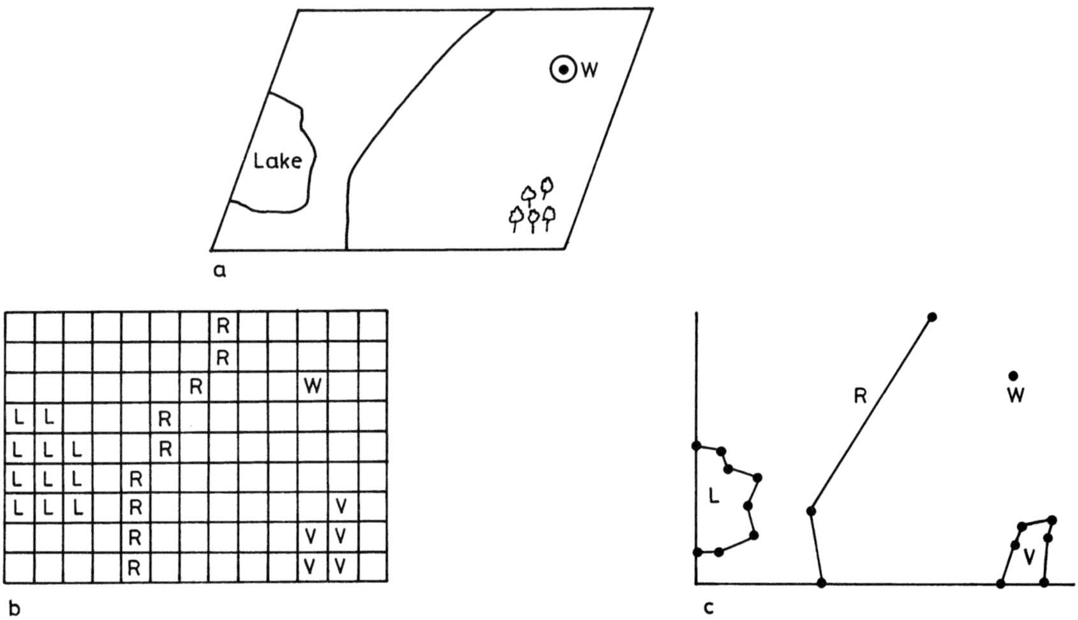

FIGURE 6.4 Basic data structures in GIS. (a) A map and the same in (b) raster format and (c) vector format.

hand, raster structure is simple, easy to handle and understand and suitable for performing most GIS functions, as well as facilitating all image processing operations. In this work, the examples of groundwater data processing where not otherwise specifically mentioned, utilize raster data structure.

6.3 GIS methodology

Broadly, a GIS comprises five main segments or stages (Figure 6.5). (1) data selection and acquisition; (2) data preprocessing; (3) data management; (4) data manipulation and analysis, and (5) data output.

6.3.1 Data acquisition

Data aquisition includes identifying data needs and locating and collecting the data sets. For identifying data needs, one requires a good understanding of the thematic problem to be tackled through GIS. Some data may be available from existing records and some other data may have to be specifically generated.

6.3.2 Data preprocessing

Preprocessing is often required to convert the data sets into a form suitable for storage in the GIS database (or data bank), and make them amenable for integrated analysis. Preprocessing can vary from simple to fairly complex operations including format conversion, structure conversion, resampling, interpolation, registration (e.g. Aronoff, 1989; Star and Estes, 1990). Main preprocessing operations in image-based GIS are: (a) data input, (b) rasterization, (c) interpolation, (d) black-and-white image display, and (e) registration (Figure 6.6).

Data input means the procedure of encoding data into a computer-readable form, or entering data into the GIS computer. Two types of input methods are used in combination: (a) keyboard entry for entering attribute data, and (b) digitizing for entering locational data.

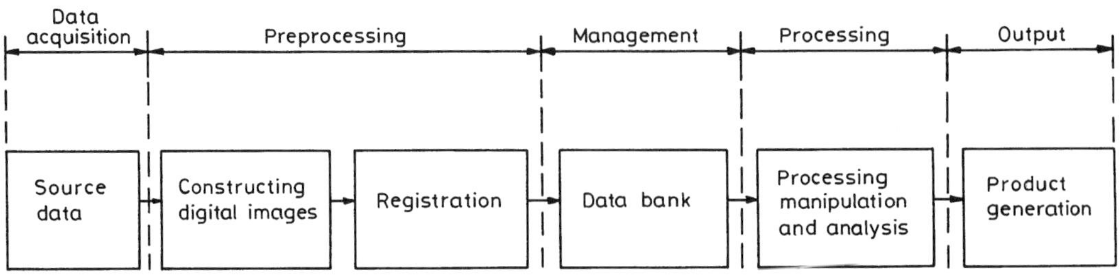

FIGURE 6.5 Main segments of a GIS.

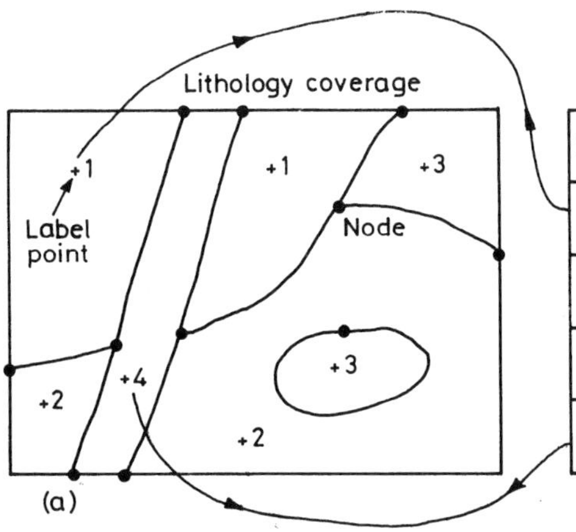

Lithology attribute table

ID	Lithology	Weathered zone	Porosity	Permeability
1	Quartzite	Thin (=100)	—	—
2	Schist	Thick(=300)	—	—
3	Granite	Medium (= 200)	—	—
4	Dolerite	Thin (=100)	—	—

(b)

FIGURE 6.6 Preprocessing for developing an image-based data bank.

Keyboard entry involves manually entering the data at a computer terminal. All attributable data are entered in this manner into tabular forms, called tables. Feature labels, i.e. numerals to identify points, lines or polygons are also entered through the keyboard. **Digitizing** involves entering positional (or locational) data into the computer, i.e. converting maps into digital data forms. This is done by either co-ordinate digitizing or by scanning which produce vector and raster data respectively. A vector to raster conversion may be necessary for some applications.

Interpolation is the process of predicting unknown values using the known values in the vicinity. It is a type of neighbourhood operation. Many GIS software packages carry modules for the following types of interpolations, (a) point-based: to estimate values at predetermined locations using points of known position and values (e.g. Thiessen polygon); (b) isoline-based: to estimate values between isolines (e.g. water-level contours); (c) areal interpolation: in this method, polygons are filled with some selected values (usually in the range 0–255 to suit image display). Often categorical data of nominal and ordinal type are handled by areal interpolation method for a relative representation. Interpolation programmes may employ a wide range of statistical methods for computation.

Image display: Employing the standard image processing systems, the rasterized and interpolated spatial data can be displayed as a black-and-white image. Typically, one image displays variation in a single parameter across the scene in shades of grey. The image can be suitably contrast-stretched and enhanced. Examples of groundwater images are given later (Figures 6.10, 6.11, 6.14).

Registration is the process of superimposing images, maps or data sets over one another with geometric congruence. It means adjusting one data layer so that it can be correctly overlaid on another data layer of the area, cell by cell. Registration is carried out using control points and resampling. The procedure is analogous to rubbersheet stretching of one data layer to suit the geometry of the other.

The above set of preprocessing procedures yields a co-registered data bank in GIS form, for further image-based data processing, interpretation and analysis (Figure 6.6).

6.3.3 Data management

The GIS software packages are built around database management systems (DBMS) which comprise a set of programs to manipulate and maintain data in a database. The DBMS approach provides numerous advantages for data processing such as efficient, controlled ordering and organization, sharing of data, data independence, centralized control, data integrity, reduction in data redundancy etc. The DBMS in GIS use relational data models, and hence are also called Relational DBMS. They provide advantages in operations such as flexible user-preferred views, search functions etc.

6.3.4 Data manipulation and analysis

Once the groundwater data such as water-table level, aquifer parameters, hydrochemical data, i.e. parameters reflecting groundwater characteristics in spatial domain, are co-registered in an image-based GIS, wide possibilities exist for data processing, enhancement and analysis. The whole gamut of standard image processing modules (Table 4.9) and GIS modules (Table 6.2) is available for data processing. Some examples are given in this treatment, and interested readers may find further information elsewhere (Jensen, 1986; Star and Estes, 1990; Gupta, 1991). Considering the organization of most commercially available software packages, the discussion is divided into two parts: (a) image processing operations, and (b) GIS analyses functions. However, it may be mentioned that any distinction between raster image processing module and raster GIS module is purely artificial.

Image processing operations

The objective of image data processing and enhancement is to render the image more interpretable, i.e. to improve the image quality for visual interpretation, and for subsequent computer analysis, if necessary. There is no standard prescription for obtaining best processing and enhancement results, which may be problem- as well as data-dependent. The various techniques of single and multiple image data enhancement available under digital image processing softwares (Table 4.9) can be applied.

A black-and-white image is typically a single parameter image, i.e. one image displays variation in a single parameter across the scene, in shades of grey. Any continuous parameter (interval or ratio type, Table 6.1) may be used for the purpose, e.g. water-table level, groundwater temperature, TDS content. The images can facilitate better interpretation of local and regional features. Also, they can be processed to generate other interesting data images, e.g. a

TABLE 6.2 Important GIS functions (modified after Aronoff, 1989)

1. Maintenance and analysis of the spatial data	– Format transformations – Geometric transformations – Editing functions etc.	
2. Maintenance and analysis of the attribute data	– Attribute editing functions – Attribute query functions	
3. Integrated analysis of spatial and attribute data	– Retrieval/classification/ measurement	– Retrieval – Classification – Measurement
	– Overlay operations	
	– Neighbouring operations	– Search – Topographic functions – Thiessen polygons – Interpolation – Contour generation
	– Connectivity functions	– Contiguity – Proximity – Network – Spread – Seek – Perspective view
4. Output formating	– Map annotation text, labels, graphic symbols, patterns etc.	

different image using topographic elevation image minus water-table level image, will produce an image showing depth of water-table. Similarly, gradient filtering of the water-table image will produce a water-table slope image.

Further, certain parameters (e.g. lithology, land use, soil type) are of categorical type measured on a nominal or ordinal scale (Table 6.1). For generating images from such data, maps are digitized, rasterized and polygons filled with certain selected DN-values (usually between 0–255). The images may be displayed as black-and-white products or in colour mode. The colour-hue axis may be selected to pertain to any parameter of interest, e.g. stratigraphic age, or permeability, or fracture density of the rocks. These images could be quite informative.

The technique of image colour coding described in sections 4.6.4 and 4.9.1 can be applied for enhancing groundwater data in image form. Single images can be enhanced by density slicing colour coding. Multiple images of groundwater data can also be combined by colour coding to generate false colour composites. On such composites, colour variations would correspond to variations in input component images.

Synthetic stereo is another interesting technique of combining and enhancing image data. It is based on the standard parallax formula used in aerial photogrammetry, the only difference being that now parallax is artificially introduced in the image-pair. A set of two images is taken; one acts as the base image and the second corresponds to the heighting image. The pixels in the base image are shifted in such a way that parallax is artificially introduced, the amount of parallax, i.e. 'heights', being introduced corresponds to the pixel values in the second image. Thus, a synthetic stereo is generated.

Several examples of image processing operation are given in Chapter Four, and some other examples related to groundwater data are shown in Figures 6.10, 6.11, 6.14, and 6.15.

GIS analysis functions

The GIS analysis functions are unique as they can concurrently handle spatial as well as attribute data. Here, the GIS analysis functions are divided into five types: retrieval, measurement, overlay, neighbourhood and connectivity (Table 6.2).

Retrieval operations
These include selective search on spatial and attribute data in such a way that the geographical locations of features are not changed, i.e. the output shows selectively retrieved data in their original geographic positions. The criteria for selective retrieval could be based on codes of attributes, or Boolean logical conditions (see later) or classification. Common examples could be: select pixels with water-table less than 1 m deep, or select pixels in a specific rock type. As the selective search can be operated on both spatial and attribute data, this becomes a powerful function in handling and processing data in the GIS environment.

Measurement operations
Commonly included in GIS software are distances between points, lengths of lines (arcs), perimeters of polygons, areas of polygons, number of points falling in a polygon and number of raster cells in each class. Sample applications in groundwater study could be: find number of wells appearing in a district, or find distance between wells, or compute density of wells. These functions are handy while dealing with exploration, planning and management of groundwater resources.

Overlay operations
These are very frequently required in almost all natural resources investigations, as often one has to deal with several data sets of the same area. Overlaying can be done in both raster and vector structures. However, overlaying in vector structures is rather tedious, for the simple reason that in this case overlaying of polygons leads to intersections and generation of new polygons of various shapes, sizes and attributes which become difficult to handle. In contrast, the overlaying operation in raster proceeds cell by cell, and therefore is relatively simple.

Overlay operations involving both arithmetic functions as well as logical type can be performed to process the multiple coverages. Overlay arithmetic functions include such computations as addition, subtraction, division and multiplication. This involves arithmetic operation of each value in a data layer by the value at the corresponding location in the second data layer (Figure 6.7). A logical or Boolean overlay operation involves finding those areas or locations where a specified set of conditions is fulfilled (or not fulfilled). Figure 6.8 displays Venn diagrams illustrating Boolean logic. An example of classification by arithmetic and logical overlay operations is given later (Figure 6.15).

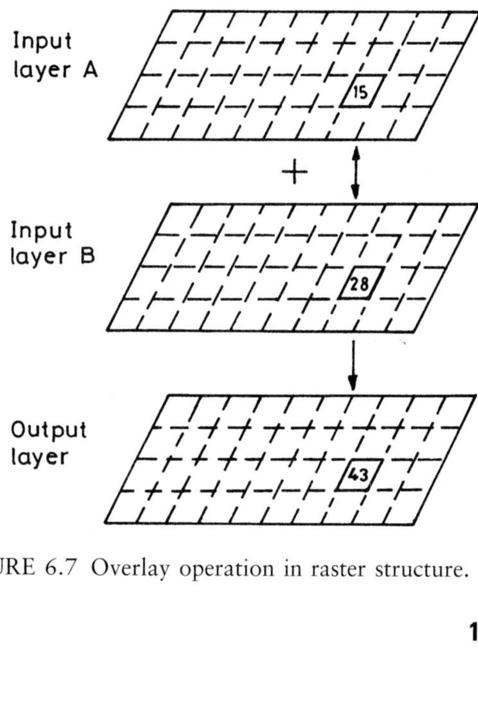

FIGURE 6.7 Overlay operation in raster structure.

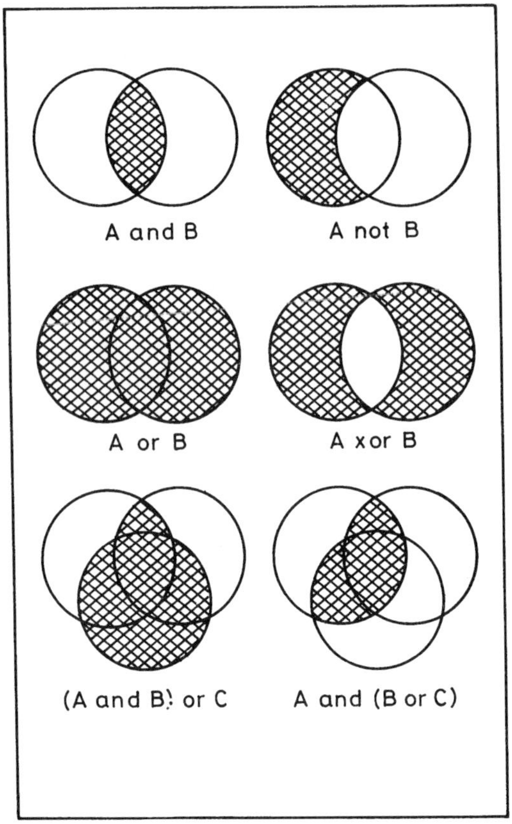

FIGURE 6.8 Concept of Boolean conditions.

Neighbourhood operations
These deal with local characteristics, or characteristics surrounding a specific target. These operations are useful in finding local variability and adjoining information. Some functions such as search, topography and interpolation are usually given in GIS software as neighbourhood operations.

(a) **Search function.** It is a very extensively used neighbourhood class of function. In this, on the basis of its neighbourhood characteristics, the target is assigned a certain value; the new value is written to a separate data layer. A sample query could be: 'select villages not having water wells within a search radius of 2 km'.

(b) **Topographic functions.** A raster data set can be represented in terms of a digital elevation model (DEM). The term topography here refers to the characteristics of such a DEM surface. The topographic functions, namely slope and aspect, are neighbourhood functions. Slope is defined as the rate of change of elevation, and aspect is the direction that a slope faces. In GIS modules, the term gradient is used only for maximum slope. Both gradient and slope are first-order derivative parameters.

(c) **Interpolation.** Interpolation is a typical neighbourhood operation. It involves predicting unknown values at given locations using the known values in the neighbourhood. An interpolation module is commonly provided in GIS packages. Besides the above, a number of image processing subroutines can also be considered as neighbourhood operations, e.g. high-pass filtering, texture transformation, image smoothing, and gradient filtering.

Connectivity operations
Connectivity operations are grouped into contiguity, proximity, network, spread and perspective-view functions.

(a) **Contiguity.** Areas possessing unbroken adjacency are classed as contiguous. What constitutes broken/unbroken adjacency in a particular case can be prescribed depending upon the problem under investigation. For example, impermeable layers and dykes etc. are important in the context of groundwater flow, as far as contiguity is concerned. Therefore, this function can facilitate search for contiguous areas in a manner that the notion of contiguity is predefined.

(b) **Proximity.** This function is based on the concept of distances between features. The notion of distance can be simple length, or based on any other computed parameter such as velocity, time, hydraulic gradient etc. A very common result of the proximity function is a buffer zone. A buffer zone is defined as an area of a specified width, drawn around the map location. Buffering can be done around points, lines or polygons. A sample example is to buffer the zone within a specified distance from the water wells (Figure 6.9).

(c) **Network.** Network functions are commonly used in analysis where resources are to be transported from one location to another. Examples of networks include canals, streams, etc. GIS packages carry modules for network analysis of varying sophistication. Some application problems could be in the field of flow functions and travel route selection. Network functions can be applied in environmental studies and pollutant dispersion investigations.

(d) **Spread.** The spread function is endowed with characteristics of both network and proximity functions. In this, a running total of the computed parameter is kept, as the area is traversed by a moving window. The rules for transport and computation of parameter can be outlined beforehand. Therefore, it is a very powerful function particularly for environmental impact assessment and pollution studies.

(e) **Perspective view.** The raster image data can be displayed as a three-dimensional surface, where the height corresponds to the value at that pixel. This generates a three-dimensional model, and forms a valuable and interesting method of enhancement, as the human mind can easily perceive shapes and forms. By defining parameters such as position of illuminating source and viewer's altitude and angle, shaded relief maps and perspective views can be generated. One more raster data set can be superimposed over the three-dimensional model by draping, e.g. a lithological map can be draped over a 3D water-table model.

(f) **Classification.** This is a procedure of subdividing a population into classes and assigning each class a name, using various statistical procedures. In vector data, each polygon may be assigned a name as an attribute. In raster structure, each cell is assigned a new numerical value for class-identification. Classification may be done using a single data set or a multiple data set (Figure 6.15).

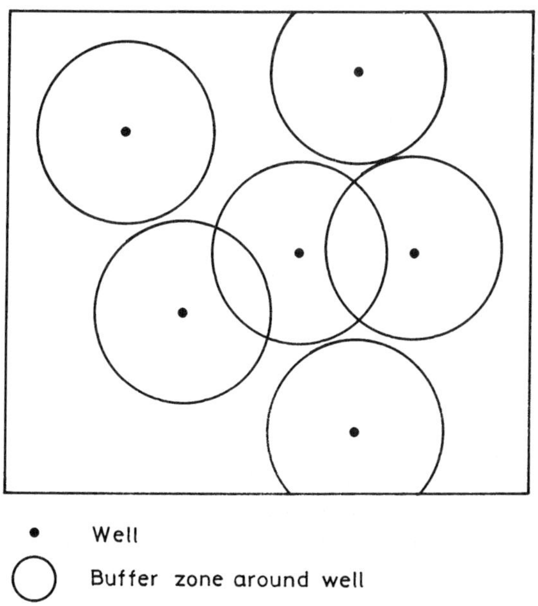

FIGURE 6.9 Buffer zone generation around points representing well locations.

6.3.5 Data output

The results of GIS analysis may be presented in three types of media: soft copy, hard copy and electronic. Soft copy output refers to the display on the computer monitor. It may be colour, monochrome, or graphic text. Soft copy output is necessary for interactive processing and display. However, it cannot be used for permanent storage, but such screen outputs can be photographed to provide permanent records. Hard copy outputs are the permanent means of display. They may use paper, photographic film or other similar materials. GIS results may also be output on electronic media such as disks and cartridges for transfer to other systems.

6.3.6 Sources of error

As GIS is used to capture, edit and manipulate spatial data sets, errors of several types may creep in at different stages and affect the data quality. Two basic categories of errors may be present: (a) inherent, and (b) operational. Inherent error is present in the source data. Operational error is introduced during the GIS working. An understanding of the types and sources of errors can lead to better job management. Although it is not possible to avoid errors altogether, they can be managed to be kept within permissible limits.

6.4 Thematic GIS applications in groundwater

Many diverse applications of GIS in groundwater studies have been reported ranging from hydrogeologic exploration, water quality and pollution, to modelling and management. A detailed treatment of each type of application would be beyond the scope of the present work. In the following paragraphs selected examples are mentioned.

6.4.1 Study of water-table behaviour and seepage pattern

The importance of the study of water-table fluctuation and flow patterns has been discussed in Chapter Three. GIS environment is very aptly suited to process the spatial data of groundwater table, topography etc. to enable interpretation of water-table behaviour and seepage patterns. The water-table data can be digitized, rasterized and interpolated to produce an image. Figure 6.10a

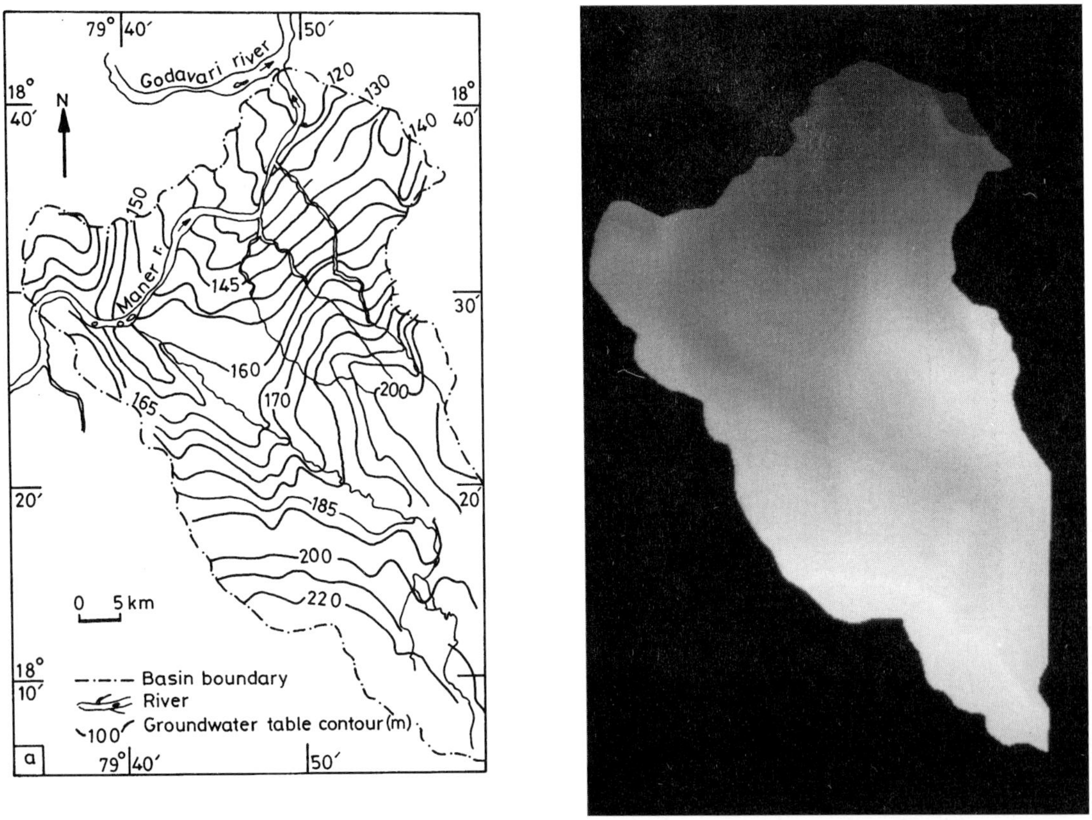

FIGURE 6.10 (a) Water table contour map of Maner river basin (India); (b) corresponding water-table image; (a: after Radhakrishna *et al.*, 1976; b: after Srivastava, 1993).

is a groundwater table contour map from which an image has been generated (Figure 6.10b). The grey tones correspond to water-table elevations – darker tones implying lower elevations and lighter tones implying higher elevations. The image provides a prompt synoptic view of the lateral groundwater flow directions. Further, a comparative study of the groundwater flow pattern indicates that the groundwater flow is mostly effluent towards the streams.

In the same manner as above, it is possible to create and display other elevation data images in black-and-white, such as topographic elevation image (from topographic data), depth to water-table image (from depth to water-table data), or depth to bedrock 1, and others.

Elevation data can be subjected to spatial filtering to generate gradient images, depicting variation in slope, pixel-to-pixel in either x or y or both directions (e.g. Jensen, 1986). Figure 6.11 is an example of a groundwater table gradient image generated by applying Sobel's operator on the water-table image (Figure 6.10b). Note that this image appears artificially illuminated from the northwest, which is due to directional computation in the Sobel's filter. The bright pixels pertain to sites of local higher gradients.

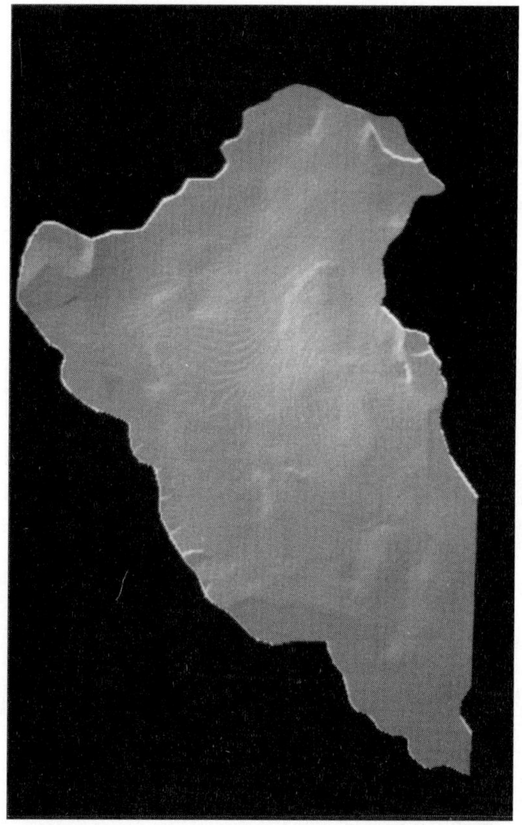

FIGURE 6.11 Groundwater table gradient image generated by applying Soble's filter on Figure 6.10(b) (after Srivastava, 1993).

6.4.2 Evaluation of well distribution in different aquifers

In a multiaquifer system it is pertinent to have information about the wells tapping different aquifers. GIS can be used to identify and display locations of wells tapping one particular aquifer or a group of aquifers (e.g. Baker, 1993). This may be useful elsewhere, viz. for water management etc. In such studies, data on location of wells, lithologic logs and depth of wells are incorporated in GIS. Spatial surfaces of the aquifer boundaries can be generated by interpolation. Then, by relating water well depth at each location with the aquifer boundaries, wells with respect to source aquifers can be searched for and classified (Figure 6.12).

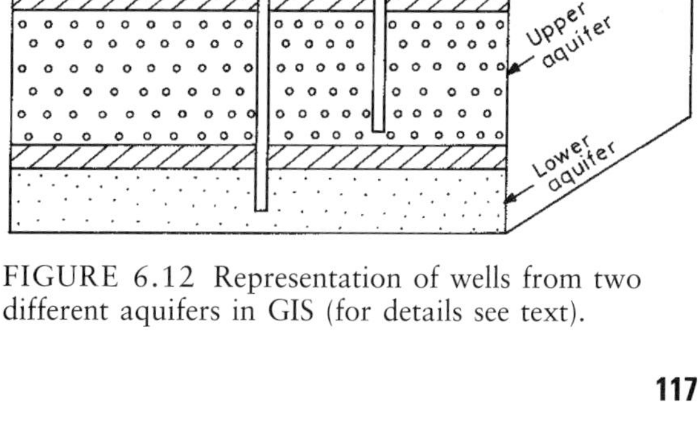

FIGURE 6.12 Representation of wells from two different aquifers in GIS (for details see text).

6.4.3. Groundwater exploration

Groundwater exploration has to utilize multiple data sets. Information on geology, topography, rainfall, landform, hydraulic parameters etc. has to be collectively interpreted. The integrated data interpretation can be done in the GIS domain in various ways. Some useful strategies could be, for example: superimposition of geophysical data image, lineament image and landform image, or superimposition of lineament image, lithology image and water-table image. Depending upon the problem, various combinations may be selected.

An interesting example of groundwater exploration involving GIS-based approach is furnished by Hansmann *et al.* (1992). In an area in Sri Lanka, the available data indicates that

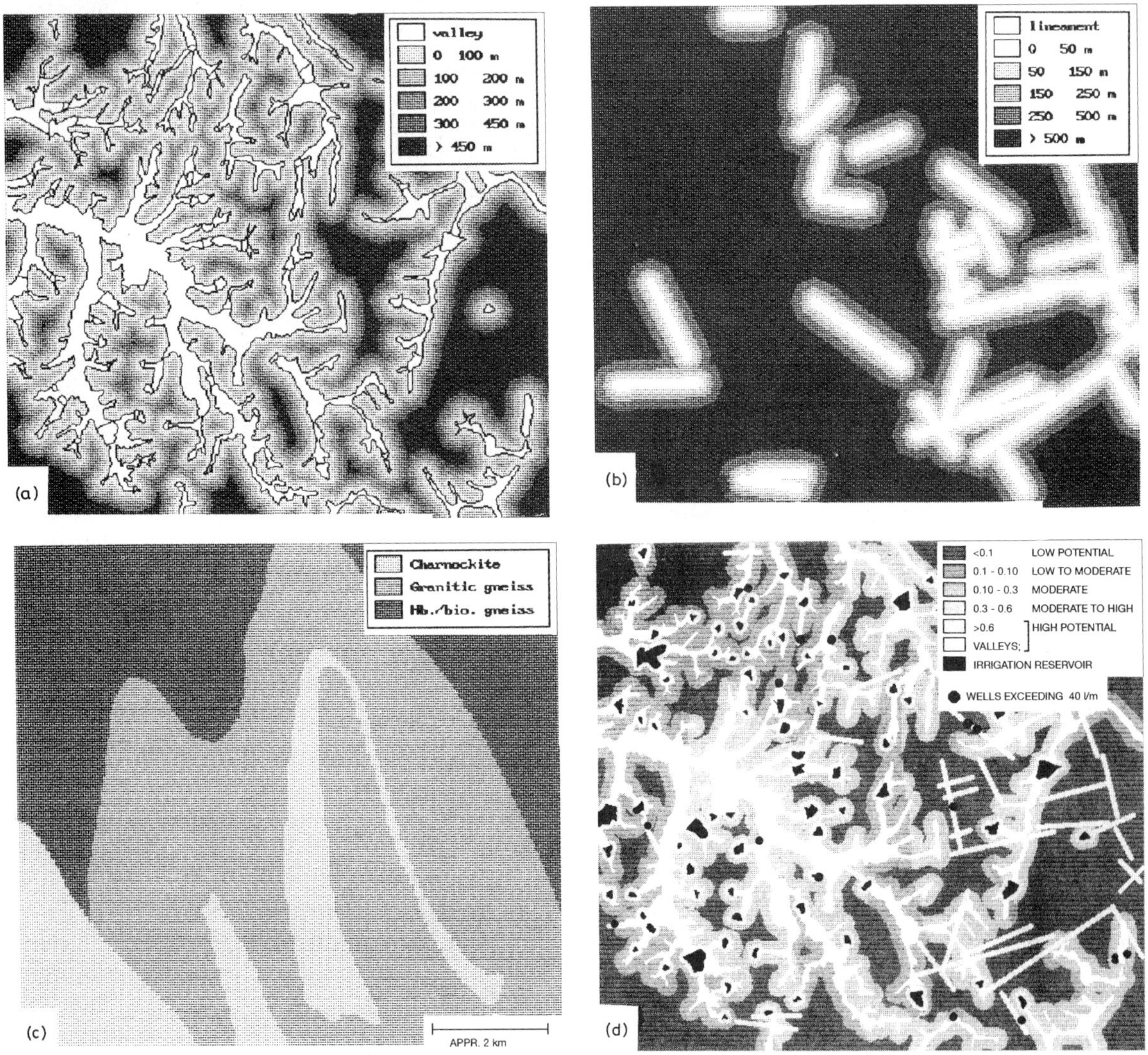

FIGURE 6.13 Raster maps of (a) distance to valley, (b) distance to lineament and (c) lithology. (d) Superimposition of the three raster data layers gives a groundwater potential map (after Hansmann *et al.*, 1992).

hydrogeology is influenced by climatic zones. Within a particular climatic zone, the well yield is related to three main hydrogeologic parameters, namely, distance to valley, distance to lineament and lithology. Raster maps of distance to valley and distance to lineament are created by first mapping valleys and lineaments on remote sensing data, and then using proximity functions in GIS (Figure 6.13). The raster map of lithology is also created as an additional data layer from a geological map. For integrating the above three data layers, Hansmann and co-workers used the statistical approach of 'weights of evidence modelling'. Integration of the data layers yields a raster map showing posteriori probability of groundwater resource at each cell in the area (Figure 6.13d).

6.4.4. Chemical quality and water use

Chemical quality of groundwater is an important attribute which controls water use (Chapter Ten). These data, being spatially variant, can be processed and analysed in the GIS environment in a highly efficient manner.

The chemical quality of groundwater is expressed in terms of various parameters such as concentration of various ions or total dissolved solids (TDS). In GIS, each of these parameters can be treated as a data layer. The data can be suitably contrast manipulated and displayed as a black-and-white product. For example, Figure 6.14 shows a TDS image generated from the analytical data. Brighter tones indicate higher TDS concentrations in groundwaters. Similar images for other chemical parameters could be generated and studied to provide a quick pictorial view of the spatial variation in the chemical quality of groundwater. Further, the chemical data images can be combined in colour mode to depict spatial variation of chemical parameters.

FIGURE 6.14 TDS image of the Maner river basin; lighter tones in the image correspond to higher TDS content (after Srivastava, 1993).

Overlay operations involving arithmetic as well as logical functions can be performed to process the multiple coverages, and the output can be used for hydrochemical classification. Figure 6.15 is an example in which four raster data sets of chemical characters, viz. TDS, Cl^{-1}, pH and SO_4^{2-} have been overlaid. Applying WHO norms and Boolean logic, groundwaters of the area have been classified with respect to suitability for drinking purposes.

GIS can also be used to relate hydrochemical quality with lithology. For example, Kalkoff (1993) made one such study in the Roberts Creek watershed, Iowa, relating stream water quality with the geology of the catchment area.

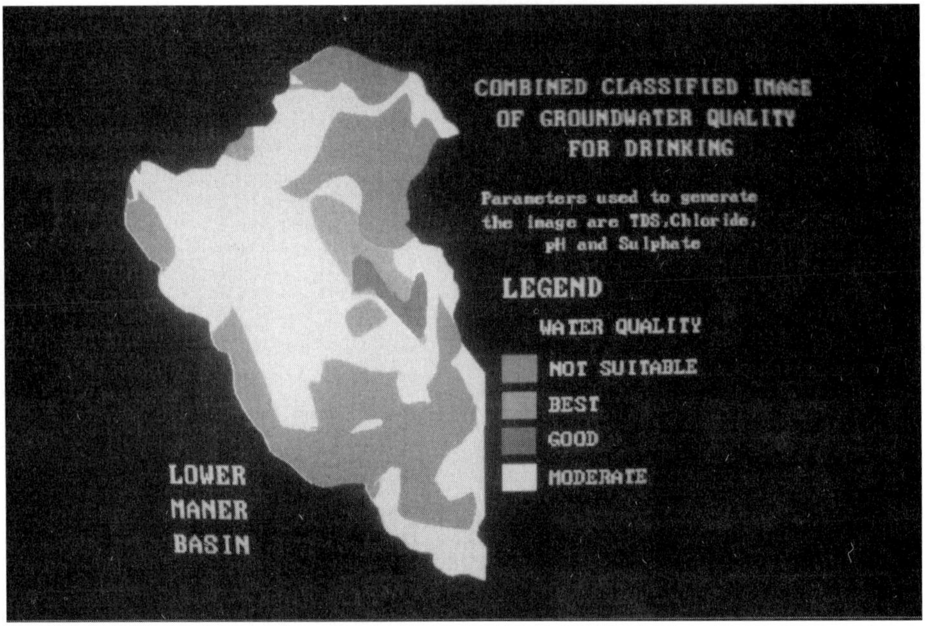

FIGURE 6.15 Classification using multiple image data sets of TDS, Cl, pH and SO_4. The groundwaters have been classified with respect to suitability for drinking purposes, applying WHO norms and Boolean logic (after Srivastava, 1993).

6.4.5. Groundwater pollution potential and hazard assessment

The GIS-based spatial model can be used to evaluate groundwater vulnerability to contamination. Most such studies utilize the raster GIS structure. The procedure involves generating relevant hydrogeological data layers, converting the parameters into ordinal numbers (using ratings and weights), and finally integrating the data sets to give a spatial property map.

The DRASTIC model developed by Aller *et al.* (1987), has been used more frequently for this purpose. It integrates parameters of depth to water-table (D), net recharge (R), aquifer media type (A), soil media character (S), topographic slope (T), impact of vadose zone (I), and hydraulic conductivity of the aquifer (C) for groundwater pollution potential evaluation. It is a model utilizing weights, ranges and ratings on the ordinal scale, to provide relative rather than absolute evaluations. Each data plane (i.e. layer) is given a (subjective) numerical value or weight, depending upon its importance. Table 6.3a shows that the parameter of depth to water-table (D) is most important and has a weight of 5; in contrast, topography is less important and has a weight of 2. In each of the data layers (i.e. thematic maps), the ranges/attributes are converted into ordinal rating numbers. Table 6.3b gives an example of converting depth to water-table (continuous) data into rating numbers, and Table 6.3c gives an example of converting aquifer media (categorical) data into rating numbers. Thus, various data layers with cell values are generated. Using an additive model the pollution potential can be calculated as follows:

$$\text{DRASTIC index} = DrDw + RrRw + ArAw + SrSw + TrTw + IrIw + CrCw \qquad (6.1)$$

TABLE 6.3a Weights for DRASTIC features

Features	*Weight*
Depth to water-table	5
Net recharge	4
Aquifer media	3
Soil media	2
Topography	2
Impact of the vadose zone	5
Hydraulic conductivity of the aquifer	3

TABLE 6.3b DRASTIC ranges and ratings for depth to water-level

Range (ft)	*Rating*
0.5	10
5–15	9
15–30	7
30–50	5
50–75	3
75–100	2
100+	1

TABLE 6.3c DRASTIC ranges and ratings for aquifer media

Range	*Rating*	*Typical rating*
Massive shale	1–3	2
Crystalline rocks	2–5	3
Weathered zones	3–5	4
Thin bedded rocks	5–9	6
Sandstone massive	4–9	6
Sand and gravel	6–9	6
Basalt	2–10	9
Karst limestone	9–10	10

where r is the cell rating and w is the layer weight. Figure 6.16 shows the working concept; an example of DRASTIC relative hydrogeologic vulnerability map is shown in Figure 6.17.

At times a distinction is made between groundwater pollution potential and hazard (e.g. Atkinson and Thomlinson, 1994). Pollution potential is derived at any given location based on local groundwater depth, hydraulic conductivity, land surface slope, and soil permeability, from the DRASTIC parameters. It is a natural risk index, not involving the effect of man-made

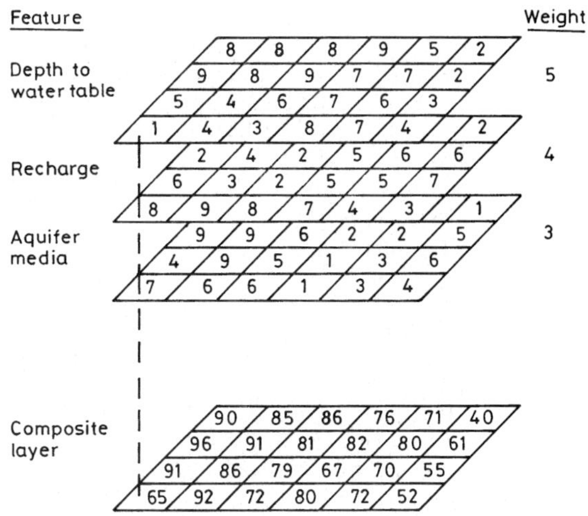

FIGURE 6.16 Raster-based scheme for calculating index for groundwater pollution potential; only some of the DRASTIC parameters are shown.

features. On the other hand, groundwater pollution hazard is considered in terms of the probability of groundwater pollution in the area. It is based on pollution potential together with society's pollution contribution in the area, given in terms of land use/cover and septic system density etc.

Halliday and Wolfe (1991) used GIS for assessing groundwater pollution potential from nitrogen fertilizers in Texas. They combined an agricultural pollution susceptibility index map with information on cropped areas, recommended nitrogen fertilizer application rates and aquifer outcrops, and finally generated a nitrogen pollution potential map.

6.4.6. Salinity hazard mapping

Dryland salinity occurs in many areas. In the northern part of Australia, salinity hazard mapping has been carried out by Tickell (1994) using GIS. The study is a futuristic one, aiming to identify areas which would be vulnerable to dryland salinity, if large-scale clearing of deep rooted vegetation were to take place for dryland

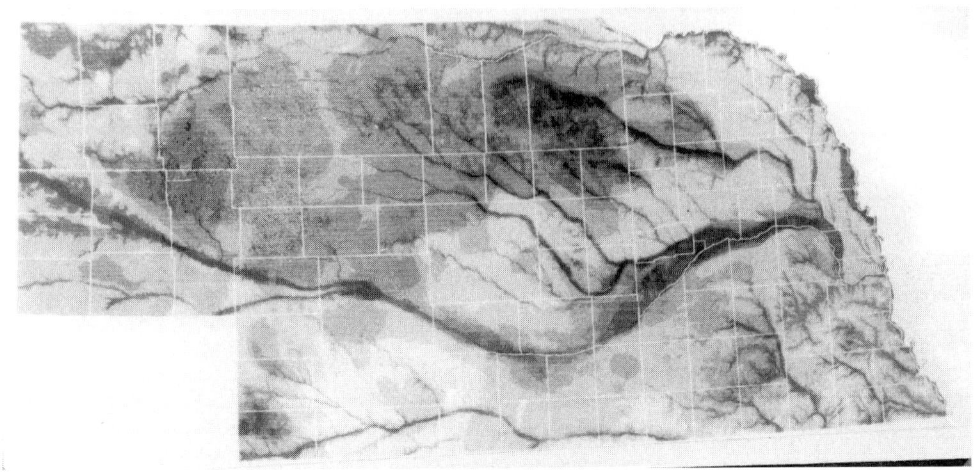

FIGURE 6.17 DRASTIC relative hydrogeologic vulnerability map of Nebraska. Increasing shades of darkness indicate greater vulnerability to groundwater contamination (after Kalinski *et al.*, 1994).

agriculture. For this study five indicators viz. groundwater salinity, vegetation, median annual rainfall, aquifer yield and presence/absence of laterite were chosen. It is considered that the dryland salinity hazard in the area is higher with higher groundwater salinity, low open vegetation, low aquifer yield, a certain range of rainfall, and presence of laterite.

Data layers (i.e. thematic maps) for each of the above five indicators were generated from field data; the maps were digitized in vector format. Ratings from each of the data layers were added to produce a salinity hazard index map which was simplified to give a map showing four categories of salinity hazards – high, medium, low, and very low (Figure 6.18).

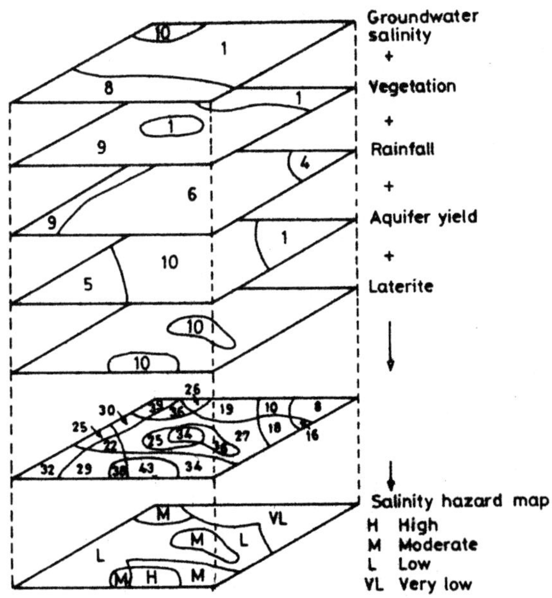

FIGURE 6.18 Scheme of salinity hazard mapping (after Tickell, 1994).

6.4.7 GIS and groundwater modelling

GIS is endowed with graphical and data handling capabilities. Hence, the GIS environment is conducive to modelling owing to the following main features (El-Kadi *et al.*, 1994): (a) information on the database is available on the same screen during data preparation; (b) it provides graphical error checking on a continuous basis; (c) it saves and analyses spatial data and results on any GIS layer, and (d) it has flexibility of overlaying and integrating maps and data.

GIS can be integratively used with numerical

groundwater modelling software (e.g. Richards *et al.*, 1993; Roaza *et al.*,1993; El-Kadi *et al.*, 1994). In groundwater modelling, one requires to create a raster/spatial mesh with aquifer parameter values assigned at each cell. With aquifer parameters being known at certain points, interpolation/extrapolation is necessary for unknown points. Solutions may require initial and boundary conditions and results to be spatially displayed. A groundwater model may be run as an option-interface in the GIS. The model could be for predicting drawdowns, recharge, contaminant transport, well-head protection area (WHPA) etc.

Contaminant transport modelling

Pollution of groundwater is a burning problem throughout the world and has been discussed in Chapter Ten. GIS can play an important role in such studies. Popular groundwater flow and contaminant transport models like MODFLOW, MODPATH and MT3D can be interfaced with GIS. MODFLOW describes 2D and 3D transient or steady-state flow in anistropic heterogeneous layered aquifers using finite difference modelling. MODPATH is used to compute path-lines in 3D steady-state groundwater flow systems, i.e. it computes position of particles at user-specified time instances. MT3D is a three-dimensional transport model to describe the solute transport. These models can be interfaced with GIS.

WHPA modelling

A capture zone represents the zone around a pumping well supplying groundwater to the well. The gradient created around a pumping well may attract contaminants if the contamination source is located within the capture zone. The delineation of WHPA utilizes various parameters such as distance, drawdown, time of travel, flow boundaries and assimilative capacities. Models for WHPA have been evolved to delineate capture zones. While interfacing with GIS, the procedure includes: (a) extraction of the necessary input data from the GIS database, (b) running the WHPA model, and (c) displaying the WHPA map generated from the model in the GIS (e.g. Figure 6.19; Rifai *et al.*, 1993).

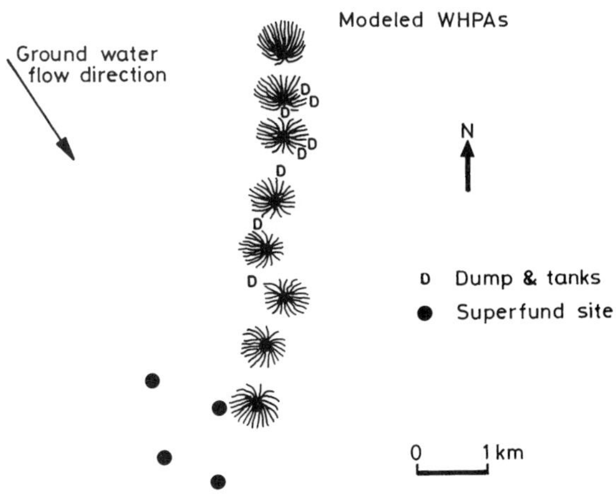

FIGURE 6.19 Typical output from modelling interface in GIS, showing the modelled well-head protection areas (WHPAs) and the distribution of dump sites in Harris County, Texas (simplified after Rifai *et al.*, 1993).

6.4.8 Concluding remarks

Some examples of groundwater data manipulation, processing and analyses in GIS environment have been given above. It is obvious that the GIS technology offers a very powerful tool for integrating a variety of georeferenced data sets, such as lithologic maps, borehole data, hydrochemical data, geophysical data, river discharge, hydrogeological data etc. This is a rapidly advancing field as new applications are being devised by different workers. In fact, in this methodology one can say that limitations of applications may be imposed only by the sophistication of the particular software package and the imagination and skill of the investigator.

Chapter seven

Principles of groundwater flow and solute transport

Although the mechanism of groundwater flow through porous media is well understood, many uncertainties exist about the flow of water and transport of solutes in fractured rocks. This has acquired greater interest as low permeability formations form potential repositories for high-level waste.

7.1 Groundwater flow

7.1.1 Laminar and turbulent flow

Groundwater flow through various rock types may be either laminar or turbulent depending on their permeability and the hydraulic gradient.

In laminar flow, also known as viscous or streamline flow, the flow lines are parallel and the velocity and direction of flow are constant. The turbulent flow is characterized by high velocities and the formation of eddies. In laminar flow the velocity of flow is proportional to the first power of the hydraulic gradient (equation 7.1).

$$V = KI \qquad (7.1)$$

But in turbulent flow, the velocity of flow will be expressed as

$$V = KI^{\alpha} \qquad (7.2)$$

where α is the degree of nonlinearity of the flow ($0.5 \leq \alpha \leq 1$).

Relatively high hydraulic gradient may give rise to turbulent flow through large pores, solution cavities and wide fractures. In laminar flow, the inertial forces are much smaller than viscous forces. The transition from laminar to turbulent flow depends on the Reynolds number (R_e) which is a dimensionless number expressing the ratio of inertial to viscous forces acting on the fluid (equation 7.3).

$$R_e = \frac{\rho V d}{\mu} = \frac{V d}{\upsilon} \qquad (7.3)$$

where ρ is the fluid density, V is the mean fluid velocity, d is the diameter of the pipe, μ is the dynamic viscosity and υ is the kinematic viscosity. In porous media, d can be replaced by the effective diameter, (d_e) of sand grains which constitutes the aquifer (section 9.2.1).

By considering porosity, η the Reynolds number, R_e can be expressed as (UNESCO, 1972)

$$R_e = \frac{1}{0.75\eta + 0.23} \frac{V d_e}{\upsilon} \qquad (7.4)$$

and by considering the intrinsic permeability (k) instead of effective diameter (d_e), R_e is given by equation (7.5) (van Golf-Racht, 1982)

$$Re = \frac{5 \times 10^{-3}}{\eta^{5.5}} \frac{V\sqrt{k}}{\upsilon} \qquad (7.5)$$

Generally, the flow regime is laminar for $R_e < 2000$ and turbulent for $R_e > 2000$ (Marsily, 1986).

In fractured rocks, it is difficult to determine the Reynolds number, since for a given type of flow it can vary greatly from one point to another along the same fracture. In fractured rocks, the diameter of the pipe in equation (7.3) can be replaced by the hydraulic diameter (D_h) which in a plane fracture is expressed by

$$D_h = 4A / \rho \qquad (7.6)$$

where A is the cross-sectional area of the fracture through which flow takes place and ρ is the outside perimeter of the cross-section area of the flow (Marsily, 1986). For a very long fracture, D_h is equal to twice its aperture, i.e. $D_h = 2a$, and R_e is given by the relation

$$R_e = \frac{2Va}{v} \qquad (7.7)$$

The walls of a natural fracture will always have a certain degree of roughness causing an additional drop in pressure and thereby influence the flow conditions. The relative roughness, R_r (dimensionless) is defined by

$$R_r = R_f/D_h \qquad (7.8)$$

where R_f is the mean height of the irregularities (asperities) in the fracture. Some measured values of R_r lie in the range of 0.002 to 0.01. However, as it is often difficult to measure roughness in the field, the effect of roughness on groundwater flow is more of academic interest.

From experimental data, van Golf-Racht (1982) showed that the transition from laminar to turbulent flow in fractures takes place at very low values of Reynolds number (≈600). According to Louis (1974), the transition from laminar ($\alpha = 1$) to completely turbulent ($\alpha = 0.5$) is quite progressive, the exponent slowly changes from 1 to 0.5 as the Reynolds number changes for instance from 100 to 2300. Lee and Farmer (1993) indicated that the critical Reynolds number commonly lies between 100 and 2300, decreasing as the relative roughness of the fracture increases. The flow, therefore, will change from laminar to turbulent condition with increase in flow velocity, pore size, sinuosity and roughness.

7.1.2 Darcy's law

In 1856, Henry Darcy, a French hydraulic engineer, investigated the flow of water through filters. His experiments demonstrated that the rate of flow, i.e. volume of water per unit time, Q is (a) directly proportional to the cross-sectional area, A, and head loss, h_L, and (b) inversely proportional to the length of the flow path, L (Figure 7.1), i.e.

$$Q \sim A \text{ and } \sim \frac{h_L}{L} \qquad (7.9)$$

By combining the above relations and introducing proportionality constant (K) Darcy's law can be written as

$$Q = KA\frac{h_L}{L} \qquad (7.10)$$

or

$$Q = KA\frac{dh}{dl} \qquad (7.11)$$

where K is the hydraulic conductivity of the porous medium and dh/dl is the hydraulic gradient.

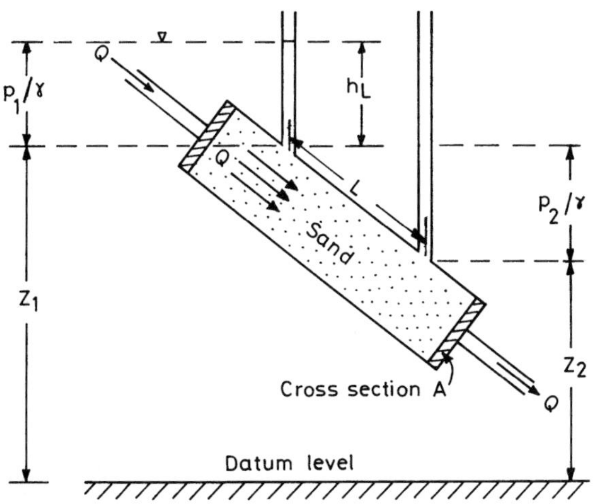

FIGURE 7.1 Flow through an inclined sand column showing pressure distribution and head loss.

Darcy's law can also be written as

$$q = V = \frac{Q}{A} = K\frac{dh}{dl} \tag{7.12}$$

where V (or q) is the specific discharge or darcy velocity. Equation (7.12) simply denotes that the specific discharge is the volume of water per unit time through a unit cross-sectional area normal to the direction of flow. Although darcy velocity, (specific discharge, q) assumes that flow occurs through the entire cross-section of the aquifer, however, in nature the actual flow takes place only through the interstitial spaces (pores). Therefore, the average interestitial velocity, $\overline{V}_a$ is

$$\overline{V}_a = \frac{Q}{\eta A} = \frac{q}{\eta_e} \tag{7.13}$$

where η is the porosity of the medium.

In order to determine the average (interstitial) velocity (or simply velocity) one should use the kinematic or effective porosity (η_e), instead of total porosity (η), therefore,

$$\overline{V}_a = \frac{q}{\eta_e} = \frac{K(dh/dl)}{\eta_e} \tag{7.14}$$

It can be seen from Figure 7.1 that

$$Q = KA\left[\left(\frac{p_1}{\gamma} + z_1\right) - \left(\frac{p_2}{\gamma} + z_2\right)\right] / L$$

where p is pressure and γ is the specific weight of water. Therefore, the head loss can be defined as energy loss or potential loss which is due to the frictional resistance through the porous medium. As velocity of groundwater movement is low, the kinetic energy of water or the velocity heads are neglected. The sum of the pressure head (p/γ) and the elevation head (z) is the piezometric head.

It may also be seen from Figure 7.1, that although the pressure (p) at point 1 is less than at point 2 but the piezometric head at 1 ($p_1/\gamma + z_1$) is more than at 2 ($p_2/\gamma + z_2$) indicating thereby that the flow takes place from a higher piezometric head to a lower one and not necessarily from higher to a lower pressure. It would mean that the flow takes place in the direction of decreasing head and is independent of the inclination of the pipe.

Darcy's law is valid only for laminar flow in porous media and hence the upper limit of its validity is decided by the Reynolds number. Experiments indicate that Darcy's law is valid as far as the Reynolds number (R_e) is between 1 and 10 (Todd, 1980). As for groundwater flow in porous media, R_e is usually less than 1, one could conclude that Darcy's law governs the flow of groundwater in porous media except in some situations where either the hydraulic gradients are very steep (as in the neighbourhood of the pumping well or spring outlets) or in rocks with large cavities, e.g. karstic formations. There may also be some lower limit for the validity of Darcy's law as some threshold hydraulic gradient is required to initiate flow especially in silty and clay formations where the water molecules are adhered to the electrically charged mineral particles. This, of course, is not of much practical interest (Bear, 1979).

7.1.3 General equation of flow

Darcy's law in a general form can be written as

$$V = -K\frac{dh}{ds} \quad (7.15)$$

where V, K and h are as defined earlier and s is the distance along the average direction of flow. If an element of saturated material has the dimensions x, y and z (Figure 7.2) then the velocity components in a rectangular co-ordinate system can be written as

$$V_x = -K_x\frac{\partial h}{dx} \quad (7.16)$$

$$V_y = -K_y\frac{\partial h}{dy} \quad (7.17)$$

$$V_z = -K_z\frac{\partial h}{dz} \quad (7.18)$$

where h is the total head under steady-state conditions, and K_x, K_y and K_z are the coefficients of hydraulic conductivty in the x, y and z directions respectively.

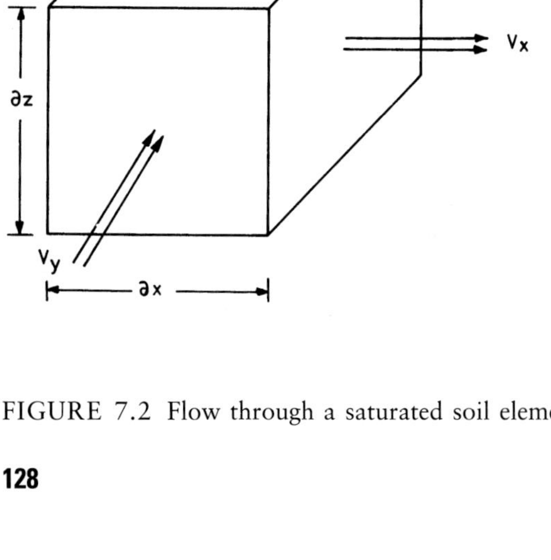

FIGURE 7.2 Flow through a saturated soil element.

Groundwater satisfies the equation of continuity. It expresses the principle of conservation of mass, i.e. the net inward flux through an elemental volume of an aquifer in the flow field must be equal to the rate at which matter is accumulating within the element (DeWiest, 1969). The continuity equation in its general form can be written as

$$\frac{\partial(\rho V_x)}{\partial x} + \frac{\partial(\rho V_y)}{\partial y} + \frac{\partial(\rho V_z)}{\partial z} + N = \rho S_s\frac{\partial h}{\partial t} \quad (7.19)$$

where ρ is the density of the fluid, N is the source/sink term and S_s is the specific storage defined later.

By eliminating ρ from both sides of equation (7.19) and using Darcy's law, we obtain equation (7.20) which represents transient flow through a saturated anisotropic medium

$$\frac{\partial}{\delta x}\left(K_x\frac{\partial h}{\partial x}\right) + \frac{\partial}{\partial y}\left(K_y\frac{\partial h}{\partial y}\right) + \frac{\partial}{\delta z}\left(K_z\frac{\partial h}{\partial z}\right) \pm W = S_s\frac{\partial h}{\partial t} \quad (7.20)$$

where W is the volume of flux per unit volume of the porous medium (a positive sign for the inflow and negative sign for the outflow).

For an homogeneous and isotropic medium, equation (7.20) reduces to

$$\frac{\partial^2 h}{\partial x^2} + \frac{\partial^2 h}{\partial y^2} + \frac{\partial^2 h}{\partial z^2} = \frac{S_s}{K}\frac{\partial h}{\partial t} \quad (7.21)$$

In a horizontal confined aquifer of thickness b, the storage coefficient, S will be equal to $S_s b$ and transmissivity $T = Kb$. Therefore, equation (7.21) reduces to

$$\frac{\partial^2 h}{\partial x^2} + \frac{\partial^2 h}{\partial y^2} + \frac{\partial^2 h}{\partial z^2} = \frac{S}{T}\frac{\partial h}{\partial t} \quad (7.22)$$

This is the partial differential equation, governing non-steady-state groundwater flow in a confined aquifer.

If the flow is steady, $dh/dt = 0$, i.e. velocity and pressure distribution do not change with time, equation (7.22) changes to

$$\frac{\partial^2 h}{\partial x^2} + \frac{\partial^2 h}{\partial y^2} + \frac{\partial^2 h}{\partial z^2} = 0 \qquad (7.23)$$

which is the Laplace equation governing for steady-state flow of groundwater in an homogeneous and isotropic aquifer.

For groundwater flow to wells, the differential equation governing the unsteady state radial flow in a confined aquifer in polar co-ordinates is expressed as

$$\frac{\partial^2 h}{\partial r^2} + \frac{1}{r}\frac{\partial h}{\partial r} = \frac{S}{T}\frac{\partial h}{\partial t} \qquad (7.24)$$

where r is the radial distance from the well to the point of observation. For steady flow, equation (7.24) reduces to

$$\frac{\partial^2 h}{\partial r^2} + \frac{1}{r}\frac{\partial h}{\partial r} = 0 \qquad (7.25)$$

Equations (7.24) and (7.25) are applied to solve problems of well hydraulics including analysis of pumping test data in homogeneous and isotropic aquifers under idealized conditions. In nature, aquifers exhibit complex situations for which necessary modifications are required. The adaptability of these equations to well hydraulics under different hydrogeological conditions is discussed in Chapter Nine.

7.2 Groundwater flow in fractured rocks

In fractured rocks, the groundwater movement mainly takes place along discontinuities, i.e. joints, fractures and shear zones. The interconnection between rock discontinuities and their spacing, aperture size and orientation decide the porosity and permeability of such rock masses. Open joints and fractures which are not filled with weathered or broken rock material form potential passage for groundwater movement but their permeability is greatly reduced when filled with clayey material such as smectite or montmorillonite. These filling materials form fracture skins which also influence the movement of solutes from the fractures into the porous matrix.

Depending on the porosities and permeabilities of the fractures and the matrix blocks, the fractured rock formations can be classified into: (a) purely fractured medium, (b) double porosity medium, and (c) heterogeneous medium (Figure 7.3). In a purely fractured medium, the porosity and permeability is only due to interconnected fractures while blocks are impervious. In a double (dual) porosity medium, both fractures and matrix blocks contribute to groundwater flow but fractures are the main contributors. In a situation when fractures are filled with clay or silty material, the fracture permeability is considerably reduced and such a medium is termed as heterogeneous.

Some of the models describing flow in a fractured medium are:

1. parallel plate model
2. double porosity model
3. equivalent porous medium model
4. discrete fracture network model, and
5. equivalent parallel plate model.

It may be mentioned that various models require knowledge about characteristic features of individual fractures and thereby account for the heterogeneity. Therefore careful mapping of fractures by surface and subsurface measurements is important. The level of details depends on the purpose for which the model is being developed. A greater accuracy is required for modelling solute transport as the heterogeneity of the fracture system greatly influences the travel time and solute concentration. It is also important to estimate the hydraulic properties

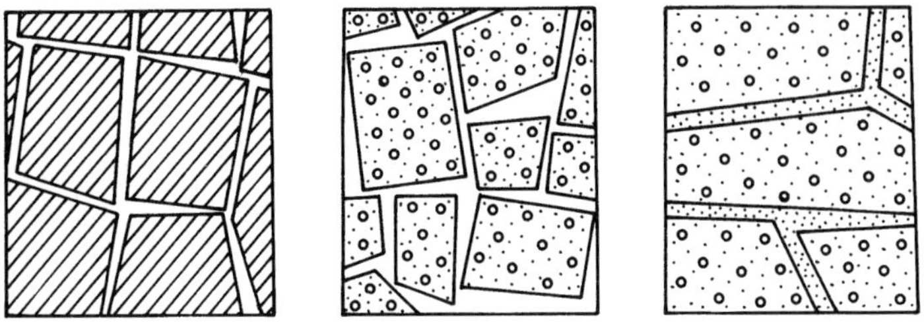

FIGURE 7.3 Schematic representation of (a) purely fractured medium, (b) double-porosity medium, and (c) heterogeneous medium (after Streltsova, 1977a).

of different fracture sets. Studies indicate that fractures parallel to the maximum compressive stress tend to be open, whereas those perpendicular to this direction tend to be closed (Anon, 1996).

7.2.1 Parallel plate model

The Darcy's law for flow in a single fracture can be written as

$$V = K_f I \tag{7.26}$$

where K_f is the hydraulic conductivity of the fracture, defined by

$$K_f = \frac{\gamma}{\mu}\frac{a^2}{12} \tag{7.27}$$

where a is the fracture aperture.

As the hydraulic conductivity (K) and permeability (k) are related by the expression

$$K = \frac{\gamma}{\mu}k \tag{7.28}$$

therefore, the permeability of the fracture, k_f, can be defined as

$$k_f = \frac{a^2}{12} \tag{7.29}$$

By combining equations (7.26) and (7.27), the average velocity ($\overline{V}_a$) in the fracture expressed by a single parallel plate model is given by

$$\overline{V}_a = \frac{\gamma a^2}{12\mu}\frac{dh}{dl} \tag{7.30}$$

Here it is assumed that the fracture walls are impermeable. If the fracture walls are permeable, the average velocity in the fracture will be (Bear, 1993)

$$\overline{V}_a = \frac{\gamma a^2}{12\mu + 2a\,\rho ql}\cdot\frac{dh}{dl} \tag{7.31}$$

where ql denotes the rate of leakage into the wall of the fracture. Equation (7.31) shows that if ql is negligible, or $\mu >> (apql)6$, then the conductivity of the fracture can be approximated by equation (7.27).

In terms of transmissivity of fracture, T_f, equation (7.29) can be written as

$$T_f = \frac{\gamma}{\mu}\frac{a^3}{12} = K_f a \tag{7.32}$$

Many researchers also define T_f as the hydraulic conductivity of fracture (Bear, 1993).

The volumetric flow rate per unit plate (fracture) width (Q_f) will be

$$Q_f = \left(\frac{\gamma a^3}{12\mu}\right) I \tag{7.33}$$

Equation (7.33) is referred as the **cubic law** which is valid for laminar flow through parallel wall fractures with smooth surfaces. In natural conditions these assumptions usually do not hold good. Under field conditions it is also difficult to define the representative distance between the fracture walls. The validity of cubic law is discussed by several researchers (Lee and Farmer, 1993). At low applied stress, when the fractures are open, the parallel plate approximation for fluid flow through the fracture may be valid. However, due to stress the contact area between fracture surfaces will increase and therefore variation in aperture should be considered (Tsang and Witherspoon, 1985).

Louis (1984) empirically defined the following five steady-state flow regimes, both laminar and turbulent, depending on various degrees of relative roughness, (R_r) fracture aperture, (a) hydraulic head gradient in the fracture plane, (I_f) and the kinematic viscosity (μ/ρ).

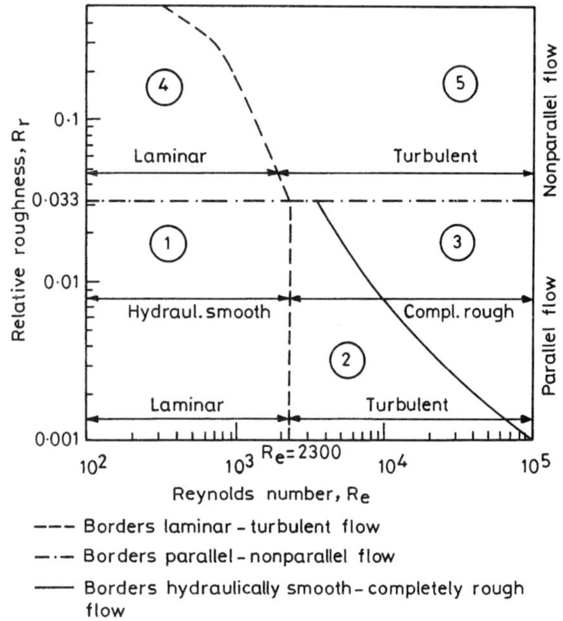

FIGURE 7.4 Definitions and range of validity of the various flow regimes in a fracture (after Louis, 1984).

Type 1: Smooth laminar

$$V = \left(\frac{\rho g a^2}{12\mu}\right) I_f \tag{7.34}$$

Type 2: Smooth turbulent

$$V = \left[\frac{g}{0.079}\left(\frac{2\rho a^5}{\mu}\right)^{\frac{1}{4}} I_f\right]^{\frac{4}{7}} \tag{7.35}$$

Type 3: Rough turbulent

$$V = \left(4\sqrt{ag}\ \ln\frac{3.7}{R_r}\right)\sqrt{I_f} \tag{7.36}$$

Type 4: Rough laminar

$$V = \left[\frac{\rho g a^2}{12\mu\,(1 + 8.8\,R_r)^{1.5}}\right] I_f \tag{7.37}$$

Type 5: Very rough turbulent

$$V = \left(4\sqrt{ag}\ \ln\frac{1.9}{R_r}\right)\sqrt{I_f} \tag{7.38}$$

The domains of validity of the various flow regimes in a fracture are shown in Figure 7.4.

If the fracture is not completely open, i.e. the fracture walls touch at places, the right-hand side of equations (7.34) to (7.38) should be multiplied by the degree of separation, F, which is defined as

$$F = \frac{\text{Open fracture surface area}}{\text{Total fracture surface area}}$$

Two types of laminar flow are distinguished by Sharp and Mani (Lee and Farmer, 1993):

1. Linear laminar: $V = KI$
2. Non-linear laminar: $V = K\,[i - \beta\,(i - i_{lim})^n]$

where K is the hydraulic conductivity of the fracture, i_{lim} is the limiting gradient for linear laminar flow and β and η are empirically determined constants. They also show that the flow rates are often not proportional to the cube of the aperture especially in rough fractures with asperities and gave the following exponent values for the relation $Q \propto a^n$.

	Rough fractures	Parallel plate
Linear laminar flow	n = 2	n = 3
Non-linear laminar	1.2 < n < 2	–
Fully turbulent flow	n = 1.2	n = 1.5

Several authors have studied the effect of variable aperture on flow through fractures. For example, Bear (1993) expressed the effective aperture, a_{eff} for a series of m discrete segments with different apertures (Figure 7.5a) as

$$a_{eff}^3 = \frac{\sum_{i=1}^{m} l_i}{\sum_{i=1}^{m} (l_i / a_i^3)} \qquad (7.39)$$

where l_i is the length of a fracture segment of aperture, a_i.

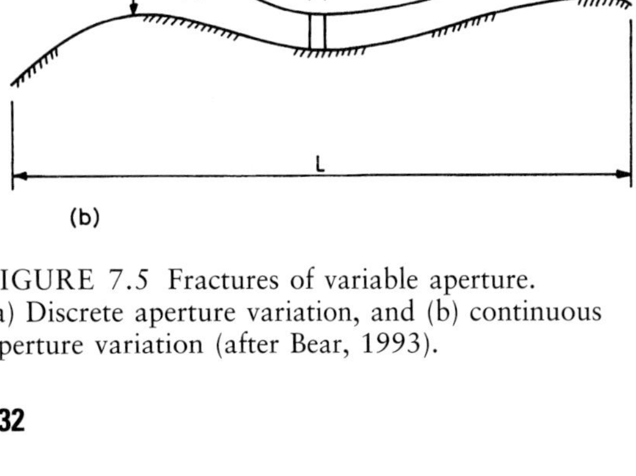

FIGURE 7.5 Fractures of variable aperture. (a) Discrete aperture variation, and (b) continuous aperture variation (after Bear, 1993).

For an aperture that varies continuously along the fracture (Figure 7.5b), i.e. $a = a(s)$, equation (7.39) is replaced by

$$a_{eff}^3 = \int_o^L \left(\int_o^L \frac{ds}{[a(s)]^3} \right)^{-1} ds \qquad (7.40)$$

where L is the total fracture length.

Flow in a multiple fracture system

The volumetric flow rate for a single fracture based on the parallel plate model is given by equation (7.33). It can be extended to various types of multiple fracture systems (Bear, 1993). In the case of a single family of m fractures of equal aperture oriented parallel to the flow direction (Figure 7.6), the total discharge Q is given by

$$Q = m \frac{a^3}{12} \frac{\gamma}{\mu} I \qquad (7.41)$$

and specific discharge, q through a cross-section having a width, L and unit length normal to the flow direction will be

$$q \cong \frac{Q}{L} = \frac{m}{L} \frac{a^3}{12} \frac{\gamma}{\mu} I = K_f I \qquad (7.42)$$

with

$$K_f = \frac{ma^3}{12L} \frac{\gamma}{\mu} \qquad (7.43)$$

If the fractures are of varying apertures, a_i, = 1, 2 ..., m, then according to Bear (1993)

$$Q = qL = \frac{1}{12\mu} \frac{\gamma}{\mu} \left(\sum_{i=1}^{m} a_i^3 \right) I = K_f L I \qquad (7.44)$$

Validity of cubic law

It can be seen from the above discussion that the cubic law is valid for fractures of small aperture and with plane surface. Natural fractures vary widely as far as planerity and surface

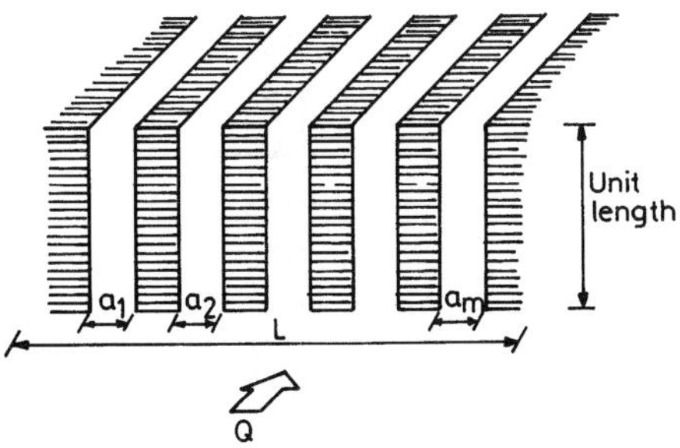

FIGURE 7.6 A set of parallel fractures (after Bear, 1993).

geometry is concerned. Bedding plane fractures in fine-grained sedimentary rocks like shales may be relatively smooth and parallel but in crystalline rocks such as granites, fracture surfaces are usually rough and the aperture varies. Further, under greater stress, fracture surfaces are also partially closed and therefore validity of the cubic law is disputed. Witherspoon *et al.* (1980) concluded that the cubic law is valid for apertures up to 4 μm opening despite local areas of contact at stresses up to 20 MPa. Deviation from cubic law in induced fractures subjected to normal stresses greater than 20 MPa is also demonstrated by Gale (1982a).

An increase in the contact area of the fracture surface will decrease the rate of fluid flow. A reduction of two or more orders of magnitude is reported when the contact area of the surface is greater than 30% as in the case of limestone and sandstone samples, where the contact area is estimated to be in the range of 40 to 70% at an applied stress of 30 MPa (Tsang and Witherspoon, 1985).

Channelling also influences the flow rate and transport of solutes due to the increase in linear velocity, more so at depths greater than 500–1000 m (Tsang and Tsang, 1987). These authors have therefore questioned the validity of the parallel plate model in nature as the channelling effect leads to preferred paths of flow due to which it may be difficult to predict the flow paths.

7.2.2 Double-porosity model

This is also known as **dual porosity model**. The behaviour of a fractured aquifer for regional groundwater investigations can best be represented by a double-porosity model. The concept of the double-porosity model was first developed by Barenblatt *et al.* (1960) and has been explained in detail by Streltsova-Adams (1978) and Gringarten (1982). The model assumes two regions – the porous block and the fracture – which have different hydraulic and hydromechanic characteristics. Three types of distribution of the matrix block are considered – horizontal slabs (strata-type), spherical blocks, and cubes (Figure 7.7). The fissured medium with horizontal fractures has been considered to be analogous to an alternate aquifer–aquitard system (Boulton and Streltsova, 1977). The block consists of fine pores which are separated from fractures; the blocks supply fluid to the fractures and act as a uniformly distributed source. Such an approach with some modifications has also been used by several researchers in analysing behaviour of fractured oil and geothermal reservoirs (Earlougher, 1977; Horne, 1990; Anon, 1996).

In the double-porosity model, the porous blocks have high primary porosity but low hydraulic conductivity while the adjacent fractures have low storativity but high conductivity. The differences in pressure between the porous blocks and the fractures lead to flow of fluid from porous blocks to adjacent fractures.

The characteristics of a double (dual) porosity aquifer system are described in terms of K_f, K_m, S_f, S_m, where subscripts f and m are for fractures and matrix respectively. The other two parameters are storativity ratio (ω) and the transmissivity ratio (λ). Storativity ratio (ω) is the ratio of fissure to total system (blocks plus fissures) storativity and can be expressed as

$$\omega = \frac{S_f}{S_f + S_m \beta} \tag{7.45}$$

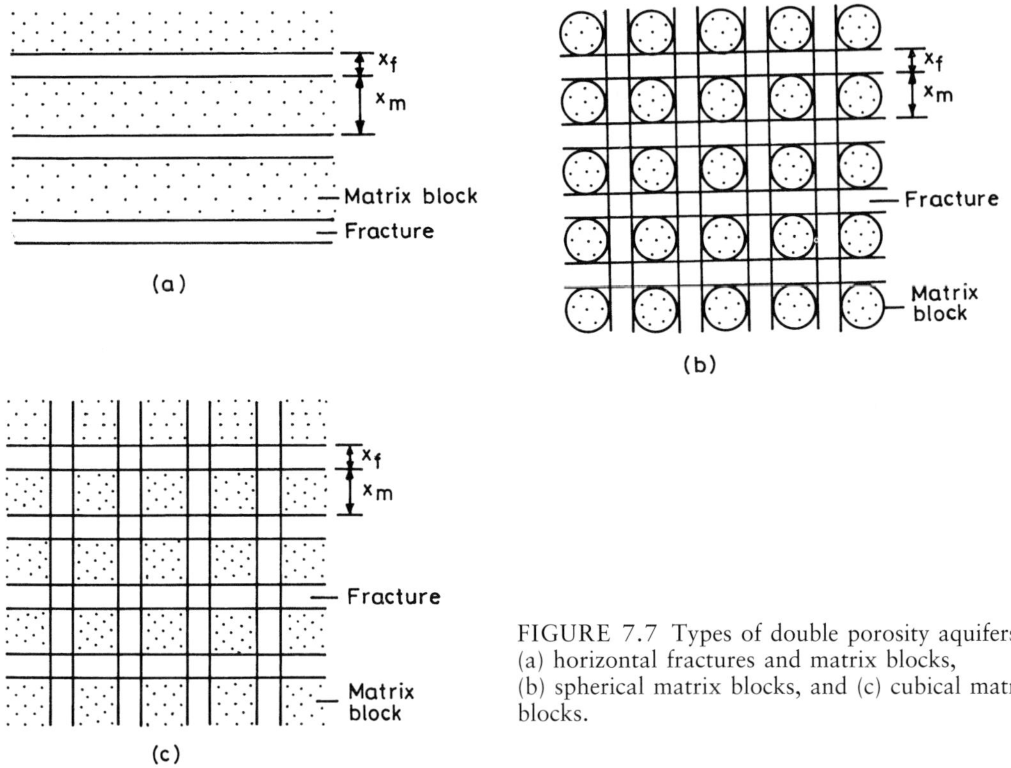

FIGURE 7.7 Types of double porosity aquifers: (a) horizontal fractures and matrix blocks, (b) spherical matrix blocks, and (c) cubical matrix blocks.

where S_f and S_m are the storativities of fracture and blocks respectively, and β is a factor depending on matrix block geometry which is taken to be equal to 1 for strata type.

The transmissivity ratio, or interporosity flow coefficient (λ) (dimensionless) is given by

$$\lambda = \alpha r_w^2 \frac{K_m}{K_f} \tag{7.46}$$

where α is the shape factor parameter depending on the geometry of the aquifer matrix and fracture (dimension reciprocal area), r_w is well radius; α is defined as

$$\alpha = \frac{A}{lv} \tag{7.47}$$

where A is the surface area of the matrix block, v is the matrix volume, and l is the characteristic length.

If the matrix blocks are in the form of horizontal slab blocks (Figure 7.7a), as may also be the case in thinly layered formations, $\alpha = 12/x_m^2$, where x_m is the thickness of a matrix block (Kruseman and de Ridder, 1990) and h_f is the thickness of the high permeability layer, i.e. fracture. When matrix blocks are in the form of cubes or spheres (Figure 7.7b, c), $\alpha = 60/x_m^2$, where x_m is the length of a side of the cubic block, or the diameter of the spherical block.

Values of ω are generally in the range of 10^{-1} to 10^{-4} (Kruseman and de Ridder, 1990). However, in a special case when $\omega = 1$, the matrix porosity will be zero and the reservoir will be characterized by single porosity instead of double porosity (Horne, 1990). Values of λ are usually very small (10^{-3} to 10^{-10}). A larger value of λ indicates lesser heterogeneity and thereby points towards a reservoir of single porosity (Horne, 1990).

Based on the continuity equation and Darcy's law, the differential equations describing the one dimensional confined flow within the fractures and matrix block can be obtained as

$$T_f \frac{\partial^2 h_f}{\partial x^2} = S_f \frac{\partial h_f}{\partial t} - V_d \quad (7.48)$$

$$T_m \frac{\partial^2 h_m}{\partial x^2} = S_m \frac{\partial h_m}{\partial t} + V_d \quad (7.49)$$

where V_d is the water transfer per unit time from blocks to fractures in a column of aquifer having unit horizontal area. h_f and h_m are the mean piezometric heads in the fracture and matrix block respectively.

According to Streltsova (1976a), the quasi-steady state transfer of water between porous blocks and fractures can be expressed as

$$V_d = \frac{K_m}{l}(h_f - h_m) \quad (7.50)$$

where l is the characteristic dimension of block.

As the hydraulic conductivity of porous blocks is much less than their storativity, the term $T_m\,(\partial^2 h_m/\partial x^2)$ may be neglected as compared to $S_m\,(\partial h_m/\partial t)$ in equation (7.49) (Barenblatt *et al.*, 1960; Streltsova-Adams, 1978). Therefore, equations (7.48), (7.49) and (7.50) can be expressed as

$$T_f \frac{\partial^2 h_f}{\partial x^2} = S_f \frac{\partial h_f}{\partial t} + S_m \frac{\partial h_m}{\partial t} \quad (7.51)$$

$$S_m \frac{\partial h_m}{\partial t} = \frac{K_m}{l}(h_f - h_m) \quad (7.52)$$

The flow from block to fissure will be time dependent, which can be expressed as

$$V_d = S_m \frac{\partial h_m}{\partial t} = K_m \frac{h_f - h_m}{l} \quad (7.53)$$

or

$$\frac{\partial h_m}{\partial t} = \frac{K_m}{S_m}\frac{h_f - h_m}{l} = \psi\,(h_f - h_m) \quad (7.54)$$

where

$$\psi = \frac{K_m}{S_m}$$

Finally, according to Streltsova-Adams (1978), the general differential equation for the confined flow in fissure to a well in a double porosity formation will be

$$T_f\left(\frac{\partial^2 h_f}{\partial r^2} + \frac{1}{r}\frac{\partial h_f}{\partial t}\right) = S_f \frac{\partial h_f}{\partial_t} + \psi S_m \int_0^\infty \frac{\partial h_f}{\partial t_{t=\tau}} \exp\left[-\psi\,(t-\tau)\right] dt \quad (7.55)$$

An identical solution was obtained by Boulton and Streltsova considering block-and-fissure units of a regular pattern (Streltsova-Adams, 1978).

The fissure-porous block flow is a time-dependent process caused by the readjustment of the block and fissure pressures, the duration of which would depend on the elastic properties of blocks and fissures, their permeability and dimensions (Streltsova, 1976b). It is therefore necessary to define, in a representative elementary volume, these different pressures, one in the fractures and the other in the matrix (blocks), as well as exchange of mass between the intergranular porosity and the porosity of the fracture (Marsily, 1986).

Due to differences in the permeabilities of fractures and blocks, flow mechanism is different during early, long and intermediate times of pumping. At the early times, flow from the matrix block is essentially zero as initially fluid is removed primarily from the fractures which have high permeability. At the long times, the fractured reservoir behaviour is equivalent to that of an homogeneous porous medium with a permeability equal to the fissure permeability. At the intermediate times, there is a transition from fracture flow to flow from fractures and matrix blocks, during which the drawdown remains more or less constant. As the flow response from fractures will be recorded first in the well and the response from blocks later, the time-drawdown data will be plotted on two separate semi-log straight lines. The characteristics of time-drawdown curves are discussed in Chapter Nine. The separation between the two straight lines will depend on storativity ratio ω and transmissivity ratio λ (Gringarten, 1982; Horne, 1990). Therefore, the response of a dual porosity system is similar to that of delayed yield in an unconfined aquifer (Boulton and Streltsova, 1977, 1978). It could, therefore, be concluded that the flow in a porous fractured rock differs from an ordinary porous medium only during the initial stages of transient flow, and only in the vicinity of the pumping well. Beyond a relatively short period, the flow regime will be identical to that of a single continuum (Bear, 1993). Methods of analysing pumping test data from double porosity aquifers are discussed in Chapter Nine.

7.2.3 Equivalent porous medium model (EPM)

This is also known as a equivalent continuum model. It is a conceptually simple and commonly used approach in estimating flow and transport in fractured media as it avoids characterization of fractures. Much research has shown that flow in a large enough volume of fractured medium can be reasonably represented by flow through a porous medium, i.e. by an equivalent continuum model. This will be true when: (a) fracture density is high, (b) apertures are constant rather than distributed, (c) orientations are distributed rather than constant, (d) larger sample sizes are tested (Long *et al.*, 1982) and (e) the interest is mainly on volumetric flow, such as for groundwater supplies. The equivalent continuum approximation is also dependent on the relative orientations of the joint set. In cases where the rock has two sets of joints, chance for equivalent continuum behaviour is found to increase as the angle between joint sets approaches 90 deg (Panda and Kulatilake, 1995). Conventional forms of the groundwater flow equations developed for granular porous media can be adopted.

7.2.4 Discrete fracture network models

Network models use fracture characteristics and heterogeneity of rock mass based on field data. The hydraulic behaviour of a discrete fracture network is a function of fracture aperture, length, density, orientation, connectivity of fractures and fracture filling material as discussed earlier. Numerical tools have been developed to distinguish between fractured systems that can be treated as EPM and those that should be treated as discrete fracture networks (Long *et al.*, 1982).

Two-dimensional and three-dimensional fracture network models have been suggested by Long *et al.* (1982) and Lee and Farmer (1993). These models evaluate flow in fractures or fracture sets with synthetic distributions of apertures, orientations, spacings and dimensions and take into account various surface roughnesses, flow channelling, and mixing phenomena at fracture intersections. The application of these theoretical models to natural systems has been limited. These models are useful where the area of interest is small – as in the study of the effect of radionuclide transport through fractures as a result of nuclear waste storage in geological formations over a long period of time. The disadvantages of the discrete modelling are that statistical information about fracture characteristics may be difficult to obtain. These models are also complex and there is no guarantee that a model reproducing the apparent geometric properties of a fracture network will capture its essential flow or transport features (Anon, 1996)

7.2.5 Equivalent parallel plate model

Sharp (1993) proposed the equivalent parallel plate model which takes into consideration, to a greater degree, geological controls more than other models. The approach is based on the assumption that the area of interest in a fractured rock mass can be divided into smaller near-homogeneous domains, each having flow characteristics different from each other. In this way, fracture characterization can be based on a direction-dependent parameter, which has been called here 'integrated fracture density of apertures function' (IFDA). The various fractures are grouped according to their strike trends. For each strike trend, field data on aperture (a) and density (N_i) for all fracture sets are collected, and their IFDA is calculated from the relation:

$$\mathrm{IFDA} = \sum N_i\, a_i^2 \qquad (7.56)$$

IFDA values are plotted graphically for each direction (Figure 7.8). In this figure the trend of each line, with double arrows, represents the strike of the dominant fracture sets and the length corresponds to the respective IFDA value. Thus, based on this characterization, the area of interest can be divided into near-homogeneous smaller domains. Each domain has its own characteristic distribution of hydraulic conductivity which will provide an input into two-dimensional finite-element models.

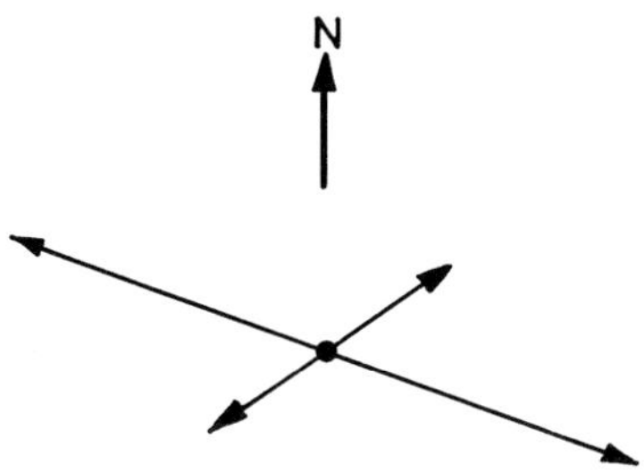

FIGURE 7.8 Graphical representation of IFDA (for details see text).

Based on the above concept, an IFDA map of a part of Bundelkhand granitic terrain in India was prepared (Figure 7.9). The different domains are demarcated based on field measurements of fracture characteristics (orientation, aperture and spacing) at 64 sites, except domains VIIb, c and d, which were mapped based on the trends of lineaments on remote sensing (aerial photographs and IRS-LISS-II) data as these areas are covered with weathered rock and soil. The trend of the IFDA tensor in the adjacent area (domain VIIIa) also helped in deciding the IFDA tensor in the domains VIIb, c and d. Figure 7.9 indicates that in this area, hydraulic tensor of the type I is by far the most extensive one, characterized by a strong NW–SE trend, though minor trends also exist in other directions.

7.3 Unsaturated flow

The zone between the water-table and the ground surface is the zone of aeration or the unsaturated (vadose) zone (see section 1.3). The mechanism of downward movement of water through this zone is of importance, both from the point of view of natural and artificial recharge to the phreatic aquifer as well as downward movement of contaminants from the ground surface. There is also an upward vertical movement of water in the unsaturated zone by evaporation and transpiration This has an important bearing on crop water requirements, drainage and soil salinization processes. The hydraulic properties of the unsaturated zone also influence the rise or fall of the water-table during recharge or abstraction of water due to pumping or drainage. Unsaturated flow is actually a two-phase flow involving air and water. However, the single phase approach to unsaturated flow behaviour holds good for all practical purposes except where the effect of entrapped air is significant.

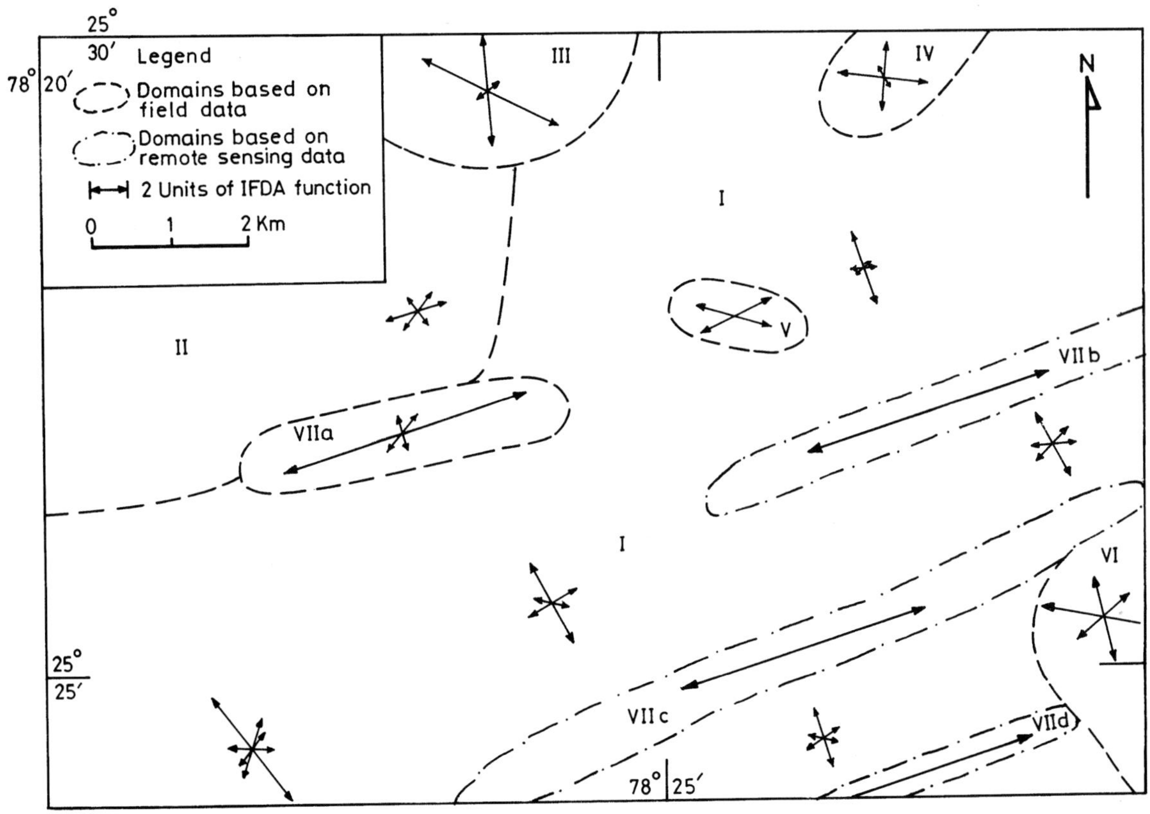

FIGURE 7.9 IFDA map of a part of Bundelkhand area, India (modified after Gupta, 1996). (Note: In this map, one unit of IFDA function = 1 μm).

7.3.1 Flow through unsaturated porous media

The **volumetric water content** (θ) in unsaturated material is defined as $\theta = V_w/V$, where V_w is the volume of water and V is the total volume of a soil or rock. It is usually reported as a decimal fraction or a percentage. For saturated flow, $\theta = \eta$ and for unsatured flow, $\theta \leq \eta$, where η is the total porosity.

In the unsaturated zone, although there will be localized pockets of water (perched water) having positive water potential, the potential (head), h, in general is negative due to the surface tension of water. The water present is subjected to capillary and adsorptive forces and energy is required to remove it. With a decrease in water content, the remaining water will have more negative water potential. As the water content decreases, the remaining water is left adhered to the surfaces of mineral grains or walls of fractures as a thin film. The flow characteristics of this thin film are also important, and depend on the thickness of the film. It is often difficult to estimate the thickness of the film and thereby the conditions under which the films will coalesce to cause saturation of the pores or the fractures.

Unsaturated flow in the vadose zone (zone of aeration) can be analysed on the basis of Darcy's law by considering the unsaturated hydraulic conductivity, (K_u), instead of normal hydraulic conductivity (K). Unsaturated hydraulic conductivity, K_u is dependent on the water content θ which is related to the (negative) pressure head, h'. Therefore, K_u can be expressed as ($K\theta$) or as K'_h depending on whether it is considered in relation to θ or h'. In unsaturated material, the presence of air in the pore spaces obstructs water movement. Therefore, the lower the value of the water content, the lower will be the unsaturated hydraulic conductivity. Based on the experimental data, the relation between K_u, K, and degree of saturation is given as (Todd, 1980)

$$\frac{K_u}{K} = \left(\frac{S_s - S_o}{1 - S_o}\right)^3 \tag{7.57}$$

where S_s is the degree of saturation and S_o is the threshold saturation (non-moving water held by capillary forces). These data, when plotted in Figure 7.10, show that K_u varies from zero at $S_s = S_o$ to $K_u = K$ at $S_s = 1$, i.e. at full saturation.

The relation between K_u/K and negative pressure, h' are S-shaped as shown in Figure 7.11. The approximate equation for this type of relation is (Todd, 1980)

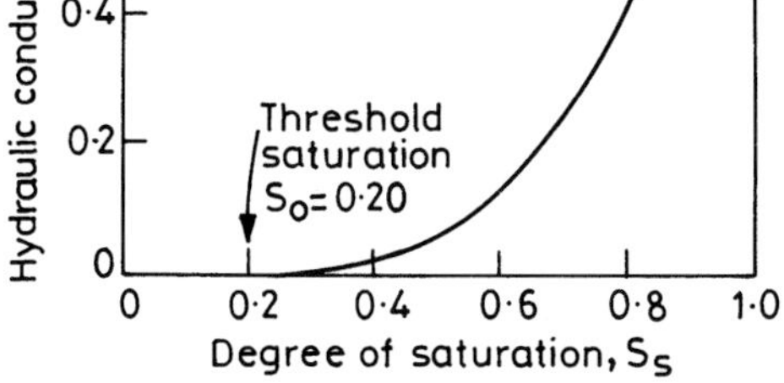

FIGURE 7.10 Ratio of unsaturated (K_u) to saturated (K) hydraulic conductivity as a function of degree of saturation (after Irmay, 1954).

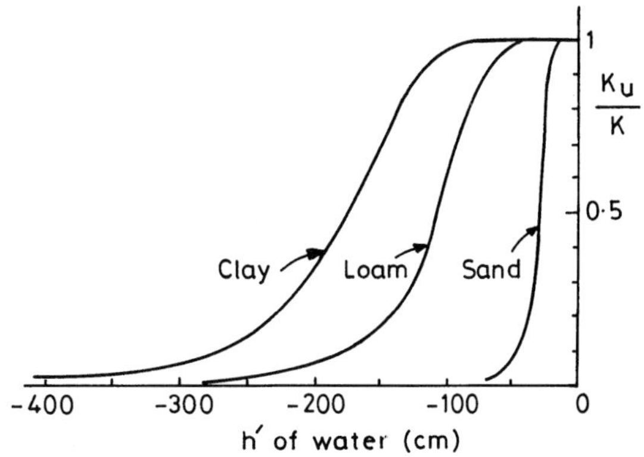

FIGURE 7.11 Relationship between hydraulic conductivity ratio (K_u/K) and soil-water pressure (h') for sand, loam and clay (after Bouwer, 1978).

$$\frac{K_u}{K} = \frac{a}{\frac{a}{b}(-h')^n + a} \tag{7.58}$$

where a, b and n are constants that decrease with decreasing grain size of the material. It can be seen that at atmospheric pressure when $h' = 0$, then $K_u = K$.

Under unsteady unsaturated flow conditions, K_u and Q continuously increase during infiltration of rain and will decrease after the cessation of recharge in the upper part of the wetted front. Evaporation and transpiration will also reduce K_u and Q in the upper part of the profile.

Darcy's law, for unsaturated flow, can be written as

$$q = -K(\theta)\, \nabla H \tag{7.59}$$

where q = volumetric flux factor; $K(\theta)$ is hydraulic conductivity tensor as a function of water content, θ; and ∇H = the total hydraulic gradient in three-dimensional space.

The hydraulic head is the sum of elevation and pressure head. However, for unsaturated conditions, the pressure head will be negative, as mentioned earlier. The flow will also be influenced by osmotic potential and temperature gradients. In addition to water content, $K(\theta)$ depends upon the total porosity, pore size distribution and the viscosity and density of the fluid. It is therefore difficult to measure $K(\theta)$ under field conditions.

The general differential flow equation for a three-dimensional unsaturated flow (equation (7.60)) was derived by L. A. Richards (Bouwer, 1978) using the analogy to heat flow in porous media, in the same way as for saturated flow.

$$\frac{\partial}{\partial x}\left(K_h \frac{\partial h}{\partial x}\right) + \frac{\partial}{\partial y}\left(K_h \frac{\partial h}{\partial y}\right) + \frac{\partial}{\partial z}\left(K_h \frac{\partial h}{\partial z}\right) + \frac{\partial K_h}{\partial z} = \frac{\partial \theta}{\partial t} \qquad (7.60)$$

The term $\partial K_h/\partial z$ is included to account for gravity flow. For two-dimensional flow in a vertical plane (stream seepage to a deep water-table, infiltration from irrigation channel etc.), the second term in equation (7.60) will be zero. For one-dimensional vertical flow (infiltration of rainfall, evaporation from water-table), the first two terms are zero. For steady flow systems, $\partial\theta/\partial t$ is zero (Bouwer, 1978).

Richard's equation is based on the assumptions that the fluid is incompressible and the flow takes place under isothermal conditions. However, the movement of water and contaminants is influenced by temperature gradients in the unsaturated zone. As the ground surface temperatures are usually higher than at depth, there will be a tendency of water vapour to move upward. This will be opposite to the movement of water which is downward due to gravity. The natural temperature gradients will also be disturbed by the disposal of radioactive waste in underground repositories. Under certain conditions, the circulation of air brought about by barometric changes will also affect the moisture content in the unsaturated zone as discussed in section 7.3.2. Therefore, although theoretical reasoning suggests that the Richard's equation does not fully represent the flow in the unsaturated zone, and may even be inaccurate in some circumstances, the errors resulting from ignoring the multiphase nature of the flow process in natural conditions in most cases will be negligible.

As most of the soils are anisotropic and heterogeneous, transport of a part of infiltrating water will be faster than the average wetting front. This **preferential flow** of water in the unsaturated zone is of importance for simulating the field water balance and therefore on the calculation of crop water use, solute transport and pollution of groundwater (Feddes *et al.*, 1988). Preferential flow, also known as bypass flow or shortcircuiting, in the unsaturated zone, may take place through macropores, caused by cracking and shrinkage of clayey soils, by plant roots or by tillage operations. Various models have been developed to simulate preferential flow by modifying the basic partial differential equation (Feddes *et al.*, 1988).

7.3.2 Flow through unsaturated fractured rock

The flow mechanism in the unsaturated unconsolidated material, which has been studied in detail in the past, is also applicable to the weathered zone in hard rocks. However, the processes affecting flow in unsaturated fractured rock are poorly understood so far. It has gained importance only in recent years, mainly for the study of migration of contaminants from the high-level repositories of radioactive waste and sanitary landfills in the unsaturated (vadose) zone. This has led to theoretical, laboratory and field investigations regarding flow mechanism in the unsaturated fractured rocks (Evans and Nicholson, 1987). One such site for radioactive waste disposal in a desert environment which has been studied in detail is the Yucca Mountain in Nevada, USA (Wang and Narasimhan, 1993; Anon, 1996). The potential repository is located in the volcanic tuff at a depth of approximately 350 m below the ground surface and 225 m above the water-table. These studies have provided useful information about the fluid flow in partially saturated, fractured porous media for predicting the migration of nuclear waste from the repository to the available environment over thousands of years.

The movement of water in unsaturated fractured rocks is essentially a problem of multiphase fluid flow (water and gas phases) as in the unsaturated porous granular medium. However, in a fractured rock there is likely to be a time-lag between the movement of fluids through the fractures and the matrix blocks. Under fully saturated conditions, the fractures will control the fluid flow while under unsaturated or partially saturated conditions, the role of fracture flow is minimal and the flow is mainly governed by matrix-block characteristics. Under unsaturated conditions, due to capillary mechanism, the fractures will drain easily as fracture apertures are usually larger than matrix pore sizes. Therefore, fractures will remain dry under partially saturated conditions while the water will be held by capillarity in the finer pores of the matrix. In such a situation where porous blocks are separated by desaturated fractures, there may not be any flow of water from one porous block to another.

However, at greater depths due to compressional stress, the walls of the fractures are pressed together forming asperities. Further, the aperture of a fracture will be smaller near the asperities and larger at other places within the fracture plane. At these places, the fractures will be partly saturated forming bridges for liquid water to move across the fractures from one porous block to an adjacent one (Figure 7.12). Borehole data show that about 12% of the fracture surfaces in the Yucca Mountain have coatings of secondary minerals, e.g. zeolite, clay and calcite, indicating that the coating area of 12% corresponds to the contact area of fractures through which inter-block flow would

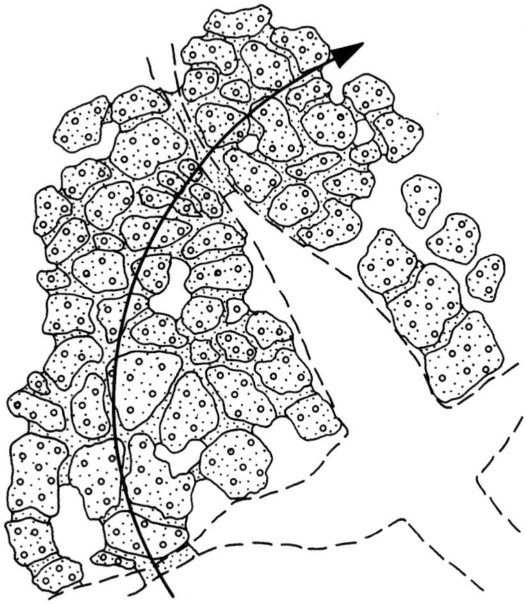

FIGURE 7.12 Schematic representation of water flow in a fractured porous medium (after Wang and Narasimhan, 1993).

take place across the fractures under partially saturated conditions (Wang and Narasimhan, 1993).

In the radioactive waste repositories, the natural temperature gradients will also be disturbed due to the heat (exceeding 100 °C) and large temperature gradients caused by the radioactive waste decay, thereby influencing the movement of water as well as contaminants both in liquid and vapour phases. The heating part will lose moisture while the condensation of water vapour will take place away from the heat source. Readers may refer to Evans and Nicholson (1987) and Pruess and Wang (1987) for experiments to study the effect of temperature gradients on the movement of water vapour, and numerical modelling of isothermal and non-isothermal flow in unsaturated fractured rock.

The convective transport of vapour in the unsaturated fractured rock is also dependent on air pressure differences which are controlled by changes in barometric pressure, thermal gradients and changes in rock saturation due to addition or removal of water. In areas of steep relief, as in the Yucca Mountain, it is observed that barometric effects will influence the air circulation and thereby drying of the rock in the upper parts of the unsaturated zone (Weeks, 1987). Therefore, the hydrological properties and vapour transport in the unsaturated zone will be affected by topographically enhanced air circulation under natural conditions. The drying of rocks would reduce the potential for deep percolation through the proposed repository site (Weeks, 1987).

7.4 Principles of solute transport in fractured media

The transport of dissolved substances (solutes) in groundwater is important when studying the movement of natural constituents due to rock water interaction, and also of contaminants introduced by man's activities as a result of domestic, industrial or agricultural wastes. The solutes could either be non-reactive (conservative), viz. chloride, or reactive substances, viz. sodium. The non-reactive substances are not subjected to any exchange or reaction while moving through the matrix, thereby excluding radioactive decay and adsorption.

The assessment of contaminant transport through fractured media has gained greater importance in recent years in view of the programmes of permanent storage of high-level nuclear fuel wastes in crystalline rocks in several countries, especially in North America and Europe (see Section 10.9).

7.4.1 Mechanism of solute transport

Similar to transport of contaminants in ordinary porous media, the main mechanisms of transport of solutes through fractured porous rocks are by advection, dispersion, molecular diffusion, decay phenomena, rock water interaction and retardation including adsorption (Figure 7.13).

As most of the fractured rock aquifers show a double porosity character, like fluid flow, the mechanism of contaminant transport depends on the relative porosities and permeabilities of the fractures and the matrix blocks. If the rock matrix is impermeable, has negligible porosity and is compact, the advective transport through the fractures will be prominent. If the matrix blocks are porous but have negligible permeability, the main mechanism in the matrix will be by molecular diffusion, and transport through the rock matrix by advection will be generally insignificant because of its low permeability. In such a case, as part of a clean-up operation, the contaminants from the fractures may be flushed out relatively quickly but the porous blocks will only slowly release the

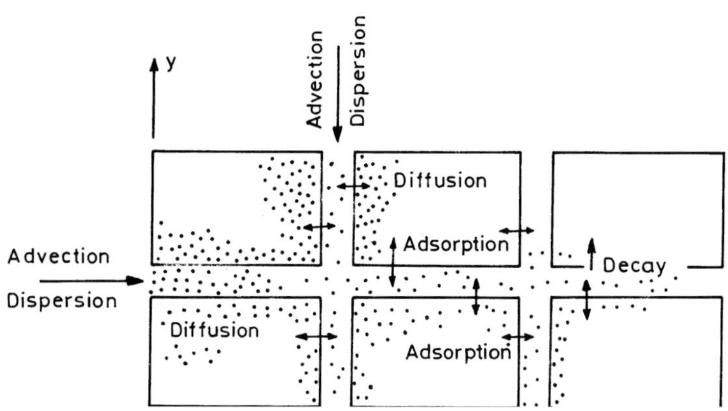

FIGURE 7.13 Schematic representation of various mechanisms of solute transport in an idealized fractured porous medium (after Germain and Frind, 1984).

contaminants into the fractures. In a third situation, when the porous block has the same permeability as the fracture, the transport of solute will take place simultaneously in the two media (fracture and matrix) by dispersion and diffusion, depending on differences in the head and the concentration of the solutes in the matrix blocks and the fractures.

Advection

The main mechanism of transport of non-reactive solutes is by advection (convection) due to which the solutes are transported at an average rate equal to the average linear-velocity, $\overline{V}_a$ of the water in the porous medium as given by

$$\overline{V}_a = \frac{q}{\eta_e} = \frac{K_x \frac{dh}{dx}}{\eta_e} \tag{7.61}$$

In fractured media, the average velocity of flow parallel to a planar fracture ($\overline{V}_a$) can be described by

$$\overline{V}_a = \frac{\gamma a^2 \, (dh/dl)}{12\mu} \tag{7.62}$$

The fractured rocks have much lower effective porosity (10^{-2} to 10^{-5}) compared to porous granular material (0.5 to 0.03). Therefore, groundwater velocities in fractured rocks are several orders of magnitude higher than in the granular rocks. This can be illustrated by computing groundwater velocities by using the modified Darcy's equation (7.63)

$$\overline{V}_a = \frac{K}{\eta_f} \cdot \frac{dh}{dl} \tag{7.63}$$

where η_f is the bulk fracture porosity.

If we assume that a fractured medium (granite) has a fracture porosity (η_f) of 10^{-4}, and hydraulic conductivity of 10^{-5} ms^{-1} and if the hydraulic gradient is 10^{-2}, as is generally the case in the field, the groundwater velocity using this equation (7.63) will be estimated to be equal to 86.4 m d^{-1}. This compared to a porous medium (say silty sand) with the same values of hydraulic conductivity and hydraulic gradient and effective porosity of 0.26, will have a velocity of about 0.03 m d^{-1} indicating thereby that the groundwater velocity in a fractured medium can be several times higher than in the porous rocks. Therefore, advection transport is much faster in fractured media than in porous rocks provided the matrix is impermeable and has negligible porosity. However, numerical studies show that tracer velocities may be less or greater than the average fluid velocity depending on the fracture roughness and contact area. The tracer velocity is found to be less than the average fluid velocity if flow is transverse to the roughness, and greater if flow is parallel to the roughness (Anon, 1996).

Dispersion

In addition to advection, the solute will have a tendency to spread by hydrodynamic dispersion which causes the dilution of the solute. During flow through a cylindrical pipe, the velocities are distributed in a parabolic form causing a faster movement of solute along the axis and a comparatively slower movement near the surface due to drag effect (Figure 7.14a). The same phenomenon would occur in porous media or fractures with more complexities due to variation in velocities as the pores and fracture openings may be of different sizes, shapes and orientations (Figure 7.14b). Dispersion will also

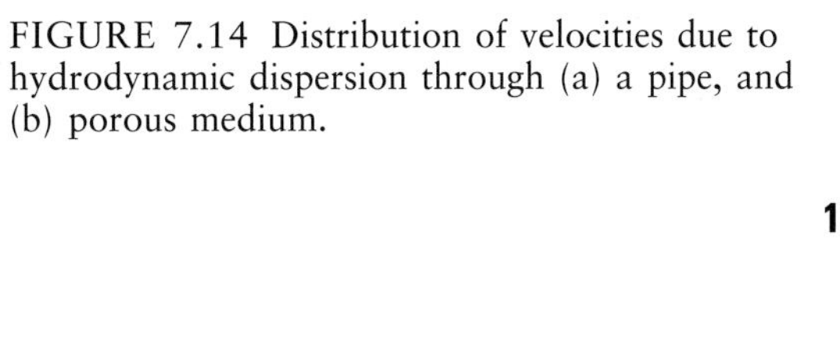

FIGURE 7.14 Distribution of velocities due to hydrodynamic dispersion through (a) a pipe, and (b) porous medium.

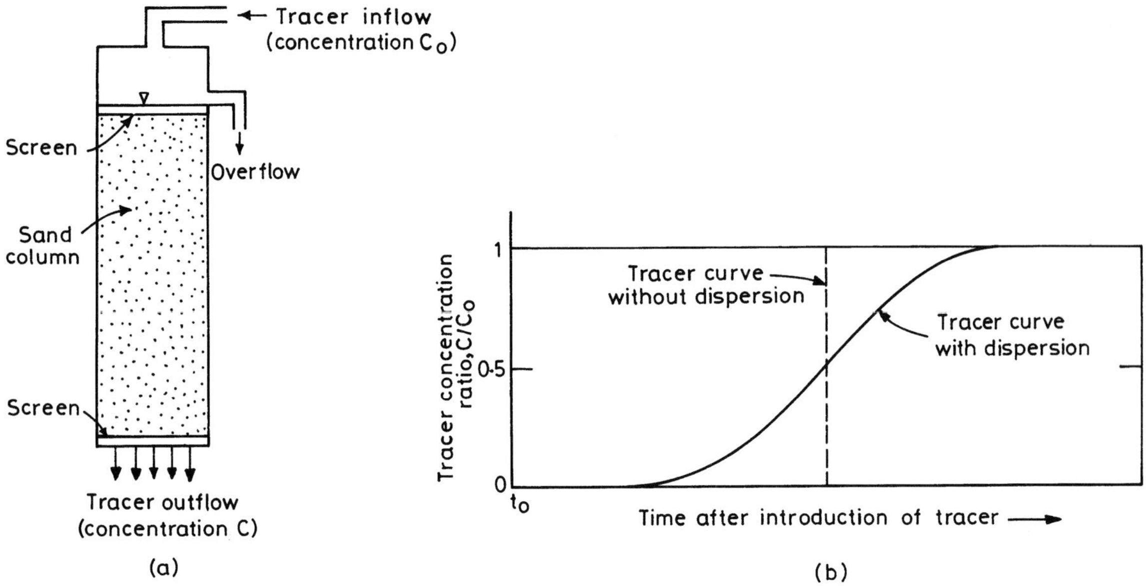

FIGURE 7.15 Longitudinal dispersion of a solute passing through a sand column: (a) experimental setup, and (b) breakthrough (dispersion) curve.

be affected by the tortuosity and connectivity of multiple flow paths. Dispersion is a non-steady irreversible process.

The dispersion phenomena can be demonstrated by simple laboratory experiments using a cylindrical sand column (Figure 7.15a). The concentration of solute will decrease with time in the direction of flow which can be demonstrated by breakthrough curves (Figure 7.15b).

Due to dispersion, the solute will spread both in the longitudinal and transverse directions. Dispersion in the direction of flow (longitudinal dispersion) is much stronger than in any direction normal to the flow (transverse dispersion) (Figure 7.16).

In fractured media, dispersion is due to heterogeneities of the velocities in a fracture (parabolic profile); also to differences in velocities from one fracture to another (because of differences in aperture sizes and tortuosity) and to intersection and channelling of fractures which cause transverse mixing and dispersion. Readers may refer to Smith and Schwartz (1993) for a detailed discussion on the effect of fracture intersection on dispersion.

In homogeneous and isotropic media, for the two-dimensional case, the equation for dispersion for a conservative tracer has the following form

$$\frac{\partial C}{\partial t} = \frac{\partial}{\partial x}\left(D_L \frac{\partial C}{\partial x}\right) + \frac{\partial}{\partial y}\left(D_T \frac{\partial C}{\partial y}\right) - V_x \frac{\partial C}{\partial x} - V_y \frac{\partial C}{\partial y} \tag{7.64}$$

where C is the relative tracer concentration ($0 < C < 1$), D_L and D_T are longitudinal and transverse coefficients of hydrodynamic dispersion respectively, V_x and V_y are the components of fluid velocities (x is the coordinate in the direction of flow and y is the coordinate normal to flow), and t is the time. Dimensions of the dispersion coefficients are L^2T^{-1}.

The longitudinal and transverse coefficients of hydrodynamic dispersion can be expressed as

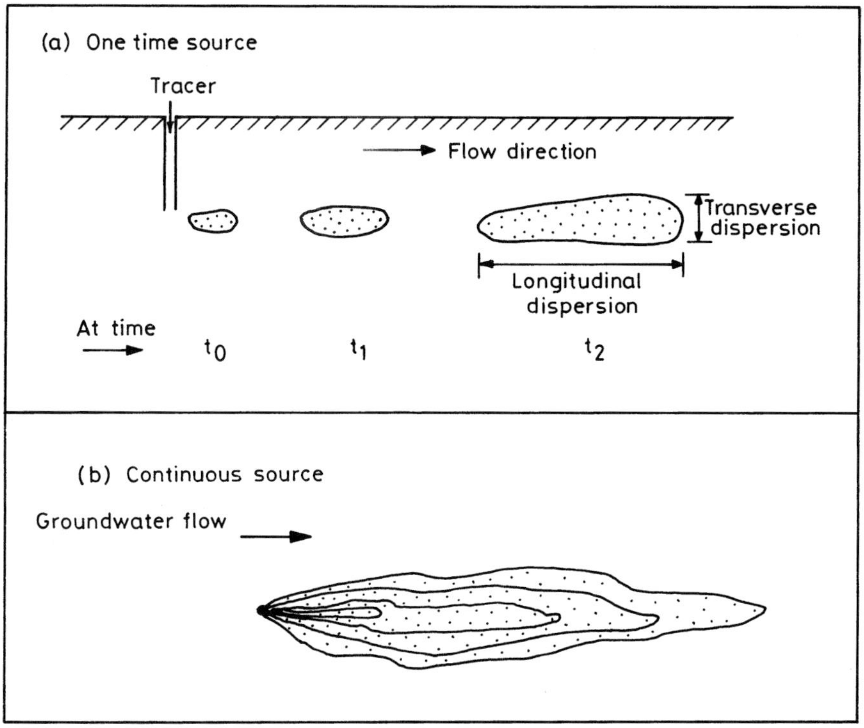

FIGURE 7.16
Longitudinal and transverse dispersion of a tracer introduced into a porous medium: (a) one-time source, and (b) continuous source.

$$D_L = \alpha_L V + D$$
$$D_T = \alpha_T V + D$$

where α_L and α_T are the characteristic properties of the porous medium known as **intrinsic dispersion coefficient** or **dispersivity** with the dimension of L, and D is the coefficient of molecular diffusion for the solute in the porous medium (L^2T^{-1}).

At low velocity, as dispersion is mainly on account of diffusion, the coefficient of hydrodynamic dispersion, D_L, equals the diffusion coefficient, D. At high velocity, mechanical mixing will be the dominant mechanism of solute transport, therefore $D_L = \alpha_L V$. Larger dispersivity value of the medium produces greater mixing of the solute front as it advances. Laboratory studies on homogeneous sandy material have given values of longitudinal dispersivity, α_L in the range of 0.1 to 10 m while the transverse dispersivity α_T values are lower by a factor of 5–20. Laboratory values are significantly smaller than those obtained from field studies (Freeze and Cherry, 1979) which is attributed to the effect of heterogeneities on field scale (Fetter, 1993). Therefore, the values of dispersivities obtained from laboratory study will not be of much help in predicting dispersion in the field. Methods of determining dispersivity in the field are given by Freeze and Cherry (1979). Field studies indicate that longitudinal dispersivity, α_L typically ranges from 10 to 100 m and transverse dispersivity, α_T is 10 to 30% of α_L (Table 7.1). The ratio of longitudinal to transverse dispersivity (α_L/α_T) in an aquifer is an important criterion which influences the shape of the contaminant plume.

Dispersion will be greater in consolidated (fractured) media than in unconsolidated media, as the pore-size distribution is wider in the consolidated rocks and thus the distribution of velocities is also wider. A greater compactness also leads to greater branching of flow

TABLE 7.1 Values of dispersivity (α_L and α_T) in some rock types estimated from field studies (after Fetter, 1988; Marsily, 1986)

Rock	*Location*	*Dispersivity (m)*
Basalt	Hanford, Washington, USA	$\alpha_L = 30$ $\alpha_T = 20$
Basalt	Snake River, Idaho, USA	$\alpha_L = 91$ $\alpha_T = 137$
Limestone	–	$\alpha_L = 152$
Alluvium	–	$\alpha_L = 10–60$ $\alpha_T = 5–50$

lines, hence greater lateral dispersion (Rasmuson and Neretnieks, 1981). It can be also noted from Table 7.1 that the fractured basalt of Idaho shows greater value of α_T as compared with α_L which is due to fracturing.

Molecular diffusion

This physical phenomenon causes movement of particles (ions or molecules) due to a concentration gradient between two neighbouring points from zones of high concentration to those of low concentration. Diffusion which is also a dispersion phenomenon is of importance only at lower velocities. Diffusion can take place even if the fluid is stationary as the movement is induced primarily under the influence of the concentration gradient. The molecular diffusion coefficient decreases with a decrease in temperature.

The mass flux of diffusing substance is proportional to the concentration gradient and is given by Fick's first law:

$$F = -D_o \frac{dC}{dx} \qquad (7.65)$$

where F = mass flux, i.e. the mass of solute per unit area per unit time; D_o = molecular diffusion coefficient; C = solute concentration; and dC/dx = concentration gradient. The negative sign would indicate movement of solute from areas of greater concentration to those of lower concentration.

In the case where concentration changes with time, Fick's second law will apply

$$\frac{dC}{dt} = \frac{Dod^2C}{dx^?} \qquad (7.66)$$

where dC/dt = change in concentration with time.

The molecular diffusion coefficient D_o of common ions (Na^+, K^+, Ca^{2+}, Cl^-, HCO_3^-, SO_4^{2-}) in water are in the range of 1×10^{-9} to 2×10^{-9} $m_2^2 s^{-1}$ at 25° C (Freeze and Cherry, 1979).

The molecular diffusion coefficient in porous media, *D*, will be small as compared with that in water due to the resistance in the porous medium by the length and tortuosity of the flow paths. The ratio of D/D_o varies from 0.1 in clayey sands to 0.7 in sands; in crystalline rocks, D/D_o is < 0.02 (UNESCO, 1980).

Molecular diffusion is a slow process. In formations of high permeability, the effects of molecular diffusion are masked by higher velocities but in low-permeability material, such as clay or silt, where groundwater velocities are low, molecular diffusion can be a significant process influencing the distribution of various ions. This also depends on the relative diffusion coefficients of ions and whether they are reactive to the matrix or not. Tritium (^3H) profiles in fractured clayey till show that tritium has moved by active groundwater circulation in fractures up to about 7 to 10 m and has travelled 1 or 2 m beyond that depth by molecular diffusion in the matrix (Ruland *et al.*, 1991).

As mentioned earlier, transport through the rock matrix by advection in fractured rocks is generally insignificant because of its low permeability. However, molecular diffusion from the fractures into the porous blocks will cause attenuation by removing solute (contaminant) mass from the fracture flow, thereby retarding the advance of contaminants along the fracture. Figure 7.17 presents the effect of matrix diffusion on concentration distribution of a non-adsorbing and adsorbing nondecaying solute migrating through a fracture in a non-porous and a porous matrix. In Figure 7.17a, as the matrix is non-porous and dispersion in the fracture is neglected, the concentration profile has a rectangular shape, but in a porous medium diffusion will be prominent which will cause greater retardation of the solute (Figure 7.17b). In tight fractured granite, diffusion would be minimal but in clays where the matrix is porous and has low permeability, solute will diffuse into the rock matrix resulting in a decrease in the concentration of the contaminant with increasing distance along the fracture. Diffusion is therefore an important process causing retardation in the movement of solutes in fractured porous rocks. Tracer study using bromide in fractured clay demonstrated the strong retardation effect due to molecular diffusion (Cherry, 1989). A retardation of about three orders of magnitude is reported from fractured clays (Germain and Frind, 1989). Figure 7.17c shows the added effect of adsorption in the surface of the fracture which will cause further retardation of the migration of contaminant in the fracture. We return to this subject later in this chapter. The presence of fracture coatings may also create a zone of diffusion having different diffusion coefficients from those in the unaltered matrix.

Migration of contaminants has been studied for various fracture openings, spacing and other fracture geometry by a number of workers (Sudicky and Frind, 1982; Vogel and Giesel, 1989; Rowe *et al.*, 1989). Large fracture openings or high velocities will reduce matrix diffusion due to increase in the solute-transported to solute-stored ratio. Studies by Grisak and Pickens (1980) and Sudicky and Frind (1982) show that the fracture spacing will have a significant effect on the advance rate of a contaminant. A small fracture spacing will result in a greater penetration distance along the fractures because of the limited capability of the porous matrix to store solute, resulting in insignificant molecular diffusion in the matrix.

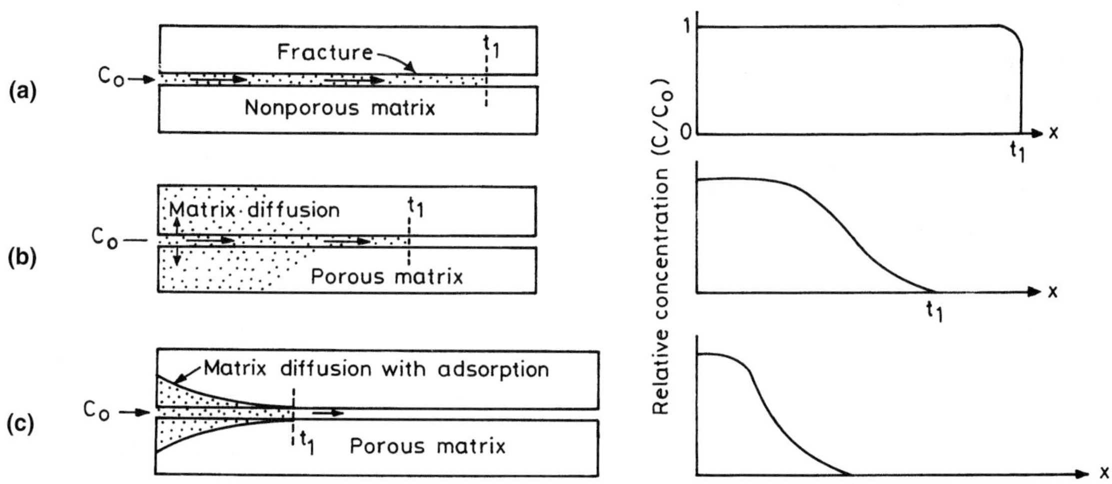

FIGURE 7.17 Effect of advection, matrix diffusion and adsorption on contaminant distribution in fractured medium. (a) Advective transport along a fracture in a nonporous matrix; (b) advective transport with matrix diffusion into porous matrix due to molecular diffusion; (c) advective transport with molecular diffusion and adsorption. Profiles of relative concentration (C/Co) of the contaminant along the fracture at time, t_1 in these cases, are also shown (modified after Vogel and Giesel, 1989).

Migration of contaminants has been also studied for various fracture geometry. A parallel fracture system was considered by Vogel and Giesel (1989). Rowe *et al.* (1989) gave a semi-analytical solution for 1D, 2D and 3D contaminant diffusion into the matrix in a orthogonally fractured system. Their study shows that in fractured shales, three-dimensional effects are minimal for very highly fractured systems (i.e. fracture spacing of 0.01 m or less) but the three-dimensional diffusion can appreciably alter the shape and extent of the contaminant plume for wider fracture spacings (e.g. 1 m).

The transport of solutes in the fractures is also affected by channelling because effective porosity and fracture surface area for diffusive flux is reduced causing a reduction in the attenuation of solute. Therefore, small-scale laboratory experiments may significantly overestimate the effective porosity and specific surface area unless network effects are taken into account. This will lead to overestimation of the capacity of retardation by means of sorption and matrix diffusion (Dverstorp *et al.*, 1992).

The effect of roughness of fracture walls on the transport of solutes is demonstrated by Raven *et al.* (1988). They have shown that wall-irregularities in fractures promote formation of immobile liquid zones (Figure 7.18). In these parts the solutes get stored in the early stages and are released later when the solute concentration in the mobile liquid decreases.

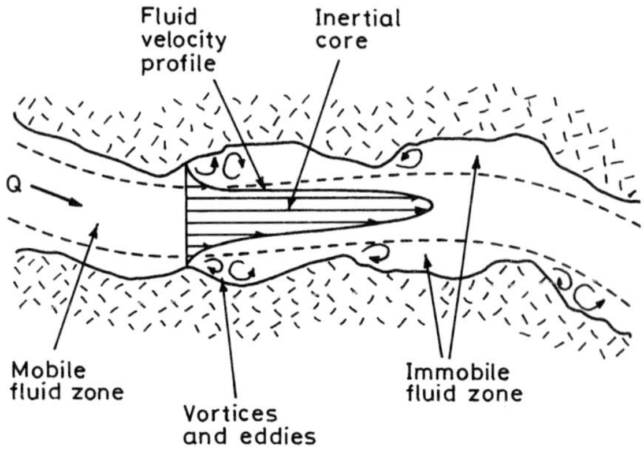

FIGURE 7.18 Zones of mobile and immobile water in a fracture (after Raven *et al.*, 1988).

Retardation

Solutes dissolved in groundwater may undergo chemical changes due to reaction with water and aquifer material by ion exchange, radioactive decay, biodegradation and other processes.

Ion exchange or adsorption reactions are of importance in the movement of contaminants in groundwater systems. The mechanism of ion exchange (adsorption) is further elaborated in section 10.5.1. Adsorption reactions are very rapid relative to the flow of groundwater; in a purely fractured rock, adsorption will take place only on the surface area of the fractures which is relatively small. However, in porous fractured rock, adsorption will also take place on the very large solid surface area within the porous blocks.The extent of adsorption depends on type of ions and composition of matrix material. Adsorption will cause retardation of contaminant front as is evidenced from column studies (Figure 7.17c). The retardation of the solute or the tracer due to adsorption will be much greater by the matrix material due to its greater surface area as compared with the adsorption on the fracture surface. In the case when two tracers are used, one reactive and the other non-reactive, the reactive tracer front will lag behind the non-adsorbed tracer which will spread out as a result of dispersion. Breakthrough curves for simultaneous injection of non-sorbing iodide and sorbing strontium followed by highly sorbing caesium in fractured granite are given in Figure 7.19. The strontium peak is lower than that of iodide which is explained by the adsorption of strontium. The secondary peak of strontium after about 1000 h is related to the injection of caesium which caused elimination of strontium. These investigations therefore demonstrate that migration of a single solute may differ from those involving simultaneous injection of several solutes having different sorbing properties.

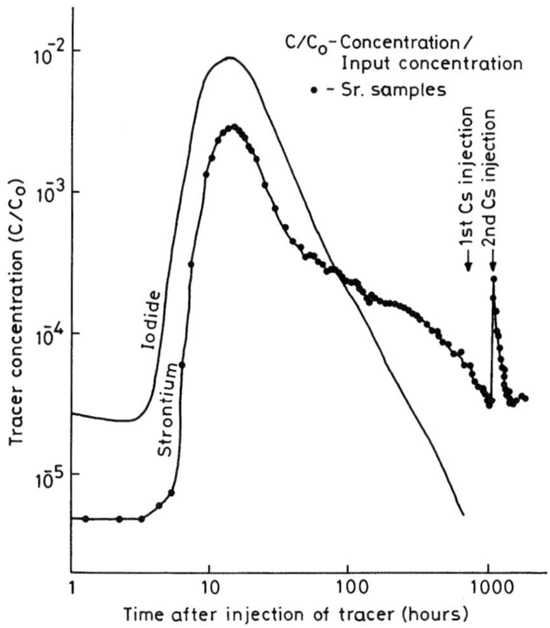

FIGURE 7.19 Breakthrough curves for iodide and strontium following simultaneous injection of a pulse of each solute (initial iodide background elevated by previous experiment) (after Heath, 1985).

Similar trends are also observed by using conservative tracer (uranin) and reactive tracers (^{22}Na and ^{85}Sr) in the crystalline (granodiorite) rock mass at the Grimsel Test Site, Switzerland (Keppler *et al.*, 1996).

Equation (7.67), known as the retardation equation, gives a simplified expression for the retardation of a contaminant (A) in an aquifer (UNESCO, 1980)

$$\overline{V}_A = \frac{\overline{V}_a}{1 + \frac{\rho_b}{\eta} K_d^A} \qquad (7.67)$$

where $\overline{V}_A$ is the average velocity of a contaminant (A); $\overline{V}_a$ is average groundwater velocity; ρ_b is the bulk density; η is the porosity; and K_d^A is the distribution (retardation) coefficient which expresses the ratio of the amount of contaminant (A) adsorbed or precipitated per gram of soil or aquifer material to the amount of the same contaminant remaining in solution, i.e. groundwater (units ml g^{-1}). Consequently, a higher value of K_d will indicate greater adsorption or precipitation of a contaminant.

Equation (7.67) illustrates that if the value of K_d for a particular contaminant increases, there must be a corresponding decrease in the average velocity of that contaminant ($\overline{V}_A$) meaning thereby its retardation relative to the average groundwater velocity. K_d values in different rock types for two important radioactive elements (strontium and caesium) which have long-lived radionuclides of great toxicity, are given in Table 7.2.

For fractured media, the retardation equation (7.67) becomes (Freeze and Cherry, 1979)

$$\frac{\overline{V}_a}{\overline{V}_A} - = \frac{1 + 2K_d^A}{a} \qquad (7.68)$$

where $\overline{V}_a/\overline{V}_A$ is known as the relative velocity and a is the fracture aperture. Equation (7.68) indicates that retardation decreases with increase in fracture aperture. This equation will be valid only when rock matrix is non-porous. In the case of porous rock matrix adjacent to fractures, a part of the contaminant will migrate by molecular diffusion from the fracture into the rock matrix thereby causing attenuation of the contamination in the fracture as discussed earlier.

The equation (7.68) assumes an uniform value of retardation coefficient which will not

TABLE 7.2 Measured field K_d values (in ml g^{-1}) for caesium and strontium (after UNESCO, 1980)

Material	*Element* *Cs*	*Sr*
Granite	34.0	1.7
Limestone	13.5	0.2
Basalt	792–9520	16–135
Sandstone	102.0	1–4
Sand	53–523	4–19
Alluvium	450–950	40.0

be true in the case of heterogeneous fractured media, due to variation in fracture aperture and orientation, which will not only influence the advective flow but also the retardation factor (Wels and Smith, 1994).

Krishnamoorthy *et al.* (1992) computed the concentration and flux of radionuclides commonly encountered in high-level waste (HLW) assuming certain parametric values in fractured granite (Figure 7.20). It was noted that various nuclides show varying trends depending on their K_d values. For example, ^{129}I reaches a peak value of 3×10^{-6} Bq ml^{-1} after about 300 yr but does not show any marked decrease due to lack of adsorption. ^{239}Pu (K_d = 500 ml g^{-1}) and ^{237}Np (K_d = 100 ml g^{-1}) reach peak concentration around 2×10^5 yr; ^{239}Pu decays faster but ^{237}Np continues to have significant concentration even after one million years due to relatively low K_d (1.7 ml g^{-1}). The variation in their peak concentrations is also explained due to differences in K_d values. For example ^{90}Sr shows a peak much earlier (around 400 yr) due to its low K_d (1.7 ml g^{-1}).

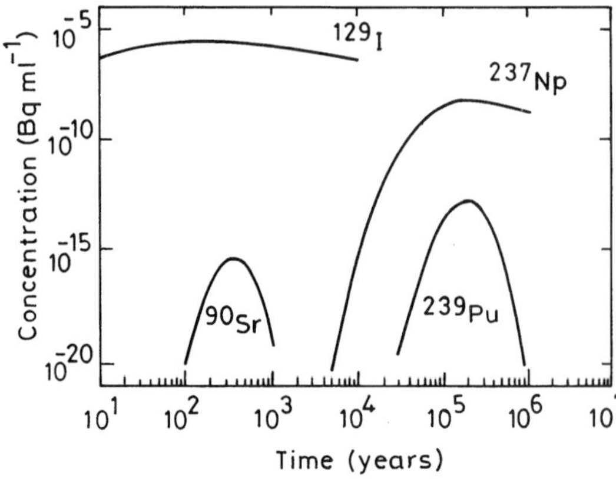

FIGURE 7.20 Concentration profiles of some typical radionuclides. Q = 1Bq, Z = 10 m, Va = 10 cm^{3}yr^{-1}, η = 0.005, a = 0.02 cm (after Krishnamoorthy *et al.*, 1992).

Further, the above studies also indicate that most of the radionuclides will decay to insignificant levels at distances of over 50 m or so. However, even this small fraction may lead to high concentrations of those radionuclides which have low K_d and a long half-life due to lower flux of water into the fractures.

Chapter eight

Hydraulic properties of rocks

8.1 Basic concepts and terminology

Hydraulic properties of water bearing formations are important as they govern their groundwater storage and transmitting characteristics. These are described below.

Porosity (η)

Porosity, η is a measure of the interstices or voids present in a rock formation. It is defined as the ratio of volume of voids, V_v to the total volume, V of the rock mass (equation 8.1)

$$\eta = \frac{V_v}{V} \tag{8.1}$$

Porosity, (η) is given as a percentage or as a decimal fraction. Porosity is of two types, primary and secondary. Primary porosity is the inherent character of a rock which is developed during its formation. Secondary porosity is developed subsequently due to various geological processes, viz. fracturing, weathering and solution activity. In unconsolidated rocks, primary porosity is of importance but in hard rocks secondary porosity is of greater significance.

Porosity is controlled by: (a) shape and arrangement of constituent grains, (b) degree of sorting, (c) compaction and cementation, (d) fracturing, and (e) solution. The geometrical arrangement of constituent grains (packing) and sorting have an important influence on porosity. Well-sorted clastic material has high porosity irrespective of grain size. In poorly sorted material, porosity is less as small-size grains occupy pore spaces between bigger grains. Compaction and cementation reduces porosity. In unconsolidated formations, porosity at deeper levels will be less due to compaction, e.g. shales have lower porosity than clays.

The fractured rock formations are made up of two porosity systems, (a) the intergranular porosity or matrix porosity (η_m), formed by intergranular void spaces, and (b) the secondary porosity developed due to fractures and solution cavities, termed as fracture porosity, (η_f). Therefore, in fractured rocks, the total porosity is the sum of matrix and fracture porosities, i.e.

$$\eta = \eta_m + \eta_f \tag{8.2}$$

The two porosities can be expressed in the conventional manner as

$$\eta_m = \frac{\text{Matrix void volume}}{\text{Total bulk volume}}$$

$$\eta_f = \frac{\text{Fracture void volume}}{\text{Total bulk volume}}$$

In the laboratory, porosity of rock samples can be estimated by immersion in liquids, density test and by gas-porosity meters (UNESCO,

1984b). In the field, geophysical logging methods, viz. resistivity, neutron and gamma methods can be used for determining porosity.

In the field, fracture porosity can be estimated from the scan-line method by the relation $\eta_f = Fa$, where F is the number of joints per unit distance intersecting a straight scan line across the outcrop and a is the mean aperture of fractures. The porosity of natural materials may range from almost zero in hard massive rocks to as much as 60% in clays (Table 8.1).

Laboratory measurements on a variety of fractured rocks show that fracture porosity η_f is considerably less than the matrix porosity. η_m is reported to generally vary from 0.1% to 8% and η_f from 0.001% to 0.01% (Lee and Farmer, 1993).

Void ratio (e)

This is generally used in soil mechanics and is expressed as

$$e = \frac{V_v}{V_s} \tag{8.3}$$

TABLE 8.1 Representative values of porosity (η), specific yield (S_y) and specific retention (S_r) of geological materials (after Morris and Johnson, 1967; Hamill and Bell, 1986)

Geological formation	*η(%)*	*S_y(%)*	*S_r(%)*
Unconsolidated deposits			
Gravel	28–34	15–30	3–12
Sand	35–50	10–30	5–15
Silt	40–50	5–20	15–40
Clay	40–60	1–5	25–45
Dune sand	40–45	25–35	1–5
Loess	45–50	15–20	20–30
Rocks			
Sandstone	15–30	5–25	5–20
Limestone, dolomite	10–25	0.5–10	5–25
Shale	0–10	0.5–5	0–5
Siltstone	5–20	1–8	5–45
Till	30–35	4–18	15–30
Dense crystalline rock	0–5	0–3	–
Fractured crystalline rock	5–10	2–5	–
Weathered crystalline rock	20–40	10–20	–
Basalt	5–30	2–10	–

where V_S is volume of mineral grains and V_v is as defined earlier. Void ratio has large numerical variation; in natural soils, where total porosity ranges from 0.3 to 0.6, the corresponding void ratio range is 0.45 to 1.5. The relation between total porosity η and void ratio e can be expressed as

$$\eta = \frac{e}{1+e}, \text{ or } e = \frac{\eta}{1-\eta} \tag{8.4}$$

Specific retention (S_r)

This is a measure of the volume of water retained by the rock material against gravity on account of cohesive and intergranular forces. It can be expressed as

$$S_r = \frac{V_r}{V} \tag{8.5}$$

where V_r is the volume of water retained. Specific retention depends on the **specific surface** of constituent mineral grains and in turn is influenced by the grain size, shape and type of clay minerals present. Specific surface is defined as the area per unit weight of the material and is expressed in m^2g^{-1}. The specific surface values of coarse grained material, such as gravel and sand, is small compared with silt and clay size fractions. Among clay minerals, non-swelling clays have specific surfaces in the range of 10 to 30 m^2g^{-1}, but swelling clays such as montmorillonite have large values of about 800 m^2g^{-1}. A similar property is the specific surface area, (S_{Sp}), defined by

S_{Sp} = Total surface area of the interstitial voids/Total volume of the medium

S_{sp} has the dimensions of L^{-1}. In fine-grained material, S_{sp} will be more, viz. in sands, S_{sp} will be of the order of 1.5×10^4 m^{-1} but in montmorillonite, it is about 1.5×10^9 m^{-1} (Marsily, 1986). These properties are of importance in the adsorption of water molecules and ions on the surfaces of mineral grains, especially on clay minerals.

Specific yield (S_y)

This is defined as the ratio of the volume of water that an unconfined aquifer will release from storage by gravity, to the total volume of fully saturated aquifer material. It is expressed either as a decimal fraction or as a percentage. Specific yield depends on the duration of drainage, temperature, mineral composition of water, grain size and other textural characteristics of aquifer material. Values of specific yield (S_y) of some common rock materials are given in Table 8.1.

Effective porosity (η_e) or kinematic porosity

This is the same as specific yield (Marsily, 1986). Fine-grained and poorly sorted materials have low effective porosity as compared with coarse-grained and well-sorted material, due to the greater retention of water on account of intergranular forces. Instead of total porosity, the effective porosity (η_e) is more important for estimating the average velocity of groundwater and transport of contaminants as discussed in sections 7.2 and 7.3.

In crystalline and other hard rocks, the size and interconnection of fractures are mainly responsible for imparting effective porosity to the rock mass. In such rocks, although total porosity may be high but due to unconnected fractures, the effective or kinematic porosity will be less. Similarly, in dolomites, which are formed as a result of diagenesis, although the rock may acquire high secondary porosity due to reduction in volume of mineral grains, its effective porosity will be less.

Hydraulic conductivity (*K*)

This is a measure of the ability of a formation to transmit water. It has the dimensions of LT^{-1} and is usually expressed in ms^{-1}.

In terms of Darcy's law (equation 7.12), *K* can be expressed as

$$K = \frac{V}{dh/dl} \tag{8.6}$$

where *K* is hydraulic conductivity, *V* is groundwater velocity, and dh/dl is hydraulic gradient.

In the USA, *K* was earlier expressed in two forms – as the Meinzer's (laboratory) coefficient of permeability, K_m and also Field coefficient of permeability (K_f), both using units of us gal. d^{-1} ft^{-2}. The main difference between the two is that K_m is expressed at a constant temperature of 60 °F (15.6 °C) while K_f is measured at the actual field temperature. As field conditions do not influence the groundwater temperature to any significant extent, the distinction between K_m and K_f has now been discarded.

The hydraulic conductivity depends both on the properties of the medium (rock material) as well as of the fluid. In sedimentary formations, grain-size characteristics are most important as coarse-grained and well-sorted material will have high hydraulic conductivity as compared with fine-grained sediments like silt and clay. Increase in degree of compaction and cementation reduces hydraulic conductivity. In hard (fractured) rocks, *K* depends on density, size and interconnection of fractures (section 8.2).

Permeability (k)

This is a more rational concept than hydraulic conductivity (*K*) as it is independent of fluid properties and depends only on the properties of the medium. Fluid properties which influence hydraulic conductivity are viscosity (μ), expressing the shear resistance, and specific weight, (γ), expressing the driving force of the fluid.

The relation between hydraulic conductivity, *K* and properties of the medium and the fluid can be expressed as

$$K = \frac{c d_e^2 \gamma}{\mu} = \frac{k\gamma}{\mu} \tag{8.7}$$

where c is a dimensionless constant also known as shape factor and d_e is effective diameter; c depends on porosity and packing etc. Equation (8.7) when substituted in Darcy's equation gives

$$k = \frac{\mu\, Q/A}{\gamma\,(dh/dl)} \qquad (8.8)$$

The value of k can be given in darcy units which in terms of equation (8.8) is expressed as

$$1\ \text{darcy} = \frac{[1\,(\text{cm}^3/\text{sec})/\text{cm}^2]\,1\ \text{cP}}{1\ \text{atm/cm}} \qquad (8.9)$$

Thus, a porous saturated medium will have a permeability of one darcy if a fluid of 1 cP viscosity will flow through it at a rate of 1 cm^3s^{-1} per cm^2 cross-sectional area under a pressure or equivalent hydraulic gradient of 1 atm per cm.

According to equation (8.9), k has the units of area. As the value of k is very small, this is also expressed in square micrometres (μm^2).

$$1\mu m^2 = 10^{-12}\ m^2$$

By substitution of appropriate units in equation (8.9), it can be shown that

$$1\ \text{darcy} = 0.987\ \mu m^2 = 10^{-8}\ cm^2 = 10^{-5}\ m\ s^{-1}\ \text{(approx.)}$$

The range of values of hydraulic conductivity, *K* and permeability, k, are given in Table 8.2. Conversion factors for the various common units of *K* and k are given in the Appendix. It could be noted from Table 8.2 that the permeability of natural materials has a wide variation. As the permeability of crystalline rocks and other tight formations is usually very small, k in such cases is usually expressed in millidarcy (md) which is approximately equal to 10^{-8} ms^{-1}. The permeability of dense unfractured rocks is usually very low, being up to 1 md, and usually below 0.001 to 0.5 md. Fractures increase the permeability above the solid rock mass by several orders of magnitude. The fracture permeability can be 100 and even 1000 md.

TABLE 8.2 Range of values of hydraulic conductivity and permeability for various types of geological materials

Hydraulic conductivity, K (ms⁻¹)	1	10^{-1}	10^{-2}	10^{-3}	10^{-4}	10^{-5}	10^{-6}	10^{-7}	10^{-8}	10^{-9}	10^{-10}	10^{-11}	10^{-12}	10^{-13}
Permeability, k (darcy)	10^5	10^4	10^3	10^2	10	1	10^{-1}	10^{-2}	10^{-3}	10^{-4}	10^{-5}	10^{-6}	10^{-7}	10^{-8}
Relative values	*Very high*		*High*			*Moderate*			*Low*			*Very low*		

Representative materials
Unconsolidated deposits
Gravel
Clean sand
Silty sand
Clay till (often fractured)
Rocks
Shale & siltstone (unfractured)
Shale & siltstone (fractured)
Sandstone
Sandstone (fractured)
Limestone & dolomite
Karst limestone & dolomite
Massive basalt
Vesicular & fractured basalt
Fractured & weathered crystalline rock
Massive crystalline rock

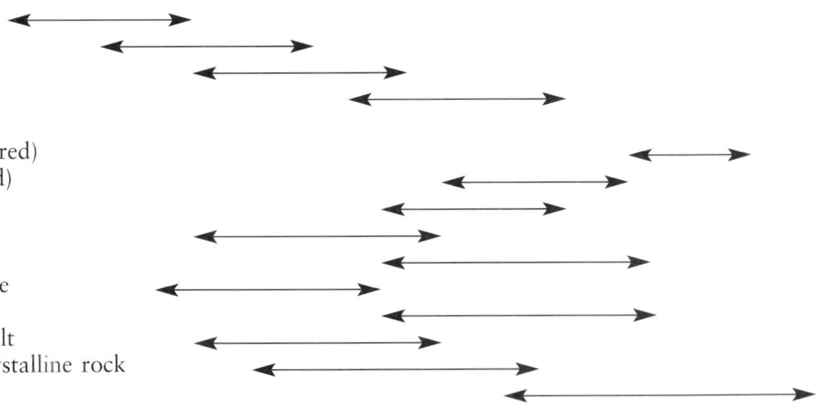

Transmissivity (T)

This is defined as the rate of flow of water at the prevailing field temperature under a unit hydraulic gradient through a vertical strip of aquifer of unit width, and extending through the entire saturated thickness of the aquifer (Figure 8.1). Transmissivity (T) has dimensions of L^2T^{-1} and is expressed in m^2d^{-1} or m^2s^{-1}. Darcy's law, in terms of T, can be written as

$$Q = TIL \tag{8.10}$$

where Q = rate of flow, I = hydraulic gradient, L = width of the flow section, measured at right angles to the direction of flow.

In the confined aquifer, T = Kb, where b is the saturated thickness of the aquifer. In the unconfined aquifer, the saturated thickness will be less than the true thickness. Here it is assumed that K is isotropic and constant across the thickness of the aquifer which may be horizontal or dipping. Transmissivities greater than

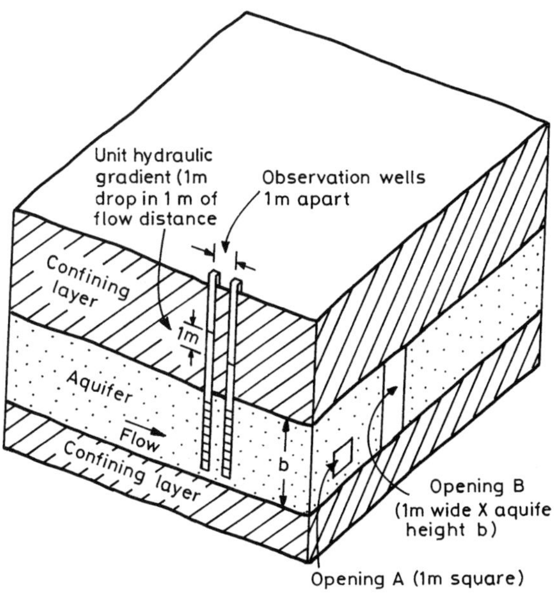

FIGURE 8.1 Diagram illustrating coefficients of hydraulic conductivity (K) and transmissivity (T). Flow of water through opening A will be equal to K and that through opening B equal to T.

1000 m^2d^{-1} represent good aquifers for groundwater exploitation. In the geothermal reservoir, transmissivity is usually expressed in terms of permeability-thickness (kb) in units of d-m (1 m^3= 10^2 d-m). We return to this subject with respect to geothermal reservoirs in Chapter Sixteen.

Storativity (S)

Storativity of an aquifer is defined as the volume of water which a vertical column of the aquifer of unit cross-sectional area releases from storage as the average head within this column declines by a unit distance (Figure 8.2). It is therefore dimensionless. In a confined aquifer, where water released from or taken into storage is entirely due to compressibility of the aquifer and water, the storage coefficient is given by S = bS_s, where S_s is the specific storage, defined later. Value of storativity in a confined aquifer is of the order of 10^{-3} to 10^{-6}. In an unconfined aquifer, storativity S, is given by $S = S_y + bS_s$. Usually $S_y \gg bS_s$, thus storativity of an unconfined aquifer for all practical purposes is regarded equal to its specific yield or effective porosity, (η_e) (Hantush, 1964). Storativity in unconfined aquifers ranges from 0.05 to 0.30.

The relation between storativity and the compressibility of the aquifer material and of water can be expressed as

$$S = \gamma \eta b \left(\beta + \frac{\alpha}{\eta} \right) \tag{8.11}$$

where β is the compressibility of water (4.7 × 10^{-10} Pa^{-1}), α is the compressibility of solid skeleton of the aquifer and S, η and b are defined earlier.

Specific storage (S_S)

This is the volume of water which a unit volume of the confined aquifer releases from storage because of expansion of water and compression of the aquifer under a unit decline

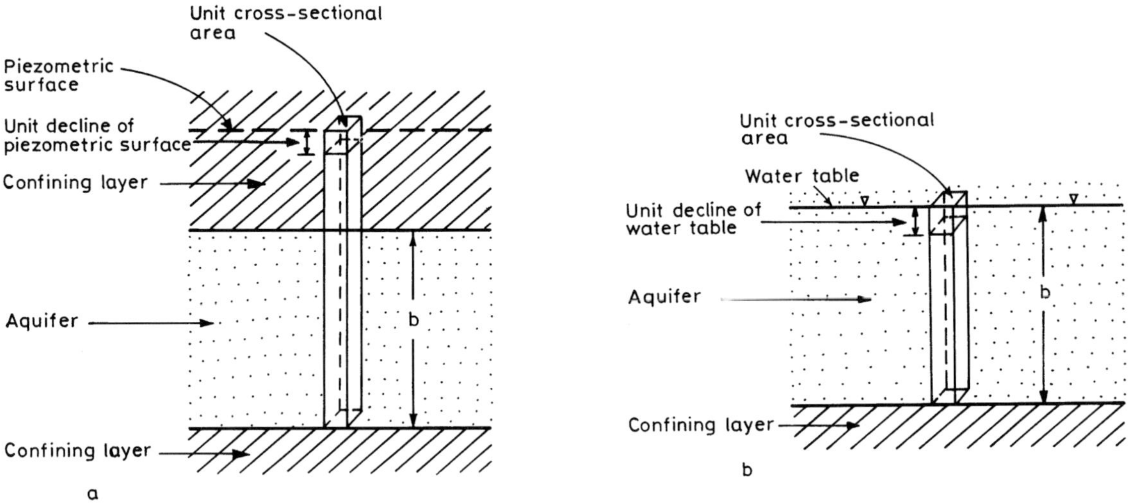

FIGURE 8.2 Diagrams illustrating coefficient of storage (S) for: (a) confined aquifer, and (b) unconfined aquifer (after Ferris *et al.*, 1962).

in the average hydraulic head. It has the dimension of L^{-1}. S_s is used exclusively in confined aquifer analysis, as in an unconfined aquifer the water released from storage is mainly due to gravity drainage and not due to the compressibility of aquifer material or of water. S_s is a more fundamental parameter than S as the latter depends upon both the specific storage and the aquifer geometry. In terms of equation (8.11)

$$S_s = \gamma\,(\alpha + \eta\beta) \qquad (8.12)$$

Therefore, S_S depends both on coefficient of compressibility (α) and porosity, (η) of the rock. As hard dense rocks have low porosity and low coefficient of compressibility, S_s is also less as compared with sands and clay formations. Some representative values of α and S_s are given in Table 8.3.

Hydraulic diffusivity (κ)

This is a single formation characteristic that couples the transmission properties, *K* and storage property, S_s or alternately T and S.

$$\kappa = T/S = K/S_s \qquad (8.13)$$

κ has dimensions of $L^2\,T^{-1}$. It is a significant property of the medium for transient flow. A comparison of measured κ in rocks of low permeability under varying stress is illustrated in Figure 8.3. The equivalent depth is also given. It could be noted that in argillaceous materials, the hydraulic diffusivity κ is generally less (10^{-9}–10^{-7} m^2s^{-1}) than in crystalline rocks, viz. gabbros and granites (10^{-7}–10^{-5} m^2s^{-1}), which appears to be due to the low S_s of crystalline rocks.

Leakage coefficient or leakance (dimensions T^{-1})

This is the property of the semi-confining (aquitard) layer. It is equal to K'/b', where K' and b' are the vertical hydraulic conductivity and thickness of the aquitard respectively.

Hydraulic resistance (C)

This is the reciprocal of leakage coefficient. It indicates resistance against vertical flow in an aquitard. It is equal to b'/K' and has the dimensions of time. If hydraulic resistance C = ∞, the aquifer is confined.

TABLE 8.3 Range in values of the coefficient of compressibility of solid material (α) and specific storage (S_s) (after Domenico, 1972; Streltsova, 1977)

Rock type	α *(Pa^{-1})*	S_s *(m^{-1})*
Dense rock	$10^{-12} - 10^{-10}$	$10^{-7} - 10^{-5}$
Fissured and jointed rock	$10^{-10} - 10^{-9}$	$10^{-5} - 10^{-4}$
Sand	$10^{-9} - 10^{-8}$	$10^{-4} - 10^{-3}$
Clay	$10^{-8} - 10^{-7}$	$10^{-4} - 10^{-2}$

Leakage factor (B)

This determines the distribution of leakage through an aquitard into a leaky aquifer. It is defined as

$$B = \sqrt{TC} \qquad (8.14)$$

Leakage factor has dimensions of length and is expressed in metres. High values of B indicate greater resistance of the semi-previous strata to leakage.

Boulton's delay index (1/α)

This is a measure of the delayed drainage of an unconfined aquifer. It has the dimensions of time. The value of $1/\alpha$ may vary from about 50 min in coarse sand to 4000 min in silt and clay.

Drainage factor (D)

Drainage factor ($D = \sqrt{T/\alpha Sy}$) is a property of unconfined aquifer. It has the dimensions of length and is usually expressed in metres. Large values of D indicate fast drainage. If $D = \infty$, the yield is instantaneous with the lowering of the water-table, i.e. the aquifer is unconfined without delayed yield.

Storativity ratio (ω)

This and the **interporosity coefficient** (λ) are the properties of fractured aquifers are described in section 7.1.4. Methods of estimating the hydraulic properties are described in Chapter Nine.

8.2 Hydraulic conductivity of fractured media

Fractures control the hydraulic characteristics of low permeability rocks, viz. crystalline, volcanic and carbonate rocks. Also in some clastic sedimentary formations, viz. sandstones, shales, glacial tills and clays, fractures form the main pathways for movement of fluids and contaminants.

In fractured rocks, a distinction can be made between hydraulic conductivity of fracture, K_f and of intergranular (matrix) material, K_m. As

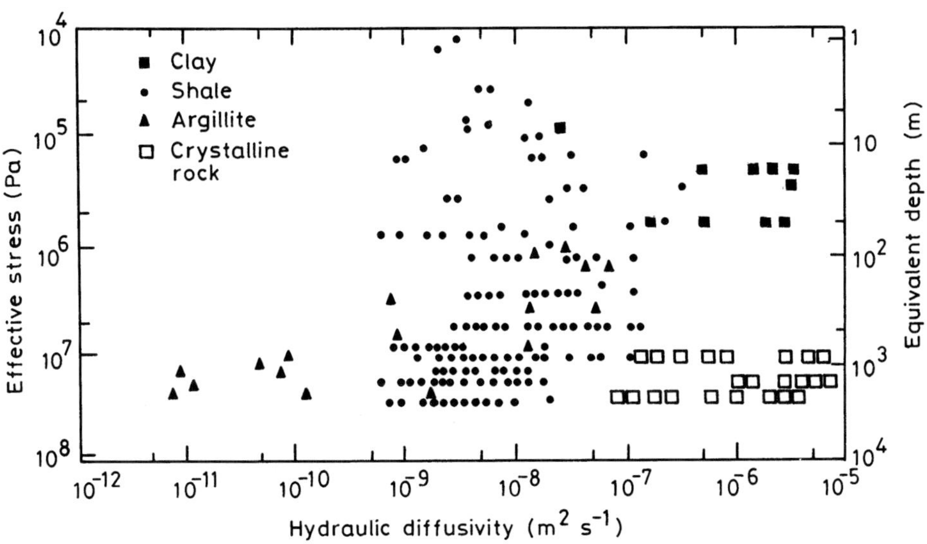

FIGURE 8.3 Measured values of hydraulic diffusivity (κ) for a variety of low-permeability rocks plotted against effective stress and the equivalent depth (after Neuzil, 1986).

fractures form the main passage for the flow of water, the hydraulic conductivity of fractured rocks mainly depends on the fracture characteristics described in Chapter Two. The role of some important parameters, e.g. aperture, spacing, stress, infilling (skin), connectivity etc. is discussed below.

8.2.1 Relationship of hydraulic conductivity with fracture aperture and spacing

The relationship between hydraulic conductivity (K_f) of a single plane fracture with aperture (a) is given by equation 8.15.

$$K_f = \frac{\gamma a^2}{12\mu} \tag{8.15}$$

The equivalent hydraulic conductivity of a rock mass, (K_s) with one parallel set of fractures is expressed by

$$K_s = \frac{a}{s}K_f + K_m = \frac{\gamma a^3}{12s\mu} + K_m \tag{8.16}$$

where s is fracture spacing. Usually K_m is very low except when the rock matrix is porous and/or fractures are filled with impervious material. Therefore

$$K_s = \frac{\gamma a^3}{12s\mu} = \frac{ga^3}{12\nu s} \tag{8.17}$$

where g is gravitational acceleration (981 cm sec^{-2}) and ν is the coefficient of kinematic viscosity which is 1.0×10^{-6} m^2s^{-1} for pure water at 20 °C.

In fractures with infillings, the hydraulic conductivity of the fracture will depend on the permeability of the filling material, assuming that this permeability is still significantly greater than that of the rock matrix.

The hydraulic conductivity of a rock mass with three orthogonal joint sets with similar spacing and constant aperture in all directions in the three dimensional space is given by equation (8.18) (Lee and Farmer, 1993)

$$K_s = \frac{2\gamma a^3}{12s\mu} + K_m \tag{8.18}$$

Figure 8.4 shows fractures with a common spacing of one joint per metre but with different apertures. In Figure 8.4a fracture apertures a_1 = 0.1 cm, and in Figure 8.4b a_2 = 0.01 cm. Substituting these values in equation (8.17) will give equivalent hydraulic conductivity of K_1 to be 8.1×10^{-4} ms^{-1} and K_2 for the second type will be 8.1×10^{-7} ms^{-1}, i.e. K_2 will be about three orders of magnitude less than that of set 1.

Figure 8.5 gives the hydraulic conductivity values of fractures with different apertures and frequencies. It can be seen that one fracture per metre with an aperture of 0.1 mm gives rock hydraulic conductivity of about 10^{-6} ms^{-1}, which is comparable to that of porous sandstone. With a 1 mm aperture and the same spacing, the hydraulic conductivity will be 10^{-3} ms^{-1}, similar to that of loose clean sand.

The equivalence between hydraulic conductivity in a fractured rock and that of porous material is depicted in Figure 8.6. As an example, the flow from a 10 m thick cross-section of a porous medium with a hydraulic conductivity of 10^{-4} m s^{-1} could be the same as from one single fracture with an aperture of about 1 mm. This demonstrates the large amount of flow which can be expected from fractures of even small openings.

A distinction is made between real or mechanical aperture (a_r) and conducting or hydraulic aperture (a_c). The real aperture is usually larger than the conducting aperture if the fracture surfaces are rough. In smooth and wide fractures, the mechanical aperture and the conducting aperture will be equal. The empirical relation between a_r and a_c can be expressed as (Lee *et al.* 1996)

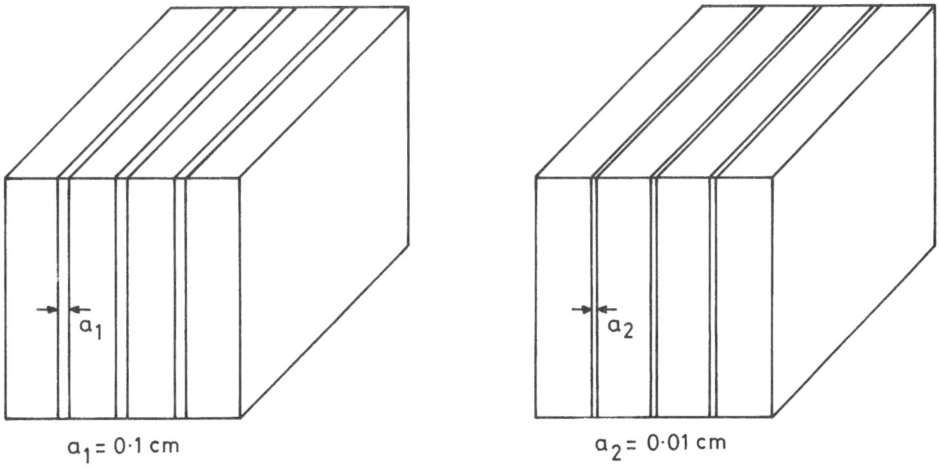

FIGURE 8.4 Two sets of fractures with same spacing but different apertures.

$$a_c = \frac{a_r^{\,2}}{JRC^{2.5}} \qquad (8.19)$$

where a_r and a_c are in μm and JRC is joint roughness coefficient, having a range from 20 (roughest surface) to 0 (smoothest surface).

Assuming a maximum initial conducting aperture (a_c) equal to 350 μm and minimum initial conducting aperture to be 50 μm and by considering other mechanical properties of similar rock types, Lee and Farmer (1993) obtained two curves showing variation in conducting aperture with depth (Figure 8.7). These curves indicate a good agreement between the range of calculated and observed values. A similar trend of variation in fracture aperture was obtained by Oda *et al.* (1989).

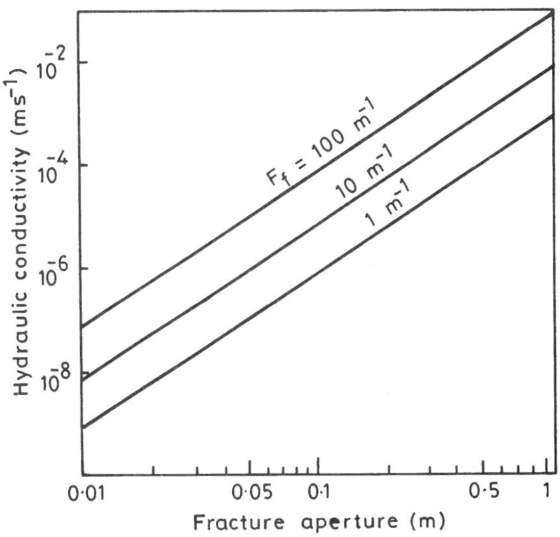

FIGURE 8.5 Variation in hydraulic conductivity with fracture frequency and conducting aperture (after Lee and Farmer, 1993).

8.2.2 Effect of stress on permeability

The mechanical behaviour and fluid flow in fractures is greatly influenced by the effective stress which is taken to be the normal stress on fracture minus the fluid pressure. Effective stress values are usually positive but in some cases as in hydrofracturing, where fluid pressure exceeds the normal stress, effective stress values will be negative.

The effect of stress on permeability is of importance in geotechnical and nuclear waste disposal studies and recovery of oil, gas and geothermal fluids from reservoirs. It is shown, both by theory and in experimental studies, that stress reduces fracture aperture and thereby the permeability of fractured rocks. Stress being a

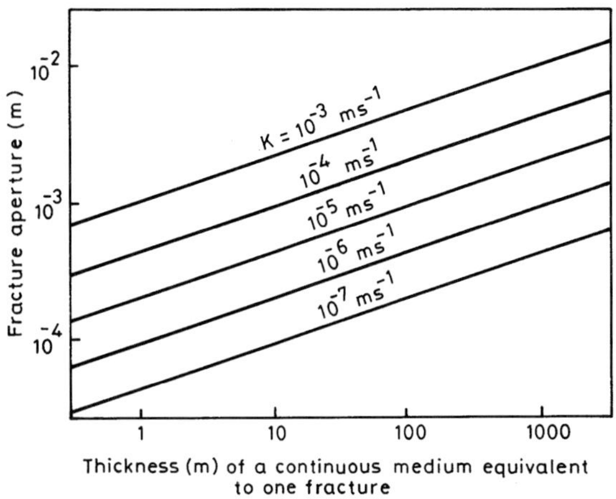

FIGURE 8.6 Comparison between the hydraulic conductivity of the porous medium and that of the fractured medium as a function of aperture (after Maini and Hocking, 1977, reproduced with permission of the Geological Society of America, Boulder, Colorado, USA).

directional phenomenon, its state determines the relative permeability of different fracture sets in a rock mass. It has been shown that fractures parallel to the maximum stress tend to be open, whereas those perpendicular to it tend to be closed.

Several experiments have been designed to estimate permeability variation in a variety of rocks under varying stress and thermo-mechanical conditions. (Brace, 1978; Gale, 1982a, b; Oda *et al.*, 1989; Read *et al.*, 1989; Jouanna, 1993; Azeemuddin *et al.*, 1995). Snow (Gale, 1982a) proposed an empirical model of the form of

$$k = ko + (K_n a^2/s)(P - Po) \qquad (8.20)$$

where k is the permeability of horizontal fractures after loading, ko is the permeability at an initial pressure Po, K_n is the normal stiffness of the fracture and a and s are defined earlier.

Lee and Farmer (1993) quoting the work of Brace *et al.* showed a decrease in permeability of Westerly Granite from 350 nd at 10 MPa pressure to 4 nd at 400 MPa pressure (1 nd = 10^{-18}m^2). In Berea sandstone, a decrease in permeability from 10^{-10}m^2 to 10^{-11}m^2 due to increase in hydrostatic pressure from ambient to 30 MPa is reported by Read *et al.* (1989). The permeability reduction from a uniaxial strain test on Berea sandstone was estimated to be 20% but it was drastic (75%) in Indiana limestone which is attributed to pore collapse (Azeemuddin *et al.*, 1995).

The effect of stress on permeability of jointed rocks depends on the direction of stress in relation to joint orientation. Laboratory studies on jointed granite show that a uniaxial stress of 12 MPa, parallel with the joint, raised fracture permeability (k_f), whereas 3 MPa, normal to the joint, decreased k_f to half of the initial value. However, even at the highest pressures (100 MPa), the permeability of the rock containing joints was at least a factor of 10^3 higher than the matrix permeability (k_m) (Brace, 1978).

Studies indicate that when normal stress is applied to a natural fracture in the laboratory, there occurs a reduction in permeability indicating fracture deformation. The change in fracture permeability is higher in the initial stages (Figure 8.8). Further, varied responses of permeability during compression and decompression are also indicated (Van Golf-Racht, 1982; Anon, 1996). After a cycle of compression and decompression, the rock permeability may either return to the original permeability (Figure 8.8), or may be reduced due to permanent deformation (Figure 8.9). Laboratory experiments also show that in a mica schist, on application of normal stress (applied perpendicular to the foliation plane), the decrease in rate of flow is greater when normal stress increases than when the stress decreases (Figure 8.10).

Although effect of changes in normal stress on fracture permeability has been studied to a large extent, there have been very few controlled studies on the effect of shear stress on fracture permeability. It is expected that shear stress will cause fracture dilation, especially in rough frac-

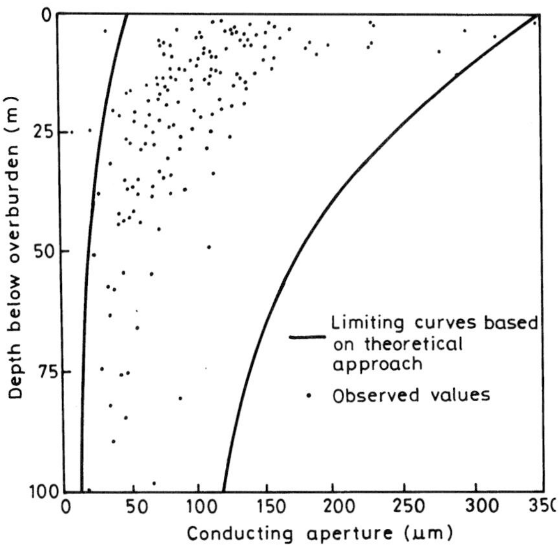

FIGURE 8.7 Relationship between conducting aperture and depth based on field data. The right-handside envelope is based on initial conducting aperture of 350 μm and lefthand side of 50 μm. JCS = 95.8 MNm^{-2}, JRC = 7.4 (after Lee and Farmer, 1993).

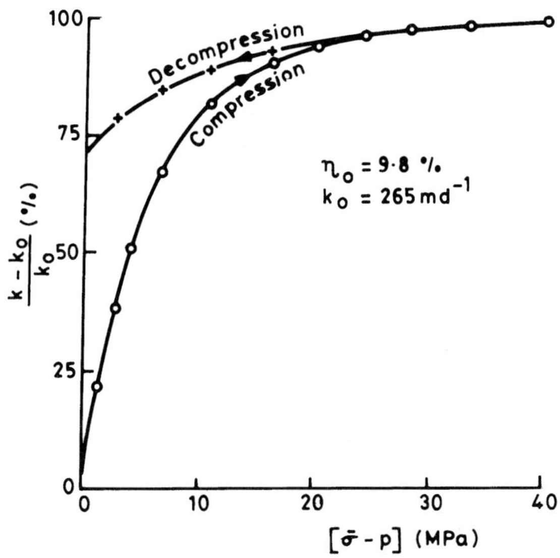

FIGURE 8.9 Change in percentage permeability with stress. Note that in this case the original permeability is not achieved after decompression (after Van Golf-Racht, 1982, with kind permission from Elsevier Science-NL, Amsterdam, the Netherlands).

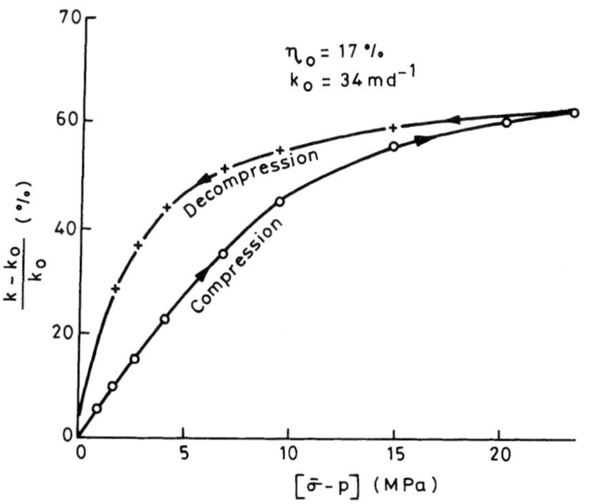

FIGURE 8.8 Change in percentage permeability with stress during compression and decompression in a fractured limestone (after Van Golf-Racht, 1982, with kind permission from Elsevier Science-NL, Amsterdam, the Netherlands).

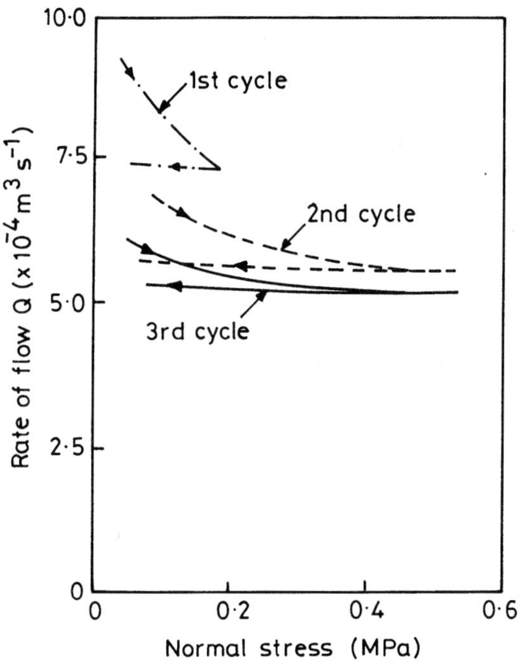

FIGURE 8.10 Effect of normal stress on rate of flow in a mica schist (after Jouanna, 1993).

tures due to displacement under low to moderate normal stress, resulting in significant changes in fracture permeability (Gale, 1982a). In rocks like granites and quartzites, with porosity less than about 5%, the dilatancy effects have been found to be quite conspicuous. In granite, the permeability increased nearly four-fold, while in sandstone the increase was about 10 to 20%, but the permeability of sand decreased considerably. The different behaviour of these materials indicates their varied response to stress (Brace, 1978). Increase in permeability (k) and specific storage (S_s) due to the growth of dilatant cracks is also demonstrated by triaxial test (Read *et al.*, 1989). On the other hand, in soft rocks, like mica schist, an increase in shear stress showed a conspicuous decrease in rate of flow along the schistosity planes (Jouanna, 1993; Figure 8.11).

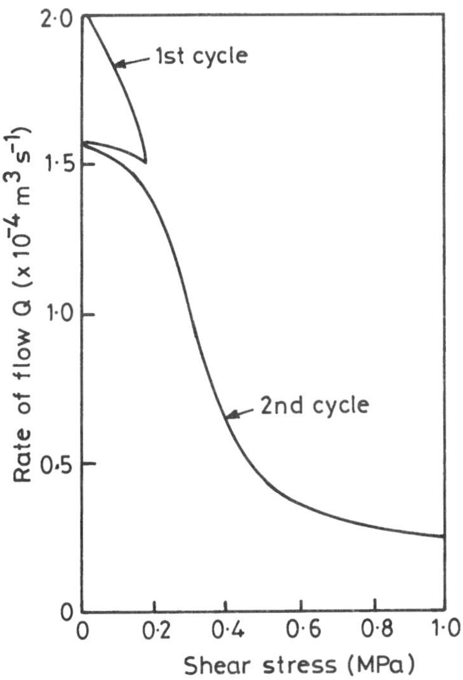

FIGURE 8.11 Influence of shear stress on rate of flow in a mica schist (after Jouanna, 1993).

8.2.3 Relationship of permeability with depth

The depth dependence of permeability can be expressed by equation (8.21) given by Black (1987)

$$k = az^{-b} \tag{8.21}$$

where a and b are constants and z is the vertical depth below ground surface. Based on the data given by Snow (1968b) about the variation in the permeability of fractured crystalline rocks with depth from the Rocky Mountains, USA, Carlsson and Olsson (1977) gave the equation (8.22)

$$K = 10^{-(1.6 \log z + 4)} \tag{8.22}$$

where K is the hydraulic conductivity in ms^{-1} and z is depth in metres. Similar empirical relations have been given by Louis (1974) and others. For example, based on tests in a number of wells in crystalline rocks in Sweden, Burgess (in Lee and Farmer, 1993) gave equation (8.23)

$$\log K = 5.57 + 0.352 \log Z - 0.978 (\log Z)^2 + 0.167 (\log Z)^3 \tag{8.23}$$

where K and Z have the same units as in equation (8.22).

Equation (8.23) can be transformed into equation (8.24) by relating stress and hydraulic conductivity on the basis of $\sigma = \gamma Z$

$$\log K = 5.57 + 0.352 (\log \sigma/\gamma) - 0.978 (\log \sigma/\gamma)^2 + 0.167 (\log \sigma/\gamma)^3 \tag{8.24}$$

The decrease in permeability with depth in fractured rocks is usually attributed to reduction in fracture aperture (Figure 8.7) and fracture spacing (Figure 8.12). A least-square fit to the Packer test data from boreholes in the granites of Stripa mine in Sweden also indicated a general decreasing trend of permeability with increasing depth (Figure 8.13a). The relation-

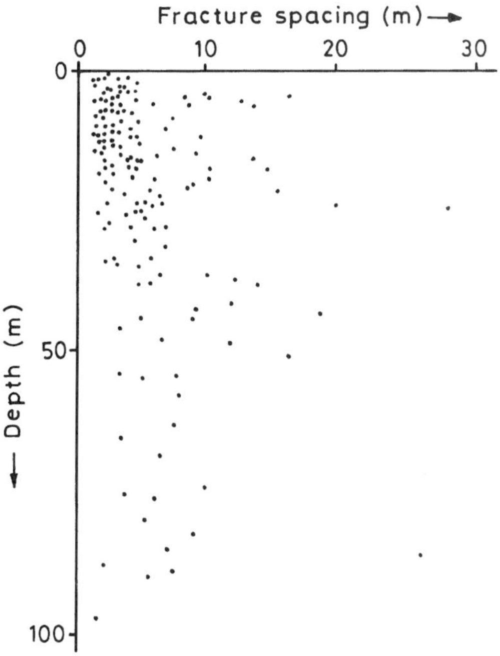

FIGURE 8.12 Variation of fracture spacing with depth (after Snow, 1968a).

ship between fracture frequency and permeability from the same area is illustrated in Figure 8.13b. A decrease in fracture aperture with depth is also reported from several other studies, e.g. recently from a radioactive waste depository in andesite rocks in Taiwan (Lee *et al.*, 1995).

Although a decrease in permeability with increasing depth is demonstrated from several places, this decrease may not be systematic, especially at greater depths (>50 m). The permeability can also vary by several order of magnitude at the same depth (Figure 8.14). Higher permeabilities at shallow depths (<50 m) can be attributed to greater influence of surficial phenomena like weathering etc. and development of sheeting joints due to unloading. Further, fractures at the same depth below the ground surface but with different orientations may be subjected to different stresses and therefore may have different permeabilities.

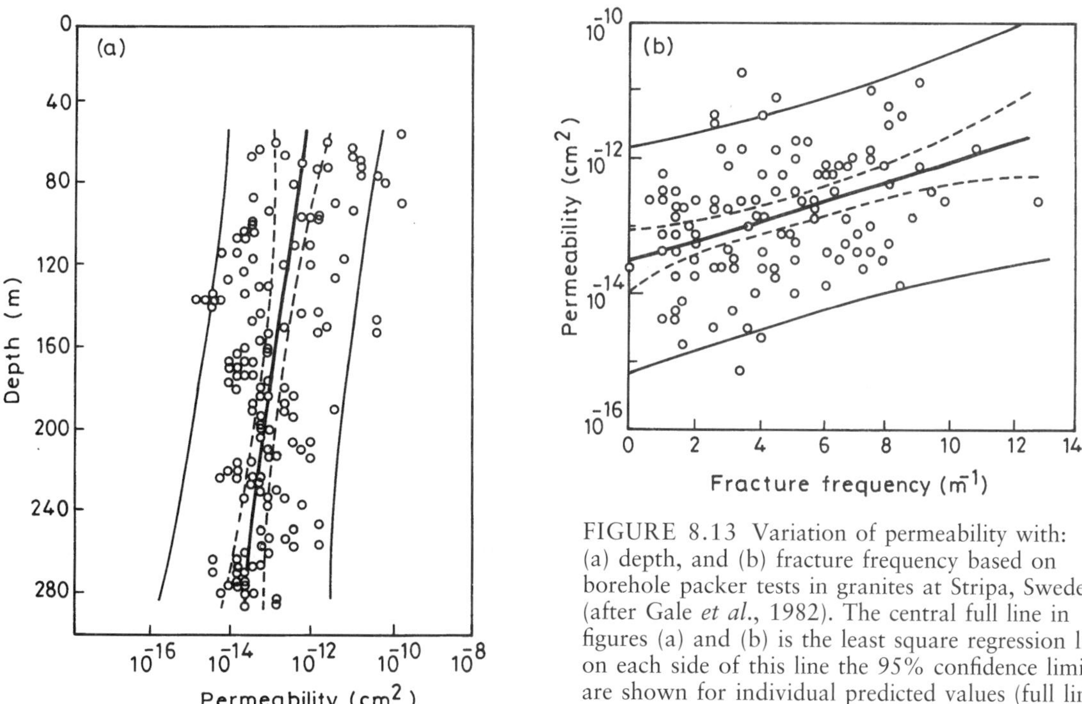

FIGURE 8.13 Variation of permeability with: (a) depth, and (b) fracture frequency based on borehole packer tests in granites at Stripa, Sweden (after Gale *et al.*, 1982). The central full line in figures (a) and (b) is the least square regression line; on each side of this line the 95% confidence limits are shown for individual predicted values (full lines) and for mean predicted values (dashed lines).

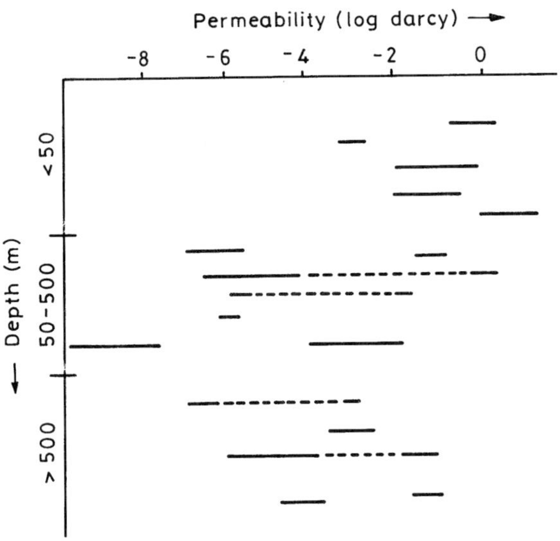

FIGURE 8.14 Ranges of permeability with depth in crystalline rocks (after Brace, 1980).

Even at depths of more than 1000 m, appreciable permeabilities are reported in fractured rocks. For example, Fetter (1988) reported higher permeabilities from fractures at depths of 664 to 1669 m in granitic rocks of Illinois, USA. Recent studies under the Continental Deep Drilling Project in Germany (Kessels and Kuck, 1995) and Hot Dry Rock (HDR) experiments in the Rhine Graben, France (Gerard *et al.*, 1996) also showed the existence of good hydraulic interconnection between adjacent boreholes through fractures even at a depth of more than 3000 m.

On the basis of the above discussion, it may be summarized that although generally in fractured rocks a decrease in permeability with depth is observed, there may be exceptions. Therefore, site specific studies are necessary.

8.2.4 Influence of temperature on permeability

Significant changes in rock temperature can occur due to natural weather conditions, viz. alternate freezing and thawing and due to man-induced changes. Formation of ice in fractures will cause extension of fractures and can also block the movement of water producing pressure build ups. The influence of temperature on rock permeabilities is important for disposal of radioactive waste and in harnessing geothermal energy. An increase in temperature will cause a volumetric expansion of the rock material leading to reduction in fracture aperture and an overall decrease in rock permeability. Studies at Stripa mine in Sweden demonstrated a reduction in permeability of granites by a factor of three when temperature was increased by 25 °C by circulating warm water. Similarly, in another experiment, a tenfold reduction in permeability was observed in a fractured gneiss when the temperature was increased by 74 °C (Lee and Farmer, 1993). On the other hand, thermal contraction causes development of new fractures during hydraulic fracturing in HDR experiments. However, doubts are created that these may not have a significant positive long-term influence on development of geothermal energy from HDR systems (Zhao and Brown, 1992; section 16.3).

8.2.5 Effect of fracture skin on permeability

Fracture surfaces are usually covered with altered or detrital material, such as clay, iron or manganese oxides. These filling materials form fracture skin, which reduce the permeability of fractures and also influence the movement of solutes from the fractures into the matrix blocks. The presence of clay in the fractures may also increase the mechanical deformability of fractures.

8.2.6 Interconnectivity of fractures

The interconnectivity of the different fracture sets is important for deciding the hydraulic continuity, which depends on fracture orientation, density, spacing and fracture size. Interconnectivity can be expressed in terms of the ratio of average fracture spacing to fracture trace length. A ratio in the range of $1/20$ to $1/50$ has been suggested to indicate continuum (Lee and Farmer, 1993). Long and Witherspoon (1985) studied numerically the effect of both the magnitude and nature of the fracture interconnection on permeability. The results showed that for a given fracture frequency, as fracture length increases, the degree of interconnection increases and thereby permeability also increases. Rouleau and Gale (in Lee and Farmer, 1993) suggested an empirical interconnectivity index I_{ij} between two fracture sets given by equation (8.25)

$$I_{ij} = \frac{l_i}{s_i} \sin \theta_{ij} \ (i \# j) \qquad (8.25)$$

where l_i is the mean trace length for set i, s_i is the mean spacing of set i and θij is the average angle between fractures of set i and j.

8.3 Anisotropy and heterogeneity

Geological formations usually do not exhibit uniformity in their texture and structure either spatially or in different directions. Accordingly, their hydraulic characteristics such as hydraulic conductivity and storativity also vary.

Anisotropy is usually a result of the rock fabric. In a porous rock consisting of spherical grains, the hydraulic conductivity will be the same in all directions and therefore it is said to be isotropic (Figure 8.15a). On the other hand in a stratified formation, the constituent mineral grains of tabular nature are usually laid with their longer axes parallel to each other. In such a case the hydraulic conductivity parallel to the bedding plane, K_h, is higher than vertical conductivity, K_v (Figure 8.15b). In fractured rocks, anisotropy is due to the presence of fractures which may have different orientations. The conductivity along the fracture will be significantly higher than that normal to the fractures (Figure 8.15c). In foliated rocks, like schists and phyllites, conductivity parallel to the foliation plane is greater than that normal to the foliation. Therefore, hydraulic conductivity is a tensorial property. The ratio of K_h to K_v may vary from 10^{-2} in fractured rocks to 10^3 in stratified sedimentary rocks, as in the former permeability is mainly due to vertical fractures

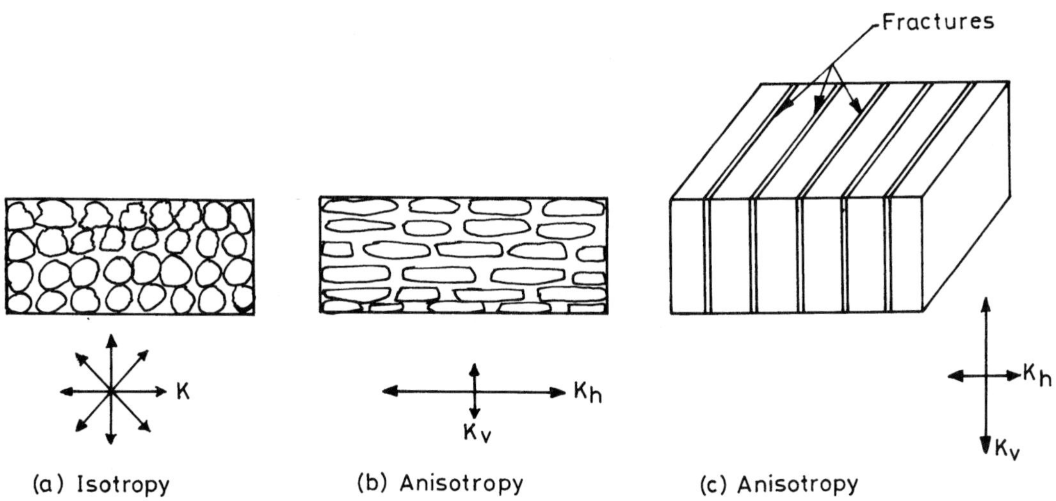

FIGURE 8.15 Isotropic and anistropic aquifers: (a) isotropic sedimentary aquifer; (b) anisotropic sedimentary aquifer, and (c) anisotropic fractured aquifer.

while in the latter the bedding planes, which form easy passage for water, are mostly horizontal. For analytical purposes, values of K_h/K_v = 1 to 10 are commonly used (Lee and Farmer, 1993).

In a sedimentary rock sequence, it may be necessary to determine the average value of hydraulic conductivity normal to bedding (K_v) and parallel to bedding (K_h). If the total thickness of the sequence is H and the thickness of the individual layers are h_1, h_2, h_3, ... h_n, with corresponding values of the hydraulic conductivity K_1, K_2, K_3 ... K_n, then K_v and K_h can be obtained by

$$K_v = \frac{[H]}{h_1/K_1 + h_2/K_2 + h_3/K_3 \cdots + h_n/K_n} \tag{8.26}$$

and

$$K_h = \frac{h_1K_1 + h_2K_2 + h_3K_3 + \cdots h_nK_n}{[H]} \tag{8.27}$$

The heterogeneity in sedimentary formations could be due to lateral variation in aquifer thickness even if the hydraulic properties such as hydraulic conductivity and specific storage remain constant. Such situations are observed in wedge-shaped aquifers (Figure 8.16a), showing lateral variation in thickness, or where alternate beds of sediments with different textures are formed under varying depositional conditions. Figure 8.16b shows a layered sedimentary formation where the individual bed has a homogeneous hydraulic conductivity K_1, K_2 ... but the entire sequence shows a heterogeneous character due to vertical variation in hydraulic conductivity. In carbonate rocks, hetrogeneity can develop due to variation in degree of solution and in crystalline rocks it is a result of varying density of aperture sizes of fractures (Figure 8.16c).

8.4 Representative elementary volume (REV)

REV is the smallest sample volume which is representative of the rock mass. The concept of REV is necessary to define the distribution of aquifer characteristics, such as hydraulic conductivity. In aquifer modelling also, it is necessary to have an idea of the minimum volume of rock which should be sampled having a representative value of rock mass properties. In unfractured homogeneous rocks, the hydraulic conductivity tends to become constant beyond a particular rock volume. This least volume is known as REV (Figure 8.17a). However, in a fractured rock mass, hydraulic conductivity value may not become exactly constant with increase in sample volume, but its variation may become insignificant (Figure 8.17b). REV

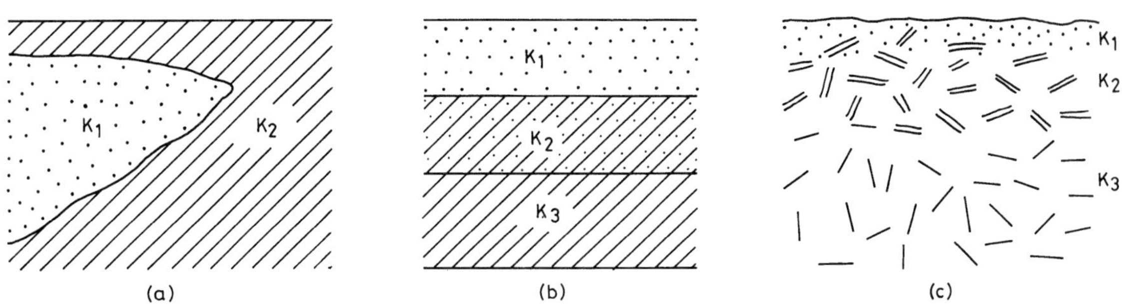

FIGURE 8.16 Heterogeneous aquifers: (a) wedge-shaped aquifer (b) layered aquifer, and (c) fractured rock aquifer ($K_1 > K_2 > K_3$).

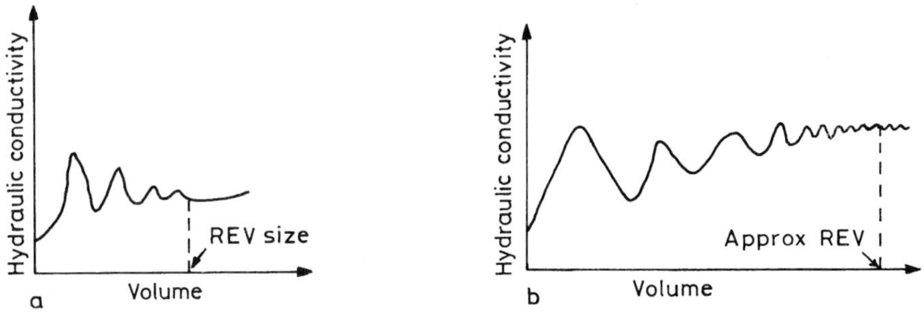

FIGURE 8.17 Representative elementary volume (REV) in: (a) unfractured rock, and (b) fractured rock.

will also depend on the flow system and the geometrical parameters of the fractured medium.

In fractured rocks, the concept of REV will be applicable when there is a constant hydraulic gradient and linear flow lines as in a truly homogeneous anisotropic medium. Further, the following criteria must be met in order to replace a heterogeneous system of given dimensions with an equivalent homogeneous system for the purposes of analysis (Long *et al.*, 1982).

(a) There is an insignificant change in the value of the equivalent permeability with small addition or subtraction of the test volume.
(b) An equivalent permeability tensor exists which predicts the correct flux when the direction of gradient in an REV is changed.

Point (a) indicates that the REV size is a good representative of the sample considering rock mass heterogeneities, and point (b) implies that the boundary condition will produce a constant gradient throughout a truly homogeneous anisotropic sample.

REV increases in size with increase in discontinuity spacing. Therefore, in order to define REV, one should have sufficient knowledge of rock discontinuities. The influence of fracture geometry on REV is shown in Figure 8.18. In

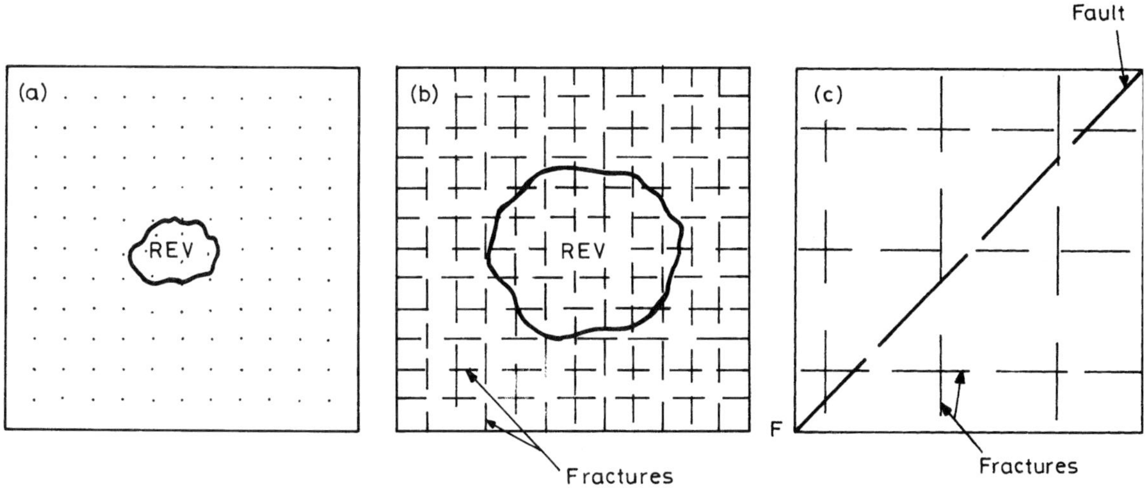

FIGURE 8.18 REV in different rock conditions: (a) homogeneous porous rock, (b) fractured rock where REV includes sufficient fracture intersections, and (c) rocks with large-scale discontinuities where REV is either very large or nonexistent.

granular rocks without discontinuities, small REV can be representative of the rock mass (Figure 8.18a), but in fractured rocks, REV should be large enough to include sufficient fracture intersections to represent the flow domain (Figure 8.18b). The size of the REV will be large compared to the size of the fractures' lengths in order to provide a good statistical sample of the fracture population. However, in the case of large scale features, such as faults and dykes, REV may not be feasible as it will be too large an area (Figure 8.18c). This implies that the concept of REV may not be true and practical in every rock mass.

Chapter nine

Methods of estimating hydraulic properties of rocks

Hydraulic properties of rock materials can be estimated by several techniques in the laboratory and in the field. The values obtained in the laboratory are not truly representative of the formation. However, the advantage of laboratory methods is that they are much less expensive and less time consuming.

9.1 Laboratory methods

Laboratory techniques are based on (a) indirect and (b) direct methods.

9.1.1 Indirect methods

In unconsolidated material, hydraulic conductivity can be determined from grain size analysis. The hydraulic conductivity of unconsolidated material is found to be related empirically to grain-size distribution by a number of investigators (A. Hazen, 1893; Krumbein and Monk, 1942; and Uma *et al.*, 1989, among others). Hazen, as far back as 1893, developed the empirical relation (equation (9.1)) between hydraulic conductivity (K) and effective diameter (d_e)

$$K = Cd_e^2 \qquad (9.1)$$

where C is a coefficient based on degree of sorting (uniformity coefficient) and packing. If K is in cm s^{-1} and d_e is in cm, the value of C in equation (9.1) ranges between 45 in very fine poorly sorted sand to 150 in coarse well sorted sand; a value of C = 100 is used as an average.

Effective diameter, (d_e), is the diameter of the sand grain (d_{10}) such that 10% of the material is of smaller size and 90% is of larger size. It can be estimated by plotting the grain size distribution curve. Uniformity coefficient, being the ratio of d_{60} to d_{10}, is a measure of the degree of sorting (Figure 9.1).

On comparison of values of K obtained from pumping tests and Hazen's formula, Uma *et al.* (1989) showed that Hazen's formula gave consistently higher values of K which is perhaps due to the reason that degree of compactness is not considered in Hazen's formula. They suggested that for unconsolidated and poorly cemented sandy material, the value of C in Hazen's formula is 6.0, for moderately cemented sandstones

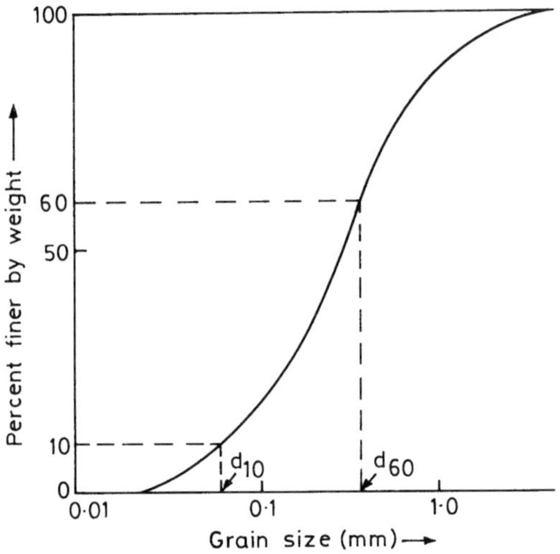

FIGURE 9.1 Grain-size distribution curve of unconsolidated material based on sieve analysis data.

3.8, and for well compacted and cemented sandstones its value is 2.0.

One of the well-known equations for determining hydraulic conductivity by indirect method is the Kozeny-Carman equation (Lee and Farmer, 1993), which has the form

$$K = \frac{1}{C_T C_o S_{sp}^2} \frac{e^3}{1+e} \frac{\gamma}{\mu} \tag{9.2}$$

where C_T is the pore tortuosity factor, and C_o is a pore shape factor: S_{sp}, specific surface area, and e, void ratio, were defined earlier in Chapter Eight.

9.1.2 Direct method

In the laboratory, hydraulic conductivity can be measured directly in three ways: (a) steady or quasi-steady flow techniques, (b) hydraulic transient flow tests, and (c) mechanical transient flow tests. The last two methods which analyse the time dependent behaviour, can be used for the estimation of both hydraulic conductivity and specific storage. Steady-state flow techniques are applicable in rocks of high conductivity, viz. sand, sandstone etc. while transient flow tests are recommended in low permeability tight formations like crystalline rocks, carbonate rocks and shales. Mechanical flow tests are generally used in compressible media such as clays.

Steady or quasi-steady flow techniques

These involve use of various types of permeameters – constant and variable head types. In these methods the rate of fluid flow through the specimen and the hydraulic gradient across the specimen are measured. Darcy's law can be used to calculate hydraulic conductivity. The description of these methods is given in many texts on groundwater hydrology, e.g. Todd (1980), Fetter (1988).

When the conductivities are very small (of the order of 10^{-10} m s^{-1}), steady-state flow can be achieved by taking small lengths of samples in the flow direction and using large hydraulic gradients (Neuzil, 1986). Falling (variable) head permeameters are also used for estimation of hydraulic conductivity of both coarse-grained and tight formations. Some researchers have used a closed reservoir instead of an open standpipe which is advantageous for clayey soil (Neuzil, 1986).

In low permeability formations, e.g. hard rocks, use of gas permeants can be advantageous; gas permeameters are more commonly used in the oil industry. Compressed air is a satisfactory fluid in most cases. Other gases such as nitrogen, oxygen, hydrogen, helium and carbon dioxide have also been used in special cases. The advantage of using gases is that they have low viscosity and they do not react significantly with the rock material in the dry state. Therefore, problems of swelling of clay minerals, bacterial growth and other chemical changes which can considerably affect permeability, are avoided.

Hydraulic transient flow test

The transient groundwater flow equation in one dimension can be written as

$$K = \frac{\partial^2 h}{\partial z^2} = S_s \frac{\partial h}{\partial t} \quad (9.3)$$

Transient tests are used for estimating very low conductivities (10^{-10} to 10^{-17} m s^{-1}). Neuzil (1986) performed these experiments on shale specimens for estimating conductivity, K and specific storage, S_s. In most of the cases a constant lateral and longitudinal load on the specimen was applied. In order to study the effect of rock deformation on permeability and specific storage, experiments based on transient pressure pulse method were designed by Read *et al.* (1989). Trimmer *et al.* (1980) used transient technique for determining permeabilities of intact and fractured granites and gabbros in the laboratory under high confining and pore-water pressures, to have an understanding of fluid flow behaviour in igneous rocks at great depths from the point of view of their suitability as host rocks for radioactive waste disposal.

Mechanical transient flow test

Permeability and specific storage can also be determined from transient mechanical behaviour of the specimen due to drainage of pore water as a result of loading. In soil mechanics, such a type of test (consolidation test) is used for low permeability compressible media. The values of permeability obtained by consolidation test and hydraulic test, vary considerably in the case of highly deformable media such as clays. A variety of testing methodologies have been developed (Freeze and Cherry, 1979).

Comparison of laboratory and field methods

The values of permeability determined in the laboratory from core samples, especially of fractured rocks, are usually several orders of magnitude lower than those existing under natural conditions (Clauser, 1992; Sanchez-Vila *et al.* 1996). The greatest variation is observed in karstified carbonate rocks (Rovey and Cherkauer, 1995). Such a difference could be due to the following reasons.

(a) Core (specimen) length. If the fracture spacing is more than the core length, the measured permeability will be representative of the matrix only.
(b) Fracture orientation. The permeability estimates are considerably influenced by the orientation of fractures in relation to flow direction. The radial flow to a well during field test will be quite different from the linear flow through the sample examined in the laboratory.
(c) Aperture size. As permeability is dependent on aperture size, an estimate from core samples obtained from deeply buried rocks will not be representative of *in situ* condition.
(d) Duration of testing. The values of permeability are found to decrease with time during extended testing in the laboratory. This is attributed to clogging of pore spaces by finer particles, swelling of clay minerals and other chemical reactions between permeant and pore fluids.

Therefore, as compared to laboratory methods, field or *in situ* methods provide a better estimate of hydraulic characteristics of rock formations as a larger volume of the material is tested. Table 9.1 gives a list of commonly used field methods. The choice of a particular method depends on the purpose of study and scale of investigations. For small-scale problems, as in geotechnical investigations, seepage of water to mines and contaminant transport problems, especially in fractured rocks, packer tests, slug tests and tracer tests are preferable. In cases of groundwater development and management on a regional scale, pumping test methods should be used. The choice is also governed by practical limitations or expediencies. The applicability of these methods in geothermal reservoirs is discussed in Chapter Sixteen.

TABLE 9.1 Field test methods for the estimation of hydraulic characteristics of aquifers (modified after UNESCO, 1984a)

Purpose of investigation	*Size of area under investigation*	*Distribution of fractures*	*Test method*
Geotechnical investigations; mine drainage; waste disposal, etc.	a few km^2	Random	Packer (Lugeon) test; slug test, tracer injection test
		Systematic fractures of 1, 2 or 3 sets	Modified packer test; cross-hole hydraulic test; tracer injection test
Groundwater development; water resources investigation	>100 km^2	Random and closely interconnected	Pumping test
Geothermal and petroleum reservoirs	a few km^2	Random	Well interference test; tracer injection test

9.2 Packer tests

A packer test, also known as an injection test, is used in an uncased borehole to determine the hydraulic conductivity of the individual horizon by isolating it with the help of packers (Figure 9.2). This method is widely used for estimating the hydraulic characteristics of fractured rocks in various geotechnical and waste disposal investigations (Louis, 1974; Black, 1987; Levens *et al.*, 1994, among others). In a packer test, water is injected under pressure into the test section. The borehole should be flushed beforehand to remove coatings from the wall of the borehole in order to get reliable results. The following three types of packer test could be used in fractured rocks, depending on details of the information required.

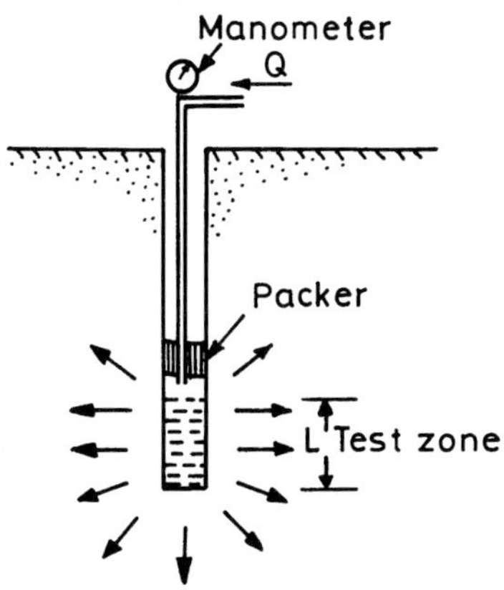

FIGURE 9.2 Standard Lugeon test with single packer.

(a) Standard Lugeon test, which gives average hydraulic conductivity.
(b) Modified Lugeon test, which gives directional hydraulic conductivity on the basis of relative orientation of the test hole to the system of fractures.
(c) Cross-hole hydraulic test.

9.2.1 Standard Lugeon test

The Lugeon method of testing was introduced by Maurice Lugeon, a French engineer, mainly for rock grouting in geotechnical work. It is a relatively low cost method especially for determining variations in hydraulic conductivity with depth and also in different strata. The test is made either in a completed borehole, or as the hole advances during drilling.

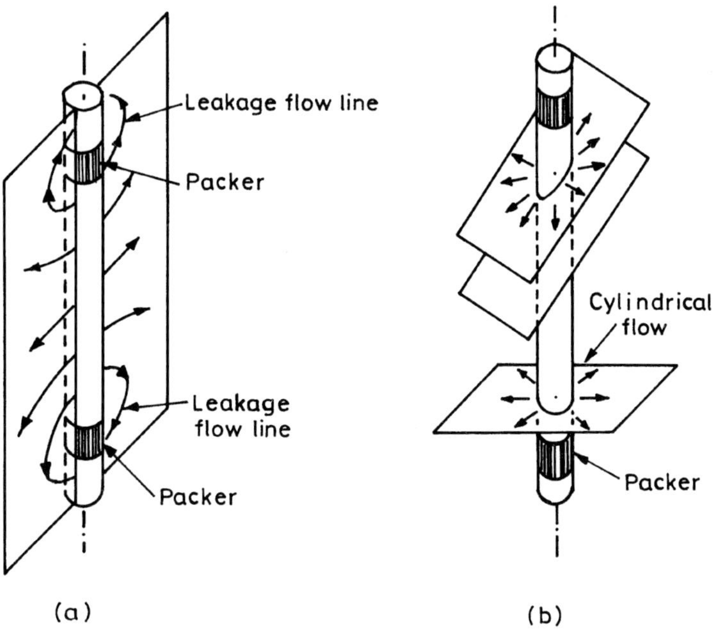

FIGURE 9.3 Influence of direction of fractures on the flow during Lugeon tests; (a) fractures parallel to the borehole, (b) fractures not parallel to the borehole.

The Lugeon test can be carried out by using either one or two packers. In the single-packer method, the packer is placed at a selected distance above the bottom of the hole (Figure 9.2). After the test is over, drilling can be further resumed and the test can be repeated in deeper horizons. The single-packer method is recommended when the rock mass is weak and intensely jointed and there are chances of the hole collapsing. In a completed borehole, two packers can be used to isolate the required section (3 to 6 m long) of the hole from the rest of it. Tests can be carried out to depths of up to 300 m.

Water is injected under pressure into the test section with increasing pressure from 0 to 10 bars (0 to 1 MPa) and then it is decreased from 1 to 0 MPa, in fixed steps at prescribed time intervals. The time interval is commonly 10 minutes. The flow rate of water in the borehole is measured under a range of constant pressures. Testing at pressures as high as 1 MPa is questionable in estimating permeability, as such high pressures are likely to increase the permeability locally by inducing new fractures and widening existing ones. It also enhances the possibilities of turbulence.

The flow pattern around the test zone depends on the orientation of the fractures in relation to the axis of the borehole (Figure 9.3). The flow rates will be very high when the intercepted fracture is parallel to the borehole.

In a test where the flow is cylindrical and an observation well (piezometer) is used (Figure 9.4), hydraulic conductivity (K) can be determined from equation (9.4) which is similar to the Thiem's equation (9.22)

$$K = \frac{Q/L}{2\pi\,(h_o - h)} \ln \frac{r}{r_w} \tag{9.4}$$

where K is hydraulic conductivity perpendicular to the axis of the borehole, in (ms^{-1}), Q is rate of inflow, in (m^{3}s^{-1}), L is thickness of the test zone, in (m), h_o and h are the piezometric heads in (m) measured in the test well and at distance (r) in the observation well, and r_w is radius of the test borehole, in (m).

In order to reduce the cost of the test, very

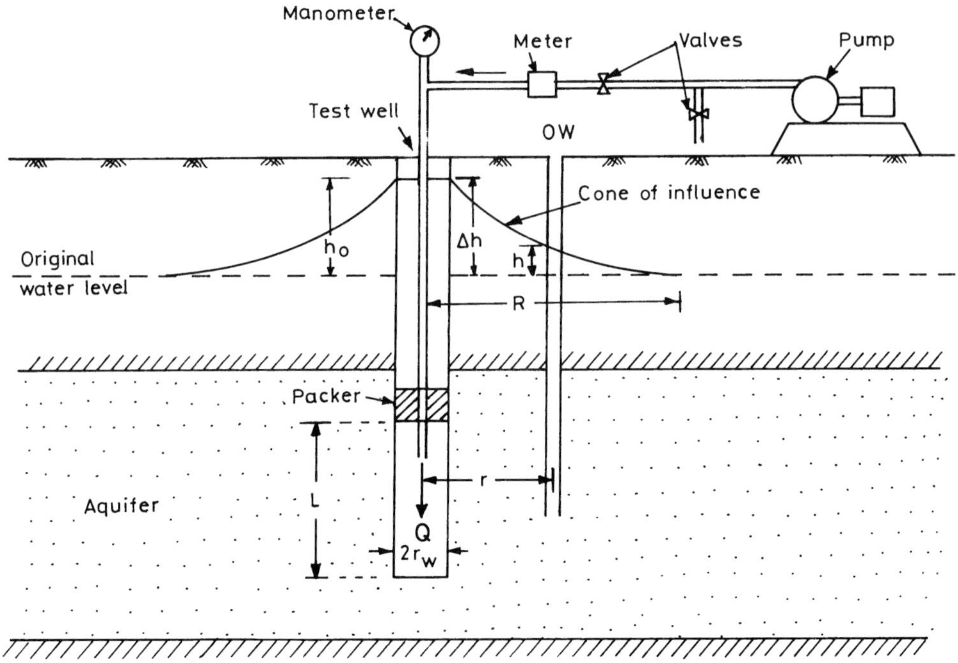

FIGURE 9.4 Standard Lugeon test – experimental setup and definitions necessary to interpret test results.

often piezometers are not installed. For such conditions, equation (9.4) becomes

$$K = \frac{Q/L}{2\pi \Delta h} \ln \frac{R}{r_w} \qquad (9.5)$$

where Δh is the hydraulic head, in (m) in the test well causing flow and R is the radius of influence, i.e. distance (m) at which initial water-level conditions do not change due to injection. Errors in the evaluation of R do not affect test results very much as $R >> r_w$. Therefore,

$$\frac{\ln (R/r_w)}{2\pi} = \text{constant} \qquad (9.6)$$

As ln R/r_w does not vary well, it can be assumed that (UNESCO, 1984b)

$$\ln (R/r_w) \simeq 7 \qquad (9.7)$$

Therefore, for numerical computations, equation (9.5) can be rewritten as

$$K = 1.12 \times \frac{Q/L}{\Delta h} \qquad (9.8)$$

where units are the same as in equation (9.4).

Some of the important rules which should be followed in conducting a packer test are as follows.

1. The test should be carried out in a saturated zone.
2. Each test at each pressure is continued until steady-state conditions are reached.
3. Pressure is measured in piezometres around the test hole. Hydraulic conductivity can also be expressed in a Lugeon unit (Lu). Lugeon is defined as the rate of flow of water in litres per minute under a pressure of water injection of 1 MPa per metre length of the tested material, i.e.

1 Lu = 1 l min^{-1} m^{-1} under 1 Mpa
or 1 Lu = 1 × 10^{-7} ms^{-1} (approx.).

It is assumed that the test section intercepts a number of fractures. Therefore, the estimated value of K represents an average value of hydraulic conductivity of the rock mass in the plane perpendicular to the borehole. The conductivity of individual fracture K_f can be expressed by

$$K_f = \frac{K}{N} \tag{9.9}$$

where N is the total number of fractures in the test section as determined from borehole core logs. It is assumed that fracture apertures are constant.

A comparison of transmissivity values determined from packer test and those obtained from the pumping test indicates that although the two values are correlatable, the values obtained from the pumping test tend to be higher (Figure 9.5). This appears to be due to the reason that in a packer test only the properties in the immediate vicinity of the borehole are reflected, but the pumping test covers a larger volume of rock mass around the borehole. Therefore, pumping tests are preferred to packer tests in order to obtain representative values of hydraulic parameters for the purpose of groundwater development.

In a standard packer test, the flow rate (Q/L) is plotted against applied pressure (P) as a flow-pressure curve. The flow-pressure curve will show different characteristics, depending on the permeability of the formation and changes brought about by water injection during the test (Figure 9.6). A typical flow-pressure curve shows a lower flow rate during the increase in pressure as compared with values during the fall in pressure (Figure 9.6a). A similar pattern (Figure 9.6b) is also expected from the cleaning of the existing fractures or development of new fractures due to hydraulic fracturing during the test. A reverse situation might also be observed due to the clogging of the fractures during the test (Figure 9.6c).

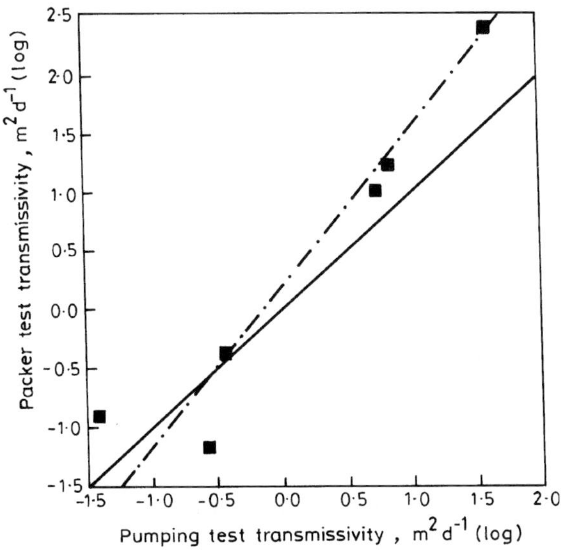

FIGURE 9.5 Comparison of transmissivity values obtained from Lugeon tests and pumping tests; (after Karundu, 1993).

9.2.2 Modified Lugeon test

In the standard Lugeon test, the borehole is vertical irrespective of the position and orientation of fractures in the rock mass. This is the only alternative when fractures are randomly distributed and are very large. Under such conditions the rock mass can be considered continuous. When there are well-defined fracture sets as in Figure 9.7 (F_1, F_2 and F_3), it is necessary to determine the hydraulic conductivities of each fracture set separately, i.e. K_1, K_2 and K_3 especially in geotechnical and contaminant transport problems. This can be achieved by having separate test holes for separate fracture sets; keeping the orientation of boreholes perpendicular to the considered fracture set.

A triple hydraulic probe to measure the directional permeabilities was proposed by Louis

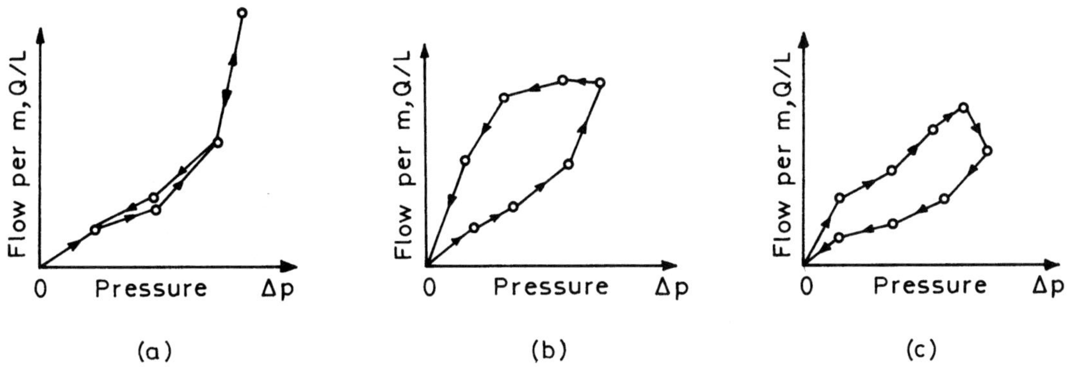

FIGURE 9.6 Characteristic flow-pressure curves for packer (Lugeon) test: (a) normal type with reversible cycle; (b) cleaning of existing fractures or development of new fractures; (c) reverse type due to the clogging of fractures.

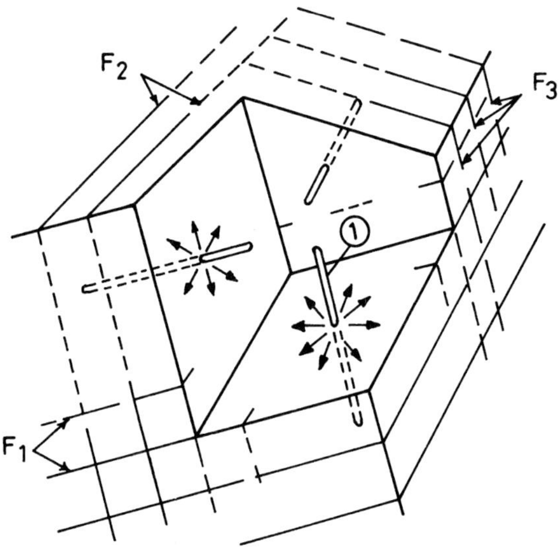

FIGURE 9.7 Modified Lugeon test in rock mass with three sets of fractures. Borehole (1) is parallel to F_2 and F_3 for testing the fracture set F_1.

(1974). In this method records are taken in three appropriate directions and measurements are made at different pressures. In order to measure the hydraulic heads during the test in the vicinity of the testing area, piezometers are installed. Four piezometric measurements enable the determination of the three-dimensional distribution of the anisotropic permeabilities.

9.2.3 Cross-hole tests

These are specially designed field tests to determine three-dimensional properties in fractured rocks from an array of randomly oriented boreholes. This method enables the computation of all three principal values of the hydraulic conductivity tensor, in three directions as well as the corresponding principal directions (Black, 1987). Cross-hole tests are usually carried out in HDR experiments and assessment of potential repositories for radioactive waste disposal for estimating relative permeabilities and interconnection between various fracture sets. A cross-hole project was implemented at Stripa mine, Sweden, for characterization of a high-level waste repository where a fan-shaped array of boreholes, 200 to 250 m long, were drilled below the ground emanating from the end of a drift, for geophysical and hydrogeological testing (Figure 9.8).

A polar plot of directional permeabilities would indicate whether the rock mass behaves as an anisotropic continuum. An ellipsoidal configuration of directional permeabilities would support continuum assumption. In the Stripa mine, Sweden, the plotted values of hydraulic diffusivity (ratio of permeability to specific storage) were not strictly ellipsoidal, indicating a limited interconnection of discrete fractures (Black, 1987).

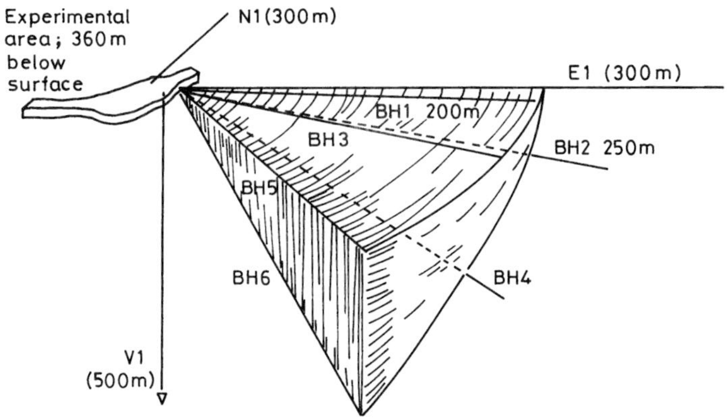

FIGURE 9.8 General layout of the cross-hole programme borehole array at Stripa, Sweden. The fan-shaped array of six boreholes (BH1–BH6) lies within an orthogonal array of three existing boreholes, E1, (east), N1 (North) and V1 (vertical) (after Black and Holmes, 1985).

Some of the problems in a packer test are (Louis, 1974):

1. Turbulence effect
2. Deformation of the medium due to high injection pressure
3. Influence of other fracture sets, i.e. effect of F_2 and F_3 when test hole is perpendicular to F_1
4. Entrance losses.

9.3 Tracer injection tests

Tracer tests are a valuable tool for determining flow characteristics, viz. groundwater flow direction, flow rates and basin boundaries in fractured rocks because the interpretations of these tests do not assume the continuum approach. The *in situ* measurement of hydraulic conductivity can also be made by using a tracer. The tracer used for such a purpose should be inexpensive, easy to detect and safe. A variety of tracers, non-radioactive and radioactive, can be used (section 3.5). Tracer tests are also carried out to determine the solute transport properties, viz. dispersivity and adsorption. Adsorptive tracers are used to investigate the sorptive properties of the fractures and rock matrix.

9.3.1 Groundwater velocity method

The procedure involves injection of a tracer in a well and its time of arrival is noted in a grid of wells located downgradient from the injection point. The average interstitial velocity, $\overline{V}$ is given by the relation

$$\overline{V} = \frac{K}{\eta_e}\frac{h}{L} \tag{9.10}$$

where K is the hydraulic conductivity, h is the difference in height between water-levels in the injection and observation wells, η_e is effective porosity and L is the distance between the observation wells. $\overline{V}$ can also be obtained from

$$\overline{V} = \frac{L}{t} \tag{9.11}$$

where t is the time of travel of tracer over the distance involved, L. Hence

$$K = \frac{\eta_e L^2}{ht} \tag{9.12}$$

Such tests are also helpful in establishing interconnection of fractures and flow continuity between both the injection and observation wells. In addition to selecting a suitable tracer, the tracer injection procedure is also important. Injection can be at the wellhead (open-hole testing) or by down-hole injectors in zones isolated by packers. In the latter case, selected fracture zones can be investigated as in hot dry rock

(HDR) experiments. In fractured geothermal reservoirs, tracers are especially useful in determining the preferential nature of permeability and flow paths (Grant *et al.*, 1982). In low permeability fractured rocks, there are problems in collecting water samples which are representative of *in situ* conditions. In such cases it is necessary to purge the borewell water by pumping for a period of time before collecting the sample (Anon, 1996).

9.3.2 Point dilution method

A point dilution method is used to determine horizontal average linear velocity of groundwater in the vicinity of a borehole. The method is based on the assumption that the aquifer is homogeneous and isotropic and the flow is horizontal. Packers are used in the well screen to isolate the portion of the aquifer to be tested. A tracer, viz. ^{3}H, ^{131}I or common salt (NaCl) is introduced into the isolated part of the well segment which gets diluted with time due to mixing with groundwater. Measurements of tracer concentration with time are made. Analysis of the resulting dilution curve gives the groundwater velocity.

The average bulk velocity across the centre of the well, ($\overline{V}^*$) is computed by using equation (9.13)

$$\overline{V}^* = \mathrm{I} - \frac{W}{At} \ln\left(\frac{C}{C_o}\right) \tag{9.13}$$

where W is the volume of the well segment where tracer is released, A is the vertical cross-sectional area through the centre of the isolated segment of the well screen, C_o is the initial concentration of the tracer at $t = 0$, and C is the concentration of tracer in water at time, t.

The average linear velocity of groundwater in the aquifer $\overline{V}$ is obtained by the relation

$$\overline{V} = \frac{\overline{V}^*}{\eta_e \alpha} \tag{9.14}$$

where α is a factor that depends on the well and aquifer characteristics. The usual range of α for tests in sand or gravel formations is from 0.5 to 4 (Freeze and Cherry, 1979). After $\overline{V}$ is known and hydraulic gradient (I) is determined from field observations, K can be calculated from equation (9.15) using Darcy's law.

$$K = \frac{\eta_e \overline{V}}{I} \tag{9.15}$$

In fractured media, the average bulk velocity $\overline{V}$ is given by equation (9.16) (Louis, 1977).

$$\overline{V} = \frac{\pi r_o L}{4\pi N a} \ln \frac{C/C_o}{t} \tag{9.16}$$

where r_o = radius of the test section, L = length of the test section, N = number of conducting fractures, a = average aperture of conducting fractures; C, C_o and t are as defined earlier.

The average hydraulic conductivity of the fractures ($\overline{K}_f$) intersecting the tested section is obtained from equation (9.17)

$$\overline{K}_f = \frac{\overline{V}}{I} \tag{9.17}$$

Other tracer test methods are: (1) divergent flow tracer test, and (2) convergent flow tracer test. In a divergent flow test, water with a tracer is injected into a well either as a pulse or by a step increase and the tracer concentration is monitored at one or more wells in the vicinity of the recharging well. In convergent flow tracer test, water is pumped from a well. Later, a tracer is injected into an adjacent injection well and the arrival of the tracer in the pumping well is monitored by sampling the pumped water.

Tracer methods may not give precise values of hydraulic conductivity due to diffusion and dispersion (Chapter Seven). The reader may refer to Anon (1996) for greater details. Care should also be taken that the injection and observation wells are properly located and are not far away from each other.

9.4 Slug tests

A slug test involves sudden injection of a known volume (or slug) of water into a well and measurement of fall of water-level with time. Alternatively a known volume of water can be withdrawn and the rise of water-level noted. Slug tests are suitable in rocks of low permeability and in weathered hard rocks. Slug tests can also be used in fractured rocks when fractures are uniformly distributed.

The advantage of a slug test is that it is cheap, as it requires less equipment and manpower and the duration of the test is also relatively short. It can be carried out even if nearby wells are being used. However, slug tests cannot replace the conventional pumping test methods as the latter have several additional advantages.

The duration of the slug test depends on the permeability of rocks – it has to be larger in low permeability media. The well radius also influences the duration of the slug test. In low permeability formations, it is necessary to have boreholes or standpipes of small diameter in order to reduce the duration of the test.

Cooper, and Bouwer and Rice, methods are commonly used for the analysis of slug test data. Cooper's method is described below. Readers may refer to Kruseman and de Ridder (1990) for the Bouwer and Rice method. The Bredehoeft and Papadopulos method which is especially suited to low permeability rocks is also given in this text.

9.4.1 Cooper's method

Cooper *et al.* (1967) developed a curve-matching method for the slug test in fully penetrating wells in confined aquifers. This method does not require any pumping or observation well. The sudden injection of a known volume of water, V in a well of diameter $2r_w$ will cause an immediate rise of water-level, h_o in the well (Figure 9.9) which can be expressed as

$$h_o = \frac{V}{\pi r_c^2} \qquad (9.18)$$

The change in water-level with time is given by equation (9.19)

$$\frac{h_t}{h_o} = F(\alpha, \beta) \qquad (9.19)$$

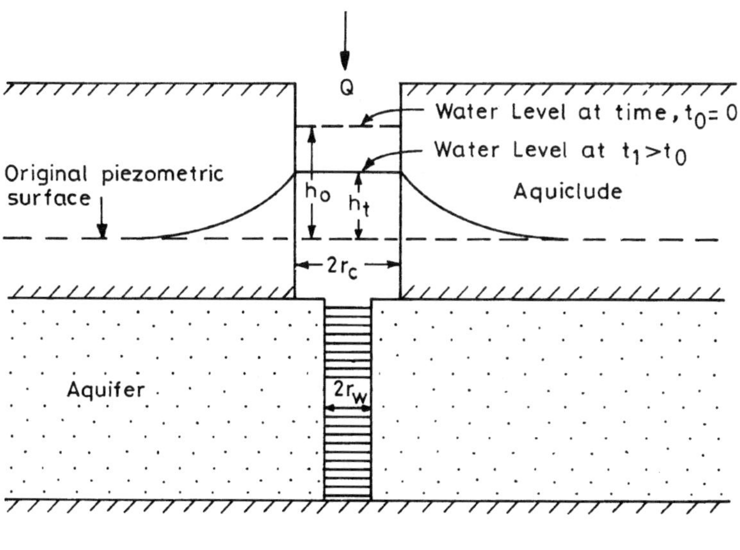

FIGURE 9.9 Slug test in a fully penetrating well of finite diameter in a confined aquifer.

where

$$\alpha = \frac{r_w^2 S}{r_c^2} \quad (9.20)$$

$$\beta = \frac{Tt}{r_c^2} \quad (9.21)$$

h_o, h_t, r_c and r_w are as shown in Figure 9.9.

Values of function F (α, β) against α for different values of α and β are plotted on semi-log paper as a family of type curves (Figure 9.10). The value of h_o is calculated from equation (9.18) based on the volume of water injected into or withdrawn from the well. The ratio of h_t/h_o is calculated for different values of t.

A field-data curve is prepared by plotting values of log h_t/h_o against corresponding values of t. The field-data curve is superimposed over the family of Cooper's type curves (Figure 9.10) to find the best match of the field-data plot with one of the type curves. The value of β for the corresponding type curve is noted. The value of t for the corresponding value of $\beta = 1$ is read from the data curve; T is calculated by substituting values of t, r_e and $\beta = 1$ in equation (9.21). The value of S can be determined from the value of α of the type curve with which the data curve is matched using equation (9.20).

The type curves particularly for small values of α are quite similar in shape and therefore it is difficult to select the type curve for an unique match. This may result in large errors in calculating S. However, T can be determined with greater accuracy but it will be representative of the formation in the immediate vicinity of the well.

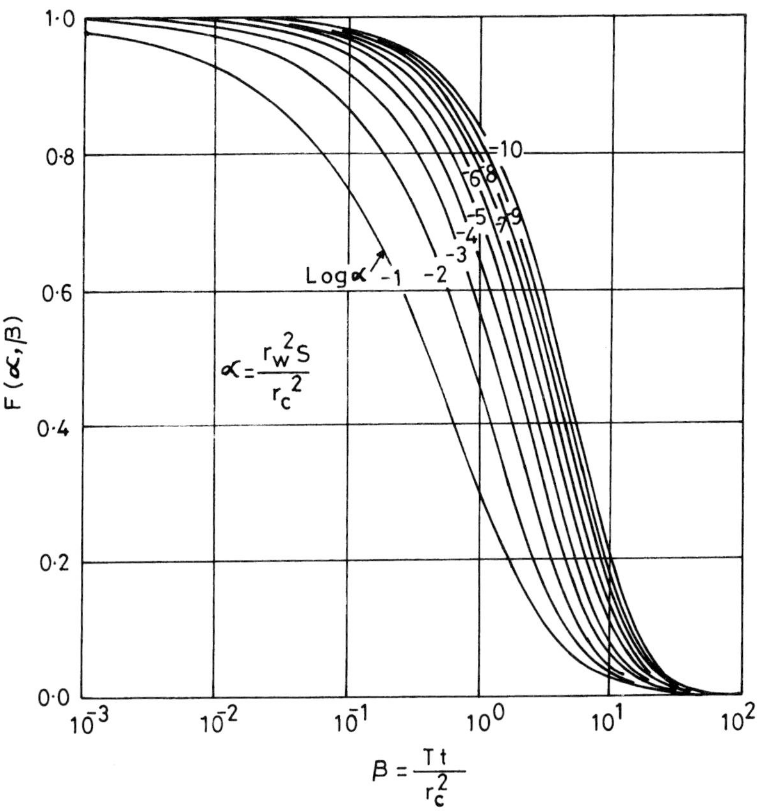

FIGURE 9.10 Family of Cooper's type curves (after Papadopulos *et al.*, 1973).

9.4.2 Bredehoeft and Papadopulos method

A modified slug test, i.e. pressure pulse method was suggested by Bredehoeft and Papadopulos (1980) in which the test well is filled with water to the surface, and suddenly pressurized with an additional amount of water. The water is then shut-in and the fall in pressure is noted with time. Packers are used to isolate the intake part of the open hole. This method is advantageous in reducing the time of test which is the main problem in tight formations. Bredehoeft and Papadopulos (1980) showed that a formation of $T = 10^{-11}\ m^2s^{-1}$ and $S = 4 \times 10^{-4}$ can be tested adequately in a few hours using the pressure pulse method, whereas years may be required by the conventional slug-testing methodology. Therefore the Bredehoeft and Papadopulos method is feasible for *in situ* estimation of hydraulic properties of tight formations, viz. shales and massive crystalline rocks having low hydraulic conductivities of the order of $10^{-13} - 10^{-16}\ ms^{-1}$. Bredehoeft and Papadopulos also argued that for $\alpha > 0.1$, type curve of F (α, β) versus the product, $\alpha\beta$ are more suitable for analysing test data than the Cooper's type curves (Figure 9.10) where F (α, β) is plotted against β only. The same procedure of matching the field-data curve (h_t/h_o versus t) plotted on semi-logarithmic paper with the type curve F (α, β) against $\alpha\beta$ can be adopted for estimating either T and S separately or the product TS.

This method is based on the assumptions, among others, that: (1) the flow in the tested interval is radial, and (2) the hydraulic properties of the tested interval remain constant throughout the test. In many cases, the first assumption holds good as the ratio of K_h/K_v is quite high. In order to satisfy the second assumption, especially in the fractured rocks, the fluid pressure in the tested interval should be kept low to avoid the formation of new fractures. This can be taken care of by initial pulses of 1-10 m (10 – 100 kPa).

Neuzil (1982) argued that the Bredehoeft–Papadopulos method does not assure the approximate equilibrium condition, i.e. equal hydraulic head in the well and the formation at the beginning of the test. Therefore, the test data will not give correct values of K. To avoid this problem, Neuzil (1982) suggested a modified test procedure to ensure near-equilibrium condition at the start of the test and also the use of two packers and two transducers to monitor pressures both below the lower packer and between the packers to detect leakage.

Black *et al.* (1983) demonstrated that the conventional slug test in fractured rocks can lead to large errors in the estimation of specific storage, but errors in the determination of hydraulic conductivity will be small. They also suggested a sinusoidally varying pressure test using two boreholes which will provide data about directional hydraulic diffusivity.

The use of nitrogen gas, instead of water, as permeant is suggested by Kloska *et al.* (1989) for determining permeability of unsaturated tight formations. This is advantageous in unsaturated or less saturated low permeability rocks as nitrogen will not change the rock properties.

9.5 Pumping tests

9.5.1 Introduction

In a pumping test, a well is pumped at a known constant or variable rate. As a result of pumping, water-level is lowered and a cone of water-table depression in an unconfined aquifer and cone of pressure relief in a confined aquifer is formed. The difference between the static (non-pumping) and pumping water-level is known as drawdown. As pumping advances, the cone of depression expands until equilibrium conditions are established when the rate of inflow of water from the aquifer into the well equals the rate of pumping. The distance from the centre of the pumped well to zero drawdown point is termed as **radius of cone of depression** (R). In an ideal uniform, isotropic and homogeneous aquifer, the cone of depression will be symmetrical and the contours of equal drawdown will be circular or near-circular. In contrast, in a fractured aquifer, due to anisotropy, the plotted drawdown cone, based on data from a number of observation wells, will be linear, highly elongated or irregular; the longer axis will be parallel to the strike of fracture conducting water. The slope of the cone of depression and its radius (R) depend on the type of aquifer and its hydraulic characteristics. In aquifers of high transmissivity, the gradient of the cone of depression is less and its radius is large, as compared with aquifers of low transmissivity (Figure 9.11).

The speed of propagation of the cone is inversely proportional to its storativity (S). In an unconfined aquifer as the storativity is high, being of the order of $10^{-1} - 10^{-2}$, corresponding to the specific yield, the spread of cone is very slow and therefore it would take a long time to reach the edges of the aquifer. In contrast to this, in confined aquifers the storativity is very small, of the order of $10^{-3} - 10^{-6}$. Hence the cone extends more rapidly in a confined aquifer than in an unconfined aquifer (Table 9.2).

TABLE 9.2 Spread of the cone of depression in metres (after Schoeller, 1959). ($T = 1.25 \times 10^{-3}$ m^2s^{-1})

	1 min	*1 h*	*1 day*	*10 days*	*100 days*
Unconfined aquifer (S = 0.2)	0.91	7.11	34.8	110	348
Confined aquifer ($S = 1 \times 10^{-4}$)	41	318	1558	4930	15580

9.5.2 Planning a pumping test

Selection of site

Pumping tests are expensive and therefore should be carefully planned. The cost depends on number of observation wells and duration of the test. Before conducting a pumping test, it is necessary to know the geological and hydrological conditions at the test site. Subsurface lithology and aquifer geometry are of help in the proper interpretation of pumping test data. Existing wells in the area can provide important information about the subsurface lithology. A preliminary estimate of the transmissivity of the aquifer can be made from subsurface lithology and aquifer thickness. At places, where subsurface geological information

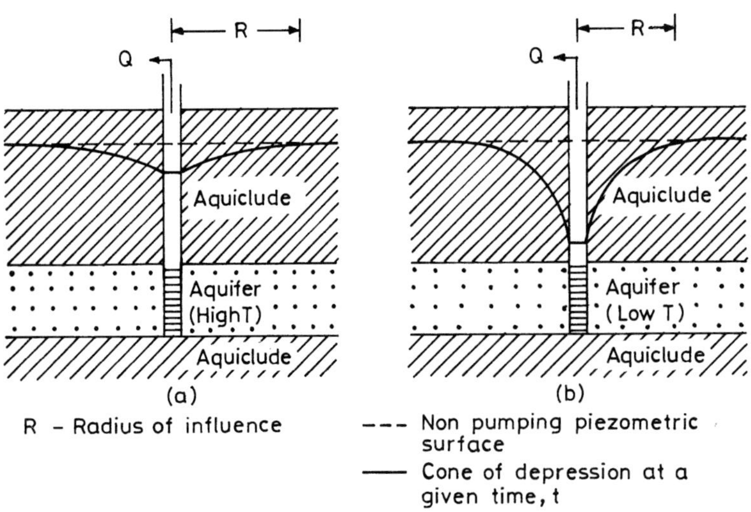

FIGURE 9.11 Comparison of cones of depression formed due to pumping in a confined aquifer of, (a) high transmissivity, and (b) low transmissivity.

is not directly available, geophysical methods can be used to ascertain the lithology (Chapter Five). The presence of hydrological boundaries in the form of rivers, canals, lakes or rock discontinuities (faults and dykes etc.) should also be noted. In the absence of this information, the test data is liable to be interpreted in different ways, viz. the effect of a recharge boundary on the time-drawdown curve may almost be the same for a leaky aquifer or an unconfined aquifer. This aspect is further elaborated in the latter part of this chapter.

The test site for a confined aquifer should not be close to a railway track or highway to avoid effect of loading which produces water-level fluctuations. The site should also be away from existing discharging wells to avoid well interference.

Design of pumping and observation wells

The well field for a pumping test consists of one pumping well and one or more observation wells. The pumping well should be at least of about 150 mm diameter so that a submersible electric pump can be installed. The diameter of the observation wells should be able to accommodate an electrical depth sounder or some other water-level measuring device, viz. automatic water-level recorder.

In order to avoid entry of pumped water into the aquifer, it should be conveyed by pipes or lined drains to a distance of 200 to 300 m away from the discharging site.

The pumping well should tap the complete thickness of the aquifer to avoid an effect of partial penetration. In thick aquifers, where partial penetration cannot be avoided, necessary correction in the drawdown data should be made. Observation well(s) should have the same depth and tap the same aquifer(s) as the pumping well. However, in some cases, where the interconnection between aquifers due to leakage is suspected, some observation wells should also tap the overlying and underlying aquifers (Figure 9.12a). This is also required in fractured rock aquifers to know the vertical component of flow (Figure 9.12b).

The length and position of the screen in the pumping well is important as this will affect the amount of drawdown. Partial penetration causes lengthening of flow lines and thereby excessive drawdown (Figure 9.13). In unconfined aquifers, as pumping causes dewatering of the aquifer, the well screen should be put in the lower one-half or one-third of the aquifer where the flow (stream) lines will be mainly horizontal, otherwise the observation well may be

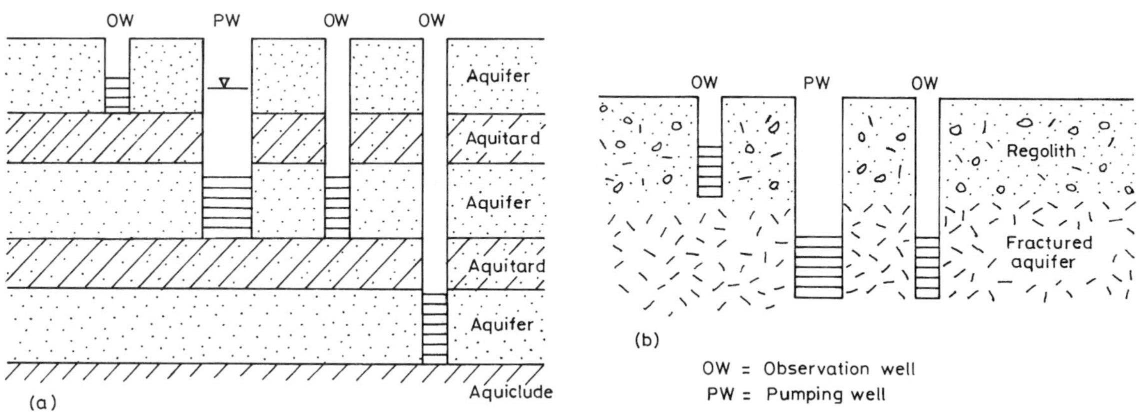

FIGURE 9.12 Layout of pumping well (PW) and observation wells (OW) in: (a) multilayered aquifer system, and (b) fractured aquifer.

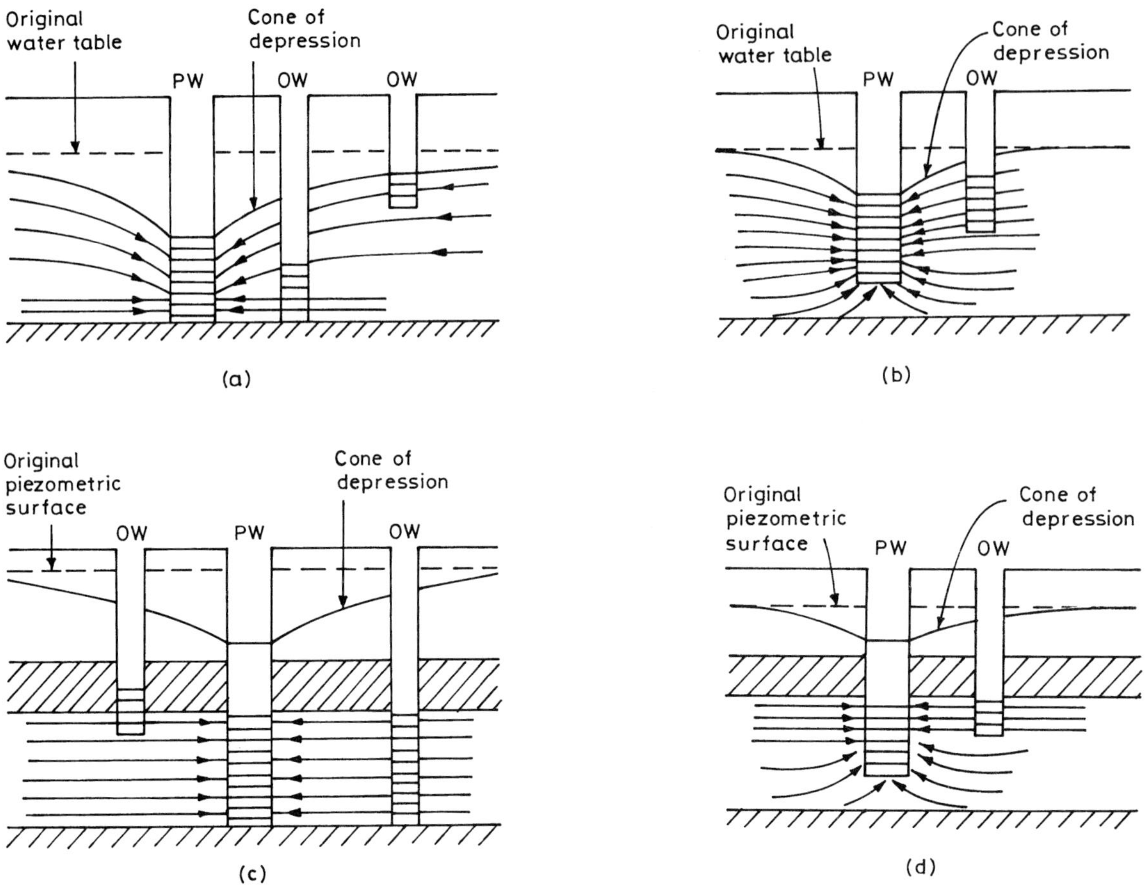

FIGURE 9.13 Positioning of screens in observation wells; (a) fully penetrating pumping well in an unconfined aquifer; (b) partially penetrating pumping well in an unconfined aquifer; (c) fully penetrating pumping well in a confined aquifer, and (d) partially penetrating pumping well in a confined aquifer.

located at distances one-and-a-half times the saturated thickness of the aquifer (Figure 9.13a). The desirable position of the screen in a confined aquifer is shown in Figures 9.13c and d. In heterogeneous aquifers, a piezometer nest is recommended for recording of water-levels in different piezometers at different depth horizons (Figure 9.14).

The number of observation wells depends on the purpose of study and financial considerations. It is desirable to have at least one observation well but three observation wells at different distances from the pumping well are better. This will help in analysing the test data by both the time-drawdown and distance-drawdown methods and identify the boundary conditions. In order to assess the anisotropy and heterogeneity of the aquifer, it is also necessary to have additional observation wells in other directions. Some of the existing wells can also be used as observation wells to minimize the cost.

Well spacing

The spacing of observation wells from the pumping well depends on the well penetration,

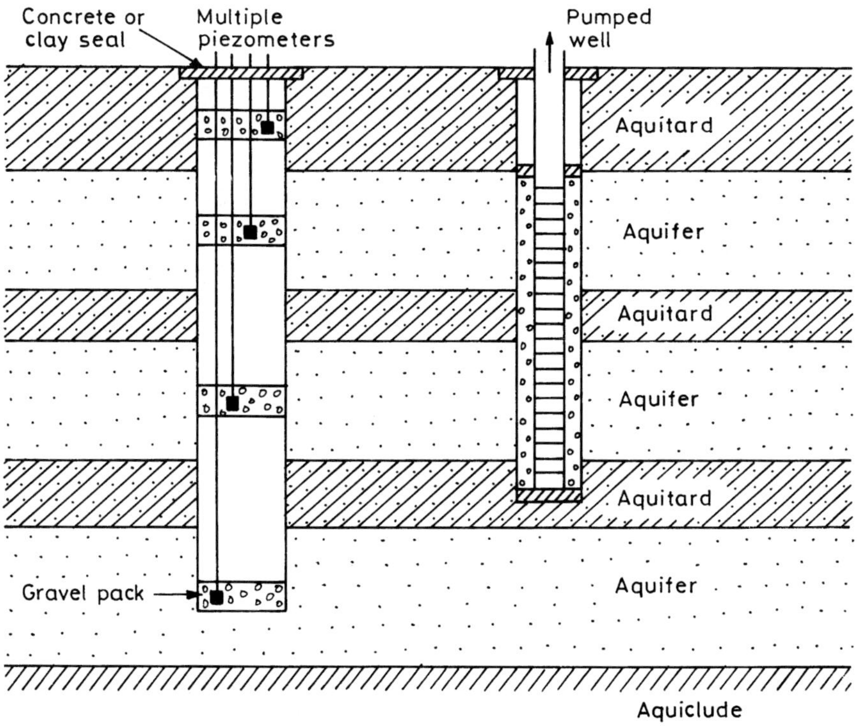

FIGURE 9.14 Schematic section of a cluster of piezometers in a multilayered aquifer system.

type of the aquifer, its hydraulic characteristics and rate and duration of pumping (Table 9.3). In an unconfined aquifer, as the radius of the cone of depression is comparatively small, the observation well should be within a distance of 100 m from the pumping well. In a confined aquifer, this distance could be up to 200 or 250 m. For semi-confined aquifers, intermediate values may be taken. Transmissivity also influences the shape and size of the cone of depression as described earlier. In aquifers with high transmissivity, observation wells can be put farther away from the pumping well than in aquifers with low transmissivity. One observation well should be selected which is outside the cone of depression of the well field. This helps to assess the effect of atmospheric or other natural causes of water-level fluctuations so that necessary correction in the drawdown data may be made.

Installation of observation wells in fractured rocks depends on whether the fractures are discrete or interconnected and also the permeabilities of rocks on the two sides of the fractures (matrix blocks) and whether fractures are open or filled with impervious material (Kruseman and de Ridder, 1990).

Duration of test

The duration of the pumping test depends on the type of the aquifer and the purpose of the test. Usually the test is continued till the water-level is stabilized so that both the non-steady and steady-state methods of analysis can be used for computing aquifer parameters. The duration of the constant rate pumping test in confined, leaky and unconfined aquifers could be about 24 h, 20 h and 72 h respectively (Hamill and Bell, 1986; Kruseman and de Ridder, 1990). In fractured rocks of low hydraulic conductivity, the pumping test can be only of a shorter duration (10 to 15 h).

TABLE 9.3 Suggested spacing of the observation wells (after Hamill and Bell, 1986).

Aquifer and well condition	*Minimum distance from pumped well to nearest observation well*	*General distance within which observation wells should be located*
Fully penetrating well in unstratified confined or unconfined aquifer	1 to 1.5 times aquifer thickness	20 to 200 m in confined aquifer, 20 to 100 m in unconfined aquifer
Fully penetrating well in a thick or stratified confined or unconfined aquifer	3 to 5 times aquifer thickness	100 to 300 m in confined aquifer, 50 to 100 m in unconfined aquifer
Partially penetrating well (<85% open hole) in a confined or unconfined aquifer	1.5 to 2 times aquifer thickness	35 to 200 m in confined aquifer, 35 to 100 m in unconfined aquifer

Measurements

Two sets of measurements are taken during a pumping test: (1) groundwater levels (during both drawdown and recovery phases) in pumping and observation wells, and (2) well discharge. Water-level measurements can be made by using a steel tape or electrical depth sounder. The automatic water-level recorders are better for obtaining continuous water-level records in observation wells. Fully automatic microcomputer-controlled systems are also now available for accurate recordings of water-levels.

As drawdown is faster in the observation wells close to the pumped well during the early part of the test, water-level observations should be made at an interval of ½ to 1 min for the first 10 min of pumping. The time interval of measurements can later be increased to 2, 5 and 10 min. In tests of longer duration extending for a period of one day or more, water-level measurements for the later part of the test could be taken at one hour or even longer time intervals. After the pumping is stopped, the rate of recovery in the pumped well and also in the observation wells is noted. As the rate of recovery is faster in the early part of recuperation, water-level measurements during this period should be taken at shorter time intervals; later the time interval can be increased. The water-levels measured during the test are also likely to be influenced by other extraneous reasons such as barometric pressure changes etc. Therefore, necessary corrections should be made to isolate the effect of such changes.

Well discharge can be measured by using different types of weirs, circular orifice weir being the most common (Driscoll, 1986).

9.5.3 Types of drawdown curves

The drawdown data are plotted as time-drawdown and/or distance-drawdown curves. A time–drawdown graph on a semi-logarithmic scale may show different slopes (Figure 9.15). The earliest time response is usually due to well-bore storage as most of the water at this stage comes from the well itself. The next part of the time-drawdown curve is influenced by the contact between the well and the aquifer due to 'skin effect'. At a later stage, the type of aquifer and its hydraulic characteristics (T and S) influence the rate of drawdown. Finally boundary conditions affect the trend of the time-drawdown curve. There could be various possible combinations of these effects. A proper interpretation of pumping test data would therefore depend on the recognition of these effects. The duration of the test is therefore important to recognize the various components (segments) of

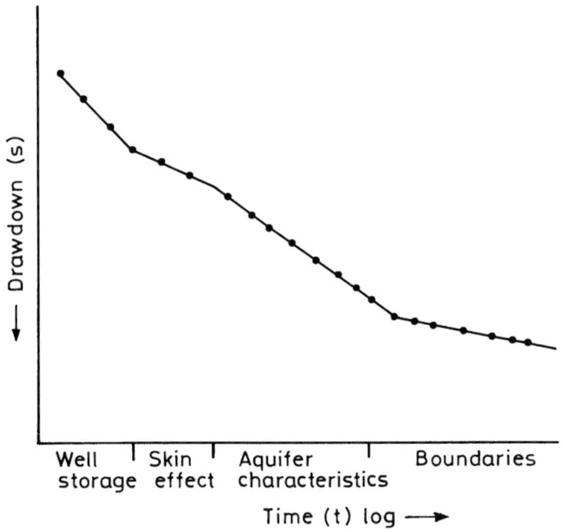

FIGURE 9.15 Drawdown response in time sequence.

a time-drawdown plot. The influence of various factors on the time-drawdown curve is discussed below.

Aquifer types

A log-log time-drawdown plot for an ideal unconsolidated confined aquifer tapped by a fully penetrating well will match with the Theis type curve (Figure 9.16a) and a semi-log time-drawdown plot will be characterized by an uniform slope (Figure 9.16a'). On the other hand, unconfined aquifers which are characterized by delayed yield, show a typical S-shape on a log-log plot with three time-segments (Figure 9.16b); the early part (first segment) conforms to the Theis type curve for a confined aquifer due to instantaneous release of water from aquifer storage, while the middle part (second segment) shows flattening due to vertical drainage. The last, or the third segment, also conforms to the Theis type curve as by that time the effect of vertical drainage becomes negligible. The semi-log plot for an unconfined aquifer is characterized by two parallel straight line slopes of segments 1 and 3 (Figure 9.16b').

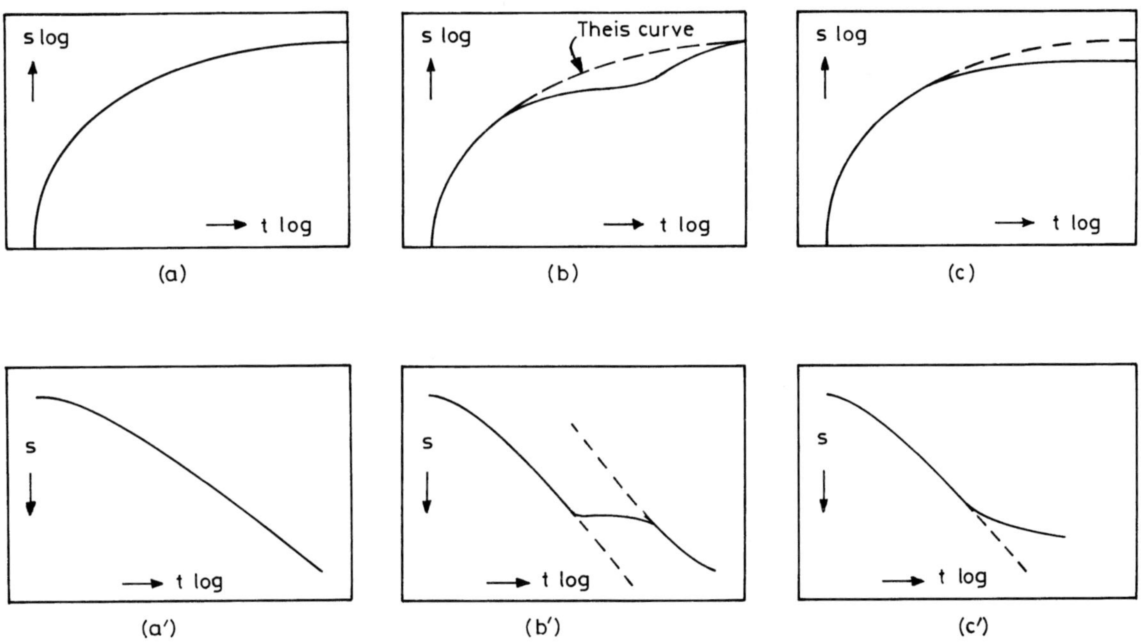

FIGURE 9.16 Comparison of log-log and semi-log plots of the theoretical time-drawdown relationship in unconsolidated aquifers; (a) and (a') confined aquifer; (b) and (b') unconfined aquifer; and (c) and (c') leaky aquifer.

The leaky aquifer shows a behaviour similar to segments 1 and 2 of an unconfined aquifer, as in this case additional water is transferred to the aquifer by leakage through the aquitard. However, segment 3 of the time-drawdown curve for an unconfined aquifer is not observed because the process of recharge by leakage continues and drawdown stabilizes after some time as is indicated by both log-log and semi-log graphs (Figure 9.16c, c').

Fractured rock aquifers

These show various types of drawdown and pressure build-up curves, depending on the nature of fractures and matrix blocks, boundary conditions and wellbore storage (Davis and DeWiest, 1966; Streltsova, 1976b; UNESCO, 1979; Gringarten, 1982; Kruseman and de Ridder, 1990). Similar trends are also observed in geothermal and petroleum reservoirs (Grant *et al.*, 1982; Horne, 1990).

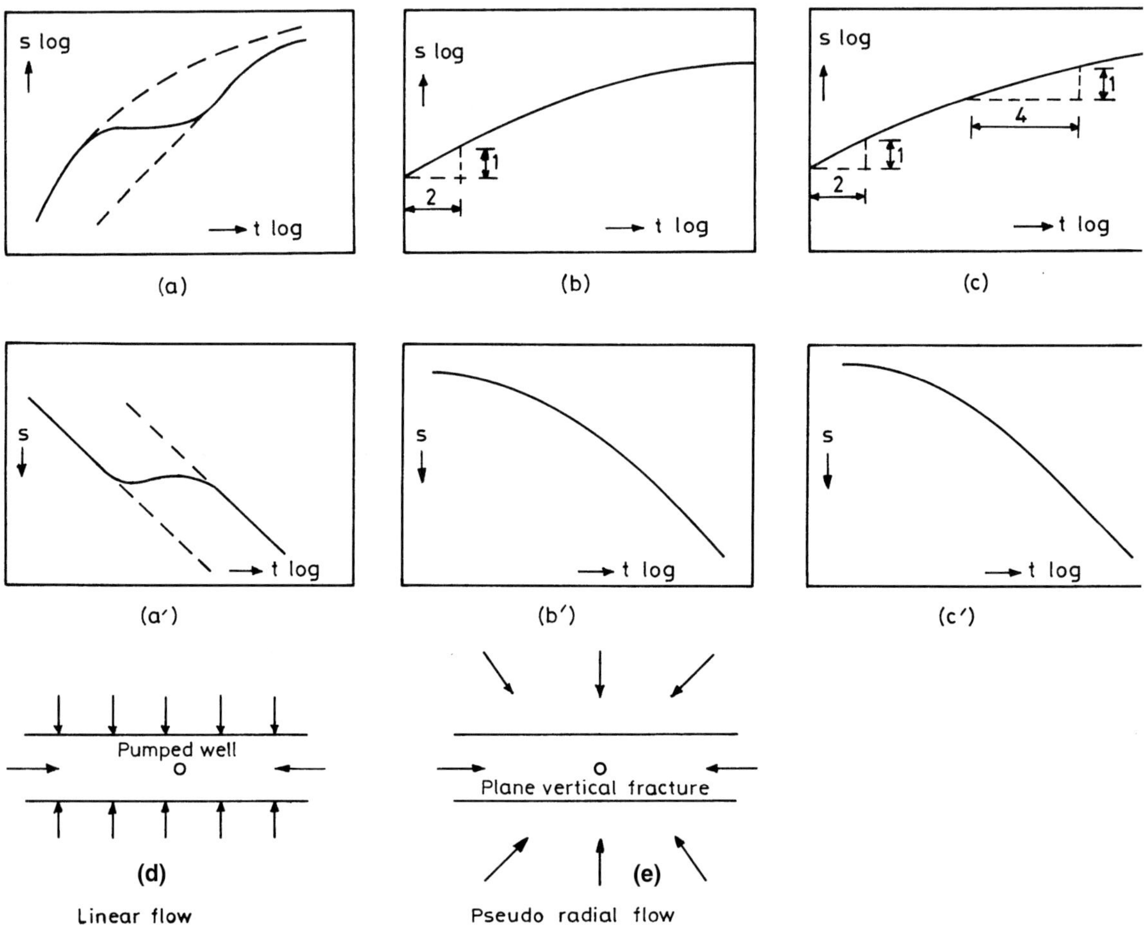

FIGURE 9.17 Comparison of log-log and semi-log plots of the theoretical time-drawdown relationship of fractured aquifers; (a) and (a') double porosity confined aquifer; (b) and (b') single plane vertical fracture; (c) and (c') fractured permeable dyke in a massive country rock of low hydraulic conductivity; (d) and (e) show linear flow and pseudo-radial flow respectively (modified after Kruseman and de Ridder, 1990).

The response to pumping in a fractured **double porosity aquifer**, having the matrix blocks divided by fractures (Figure 9.17a, a'), is similar to that of delayed yield in an unconfined aquifer. The first and third segments of the time-drawdown curve on a semi-log plot may have parallel slopes (Figure 9.17a'). The first segment is due to removal of water from storage in the fracture or in the wellbore. The second segment shows more or less constant drawdown due to a contribution from the matrix block while the third segment is due to the removal of water from storage both in the fractures and the matrix blocks.

The effect of pumping from a well in a single vertical fracture of infinite hydraulic conductivity in a confined homogeneous and isotropic aquifer of low permeability is shown in Figures 9.17b and b'. The response at early times of pumping is characterized by a ½ slope straight line on the log-log plot of drawdown against time which is explained due to linear flow Figure 9.17d. At a later time, as the flow regime changes to pseudo-radial (Figure 9.17e), the shape of the curves resembles those of (a) and (a'). Parts (c) and (c') of Figure 9.17 present the response of a pumping well in a fractured dyke of high permeability traversing a confined aquifer of low permeability and high storativity. The log-log time-drawdown plot in such a case is characterized by two straight-line segments. The first segment has ½ slope like that of the well in a single vertical fracture of infinite permeability. At intermediate times, the log-log plot shows a ¼ slope due to a contribution from the host rock. At late times, the flow in the host aquifer is pseudo-radial which is reflected by a straight-line segment in the semi-log plot (Kruseman and de Ridder, 1990).

Some anomalous trends showing greater drawdown in wells away from the pumping well as compared to those in the vicinity have also been reported from fractured rocks (Streltsova-Adams, 1978; Gringarten, 1982). Also there is field evidence of a rise of water-level near the pumping well. These anomalous behaviours in literature are referred as 'reverse water-level fluctuation' or **Noorbergum effect** which can be explained by the following phenomena (Streltsova, 1976b).

(1) Delayed release of water from storage in the porous matrix of a double porosity medium in response to pumping resulting in local increase of water-level in the fracture system.
(2) Decrease of the storage capacity of fractures due to their deformation as a result of pumping.
(3) Additional recharge of fractures due to the recycling of pumped water discharged on the ground surface.

Skin effect

The concept of skin effect or 'skin factor' on head loss in a well was first introduced in the oil well industry, where due to mud filtrate at the well face, greater well loss was observed (Horne, 1990). The idea was extended later to water wells also. The 'skin effect' will be positive when the effective well bore radius and permeability is reduced due to mud filtrate. Conversely, when the effective well bore radius is increased due to acidization or other methods of well stimulation (section 15.4), the skin effect will be negative. In fractured rocks, the skin effect will be negative due to good contact between the well and the water bearing fractures but when there is a clay coating on the walls of the fractures, the skin effect will be positive resulting in greater drawdown.

Several analytical and digital techniques are available for analysing pumping test data under different hydrogeological conditions. The advantage of numerical methods is that they can take care of complex hydrogeological conditions. The graphical procedures are subjective and the number of type curves also become prohibitively large when the number of parameters exceeds three.

Graphical methods, although cumbersome, have an advantage as one can visualize the extent to which the field data match with the assumed conditions and the investigator can be selective in choosing the best part of the data which would give reliable results under given hydrogeological conditions. In this text (sections 9.6–9.8) only graphical methods of pumping test data analysis are described. For computer assisted methods, the reader is referred to other texts (e.g. Rushton and Redshaw, 1979; Boonstra, 1989).

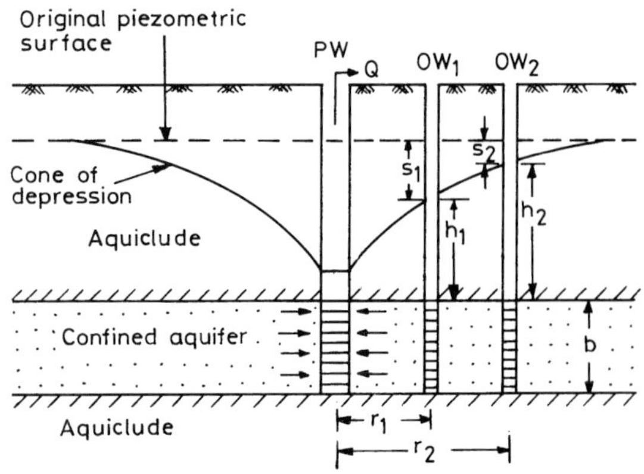

FIGURE 9.18 Schematic cross-section of a pumped confined aquifer.

9.6 Constant discharge tests

The methods described in this section were originally developed for unconsolidated aquifers but can also be applied to fractured rocks if the fractures are interconnected.

9.6.1 Confined aquifers

The pumping test data from a confined aquifer can be analysed both by steady state and unsteady state methods. The steady-state equation (7.25) for groundwater flow in a confined homogeneous aquifer is given in section 7.1.3.

Thiem's equilibrium method

Thiem (in Wenzel, 1942) was one of the first workers to use drawdown data for estimating T from unconfined and confined aquifers. Thiem's equation based on steady-state flow for a confined aquifer can be written as

$$Q = \frac{2\pi T\,(h_2 - h_1)}{\ln\,(r_2/r_1)} \tag{9.22}$$

or

$$Q = \frac{2\pi T\,(s_1 - s_2)}{2.30\,\log\,(r_2/r_1)}$$

where Q is well discharge in m^3d^{-1}, T is transmissivity in m^2d^{-1}, s_1 and s_2 are drawdowns in metres in the two observation wells located at distances r_1 and r_2 in m respectively from the pumping well (Figure 9.18).

After the system has reached steady-state (equilibrium) condition, the drawdown data from different observation wells are plotted on semi-logarithmic paper (drawdown, s_1 and s_2 on arithmetic scale and distances, r_1, r_2 on logarithmic scale). The slope Δs of this line is determined and is substituted in equation (9.23) for estimating T.

$$T = \frac{2.30Q}{2\pi\Delta s} \tag{9.23}$$

where Δs is the difference of drawdown per log cycle of r.

Theis non-equilibrium type curve method

The differential equation (7.24) for unsteady flow in plane polar coordinates is given in section 7.1.3. Based on the analogy between flow of water in an aquifer and flow of heat in an equivalent thermal system, Theis (1935) obtained equation (9.24)

$$s = \frac{Q}{4\pi T}\int_u^\infty \frac{e^{-y}}{y}\,dy \tag{9.24}$$

where

$$u = \frac{r^2 S}{4Tt} \quad (9.25)$$

s = drawdown in metres in a piezometer or observation well at distance, r in metres from the pumping well, S is storativity of the aquifer (dimensionless), t is time since pumping started, in days; Q and T are as defined earlier.

Although the Theis equation is based on several assumptions, not all of which hold good under natural field conditions, it has been very useful in solving many groundwater flow problems. The exponential integral in equation (9.24) is written symbolically as W(u) which is generally read as well function of u or Theis well function. Equation (9.24) can therefore be written as

$$s = \frac{Q}{4\pi T} \, W(u) \quad (9.26)$$

where

$$W(u) = -0.5772 - \ln u + u - \frac{u^2}{2.2!} + \frac{u^3}{3.3!} - \frac{u^4}{4.4} \quad (9.27)$$

where $u = \dfrac{r^2 s}{4Tt}$ as already indicated. Values of W(u) for various values of u are given in several standard text books on groundwater hydrology, e.g. Todd (1980).

From equations (9.24) and (9.25), it can be shown that if s is known for one value of r and several values of t or for one value of t and several values of r, and if Q is known, then T and S can be determined.

The first step in this method is to plot on double-log paper, the Theis type curves of W(u) versus u, W(u) versus $\sqrt{u}$ and W(u) versus $^1/_u$ (Figure 9.19). The field data curve can be matched with one of the Theis type curves as given in Table 9.4. A reverse type curve [W(u) versus $^1/_u$] can be matched with a data curve of s versus $^t/_{r^2}$ or if only one observation well is available then s is plotted against t. This will save computation time. If drawdown data are available for one value of t and several values of r, then a distance–drawdown curve (s versus r) is plotted which is matched with a type curve of W(u) versus $\sqrt{u}$. As an example, the matching of the Theis type curve, W(u) versus u with the field data curve, $^{r^2}/_t$ versus s is shown in Figure 9.20. An arbitrary match point is selected anywhere on the overlapping portion

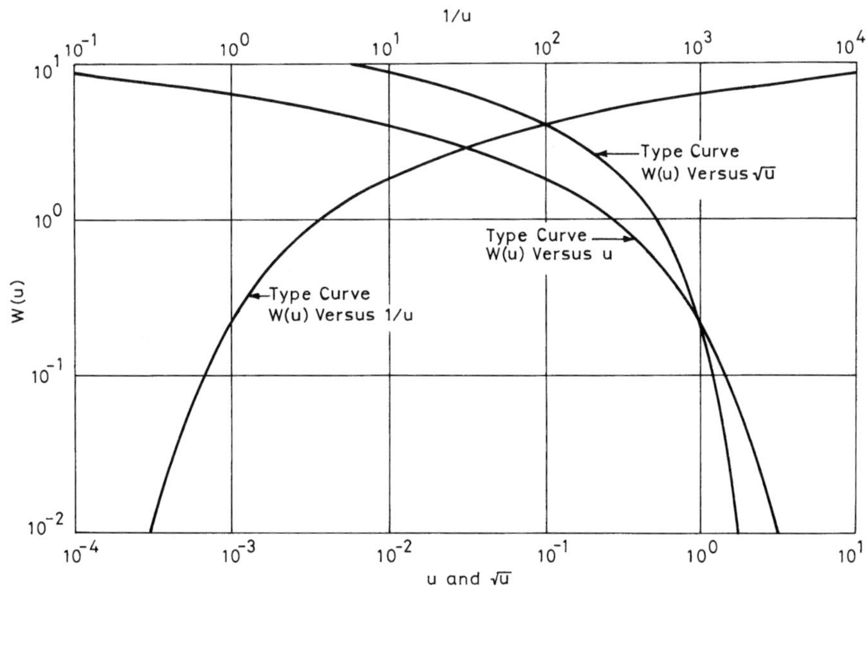

FIGURE 9.19 Theis type curves W(u) versus u, W(u) versus √μ, and W(u) versus 1/u (after Warren, 1963).

TABLE 9.4 Various combinations of type curves and data curves in Theis method of analysis

Type curve	*Data curve*
W(u) versus u	s versus r^2/t or s versus 1/t
W(u) versus l/u	s versus t/r^2 or s versus t
W(u) versus √u	s versus r

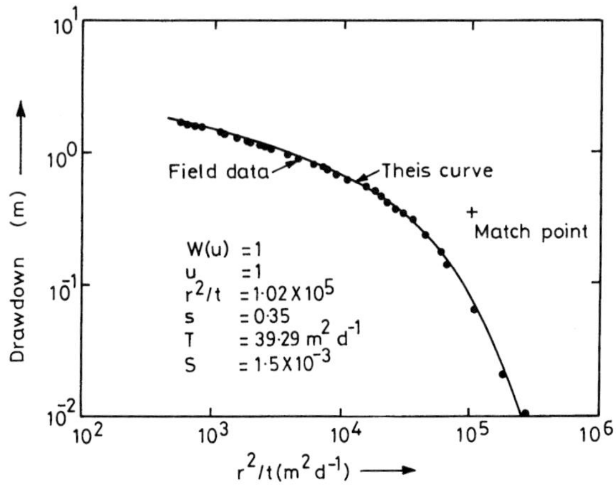

FIGURE 9.20 Analysis of pumping test data by Theis type curve method for a confined sandstone aquifer (after Radhakrishna *et al.*, 1979).

of the sheets and the coordinates of this point are noted. These data on substitution in equations (9.26) and (9.25) will give values of *T* and *S* respectively.

Jacob's method

For values of u < 0.01, equation (9.26) reduces to equation (9.28)

$$s = \frac{Q}{4\pi T}\left(\ln \frac{4Tt}{r^2 S} - 0.5772\right) \qquad (9.28)$$

In decimal logarithms, equation (9.28) reduces to

$$s = \frac{2.30Q}{4\pi T} \log \frac{2.25Tt}{r^2 S} \qquad (9.29)$$

which is known as the Jacob or Theis–Jacob equation.

According to equation (9.29), the graph of drawdown, *s* versus log *t*, will be in the form of a straight line. Equation (9.29) can be rewritten as

$$T = \frac{2.30Q\,(\log t_2/t_1)}{4\pi\,(s_2 - s_1)} \qquad (9.30)$$

or

$$T = \frac{2.30Q}{4\pi \Delta s} \qquad (9.31)$$

where Δs is the change in drawdown over one log cycle of time.

The storativity, *S* can also be determined from the same semi-log plot by extending the time-drawdown curve to zero drawdown axis. Solving for storativity, *S*, the equation in its final form becomes

$$S = \frac{2.25Tt_o}{r^2} \qquad (9.32)$$

where t_o is the time intercept in days for $s = 0$.

Theis recovery method

After pumping is stopped, the water-levels in pumping and observation wells will start rising. This is known as the recovery or recuperation phase. Theis equation for recovery phase can be written as

$$s' = \frac{Q}{4\pi T}\,[\mathrm{W(u)} - \mathrm{W(u')}] \qquad (9.33)$$

where

$$\mathrm{u} = \frac{r^2 S}{4Tt'} \quad \mathrm{u'} = \frac{r^2 S}{4Tt'} \qquad (9.34)$$

Q, *T*, *S* and *r* are defined earlier, *t* is the time in days since pumping started, *t′* is the time in days since pumping stopped and *s′* is the residual drawdown. As in Jacob's equation, for small values of r and large values t', equation (9.33) can be written as

$$s' = \frac{2.30Q}{4\pi T}\log\frac{t}{t'} \qquad (9.35)$$

Therefore a plot of residual drawdown, s' on arithmetic scale versus t/t' on logarithmic scale should form a straight line. Equation (9.35) can be rewritten as

$$T = \frac{2.30Q}{4\pi \Delta s'} \qquad (9.36)$$

where $\Delta s'$ is the change in residual drawdown per log cycle of t/t'.

9.6.2 Leaky (semi-confined) aquifers

In nature, truly confined aquifers are rare as the confining layers are not completely impermeable. Therefore, a certain amount of water is contributed to the pumped aquifer due to leakage through the aquitard. Such aquifers are known as leaky or semi-confined aquifers.

Pumping from a fully penetrating well in a leaky aquifer draws water from storage in the leaky aquifer as well as from storage in the aquitard and the water-table aquifer. The flow in the leaky aquifer is radial (horizontal) towards the well but the flow through the aquitard is vertical due to head difference between the two aquifers (Figure 9.21). As pumping advances, the cone of depression expands and the head difference between the two aquifers increase leading to a greater amount of leakage. The amount of leakage also depends on the vertical hydraulic conductivity of the aquitard (K'). A steady-state situation may arise when the rate of withdrawal becomes equal to the amount of water contributed from the pumped aquifer plus that contributed by leakage from the adjacent aquifer(s).

Earlier it was assumed that during pumping the hydraulic head in the unpumped aquifer

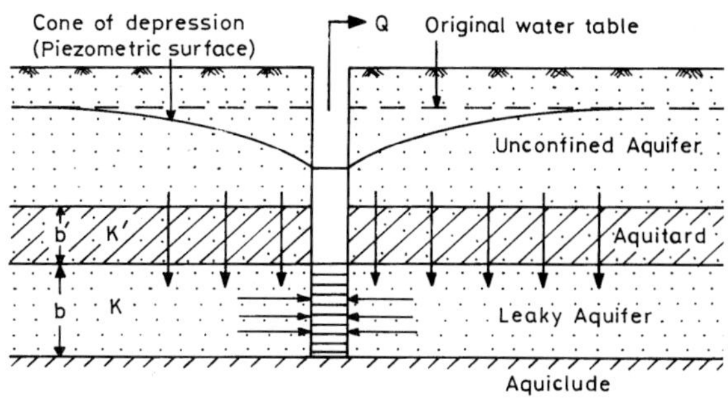

FIGURE 9.21 Schematic cross-section of a pumped leaky aquifer.

remains constant, and that the rate of leakage into the pumped aquifer is proportional to the hydraulic gradient across the aquitard (Hantush and Jacob, 1955). The first assumption will only hold good if there is a constant source of recharge to the unpumped aquifer which may not be possible. The second assumption which ignores the effects of storage capacity of the aquitard is justified when the flow has reached steady-state condition. Under unsteady-state condition, the effect of aquitard storage cannot be neglected. Therefore, the effects of aquitard storage as well as decline of hydraulic head in the unpumped aquifer were also considered by later workers (Freeze and Cherry, 1979). A review of evaluation of aquifer test data in leaky aquifers is given by Walton (1979) and Kruseman and de Ridder (1990).

Walton's method

The unsteady-state flow in a leaky aquifer without water released from storage can be given in an abbreviated form by equation (9.37)

$$h_o - h = s = \frac{Q}{4\pi T}\int_u^{\infty}\frac{1}{y}\exp\left(-y - \frac{r^2}{4B^2 y}\right)dy \qquad (9.37)$$

or

$$s = \frac{Q}{4\pi T}\,\mathrm{W}\,(\mathrm{u}\cdot r/\mathrm{B}) \qquad (9.38)$$

where

$$u = \frac{r^2 S}{4Tt} \tag{9.39}$$

$$r/B = \frac{r}{\sqrt{T/(K'/b')}} \tag{9.40}$$

or

$$B = (Tb'/K')^{\frac{1}{2}} \tag{9.41}$$

where B is the leakage factor, K' is the vertical hydraulic conductivity of the aquitard, and b' is the thickness of the aquitard (Figure 9.21).

It can be seen that for large values of B, r/B would tend to be zero and therefore equation (9.38) will approach the Theis equation (9.26).

$W(u.{}^{r}/_{B})$ is read as the well function for leaky aquifers. Values of $W\ (u.{}^{r}/_{B})$ in terms of the practical range of u and ${}^{r}/_{B}$ have been plotted on a double-log paper to provide a family of type curves for leaky aquifers (Figure 9.22). Pumping test data (s versus t) are plotted on another log-log paper which are matched with one of the leaky aquifer type curves for a particular value of r/B and a match point is selected. The coordinates of the match point $W(u.{}^{r}/_{B})$ and ${}^{1}/_{u}$ are read from the type curve sheet and s and t from the data curve sheet. The values of Q and match point coordinates, i.e. $W(u.{}^{r}/_{B})$ and s are substituted in equation (9.38) to calculate T. S is obtained from equation (9.39). The value of ${}^{r}/_{B}$ of the type curve with which the data curve was matched is substituted in equation (9.40) to determine K' and also leakage, ${}^{k'}/_{b'}$ or its reciprocal, hydraulic resistance, C from the relation $B = \sqrt{TC}$. A straight line method for estimating hydraulic properties of leaky aquifers is given by Hantush (1956, 1964).

9.6.3 Unconfined aquifers

Pumping from an unconfined aquifer causes dewatering of the aquifer resulting in a vertical component of flow in addition to the radial flow (Figure 9.23). This is in contrast to a confined aquifer where there is no dewatering of the aquifer and the water is obtained due to compaction of the aquifer and expansion of water.

The time-drawdown curve in response to pumping from an unconfined aquifer is characteristically S-shaped due to delayed yield as described in section 9.5.3. In an unconfined aquifer, if drawdown, s, is small as compared to its saturated thickness, b, the vertical component of flow can be neglected and the Theis equation can be used to determine aquifer characteristics. However, when drawdowns are significant, the vertical component of flow cannot be ignored and therefore the use of Theis equation is not justified. In such conditions where gravity drainage is considerable, methods for

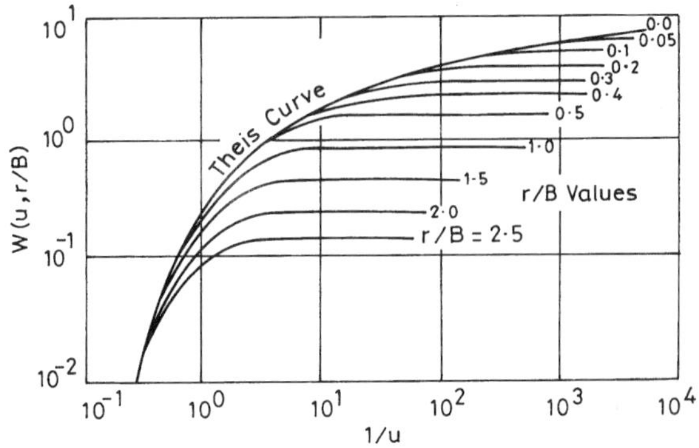

FIGURE 9.22 Family of Walton's type curves of W(u,r/B) versus 1/u for different values of r/B for a leaky aquifer (after Walton, 1962).

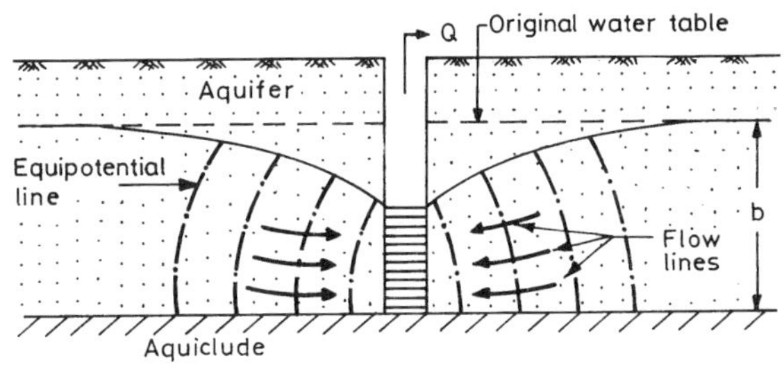

FIGURE 9.23 Schematic cross-section of a pumped unconfined aquifer.

analysing pumping test data based on the concept of delayed yield are given by Boulton (1963) and Neuman (1975).

Boulton's method

Boulton (1963) gave equation (9.42) for drawdown in an unconfined isotropic aquifer with delayed drainage

$$s = \frac{Q}{4\pi T} W\left(U_{AY} \frac{r}{D}\right) \qquad (9.42)$$

where $W(U_{AY}\ {}^{r}\!/\!_{D})$ is called the 'well function of Boulton'. Under early-time conditions, describing the first segment of the time-drawdown curve, equation (9.24) can be written as

$$s = \frac{Q}{4\pi T} W\left(U_{A} \frac{r}{D}\right) \qquad (9.43)$$

where

$$U_A = \frac{r^2 S_A}{4Tt} \qquad (9.44)$$

and S_A = early time storage coefficient.

Under later-time conditions for the third segment of the time-drawdown curve, equation (9.43) reduces to

$$s = \frac{Q}{4\pi T} W\left(U_{y} \frac{r}{D}\right) \qquad (9.45)$$

where

$$U_y = \frac{r^2 S_y}{4Tt} \qquad (9.46)$$

and S_y = specific yield. D is called the drainage factor, which is defined as

$$D = \sqrt{\frac{T}{\alpha\, S_y}} \qquad (9.47)$$

D has the dimensions of L, $\frac{1}{\alpha}$ is the 'Boulton delay index' which is an empirical constant having the dimensions of time. It is used to determine the time at which the delayed yield ceases to affect the drawdown. The procedure for the use of Boulton's method involves matching the field data curve in two steps, with one of the Boulton's type curves (Figure 9.24) for computing aquifer properties (Kruseman and de Ridder, 1970). Prickett (1965) was among the first workers to demonstrate the applicability of Boulton's type curve method to unconfined aquifers. An example of field data analysis using Boulton's type curve method in a granitic aquifer is given in Figure 9.25.

Neuman (1972) showed that Boulton's method can be used only for large values of pumping time. In general, the limitation of Boulton's method becomes more severe as the ratio of horizontal permeability to vertical permeability increases, the thickness of the aquifer increases and the distance from the pumping well decreases. Neuman (1975) suggested an alternative method as described below.

Neuman's method

Neuman's method for determining aquifers characteristics of anisotropic unconfined aquifers does not involve such semi-empirical quantities as Boulton's delay index, $\frac{1}{\alpha}$. It also takes into account aquifer anisotropy. Neuman's drawdown equation can be written as

$$s = \frac{Q}{4\pi T} W\,(U_{A'}\, U_{B'}\, \beta) \qquad (9.48)$$

where W (U_A, U_B, β) = well functions for water-table aquifer with fully penetrating wells having no storage capacity (dimensionless). Under early time conditions equation (9.48) reduces to

$$s = \frac{Q}{4\pi T} W\,(U_{A'}\, \beta) \qquad (9.49)$$

where

$$U_A = \frac{r^2 S_A}{4Tt} \qquad (9.50)$$

S_A = Elastic early-time storativity.

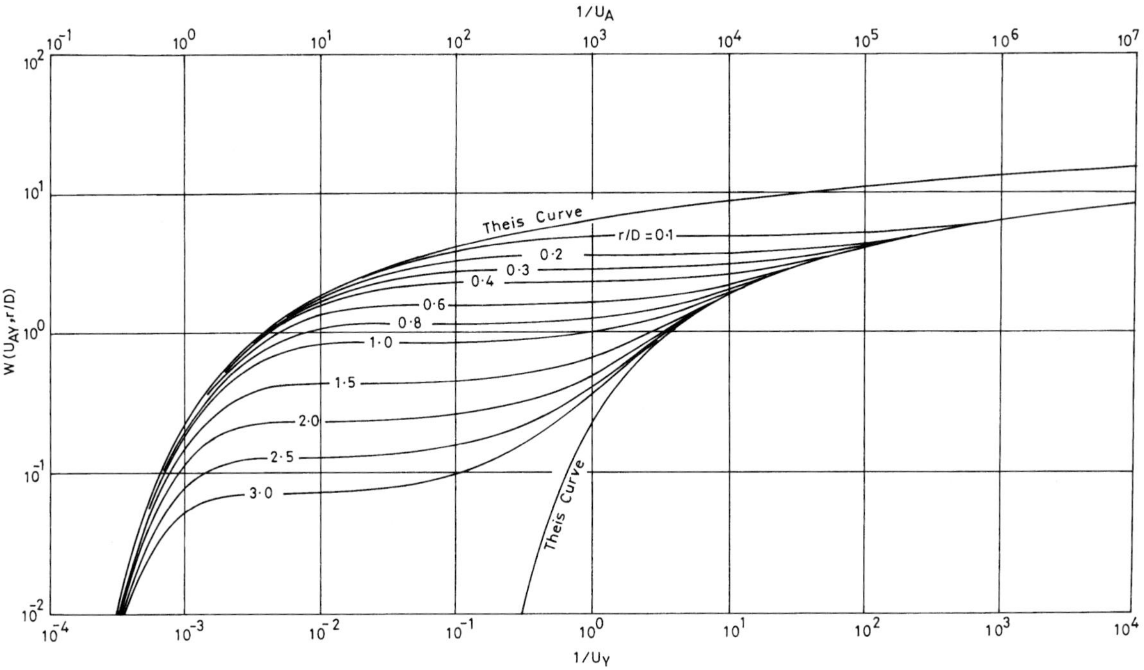

FIGURE 9.24 Family of Boulton's type curves for an unconfined aquifer (after Prickett, 1965, reproduced by permission of the journal of *Ground Water*).

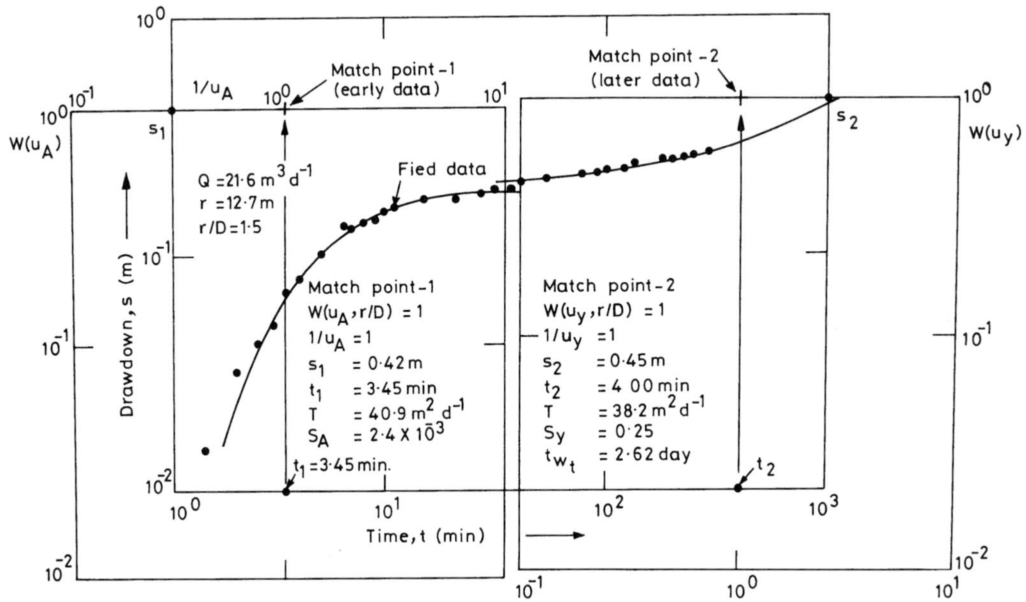

FIGURE 9.25 Analysis of pumping test data by Boulton method from a granitic aquifer in Lower Maner Basin, Andhra Pradesh (after Radhakrishna *et al.*, 1979).

Under late-time conditions, the third segment of the time-drawdown curve, equation (9.49) reduces to

$$U_B = \frac{r^2 S_y}{4Tt} \qquad (9.51)$$

Neuman's parameter β is defined as

$$\beta = \frac{r^2 K_v}{b^2 K_h} \qquad (9.52)$$

where K_v and K_h are the vertical and horizontal hydraulic conductivities of the aquifer respectively and Q, T, R, t, S_y, b are as defined earlier.

For an isotropic aquifer $K_v = K_h$. Therefore,

$$\beta = \frac{r^2}{b^2}$$

Figure 9.26 presents the family of Neuman's type curves of W (U_A, U_B, β) vesus $1/U_A$ and $1/U_B$ for various values of β.

The aquifer characteristics can be determined by type-curve and straight-line methods (Kruseman and de Ridder, 1990). Neuman (1975) showed that contrary to the assumption of Boulton, α is not a characteristic constant of the aquifer but decreases linearly with the logarithm of r.

The difference between Boulton's theory and that of Neuman (as far as fully penetrating wells are concerned) is that the former only enables one to calculate α, whereas the latter enables one to determine the degree of anisotropy, as well as the horizontal and vertical hydraulic conductivities.

9.6.4 Bounded aquifers

In the various aquifer test methods mentioned above, it was assumed that the aquifer is of infinite areal extent. However, in nature such an assumption will not always hold good as the aquifers are limited by some hydrogeological boundary. Two types of boundaries are easily identified, i.e. a recharge boundary like a river or a canal which causes replenishment of the aquifer and a barrier or impermeable boundary which is a result of either thinning or termination of an aquifer against a low-permeability formation or dyke or a fault which limits movement of water (Figure 9.27).

The method of images, commonly used in the study of heat conduction in solids, has been used for the solution of boundary problems in groundwater flow. The boundary condition is

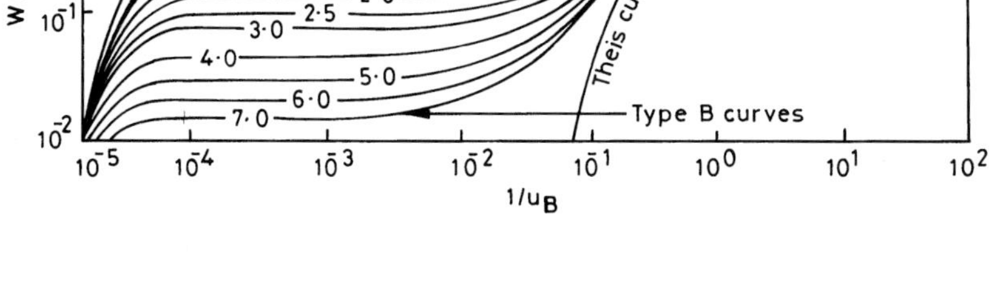

FIGURE 9.26 Neumann's type curves for unconfined aquifer.

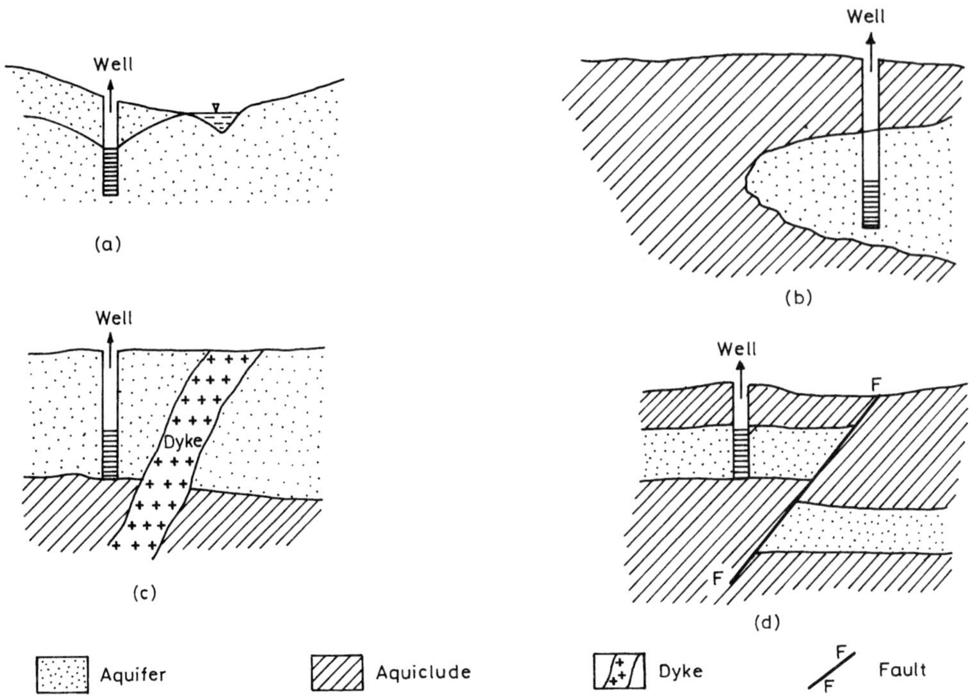

FIGURE 9.27 Diagram illustrating various types of boundaries. (a) Recharging boundary due to stream. Impermeable boundaries due to: (b) lateral termination, (c) dyke, and (d) fault.

simulated by an image well or imaginary well located at the same (equivalent) distance from the boundary but on the opposite side. Depending upon the nature of the boundary, i.e. recharging or barrier type, the flow is characterized by assuming a recharging or discharging image well. Thus the real bounded system is replaced by an equivalent hydraulic system, i.e. an imaginary system of infinite areal extent. Both the recharge and discharge image wells are located perpendicular to the boundary, but on the opposite sides at equal distances from the boundary (Figures 9.28 and 9.29). In a recharging system (Figure 9.28), the recharge image well recharges the aquifer at the same rate as the rate of withdrawal from the real well. As a result of this the build-up due to the cone of impression (recharge), and the drawdown caused due to pumping from the real well, exactly cancel each other at the recharging boundary resulting in a constant head along the line source. In cases where there are more than one boundary, more image wells are to be considered depending upon the configuration and nature of the boundaries. Several such geometric configurations are considered by Ferris *et al.* (1962), and Bear (1979), among others.

The effect of boundary conditions on the time-drawdown trend in an observation well will depend on the nature of the boundary, the time when the cone of depression intercepts the boundary and the distance of the observation well from the pumping well. On a log-log plot, the boundary effects are observed by the deviation of the field data from the Theis curve. When time-drawdown data are plotted on semi-log paper, the recharging boundary will cause a flattening of the time-drawdown curve and a barrier (impermeable) boundary will cause further steepening of the curve (Figure 9.30). The

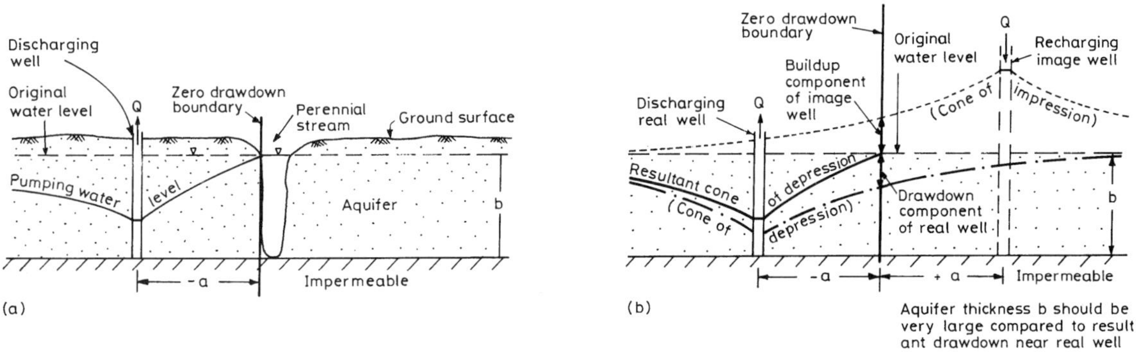

FIGURE 9.28 Schematic section of: (a) a recharging well in a semi-infinite aquifer bounded by a perennial stream, and (b) of the equivalent hydraulic system in an infinite aquifer (after Ferris *et al.*, 1962).

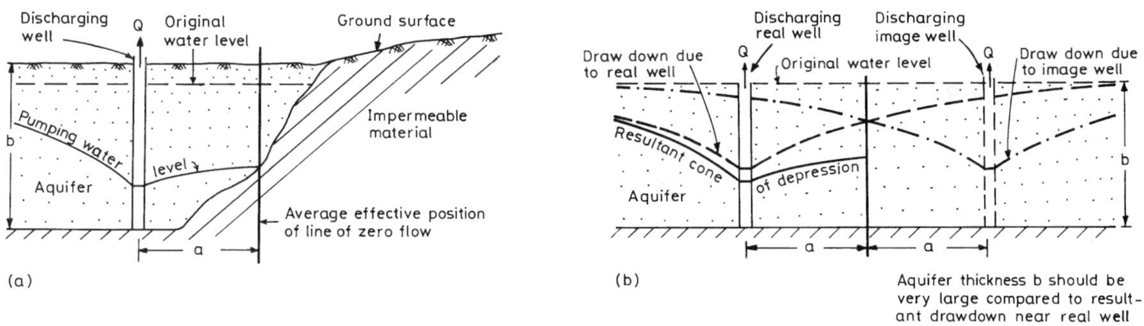

FIGURE 9.29 Schematic section of: (a) a discharging well in a semi-infinite aquifer bounded by an impermeable formation, and (b) the equivalent hydraulic system in an infinite aquifer (after Ferris *et al.*, 1963).

time of inflection of the time-drawdown curve from the normal trend (when the aquifer is of infinite areal extent) depends on the distance of the observation well from the pumping well. Note that the effect of the recharging boundary on the time-drawdown curve is the same as that for leaky aquifers or partially penetrating wells. Therefore, for a proper interpretation of pumping test data, necessary information about the hydrogeological system is necessary.

The recharging boundary, viz. rivers, canals, or lakes can be easily observed in the field but barrier boundaries are at times hidden as a buried river channel, or a subsurface dyke, or a fault. Such boundaries can be identified and mapped by the analysis of observed time-drawdown data provided that pumping is carried out for a sufficiently long time so that the cone of depression intercepts the boundary. As mentioned above, the nature of the boundary (recharging or barrier type) can first be ascertained from the trend of the time-drawdown data. The next step is to determine the distance of the image well from the observation well (r_i) for which time-drawdown data from at least three observation wells are necessary. The values of r_i from various observation wells can be estimated from the semi-log time-drawdown plot by choosing any arbitrary value of drawdown, s_w, and corresponding value of time, t_r, before the effect of the boundary influences the time-drawdown curve (Figure 9.31a). The

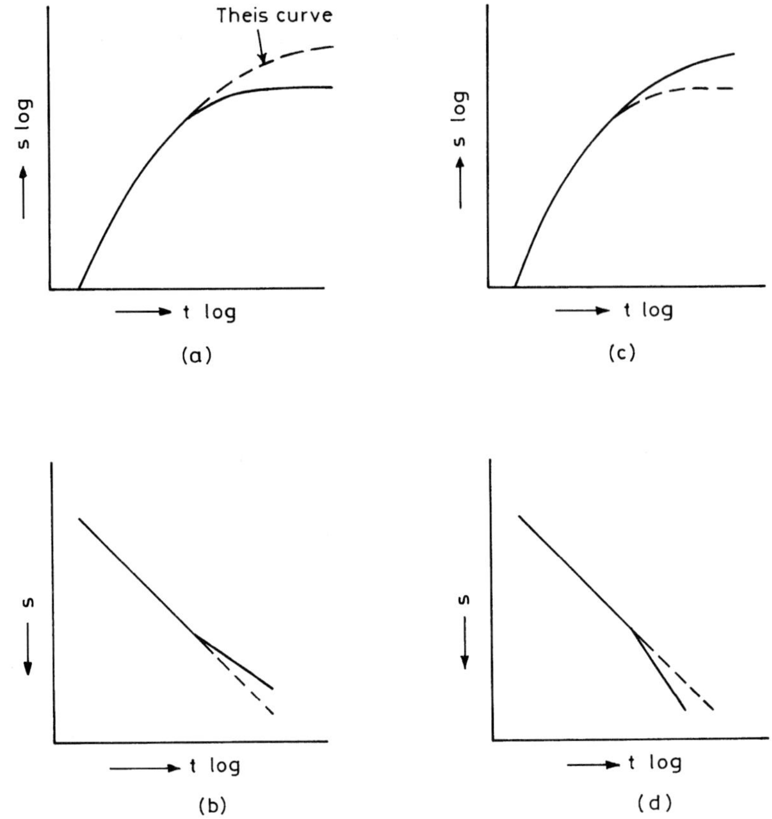

FIGURE 9.30 Effect of a recharge boundary (a and b) and impermeable boundary (c and d) on the time-drawdown curves in a confined aquifer.

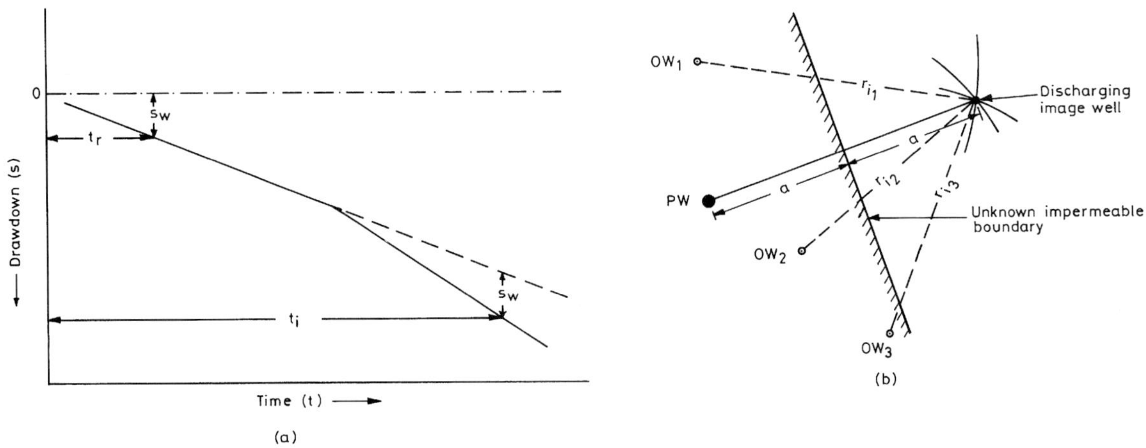

FIGURE 9.31 Diagrams illustrating application of image well theory to locate an unknown impermeable boundary; (a) time-drawdown plot showing effect of impermeable boundary, (b) location of the unknown impermeable boundary. OW = observation well, PW = pumping well.

time-drawdown curve is extrapolated beyond the time of inflection. The time intercept, t_i of an equal amount s_w of divergence caused by the image well is read from the data curve. The distance r_i, can be determined from equation (9.53)

$$r_i = r_r \sqrt{t_i/t_r} \tag{9.53}$$

where r_r is the distance of the observation well from the pumping well and r_i, t_i and t_r are as defined earlier.

After the values of r_i are computed for the individual observation wells, circles are drawn with their centres at the respective observation wells and their radii equal to the respective estimated values of r_i. The intersection of the arcs will give the location of the image well. The perpendicular bisector of the line joining the image well and the real (pumped) well will mark the strike of the boundary (Figure 9.31b).

9.6.5 Partially penetrating wells

One of the assumptions in the Theis and other methods as discussed earlier, is that the pumped well penetrates the entire thickness of the aquifer so that the flow towards the well is horizontal. However, under field conditions especially when aquifers are very thick, it is usually uneconomical to provide a screen along the entire thickness of the aquifer. Such a well which taps only a part of the aquifer thickness is known as a partially penetrating well.

In a partially penetrating well, the flow pattern differs from the radial horizontal flow which presumably exists around a fully penetrating well. Therefore, flow towards a partially penetrating well will be three-dimensional due to the vertical flow component. Accordingly, the average length of a flow line will be greater in a partially penetrating well as compared with the fully penetrating well which will cause a greater resistance to flow resulting in additional drawdown (Figure 9.13b and d). The discharge from a partially penetrating well, therefore, will be less than a fully penetrating well for the same drawdown. It would also mean that if they are pumped at the same rate, the drawdown in a partially penetrating well (s_p) will be greater than that in a fully penetrating well (s_w). The effect of partial penetration is greater near the well face and will decrease with increasing distance from the pumping well (Figure 9.32a). Anisotropy causes greater effect of partial penetration on drawdown as compared to isotropic aquifers as in the former the flow is three-dimensional. In an anisotropic aquifer, the effect of partial penetration will be negligible at a distance of $r \geq 1.5\ b \sqrt{K_h/K_v}$ and at $t > Sb^2/2T$. In a partially penetrating well, the drawdown will be greater during early time of pumping (Figure 9.32b). Methods of analysing test data from partially penetrating wells for steady and unsteady conditions in confined, unconfined, leaky and anisotropic aquifers are described by Hantush (1964) and Kruseman and de Ridder (1990).

9.6.6 Large diameter wells

The various methods described above assume that the pumping well has an infinitesimal diameter which will not be valid in large diameter (dug) wells. Therefore, these methods are not applicable to large diameter wells due to the significant effect of well storage. Dug wells are common in low permeability hard rocks of India and other developing countries. The earliest method to estimate aquifer properties from a large diameter, fully penetrating well in a confined aquifer was given by Papadopulos and Cooper (1967). Later, several other analytical and numerical solutions of unsteady flow to large-diameter wells were developed (Boulton and Streltsova, 1976; Rushton and Holt, 1981; Rushton and Singh, 1983; Herbert and Kitching, 1981; Barker, 1991; Chachadi *et al.* 1991).

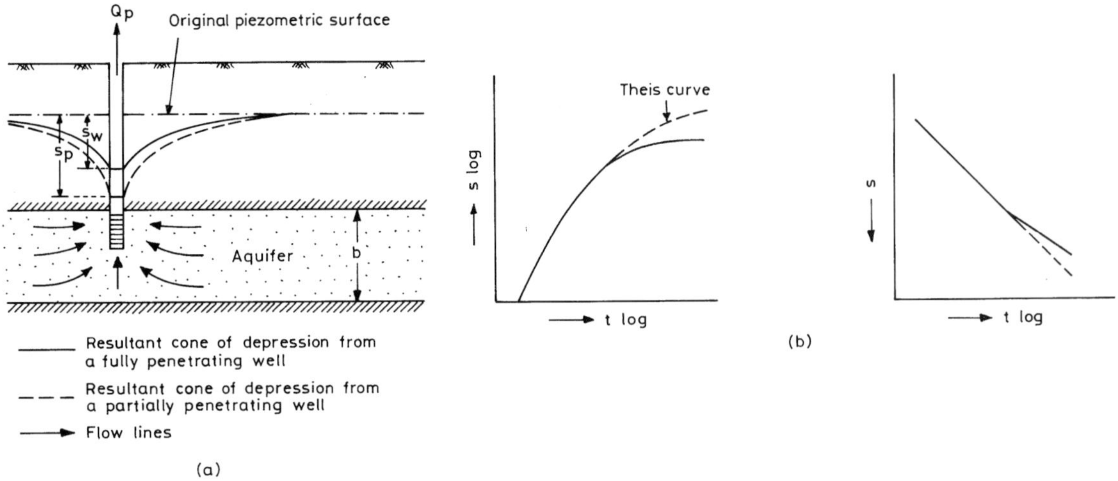

FIGURE 9.32 (a) Effect of partially penetrating well on drawdown in a confined aquifer. (b) Effect of partial penetration on the time-drawdown relationship in a confined aquifer.

Papadopulos–Cooper method

Padopulos and Cooper (1967), gave equation (9.54) to describe drawdown in a fully penetrating large-diameter well (with storage) in a confined aquifer (Figure 9.33)

$$s_w = \frac{Q}{4\pi T} F(U_w, \beta) \tag{9.54}$$

where

$$U_w = \frac{r_w^2 S}{4Tt} \tag{9.55}$$

$$\beta = \frac{r^2 w^s}{r_c^2} \tag{9.56}$$

r_w = radius of the well screen or open well, r_c = radius of the well casing over which the water-level is changing.

As in other type curve matching methods, the time-drawdown curve is plotted on a double-log paper which is matched with one of the type curves for large diameter wells (Figure 9.34). A match point is selected and the values of $F(U_w, \beta)$, $1/u_w$ and t are noted. The value of β of the type curve with which the observed data curve is matched is also noted. T is estimated from equation (9.54). Values of S can be computed by two methods: (1) by substituting the values of $1/u_w$, t, r_w and T into equation (9.55), and (2) by substituting the values of β, r_w and r_c into equation (9.56).

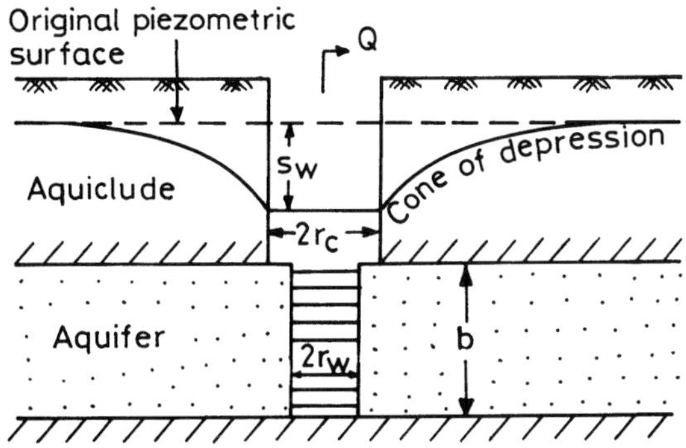

FIGURE 9.33 Schematic cross-section of a confined aquifer pumped by a large diameter well.

The main problem in curve matching is that the early part of type curves (Figure 9.34) differ only slightly in shape from each other and therefore uncertain value of β will make a significant difference in the computed value of S. Therefore, unless time of pumping is large, reliable value of S cannot be obtained, but transmissivity can be computed without such a problem.

The well storage dominates the time-drawdown curve up to a time, *t*, given by $t = (25r_c^2/T)$, after which the effect of well storage will become negligible and hence reliable values of aquifer characteristics *T* and S can be obtained. Therefore, for obtaining a representative value of S, the well should be pumped beyond this time *t*. In hard rocks, such as granites and basalts where *T* is about 10 m^2d^{-1} and r_c is 2 m, *t* will be ten days which is impractical. Therefore, the Papadopulos–Cooper type curve method may not give a correct value of *S*, but *T* values can be reliable.

The Papadopulos–Cooper method, although taking into account the effect of well storage, is applicable only to a fully penetrating abstraction well in a confined aquifer. A method for analysing test data from a partially penetrating well in an anisotropic unconfined aquifer is given by Boulton and Streltsova (1976). The Papadopulos–Cooper method also assumes constant rate of abstraction which is difficult to maintain when a large-diameter well is discharged by a centrifugal pump. A method to maintain constant discharge during the pumping test is suggested by Athavale *et al.* (1983). A curve matching technique considering the falling rate of abstraction during the test is given by Rushton and Singh (1983).

The effect of seepage face which is developed in the unconfined aquifer due to differences in the water-level outside and inside a large-diameter well was considered by Rushton and Holt (1981), Rushton and Singh (1987) and Sakthivadivel and Rushton (1989). The advantage of using recovery data over drawdown data is advocated by Herbert and Kitching (1981), Singh and Gupta (1986) and others. This is because during recovery: (a) well storage does not play an important role as all water is derived from the aquifer, and (b) well losses are small especially during the later stages of recovery.

The use of numerical methods for computing aquifer properties under complex hydrogeological conditions are advantageous. The discrete kernel approach suggested by Patel and Mishra (1983), Rushton and Singh (1987), and Barker (1991) is advantageous for estimating aquifer parameters from both drawdown and recovery data under varying rates of pumping. Chachadi *et al.* (1991) also considered storage effect both in the production and observation wells.

9.7 Variable-discharge (step-drawdown) test

In a variable-rate (step-drawdown) test, the well is pumped in three or more steps with an increase in rate of discharge, each step being of about one hour duration. Drawdown measurements in the pumped well are taken at frequent intervals.

Step-drawdown tests are useful for: (1) estimation of well performance or its productivity, (2) selection of suitable rate of pumping, and (3) estimation of hydraulic characteristics of aquifers in cases where cost factors may not permit long duration constant rate pumping tests.

The drawdown in a pumping well at a given time, s_w, includes two components, the aquifer loss (BQ) and the well loss (CQ^n) (Figure 9.35). Aquifer or formation loss (BQ) is due to the Darcy type linear flow of water in the aquifer. Well losses (CQ^n) are both linear and non-linear. As the well loss is mainly due to turbulent flow it is expressed as CQ^n, where the value of *n* may vary from 1.5 to 3.5, depending on the extent of turbulence or rate of discharge. The value of *n* is usually taken to be 2, i.e.

$$s_w = BQ + CQ^2 \tag{9.57}$$

where s_w = drawdown in pumping well, *B* =

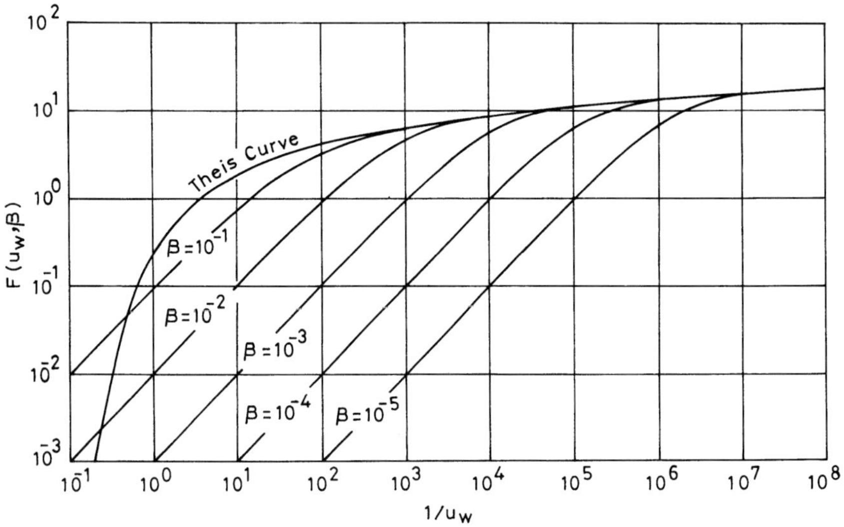

FIGURE 9.34 Family of a Papadopulos type curves for large-diameter wells.

formation loss coefficient, BQ = formation loss, C = well loss coefficient, CQ^2 = well loss.

The value of well loss coefficient, C, is a measure of the degree of well deterioration due to clogging or other reasons. Values of C between 0.5 and 1.0 min^2m^{-5} indicate mild deterioration while clogging is severe when C is between 1.0 and 4.0 min^2m^{-5}. It is then difficult to restore the original capacity (Walton, 1962). If a well does not show any well loss, it is a perfect or ideal well. Step-drawdown tests are also useful in fractured rocks for determining flow regime and in discriminating linear and non-linear head losses around a pumping well (Mackie, 1982; Mishra *et al.*, 1989).

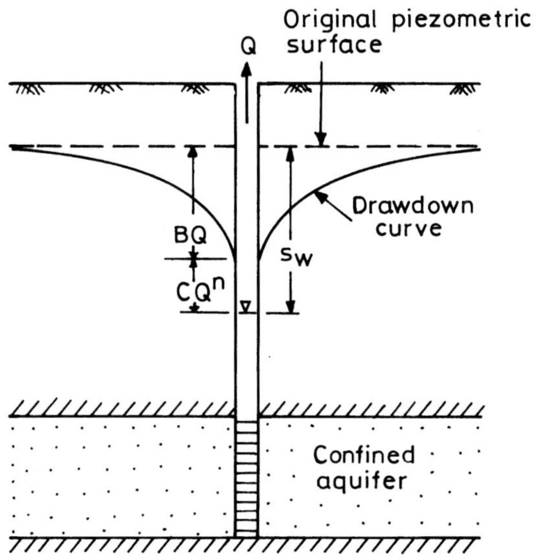

FIGURE 9.35 Schematic cross-section of a fully penetrating well in a confined aquifer showing relation of well loss (CQ^n) and formation loss, BQ to total drawdown, s_w.

9.7.1 Evaluation of well loss

Several graphical (Walton, 1962) and numerical methods (Avci, 1992) are developed for the analysis of step-drawdown test data for the estimation of aquifer parameters and well loss constant. Walton (1962) suggested a graphical method in which the time-drawdown data for successive steps are plotted on a semi-log paper; *t* on the logarithmic scale and s_w on the arithmetic scale (Figure 9.36a). Equation (9.58) can be used for determining well loss *C*.

$$C = \frac{(\Delta s_i/\Delta Q_i) - (\Delta s_{i-1}/Q_{i-1})}{\Delta Q_{i-1} + \Delta Q_i} \tag{9.58}$$

The delta term in equation (9.58) represents increments of drawdown produced by each increase in the rate of pumping (ΔQ). The incremental drawdown, Δs is determined by noting the difference between the water-level at the beginning and at the end of each step keeping the time step Δt to be the same.

For steps 1 and 2, equation (9.58) can be written as

$$C = \frac{(\Delta s_2/\Delta Q_2) - (\Delta s_1/\Delta Q_1)}{\Delta Q_1 + \Delta Q_2} \quad (9.59)$$

and for steps 2 and 3

$$C = \frac{(\Delta s_3/\Delta Q_3) - (\Delta s_2/\Delta Q_2)}{\Delta Q_2 + \Delta Q_3} \quad (9.60)$$

The value of C computed from successive steps may not be constant. An increase in the value of C in successive steps indicates well deterioration due to clogging of well screen or other reasons, while a decrease in the value of C points to the development of well during pumping.

Equation (9.57) in terms of specific drawdown, s_w/Q can be re-written as

$$\frac{s_w}{Q} = B + CQ \quad (9.61)$$

Values of specific drawdown (s_w/Q) for each step as determined from Figure 9.36a, when plotted on an arithmetic scale against Q, should give a straight line (Figure 9.36b). The well loss coefficient, C, is given by the slope of the straight line and the formation loss coefficient, B by the intercept Q = O.

9.7.2 Well characteristics

In addition to well loss coefficient, C hydraulic characteristics of the well can also be expressed in terms of well efficiency and specific capacity.

Well efficiency

Well efficiency E_w is the ratio of aquifer loss to the total drawdown in the pumped well. It can be expressed as

$$E_w = \frac{BQ}{(BQ + CQ^2)} \times 100 \quad (9.62)$$

The value of well efficiency indicates the state of the well and aquifer development. No well is 100% efficient. A progressive decline of well efficiency would indicate clogging of the well screen and surrounding aquifers.

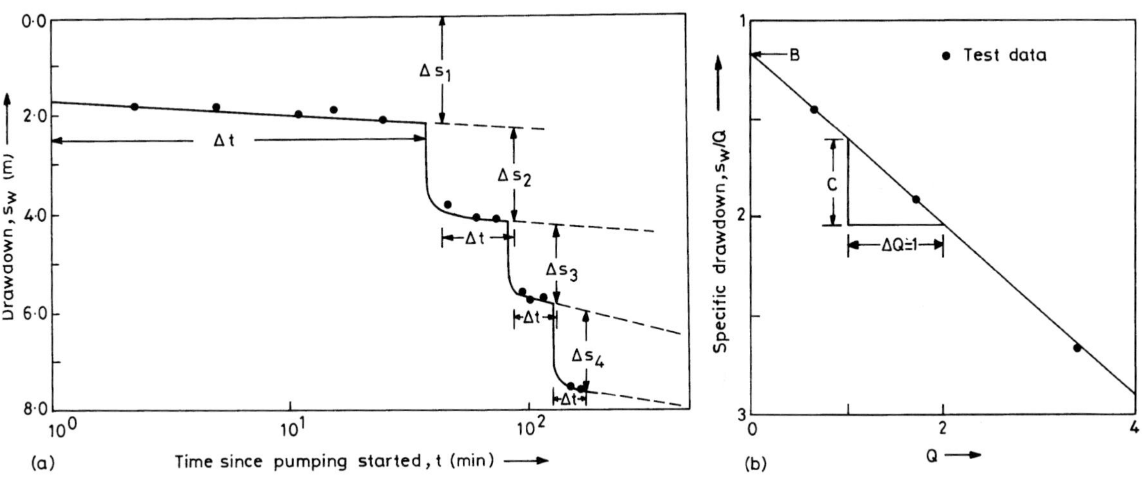

FIGURE 9.36 (a) Time-drawdown plot from step-drawdown test for estimation of C; (b) determination of B and C from graph of s_w/Q versus Q.

Specific Capacity

Specific capacity is the ratio of well yield (Q) to drawdown (s_w) in the pumping well (dimension L^2T^{-1}). It is a measure of the productivity of both the well and the aquifer; the higher the specific capacity the better the well efficiency. The specific capacity (Q/s_w) of a well discharging at a constant rate in a homogenous, isotropic confined aquifer infinite in areal extent can be expressed by equation (9.63) which is based on the non-steady state Jacob's equation (9.29)

$$\frac{Q}{s_w} = \frac{1}{(2.30/4\pi T)\log(2.25\,Tt/r_w^2 S)} \qquad (9.63)$$

Equation (9.63) assumes that: (1) the well has full penetration and receives water from the total saturated thickness of the aquifer, and (2) well loss is negligible. Considering well loss, equation (9.63) can be re-written as

$$\frac{Q}{s_w} = \frac{1}{(2.30/4\pi T)\log(2.25\,Tt/r_w^2 S) + CQ^{n-1}} \qquad (9.64)$$

Equation (9.64) indicates that specific capacity decreases with Q and t. Specific capacity also varies directly with the logarithm of $1/r_w^2$, i.e. a large increase in the radius of the well results in a comparatively small increase in the values of specific capacity.

Statistical analysis of a large number of specific capacity data from a variety of fractured rocks in Pennsylvania, USA, indicates that the main controlling factors are well diameter and duration of pumping while lithology plays a comparatively smaller role (Knopman and Hollyday, 1993).

9.7.3 Estimation of transmissivity from specific capacity data

Aquifer transmissivity can be estimated from specific capacity values to avoid the cost of a long duration pumping test (Walton, 1962, Brown, 1963). Based on equation (9.63), several graphs relating theoretical specific capacity with transmissivity for various values of S, T, and r_w are given by Walton (1962). One such plot relating specific capacity to T and S for a given value of t and r_w is illustrated in Figure 9.37. If specific capacity has been determined from a field (pumping) test, S is known a *priori* from either well-log data or water-level fluctuations, and r_w and t are known, transmissivity T can be computed by using such a graph. These graphs are based on the assumption that the well is 100% efficient and that the effective diameter is the same as the diameter of the well screen. The second assumption may hold good in consolidated rocks but in unconsolidated sediments, effective diameter is usually more than the diameter of the screen.

The other approach to estimate T from specific capacity values is based on the Thiem's equilibrium formula

$$T = \frac{2.3Q\log R/r_w}{2\pi s_w} = A\,(Q/s_w) \qquad (9.65)$$

where A is the coefficient depending on radius of influence, R and well radius, r_w; Q/s_w is the specific capacity of the well. Values of A would also depend on the hydraulic characteristics of the aquifer and degree of confinement; it can be estimated from values of r_w and R. In cases where exact values of R are not available, approximate values of R can be used as given in Table 9.5 as the value of A in equation (9.65) will not be much affected by poor estimates of R.

Based on the above approach, A was estimated to vary from 0.37 to 0.63 for the basaltic aquifer in the Deccan Traps of India (Adyalkar and Mani, 1972), and 0.39 by Fernandopulle *et al.* (1974) for the basalts of Gran Canaria in Spain. For valley fill deposits in California, USA, Huntley *et al.* (1992) reported its range from 0.9 to 1.2. For estimation of transmissivity in

TABLE 9.5 Approximate values of radius of influence, R (after UNESCO, 1972)

Type of formation	*Type of aquifer*	*R (m)*
Fine and medium-	Confined	250–500
grained sands	Unconfined	100–200
Coarse-grained sands	Confined	750–1500
and gravel-pebble beds	Unconfined	300–500
Fissured rocks	Confined	1000–1500
	Unconfined	500–1000

fractured rocks, Huntley *et al.* (1992) gave equation (9.66)

$$T = A\,(Q/s_w)^{1.18} \qquad (9.66)$$

where the value of coefficient, A will be 0.12 if both T and Q/s_w are expressed in m^2d^{-1}. Here it is assumed that well loss is negligible; this will result in the underestimation of transmissivity in alluvial aquifers due to significant well loss. In fractured rocks also, due to larger effective radius of the well and shorter duration of pumping, one will expect overestimation of transmissivity based on specific capacity values (Razack and Huntley, 1991). Further, the value of T estimated from specific capacity data applies only to the aquifer adjacent to the well which is affected by drilling and well development. Therefore, values of T estimated from specific capacity data are greater than those obtained from aquifer tests.

Well productivity

Well productivity or specific capacity index (dimension LT^{-1}) is computed by dividing specific capacity by the saturated thickness of the aquifer tapped by the well. Statistical analysis of specific capacity index data from wells screened at different depths in a multi-unit aquifer is useful in determining the relative productivity of different units or variation within the same lithologic unit (Walton, 1962, 1970). It is also useful in predicting well yields from a given thickness of the aquifer. Examples of well productivity data from different rock types are given in Chapters Eleven, Twelve, Thirteen and Fourteen.

9.8 Pumping tests in fractured rock aquifers

In areas where the fractures are closely spaced and are interconnected, conventional pumping test methods, assuming confined and leaky confined models, can be used. However, the double porosity model (section 7.22) is more representative of uniformly fractured aquifers including petroleum and geothermal reservoirs (Earlougher, 1977; Horne, 1990; Grant *et al.*, 1982).

9.8.1 Double porosity model

The flow characteristics in a double (dual porosity) aquifer were discussed earlier in section 7.2.2. In a double porosity aquifer the rock mass is assumed to consist of a number of porous blocks as well as a large number of randomly distributed, sized and oriented fractures. A method of estimating hydraulic properties of such a double porosity aquifer is described below.

Streltsova-Adams' method for confined fractured aquifer

Streltsova-Adams (1978) assumed that a confined fractured aquifer representing a double porosity model, consists of matrix blocks and fracture units in the form of alternate horizontal slabs (see Figure 7.7a). The thickness of the matrix units is greater than that of fracture units.

The drawdown distribution in the fracture (s_f) is given by equation (9.67)

$$s_f = \frac{Q}{4\pi T_f}\,W\,(u_f,\, r/B_f,\, n) \qquad (9.67)$$

where

$$u_f = \frac{r^2 S_f}{4T_f t} \qquad (9.68)$$

$$B_f^2 = \frac{T_f}{\alpha_f S_m} \qquad (9.69)$$

and

$$n = 1 + \frac{S_m}{S_f} \qquad (9.70)$$

Subscripts f and m represent properties of fractures and matrix blocks respectively, B_f is the drainage factor. Accordingly, the equation for drawdown in matrix block, s_m is also developed.

Values of the drawdown function $W(u_f, r/B_f, n)$ computed for n = 10, 100, and 1000 and various assumed values of parameters r/B_f and $1/u_f$ are given by Streltsova-Adams (1978).

Streltsova-Adams' method is based on the assumptions that: (a) fractures and blocks are compressible (b) the abstraction well is fully penetrating the fractured aquifer and receives water from it (c) pumping is at a constant rate, (d) the radius of the pumped well is vanishingly small, (e) the flow in the block is vertical, (f) the flow in the fissure is horizontal, and (g) flow in both blocks and fissures obey Darcy's law.

Based on the values of well function, and values of r/B_f and $1/u_f$, type-curves can be plotted on double-log paper. One such set of type curves for n = 10 is given in Figure 9.38. Time-drawdown data is plotted on double log paper and matched with one of the type curves. A match point is selected for which coordinate values are noted and substituted in equations (9.67), (9.68) and (9.69) to obain values of T_f, S_f, S_m and B_f. Other characteristics which can be estimated are α_f, and C_f. α_f, a characteristic of the fissure flow (d^{-1}), is given by

$$\alpha_f = \frac{T_f}{{B_f}^2 S_m} \qquad (9.71)$$

and C_f (hydraulic diffusivity of the fissure)

$$= \frac{T_f}{S_f} (m^2 d^{-1}) \qquad (9.72)$$

As an example, the time-drawdown data

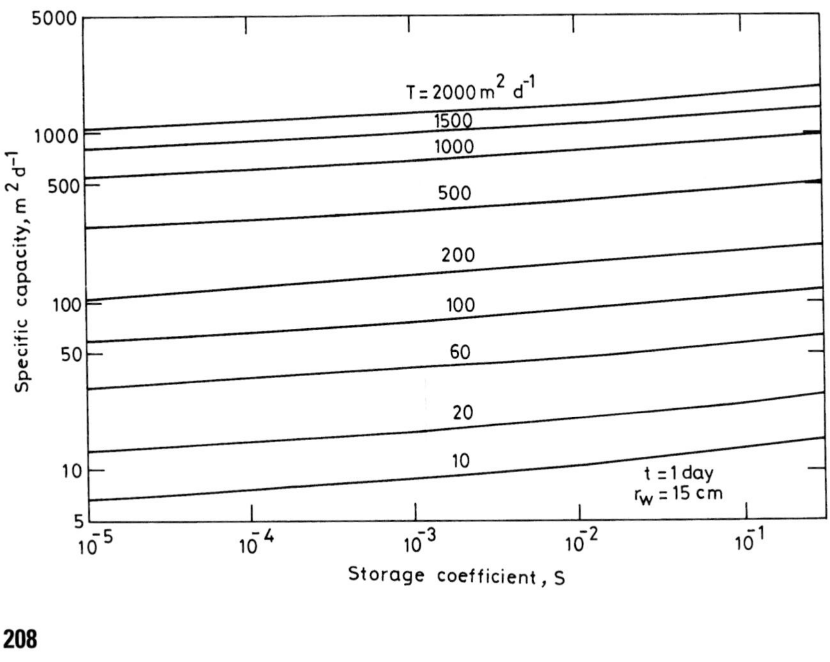

FIGURE 9.37 Relationship between specific capacity, transmissivity and storativity based on Jacob's non-equilibrium equation (after Meyer, 1963; Todd, 1980).

curve from an observation well in fractured basalt is given in Figure 9.39. The field data curve was matched with the Theis type curve, Hantush leaky aquifer type curve, and Boulton's type curves for an unconfined aquifer. However, keeping in view the hydrogeological situation and because the best match was obtained with Streltsova-Adams' type curve ($n = 10$), hydraulic parameters were computed assuming a double-porosity model (Figure 9.39).

The equations given by Streltsova-Adams for a double-porosity aquifer involve delayed yield from blocks to fissures. Hence Streltsova-Adams' type curves are identical to Boulton's type curves for the unconfined aquifer involving delayed yield in unconsolidated formations. Therefore, for proper interpretation of pumping test data, hydrogeological framework of the aquifer system should be known.

Bourdet–Gringarten's method

The hydraulic characteristics of the fractures and the matrix block in a double-porosity aquifer can also be estimated using Bourdet–Gringarten's type curve method (Bourdet and Gringarten, 1980; Gringarten, 1982). The difference between Streltsova-Adams' approach and that of Bourdet–Gringarten is that the former considered blocks and fissures of regular shape and orientation in the form of slabs, while the latter model is applicable to different geometrical shapes of blocks. Readers may refer to Kruseman and de Ridder (1990) for further details of this method.

The effect of fracture skin in a fissured double-porosity aquifer is discussed by Moench (1984) and its application to estimate hydraulic parameters of fissured rock system is given by Levens *et al.* (1994).

Single vertical fracture models

The majority of oil and gas reservoirs are hydraulically fractured to produce a single vertical fracture, to augument well production. In hot dry rock (HDR) experiments also, such fractures are induced for circulating water to deeper parts in the crust for tapping geothermal energy. Naturally occurring vertical fractures are also intersected during drilling for

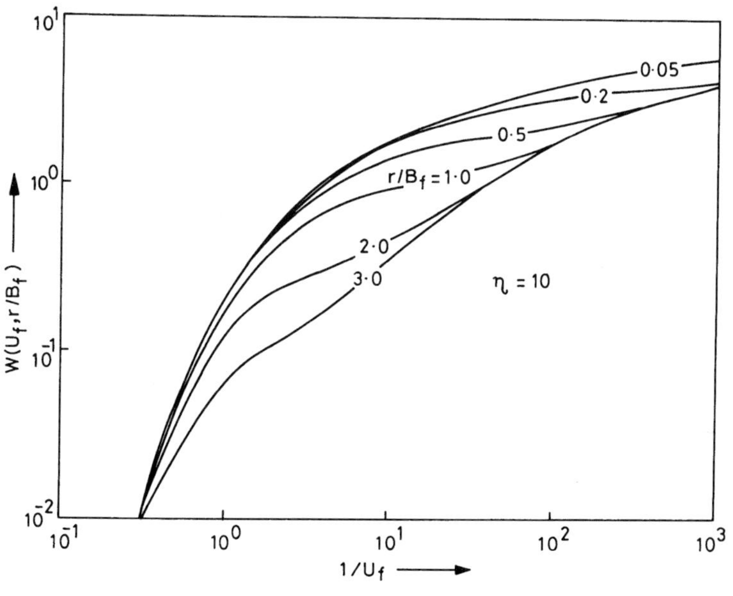

FIGURE 9.38 Streltsova-Adam's type curves for double-porosity confined aquifer (n = 10).

water and geothermal wells. Therefore, drawdown and pressure build-up in wells penetrating a single vertical fracture is of importance for estimating its hydraulic characteristics. The time-drawdown characteristics of a vertical fracture of high permeability was discussed earlier in section 9.5.3.

Gringarten–Witherspoon's method

The drawdown in an observation well as a result of pumping from a single, plane, vertical fracture in a homogeneous, isotropic confined aquifer is given by equation (9.73)

$$s = \frac{Q}{4\pi T} F\,(u_{\text{vf}}, r') \qquad (9.73)$$

where

$$U_{\text{vf}} = \frac{Tt}{Sx_{\text{f}}^2} \qquad (9.74)$$

$$r' = \frac{\sqrt{(x^2 + y^2)}}{x_{\text{f}}} \qquad (9.75)$$

x_f = half length of the vertical fracture (m), x,y = distance between observation well and pumped well, measured along the x and y axis respectively (m).

Equations (9.73) and (9.74) indicate that the drawdown in the observation well depends on its location with respect to orientation of the fracture. The observation well could be located on the x axis or y axis or along a line at an angle of 45° to the strike of the fracture (Figure 9.40).

Type curves of drawdown function $F(u_{\text{vf}}, r')$ for different values of u_{vf} and r' and location of the observation wells are given in Kruseman and de Ridder (1990). If one knows about the location of the observation well with respect to the fracture, a particular set of type curves can be selected for matching the test data plot. By matching the data curve with the type curve, coordinate values of the match point are noted and T and S are calculated from equations (9.73) and (9.74).

The various methods of analysing test data from fractures are primarily developed for hydraulically fractured petroleum reservoirs with the assumption that fractures have large hydraulic conductivity. Therefore, their application to natural fractures is limited as natural fractures will not have infinite hydraulic conductivity due to infilling. Moreover, the assumption of negligible storage may also not hold good as natural fracture zones may be quite wide.

9.8.2 Analysis of pumping test data from intrusive dykes

Dykes are intrusive bodies of igneous rocks, commonly of dolerite composition which may extend to long distances of the order of several kilometres but are of limited width. They cut across different type of rocks and at places cause fracturing of the country rocks at the contact due to baking effect. The dykes are themselves fractured into a system of joints usually of columnar type which impart secondary permeability. At other places, dykes may be comparatively massive and impermeable forming a barrier boundary.

Methods of analysis of pumping test data from fractured dykes are given by Boonstra and Boehmer (1986, 1989) and Boehmer (1993). It is assumed by these researchers that the dyke is of infinite length and has finite width and a finite hydraulic conductivity. The upper part of the dyke is less permeable due to weathering of the dyke rock into clayey material, the middle part is fractured and is permeable while the deeper part is massive and impervious as the fractures tend to die out with depth. Therefore, the middle fractured part of the dyke forms a confined horizon which is continuous with the adjacent confined part of the aquifer in the country rock (Figure 9.41).

The time-drawdown data obtained from a pumping well in a dyke is characterized by three segments as mentioned earlier (Figure 9.17). The

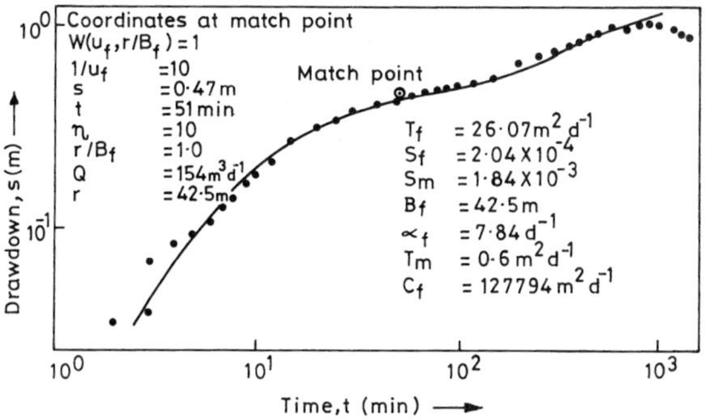

FIGURE 9.39 Time-drawdown plot of pumping test data from an observation well in fractured basalt, South India.

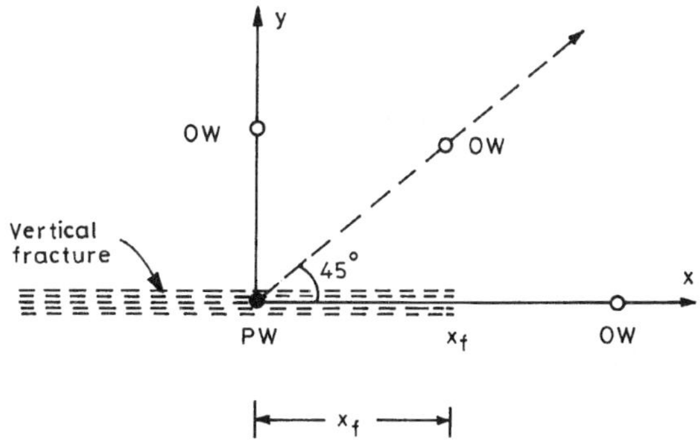

FIGURE 9.40 Plan view of a vertical fracture with a pumped well (PW) and observation wells (OW) at three different locations.

hydraulic characteristics of the dyke and the country rock can be determined by analysing time-drawdown data from observation wells in a dyke using a type-curve method developed by Boonstra and Boehmer (1986). A distance-drawdown type curve method is also suggested. They have demonstrated the applicability of their method for estimating hydraulic characteristics of a dolerite dyke of the Karoo system in the Republic of South Africa (Boehmer and Boonstra, 1987; Boonstra and Boehmer, 1989).

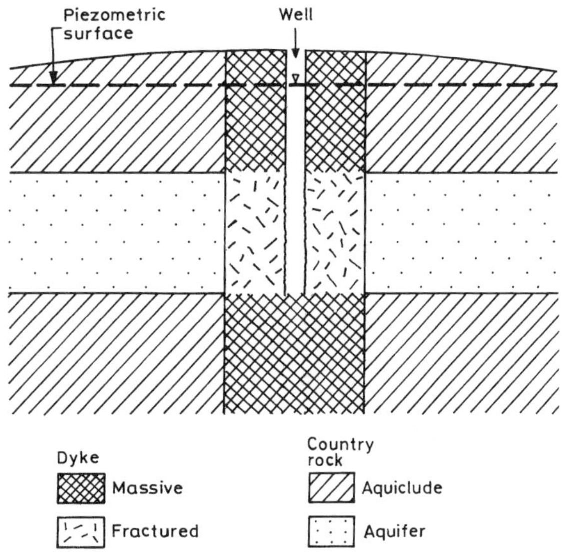

FIGURE 9.41 Composite dyke-aquifer system (after Kruseman and deRidder, 1990).

9.9 Other field methods

9.9.1 Geoelectrical methods

These are indirect field methods based on electrical properties of aquifers. However, these methods cannot replace other direct field methods, viz. pumping tests, for obtaining representative values of aquifer properties. The geoelectrical methods are based on the analogy between the Darcy's law and the Ohm's law relating hydraulic conductivity with electrical conductivity.

In a clay-free porous formation, fully saturated with water, the formation resistivity factor (F) is given by Archie law (equation 9.76)

$$F = \rho_o / \rho_w \qquad (9.76)$$

where ρ_o is the resistivity of the water saturated formation, and ρ_w is the resistivity of brine water.

For a two-layered formation, having thicknesses h_1 and h_2 and corresponding electrical resistivities ρ_1 and ρ_2 the average longitudinal (ρl) and average transverse (l_t) resistivities are given by equations (9.77) and (9.78)

$$(h_1 + h_2)\,\rho_t = h_1\,\rho_1 + h_2\,\rho_2 \qquad (9.77)$$

and

$$(h_1 + h_2)/\rho l = (h_1/\rho_1) + (h_2/\rho_2) \quad (9.78)$$

These can be extended to an n layer section (equations (9.79) and (9.80))

$$\sigma_l = \frac{1}{\rho_l} = \frac{\sum_{i=1}^{n} h_i/\rho_i}{\sum_{i=1}^{n} h_i} \quad (9.79)$$

and

$$\sigma_t = \frac{1}{\rho_t} = \sum_{i=1}^{n} h_i \Big/ \sum_{i=1}^{n} h_i \rho_i \quad (9.80)$$

where σ_ℓ and σ_t are longitudinal and transverse conductivities respectively.

Based on the above approach, Maillet (1947) introduced the concept of transverse unit resistance (T_R) and longitudinal unit conductance (S_c) which can be expressed as

$$T_R = \sum_{i=1}^{n} \rho_i h_i \quad (9.81)$$

and

$$S_c = \sum_{i=1}^{n} h_i/\rho_i \quad (9.82)$$

T_R and S_c are known as **Dar Zarrouk variable** and **Dar Zarrouk function** respectively. These parameters have been used for the estimation of hydraulic conductivity (K) and transmissivity (T) by many researchers, e.g. Kelly (1977), Sri Niwas and Singhal (1981, 1985) and Huntley (1986).

Sri Niwas and Singhal (1981) proposed the following empirical relation (equation (9.83)) relating T with T_R for a homogeneous isotropic aquifer having groundwater of uniform quality.

$$T = \alpha T_R \quad (9.83)$$

where $\alpha = K\ \sigma$ (a constant)

For incorporating effects of variation in quality of groundwater, equation (9.83) was later modified by Singhal and Sri Niwas (1983) as

$$K = \alpha' \rho' \quad (9.84)$$

where

$$\alpha' = K\sigma' \quad (9.85)$$

and

$$\sigma' = \frac{\sigma \rho_w}{\bar{\rho}_w} \quad (9.86)$$

$$\rho' = \rho \left(\frac{\bar{\rho}_w}{\rho_w} \right) \quad (9.87)$$

$\bar{\rho}$w is the average water resistivity, σ' is modified conductivity and ρ' is modified resistivity. $K\sigma'$ is a constant for an aquifer.

Therefore, the relation between transmissivity (T) and hydraulic conductivity (K) in terms of T_R will be

$$T = (K\sigma')\, T_R \quad (9.88)$$

The value of ($K\sigma'$) is area specific depending on rock type and water quality. It was estimated to be 0.5–1.15 for alluvial formations of North India (Sri Niwas and Singhal, 1985). Vertical electrical sounding data have also been used recently to estimate groundwater flow parameters in fractured rocks (Cassiani and Medina, 1997).

9.9.2 Spring discharge method

Spring discharge data can also be used to estimate transmissivity as described in section 13.6.

Chapter ten

Groundwater quality and contamination

The quality of water is as important as its available quantity. Rain and snow are the purest forms of water which undergo many complex chemical changes after coming in contact with soil and other rock materials. Man's activities also have a considerable influence on water quality.

The problem of groundwater contamination is widespread especially in developing countries due to lack of proper sanitary conditions and piped water supply. In these countries, a large number of people still use water from shallow dug wells or ponds and rivers. The WHO estimated that in 1980, about 1320 million (57%) of the developing World (excluding China) were without a clean water supply, while 1730 million (75%) were without adequate sanitation. Keeping this in view, the United Nations declared the decade 1981–1990 as the International Decade of Drinking Water and Sanitation. In India also, the Federal Government has launched a programme of Drinking Water Mission to provide safe drinking-water.

10.1 Expressing water analysis data

The concentration of the dissolved inorganic constituents in water is expressed in ionic form, while those salts which occur in undissociated or colloidal form are reported as oxides or as an uncombined element. For example, the sodium concentration is expressed as cation Na^+, iron is reported as the element Fe, and silicon as oxide SiO_2.

The concentrations of dissolved salts or ions in groundwater are usually expressed in parts per million (ppm) by weight. The parts per thousand unit is commonly used in reporting the composition of sea water. Parts per billion (ppb) or parts per trillion (ppt) are used in reporting the concentration of trace elements. In weight-per-volume units, the concentration of ions is expressed in milligrams per litre (mg l^{-1}). The ppm and mg l^{-1} units are numerically almost the same, if the density of water is nearly 1.0, and the concentration of dissolved solids is less than 7000 mg l^{-1}. In highly mineralized water, the conversion can be made by equation (10.1).

$$\text{Parts per million (ppm)} = \text{Milligram per litre/ Specific gravity of the water} \tag{10.1}$$

A unit which is more convenient for geochemical studies, is equivalent per million (epm) or milligram equivalents per litre (meq l^{-1}). The concentration of an ion in meq l^{-1} can be determined by multiplying mg l^{-1} value with the reciprocal of its combining (equivalent) weight. The concentration of an ion in meq l^{-1} is designated by putting symbol gamma (γ) before the chemical symbol, viz. γ Ca. This unit gives a better idea of the chemical character of water and is also of help in ascertaining the accuracy and completeness of chemical analysis.

The accuracy of chemical analysis of water samples can be checked by calculating the cation–anion balance, as the sum of the major cations should be equal to the sum of major anions expressed in meq l^{-1}. If it is not so, the analysis is either erroneous or incomplete. The percentage error i.e. 'ion-balance error' (e), can be determined by equation (10.2) (Matthess, 1982)

$$e = \frac{\Sigma\gamma c - \Sigma\gamma a}{\Sigma\gamma c + \Sigma\gamma a} \qquad (10.2)$$

where γc represents the cation sum and γa the anion sum in meq l^{-1}. In general the value of e should be less than 5% and certainly less than 10%. In this procedure, colloidal or suspended matter such as Al_2O_3 and SiO_2 is not included.

10.2 Isotopic composition

Natural water contains mainly hydrogen (H) of mass 1(^1H) and oxygen (O) of mass 16 (^{16}O). In addition, it also contains small amounts of deuterium (^2H), tritium (^3H), and isotopes of oxygen (^{17}O and ^{18}O). The relative abundance of these isotopes is given in Table 10.1. Of the six isotopes of hydrogen and oxygen (^1H, ^2H, ^3H, ^{16}O, ^{17}O and ^{18}O), five are stable, while tritium (^3H) is radioactive with a half-life of 12.3 y. The various isotopes of hydrogen and oxygen combine to form 18 types of water molecules. The common water molecules are $^1H_2{}^{16}O$, $^2H_2{}^{16}O$, $^3H_2{}^{16}O$, $^1H_2{}^{17}O$ and $^1H_2{}^{18}O$. Water with isotopic composition $^1H_2{}^{16}O$ and molecular weight of 18 forms 99.8% of the total water on the Earth while the proportion of heavy water $^2H_2{}^{16}O$ is only about 0.3‰ (Matthess, 1982).

Isotopic analysis of water is of importance in hydrology as it can provide the history and origin of water. For example, surface water with an abundance of heavy isotopes would indicate

TABLE 10.1 Relative abundance of hydrogen and oxygen isotopes in natural waters (after UNESCO, 1984b)

Isotope	*Rleative abundance, (%)*	*Half-life*
^1H	99.985	stable
^2H	0.015	stable
^3H	Trace	12.3 yr
^{16}O	99.76	stable
^{17}O	0.04	stable
^{18}O	0.20	stable

partial evaporation. Radioactive isotopes like tritium are of help in knowing the relative ages and rate of recharge of groundwater. Some of the applications of isotopes in groundwater hydrology are given in sections 3.5 and 17.1.3. Readers may also refer to IAEA (1983).

10.3 Dissolved constituents

The relative abundance of various elements in groundwater mainly depends upon their chemical mobility. The mobility of an element in the hydrosphere is determined by the solubility of its various compounds, the tendency of ion towards adsorption and base exchange, and the degree to which it is bound in the biosphere. The different behaviour of various elements from the point of view of their hydrochemical mobility is given in Table 10.2 where a comparison of the distribution of common elements in igneous and sedimentary rocks is made with their distribution in fresh groundwater and sea water.

Silicon, aluminum and iron which are most abundant in igneous rocks have low mobility in the hydrosphere. Chlorine which is relatively scarce in the Earth's crust is very mobile and widespread in the hydrosphere. Sodium is more mobile than potassium although both occur in almost equal amounts in the igneous rocks, which form primary source material (Table 10.2).

The various chemical constituents, based on

TABLE 10.2 Average chemical composition (mg l^{-1}) of three major rock types, sea water and groundwater. Relative mobility of common elements is also indicated (after Chebotarev, 1950; Hem, 1989)

Constituent	*Igneous rocks*	*Sandstone*	*Carbonates*	*Average sea water*	*Groundwater*	*Relative mobility*
SiO_2	285 000	359 000	34	6.4	1–30	0.20
Al	79 500	32 000	8970	<2.0	1–2	–
Fe	42 200	18 600	8190	0.01	0–5	–
Ca	36 200	22 400	272 000	400	10–200	3.00
Mg	17 600	8 100	45 300	1350	1–100	1.30
Na	28 100	3 870	393	10 500	1–300	2.40
K	25 700	13 200	2390	380	1–20	1.25
Sr	368	28	617	8	<10	–
HCO_3	–	–	–	142	80–400	–
SO_4	–	–	–	2700	10–100	57.00
Cl	200	150	–	19 000	1–150	100.0
F	715	220	112	1.3	0.1–2	–
Br	2.4	1.0	6.6	65	<0.5	–
B	7.5	90	16	4.6	<2	–
TDS	–	–	–	35 000	100–1000	–

their concentration in water, can be classified as: (1) major constituents (>1 mg l^{-1}), viz. Ca, Mg, Na, K, HCO_3, SO_4, Cl, NO_3 and SiO_2; (2) minor constituents (1 to 0.1 mg l^{-1}) viz. Fe^{++}, Mn, F, B; and (3) trace elements (<0.1 mg l^{-1}), viz. Al, As, Hg, Pb, Cr, Zn and other heavy metals. The other constituents are pesticides and radioactive materials. The source of various ions is given in Table 10.3.

10.4 Graphical presentation of chemical data

The water analysis data, for their proper evaluation can be plotted in: (a) hydrochemical maps; (b) hydrochemical diagrams, and (c) hydrochemical sections.

A brief description of these is given below. Readers may refer to Zaporozec (1972), Freeze and Cherry (1979), Matthess (1982), and Lloyd and Heathcote (1985), for a detailed account of various methods of plotting water quality data.

10.4.1 Hydrochemical (isocone) maps

In hydrochemical maps, the spatial variation in the hydrochemical characteristics of groundwater can be depicted by plotting contours (isocones) of either equal concentration of each ion or of equal ionic ratios in mg l^{-1} or meq l^{-1}. Necessary information about the lithology and structural features of rocks as well as the groundwater regime is necessary for their proper interpretation. Isochlor maps, depicting distribution of the chloride ion, are useful to determine the direction of groundwater flow as normally the chloride concentration is likely to increase in the flow direction. Isochlor and isoconductivity maps are also used in the study of sea water intrusion in coastal aquifers (section 17.6).

Equal ratio maps provide useful information about the relative dissolution of aquifer material, ion exchange and the mixing of waters of different qualities. Schoeller (1959) suggested equal ratio contour maps for $\gamma SO_4/\ \gamma Cl/\gamma Mg/\gamma Ca$ and other ionic combinations, where concentration of individual ions is expressed in meq l^{-1} (Figure 10.1). The ratio of $\gamma Mg/\gamma Ca$ in

TABLE 10.3 Sources of ions in water

Major ions (> 1 mgl^{-1})	*Source*
Calcium Ca^{2+}	Carbonates, gypsum
Magnesium MG^{2+}	Olivine, pyroxene, amphiboles
Sodium Na^{+}	Clays, feldspars, evaporites, industrial waste
Potassium K^{+}	Feldspar, fertilizers, K-evaporites
Bicarbonate HCO^{3-}	Soil and atmospheric CO_2, carbonates
Chloride Cl^{-}	Windborne, rain water, sea water and natural brines, evaporite deposits; pollution
Sulphate SO_4^{2-}	Gypsum and anhydrite, sea water, windborne, oxidation of pyrite
Nitrate NO^{3-}	Windborne, oxidation of ammonia or organic nitrogen, contamination
Silica, SiO_2	Hydrolysis of silicates
Minor ions (1 to 0.1 mgl^{-1})	
Iron Fe^{2+}	Oxides and sulphides, e.g. hematite and pyrite; corrosion of iron pipes
Manganese Mn^{2+}	Oxides and hydroxides,
Boron B	Tourmaline, evaporites, sewage, sea water
Fluoride F^{-}	Fluorine-bearing minerals, viz. flourite, biotite
Trace elements (< 0.1 mgl^{-1})	
As	Arsenic minerals, e.g. arsenopyrite, arsenic insecticides
I	Marine vegetation, evaporites
Zn	Sphalerite, industrial waste
Heavy metals (Hg, Pb, Cd, Cr)	Industrial waste and igneous rock weathering, under mild reducing conditions
Radioactive elements e.g. U, Ra, etc.	Uraniferous minerals, nuclear tests and nuclear power plants.

water is particularly useful in the study of water chemistry from carbonate rocks. In limestone and chalk aquifers $\gamma Mg/\gamma Ca$ ratio is usually in the 0.5–0.7 range and in dolomite aquifers it is in the 0.7–1.1 range. In olivine bearing (Mg-rich) silicate rocks, such as basalts, $\gamma Mg/\gamma Ca$ ratio in groundwater is greater than 1. The $\gamma Cl/\gamma HCO_3$ and $\gamma SO_4/\gamma Cl$ ratios have been used as indicators of sea water intrusion in coastal aquifers. The $\gamma HCO_3/\gamma Cl$ ratio is useful in determining the direction of groundwater movement and distance of a given water sample from its initial recharge area as this ratio will decrease in the direction of flow due to greater dissolution of Cl. Figure 10.2 illustrates an example of increase in TDS (total dissolved solids) and SO_4^{2-} and decrease in HCO_3^-/Cl^- ratio in the direction of groundwater flow. The ratio of sodium to total cations is used in studying the effect of ion exchange in sediments having cation exchange properties. The ratio of Br/Cl, I/Cl and B/Cl are recommended for determining the origin of deep oil field waters (Zaporozec, 1972). Br/Cl ratio is also a reliable parameter for identification of salinization sources (Vengosh and Rosenthal, 1994) and flow rates (Davis *et al.*, 1998). GIS methods using colour coding can be used to show the distribution of various ions (section 6.4.4.).

10.4.2 Hydrochemical pattern diagrams

The purpose of hydrochemical pattern diagrams is to depict the absolute or relative concentration of different cations and anions in terms of either mg l^{-1} or meq l^{-1}. These include the Collins bar diagram, Stiff diagram, radial diagram and circular diagram. These diagrams are useful in comparing waters from the different shapes/patterns of their plots; when plotted on maps, they will show spatial variation in water quality.

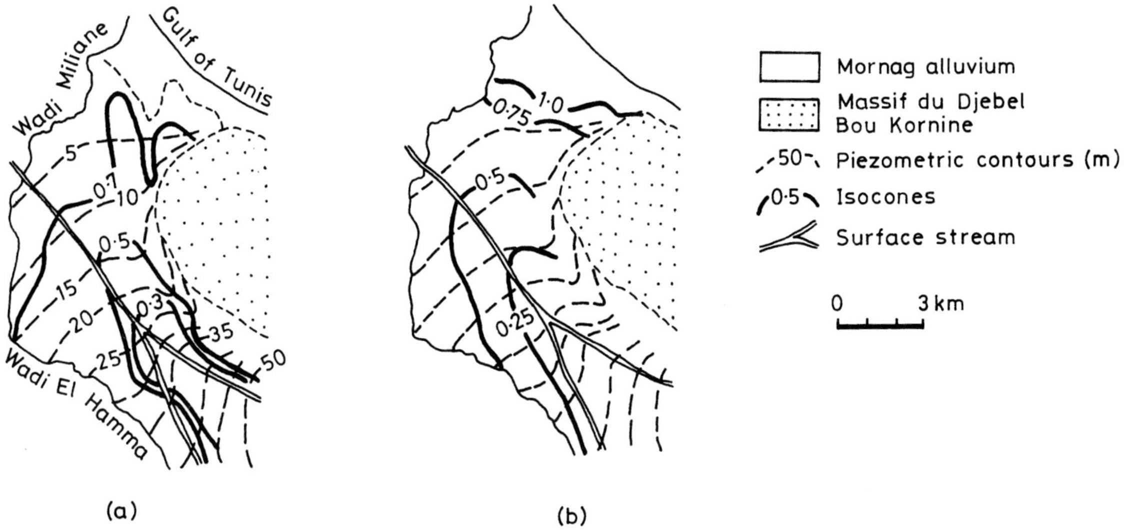

FIGURE 10.1 Equal ratio isocone maps of the Mornag alluvial aquifer, Tunisia: (a) map of SO_4^{2-}/Cl^{-1} ratios; (b) map of Mg^{2+}/Ca^{2+} ratios. Piezometric contours are also shown (after Schoeller, 1959).

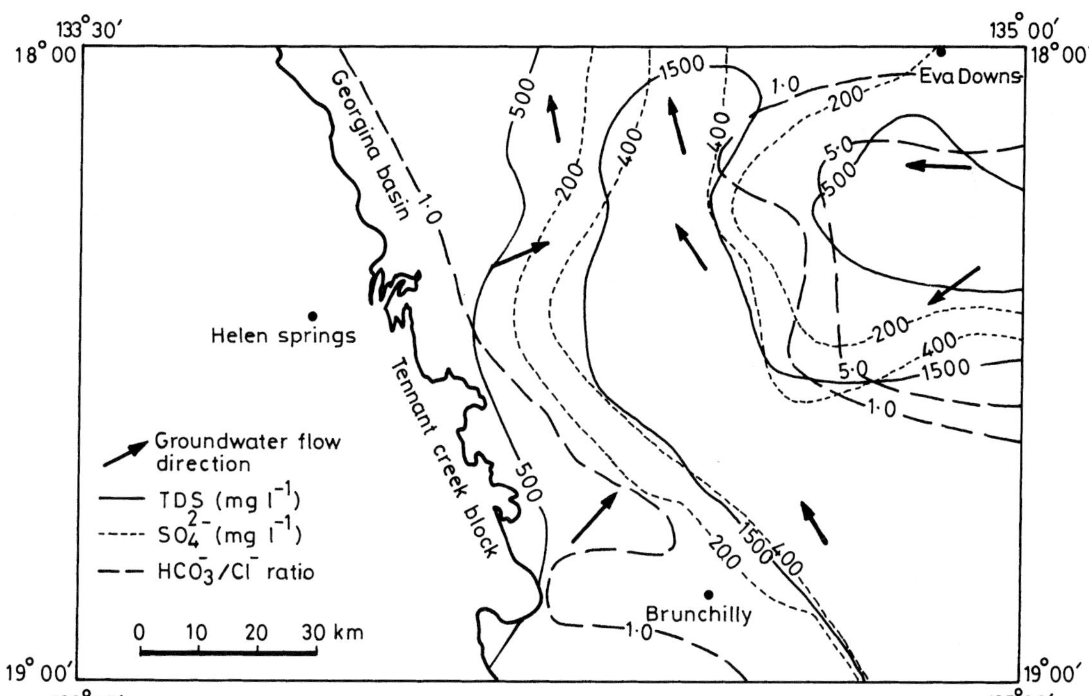

FIGURE 10.2 Isocones of TDS, SO_4^{2-} and HCO_3^-/Cl^- in the Palaeozoic limestone aquifer of Georgina Basin, Northern Territory, Australia (redrawn after Verma and Jolly, 1992).

TABLE 10.4 Chemical analysis data of two groundwater samples (after Hem, 1989)

Chemical Constituent	*Sample 1* mg l^{-1}	meq l^{-1}	meq l^{-1} (%)	*Sample 2* mg l^{-1}	meq l^{-1}	meq l^{-1} (%)
Ca^{2+}	37	1.85	6.0	49	2.45	21.8
Mg^{2+}	24	1.97	6.5	18	1.48	13.2
$Na^{+}+K^{+}$	611	26.58	87.5	168	7.29	65.0
HCO_3^-	429	7.03	23.2	202	3.31	29.6
SO_4^{2-}	1010	21.03	69.2	44	0.92	8.2
Cl^-	82	2.31	7.6	246	6.94	62.2
TDS	1980	–	–	649	–	–

The chemical analysis data of two water samples as given in Table 10.4 have been plotted in various hydrochemical diagrams (Figures 10.3 to 10.6 and Figure 10.8).

Collins' bar diagrams

These are vertical bar diagrams. Each sample is represented by two bars, one for cations and the other for anions (Figure 10.3). The height of each bar is proportional to the total concentration of cations or anions in meq l^{-1}. As the sum of the cations and sum of the anions expressed in meq l^{-1} should be equal, the height of the two bars for each sample should be the same. The concentration of cations and anions can be plotted either in absolute values (Figure 10.3) or the percentage of the total epm.

Stiff diagrams

In Stiff's diagram, the analytical data are plotted on three or four horizontal parallel axes equidistant from each other (Figure 10.4). These diagrams, like other pattern diagrams, have a disadvantage as separate diagrams are to be prepared for each analysis. However, they are useful in visualizing the differences in the distribution of cations and anions from their varying patterns.

Schoeller's diagram

A semi-logarithmic plot, in which on the abscissa (on arithmetic scale), the various

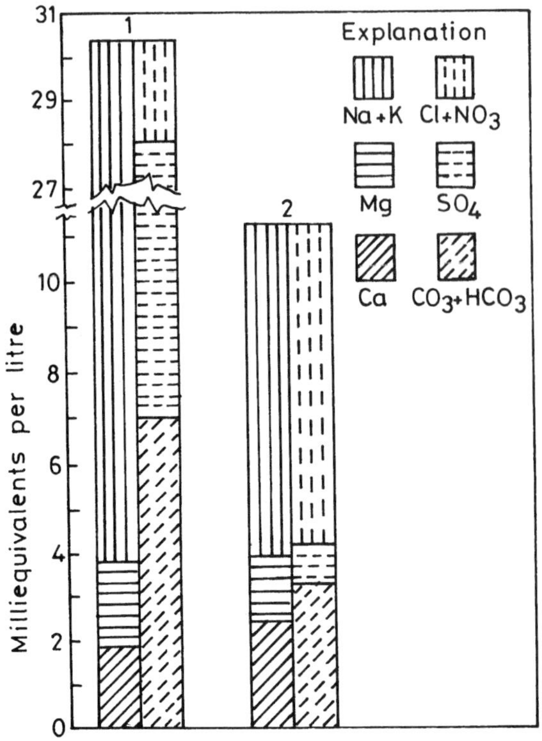

FIGURE 10.3 Representation of chemical analysis data (meql^{-1}) by Collins bar diagrams (after Hem, 1989).

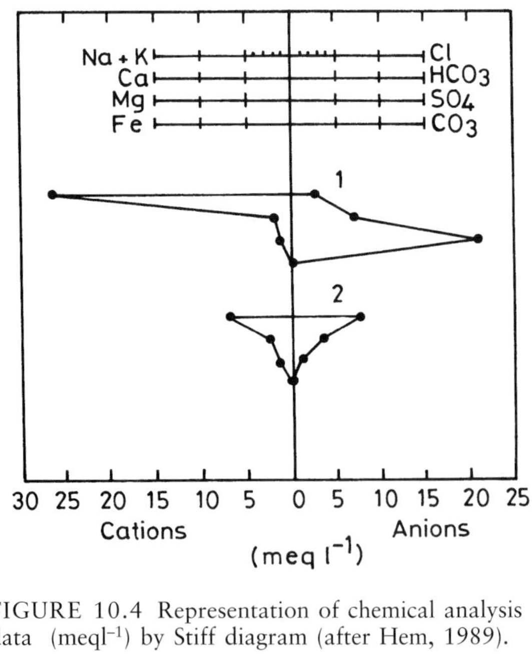

FIGURE 10.4 Representation of chemical analysis data (meql^{-1}) by Stiff diagram (after Hem, 1989).

cations and anions are arranged in the order Ca, Mg, Na, Cl, SO_4 and HCO_3, from left to right at equidistance. The concentration of each of these ions in mg l^{-1} or meq l^{-1} is plotted along the ordinate in logarithmic scale. The plots are joined by straight lines (Figure 10.5). These diagrams have an advantage over the other diagrams described earlier, as more than one analysis can be plotted on the same diagram and their hydrochemical characteristics can be compared. It is also possible to compare the ratios of the various elements from the slope of the lines joining the plots of the adjacent cations or anions, viz. $\gamma Mg/\gamma Ca$, $\gamma SO_4/\gamma Cl$ and others (Schoeller, 1959). Waters of similar composition will plot as near-parallel lines.

Hill-Piper diagram

The Hill-Piper tri-linear diagram, which is used extensively, was first conceived by Hill and later improved by Piper (1953). It has two triangular fields, one for cations and the other for

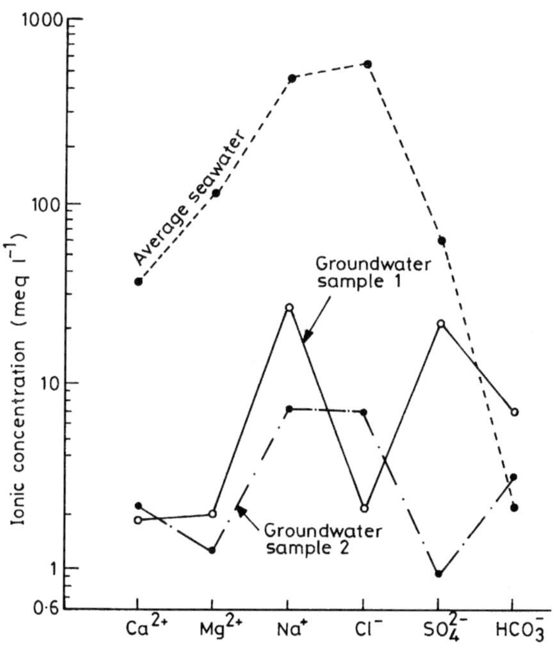

FIGURE 10.5 Chemical analyses (meql^{-1}) represented by Schoeller's semilogarithmic plotting.

anions, and a central diamond-shaped field (Figure 10.6). In the two triangular fields, the concentrations of cations and anions are plotted as percentages in meq l^{-1} so that the total of cations (Ca^{2+}, Mg^{2+}, and Na^{+} + K^{+}) and anions (Cl^{-}, SO_4^{2-} and CO_3^{2-} + HCO_3^{-}) are each considered as 100%. The respective cation and anion plots are projected into the central diamond-shaped field showing the overall chemical characteristics of the water sample in terms of certain combinations of cations and anions. The total salinity can be depicted by drawing circles with radii in proportion to the logarithmic values of totally dissolved solids (Figure 10.6).

The Hill–Piper system has an advantage of plotting analytical data of a large number of samples on one diagram. It is also used for the classification of water samples into various hydrochemical types depending on the relative concentration of major cations and anions (Figure 10.7, Table 10.5). It also has an advantage of demonstrating changes in water quality due to mixing of water of two extreme types, viz. mixing of fresh groundwater and sea water in coastal aquifers and also the effect of base exchange etc.

One of the main disadvantages of the Hill-Piper diagram is that it shows the relative concentration of different ions and not their absolute concentration.

Durov's diagram

This is also a type of tri-linear diagram, suggested by Durov (Zaporozec, 1972, Lloyd and Heathcote, 1985). In this diagram, the concentration of major cations and anions in percentage meq l^{-1} is plotted in two separate triangles. The sample points on the two triangles are projected to a central square field which represents the overall chemical character of the sample. The concentration of any other chemical characteristics, viz. TDS, EC, pH etc. can also be shown by extending the point from the central square field to the adjacent one or two scaled

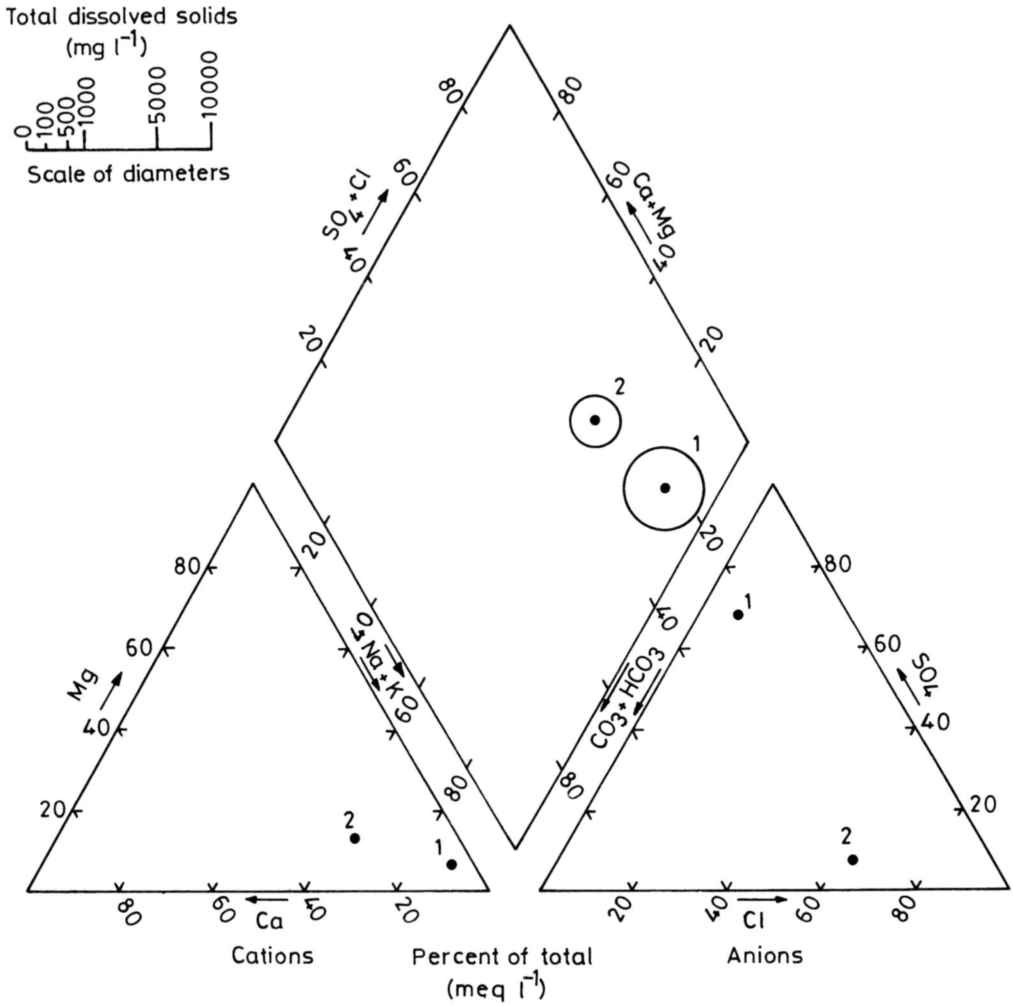

FIGURE 10.6 Hill-Piper diagram showing plotting of analytical data (after Hem, 1989).

rectangular fields as shown in Figure 10.8. Durov's diagram can also be used for plotting the concentration of minor ions. Burdon and Mazloum (1961) and Lloyd and Heathcote (1985) have given expanded versions of Durov's diagram which have an advantage over the Hill–Piper diagram in that they provide a better display of different types of water and also some hydrochemical processes such as ion exchange, simple dissolution and mixing of waters of different qualities.

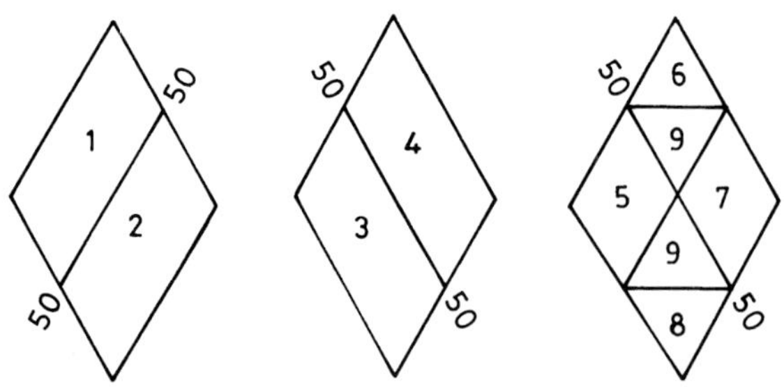

FIGURE 10.7 Subdivisions of the diamond-shaped field of Hill-Piper diagram showing various classes of water.

TABLE 10.5 Chemical characteristics of groundwater plots in different areas of the Hill–Piper diagram (after Piper, 1953)

Area 1:	Alakaline earths exceed alkalis
Area 2:	Alkalis exceed alkaline earths
Area 3:	Weak acids exceed strong acids
Area 4:	Strong acids exceed weak acids
Area 5:	Carbonate hardness (secondary alkalinity) exceeds 50% – that is, chemical properties of the groundwater are dominated by alkaline earths and weak acids
Area 6:	Non-carbonate hardness (secondary salinity) exceeds 50%
Area 7:	Non-carbonate alkalinity (primary salinity) exceeds 50% – that is, chemical properties are dominated by alkalis and strong acids, ocean waters and many brines plot in this area, near its right-hand vertex.
Area 8:	Carbonate alkali (primary alkalinity) exceeds 50% – that is, here are plotted groundwaters which are inordinately soft in proportion to their content of dissolved solids.
Area 9:	No one cation–anion pair exceeds 50%

Hydrochemical sections

These are plotted to show the variation in concentration of several hydrochemical characteristics in the direction of groundwater movement. This type of section can also be of help in depicting the changes in water quality due to mixing of sea water and freshwater in coastal aquifers. Figure 10.9 shows one such section from the carbonate rock aquifer in coastal parts of Saurashtra, India.

Software packages which are now available have greatly facilitated the plotting and interpretation of hydrochemical data.

10.5 Modification of chemical characters of groundwater

In addition to dissolution of rock material, the other processes which bring out chemical changes in the composition of groundwater are

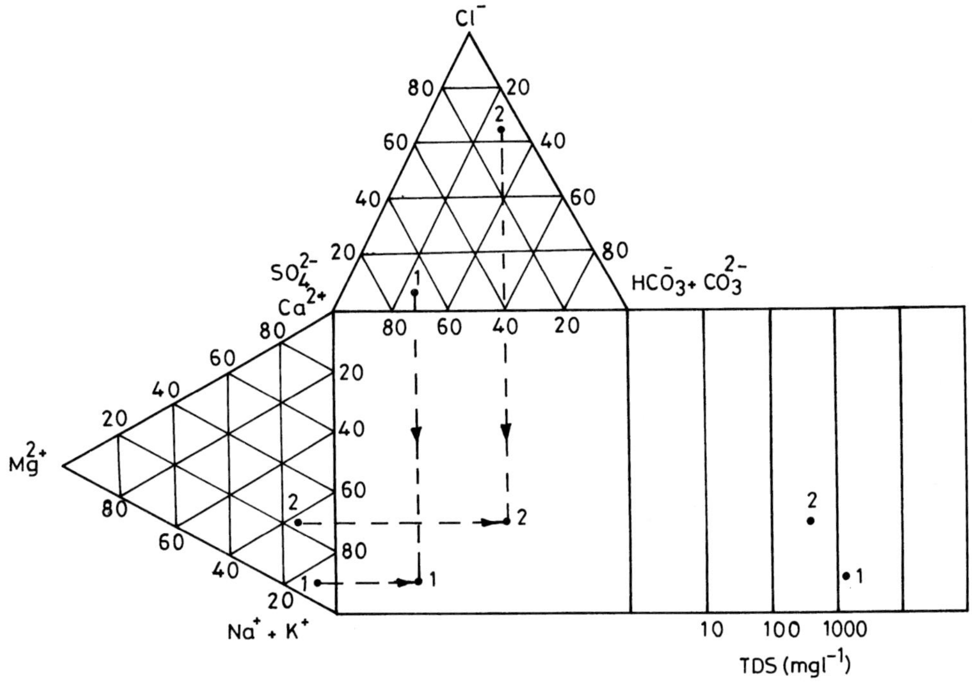

FIGURE 10.8 Durov's diagram showing plotting of analytical data.

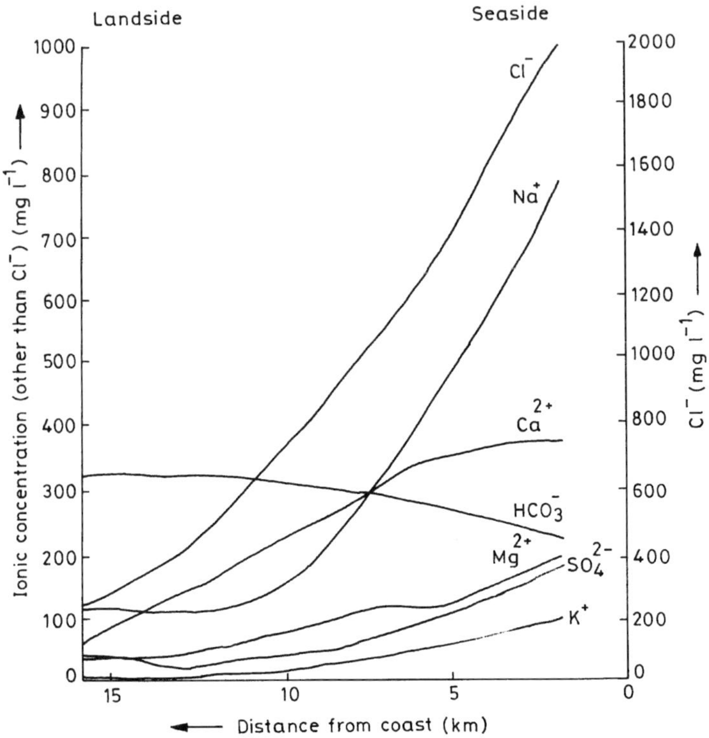

FIGURE 10.9 Hydrochemical section of the Miliolite limestone across the coast of Saurashtra, India (after Nair, 1981).

ion exchange, membrane filtration and sulphate reduction. Climate and length of flow trajectory also affect water quality.

10.5.1 Ion exchange

Groundwater when coming into contact with rock matrix and organic material has a tendency to exchange ionic constituents leading to adsorption of cations on mineral surfaces. Most of the exchange reactions involve cations and therefore this is referred to as cation or base-exchange. The common natural materials which act as the most important adsorbants are clay minerals. In addition, glauconite, rock forming silicates like zeolites, micas, feldspathoids, and organic material like humus also form adsorbants.

The cation-exchange capacity (CEC) of a material is defined as the number of exchangeable ions in milliequivalents per 100 g of solids at pH = 7. The CEC of some commonly occurring minerals and other natural materials is given in Table 10.6.

The cation exchange capacity increases with decrease in the particle size. Colloidal-sized particles (10^{-3} to 10^{-6} mm) have a greater tendency to ion exchange because they have a large electrical charge relative to their surface areas. As the nature of the surface charge is a function of pH, the tendency of adsorption of cations or

TABLE 10.6 Cation exchange capacity of some natural materials (after Matthess, 1982)

Mineral/Rock	*Cation exchange capacity (meq/100 g)*
Tuff	32–49
Kaolinite	3–15
Glauconite	10–200
Montmorillonite	70–100
Zeolites	100–600
Feldspathoids	400–1000
Organic substances in soils	150–500

anions will also depend on the pH of the solution. The exchange capacity also depends on the ion characteristics. The affinity for adsorption is greater in ions with higher valencies, i.e. it is more difficult to remove ions of higher valency than those of lower valency. For ions of the same valency, the affinity for adsorption increases with increase in the atomic number and the ionic radius. The normal order of preference for adsorption is:

$C_s^+ > Rb^+ > K^+ > Na^+ > L;^+$
Stronger → Weaker
$Ba^{2+} > Sr^{2+} > Ca^{2+} > Mg^{2+}$

The concentration of the adsorbed mole fractions at the initial condition and the concentration ratio of the two ions in solution are also of importance to determine the direction in which an ion exchange reaction will take place. The equilibrium between the quantity of the substance (S) adsorbed on the solids and the concentration of this substance in solution (C) can be described by Freundlich's isothermal equation (10.3)

$$S = K_d C^b \tag{10.3}$$

where K_d and b are coefficients that depend on the solute species and nature of the medium. Equation (10.3) shows that an increase in the concentration of a solution will raise the adsorbed quantity, and a decrease in the concentration will lead to desorption.

The effect of cation exchange on groundwater quality and movement of contaminants has been cited widely in groundwater literature (Back and Hanshaw, 1965; Freeze and Cherry, 1979; Hem, 1989; Marsily, 1986). This aspect was also discussed earlier in section 7.4.1.

Base exchange also changes soil characteristics. The replacement of Ca^{2+} by Na^+, because of the exchange process in clays, decreases the permeability of the soil drastically due to swelling. This can happen in areas where saline or brackish water is used for irrigation, resulting in a decrease of agricultural productivity due to poor drainage.

10.5.2 Membrane effects

In stratified sedimentary rock formations, the aquifers are often interbedded with relatively impermeable clay and shale beds which serve as a semi-permeable membrane for the selective movement of ions under higher confining pressures. A semi-permeable membrane is defined as one that restricts or prevents the passage of charged species (cations and anions) while allowing relatively unrestricted flow of the water molecule (Back and Hanshaw, 1965). The concentration of solutes on the input side of the membrane therefore increases relative to the concentration on the output side; thereby, the input side gets saltier and the receiving aquifer gets fresher. This ion expulsion effect is commonly referred to as **salt filtering, ultrafiltration** or **hyperfiltration** (Freeze and Cherry, 1979; Matthess, 1982). Divalent cations are usually filtered more easily than monovalent cations due to differences in ionic size and charge; temperature also influences the rate of salt filtering. Clays with higher cation exchange capacity, like montmorillonite, have higher ion-filtering efficiencies compared to kaolinite. Laboratory experiments also indicate that salt filtering is more effective in sedimentary formations at depths greater than 500 to 1000 m. The formation of deep-seated brines and the occurrence of saline waters in non-marine sediments and those devoid of evaporite deposits, has been explained by the process of salt filtering (Back and Hanshaw, 1965; Neuzil, 1986).

Sulphate reduction

The reduction of SO_4^{2-} to HS^- is a common feature in groundwater which takes place in the presence of bacteria (equation 10.4).

$$SO_4^{2-} + CH_{4(\text{bacterial})} \rightarrow HS^- + H_2O + HCO_3^- \tag{10.4}$$

Therefore, waters which have undergone sulphate reduction are characterized by the presence of HS^- and high content of HCO_3^- as is reported from several oil fields.

Sulphate-reducing bacteria are also reported from the shallow groundwater in coastal parts of Orissa, India. The H_2S released from the reduction of SO_4^{2-} gives a bad odour and also attacks iron pipes and other parts of the well assembly, thereby increasing the concentration of ferrous ion in water.

Sulphate-reducing bacteria have also been reported from even greater depths (> 500 m) in petroleum deposits. Some bacteria can withstand pressures above 1700 bars (Matthess, 1982). Sulphate reduction can also take place by hydrogen released through the decomposition of organic matter by anaerobic bacteria (equation 10.5)

$$CaSO_4 + 4H_2 \rightarrow 4H_2O + CaS \qquad (10.5)$$

10.6 Hydrochemical zoning and hydrochemical facies

10.6.1 Hydrochemical zoning

On a regional scale, groundwater exhibits both lateral and vertical variation in its chemical characteristics which is a result of variation in the lithology of the aquifer, climatic variations and length of trajectory. Schoeller (1959) has identified three types of zonations namely geological zonation, vertical zonation and climatic zonation. The geological zonation is a result of variation in the mineralogical composition of aquifer material.

Vertical and lateral variation (zonation) in a groundwater flow system is a result of greater dissolution leading to an increase in total dissolved solids and different ions along its flow paths. Therefore, the groundwater in recharge areas and at shallow depths will have a lesser concentration of dissolved solids than water in the deeper zones. Such a variation is described from Russia by Siline-Bektchourine (Schoeller, 1959) and by Chebotarev (1950) and Domenico (1972) from Australia. As per this scheme, the following changes in the anion species in groundwater are reported with increase in the length of trajectory and increase in residence time:

$$HCO_3^- \rightarrow HCO_3^- + SO_4^{2-} \rightarrow SO_4^{2-}$$
$$+ HCO_3^- \rightarrow SO_4^{2-} + Cl^- + Cl^- SO_4^{2-} \rightarrow Cl^-$$

The decrease in redox potential of groundwater along its flow path is due to a decrease in dissolved oxygen at greater depths (Freeze and Cherry, 1979).

As the climate influences the rainfall and evaporation etc., the chemical evolution of groundwater will also depend on these climatic factors. This is the reason that groundwater in arid and semi-arid regions is more saline than in humid and temperate regions. Temperature, along with humidity, also controls the activity of soil micro-organisms and the organic matter which produces CO_2. This is an essential factor in the dissolution of silicate and carbonate minerals in the rocks. Schoeller (1959) traced the evolution of the chemistry of groundwater in the former USSR, from the temperate down to the equatorial regions, and has shown variations in total dissolved solids and various cation and anion species.

The lateral and vertical sequence of various types of water is also modified by the palaeoclimatic conditions. There are several examples where, due to palaeoclimatic conditions, fresh groundwater occurs at varying depths especially in arid and semi-arid regions. Certain chemical processes like base exchange and sulphate reduction also cause reversal in the above sequence.

10.6.2 Hydrochemical facies

The concept of hydrochemical facies as given by Back (1961) is based on a similar concept

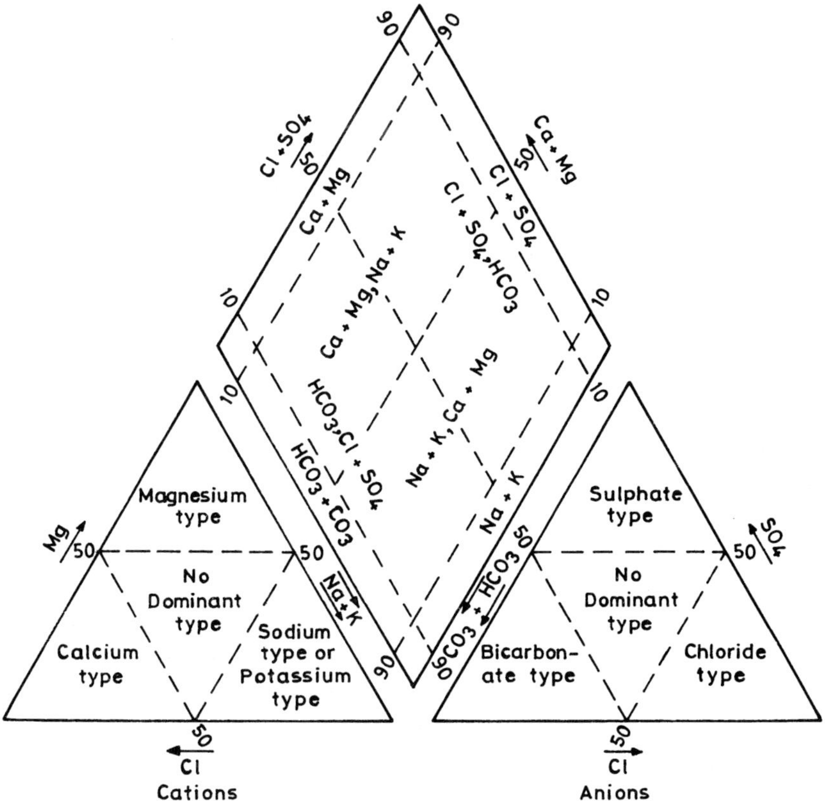

FIGURE 10.10 Hydrochemical facies diagram. The ionic concentration is percentage of total meql^{-1}

TABLE 10.7 Classification of hydrochemical facies (after Back, 1966)

	Percentage of constituents, (meq 1^{-1})			
	Ca+Mg	*Na+K*	*HCO_3+CO_3*	*C1+SO_4*
Cation facies				
Calcium-magnesium	90–100	0<10		
Calcium-sodium	50–90	10<50		
Sodium-calcium	10–50	50<90		
Sodium-potassium	0–10	90–100		
Anion facies				
Bicarbonate			90–100	0<10
Bicarbonate-chloride-sulphate			50–90	10<50
Chloride-sulphate-bicarbonate			10–50	50<90
Chloride-sulphate			0–10	90–100

of lithofacies commonly used in geology which states that there will be a definite trend in the formation of minerals under a given geological environment. Similarly, the development of a particular hydrochemical facies is controlled mainly by the lithology of the aquifer and its distribution is controlled by the groundwater flow pattern.

Piper's diagram forms the basis of classification of waters into various hydrochemical facies

(Figure 10.10). The characteristics of various hydrochemical facies are given in Table 10.7. The difference between the Piper's diagram and hydrochemical facies diagram is that instead of giving equal increments to various variables, the hydrochemical facies are distinguished into 0–10%, 10–50%, 50–90% and 90–100% domains. The areal distribution of various facies can be shown with the help of maps, fence diagrams and cross-sections (Back, 1961). The concept of hydrochemical facies has been used for explaining the changes in water quality from recharge to discharge areas and along the flow paths in a given lithology (Back and Hanshaw, 1965; Back, 1966; Toth, 1966). Some of the main conclusions are:

(a) Bicarbonate content is low in recharge areas and high in discharge areas.
(b) Sulphate content decreases in the direction of flow and bicarbonate increases, as a result of sulphate reduction.
(c) The ratio of sulphate to chloride decreases in the direction of flow.

10.7 Quality criteria for various uses

Groundwater is mainly used for drinking, irrigation and industrial purposes. Therefore, quality criteria depend on the use of water for a particular purpose, and quality standards have to be maintained in water supply for different uses to avoid deleterious effects.

10.7.1 Domestic use

Groundwater forms an important source of water for drinking and other domestic purposes. Therefore, groundwater, in general, is safer for use than surface water especially from the point of view of bacterial pollution; but the chemical composition of water is also important, as certain chemical constituents become toxic beyond a particular concentration, although they may be beneficial in lower amounts. This can be expressed by the dose response curve (Figure 10.11). The threshold value will vary for different constituents.

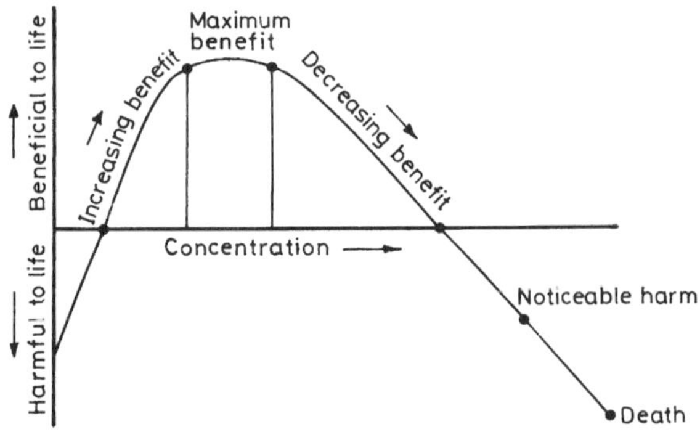

FIGURE 10.11 A general dose-response curve for human consumption.

Prescribed standards for drinking-water vary from country to country, depending upon economic conditions, climate, food habits and geographic location. There is also conflicting evidence with respect to safe limits for certain constituents.

Quality criteria for drinking-water normally are based on a water intake of two litres per person per day. The optimum desirable concentration of various constituents depends on age, food habits and climate. Therefore, exact drinking-water standards could vary from one individual to another. Drinking-water standards as prescribed by WHO (1984), European Economic Community (EEC) and Indian Standard Institution (ISI) are given in Table 10.8. The importance of some of the constituents in water quality standards for drinking purposes is discussed below.

TDS

The potability of water in terms of TDS is suggested as follows (WHO, 1984):

Excellent	< 300 mg l^{-1}
Good	300–600 mg l^{-1}
Fair	600–900 mg l^{-1}
Poor	900–1200 mg l^{-1}
Unacceptable	>1200 mg l^{-1}

TABLE 10.8 Upper limits of various constituents (mg l^{-1}) in drinking-water

Parameter	*WHO (1984)*	*EEC (Lloyd and Heathcote, 1985)*	*ISI (1983)*
TDS	1000	1500	500
pH	6.5–8.5	6.5–9.0	6.5–8.5
Total hardness ($CaCO_3$)	500	2–10 meq l^{-1}	300
Ca	75–200	75–200	75
Mg	–	50	30
Na	200	150	No limit
Cl	250	250	250
SO_4	400	250	150
NO_3	45	50	45
Fe	0.3	0.2	0.3
F	1.5	1.5	0.6–1.2
Pb	0.05	0.05	0.1
Hg	0.001	0.001	0.001
Zn	5.0	5.0	5.0
Cu	1.0	0.05	0.05
Cd	0.005	0.005	0.01
As	0.05*	0.05	0.05
Cr (hexavalent)	0.05	0.05	0.05
CN (Cyanides)	0.05	0.05	0.05
Overall α radioactivity	3 pCi l^{-1} (0.1 *Bq*l^{-1})	3pCi l^{-1}	1 pCi l^{-1}
Overall β radioactivity	30 pCi l^{-1} (1Bq l^{-1})	30 pCi l^{-1}	10 pCi l^{-1}
Coliforms	No fecal coliforms	No coliforms in 95% of sample	100 coliforms per litre

*Recently reduced to 0.01 mg 1^{-1}
EEC – European Economic Community
ISI – Indian Standards Institution

Hardness

The standards of hardness for drinking-water supply vary, viz. the World Health Organisation (WHO) suggested an upper limit of 500 mg l^{-1} of $CaCO_3$. The American Water Works Association has prescribed an upper limit of 80 mg l^{-1} of $CaCO_3$ for 'ideal' water (Hem, 1989). As per the standards prescribed by the European Economic Community, the minimum required concentration of hardness is 2–10 meq l^{-1} of $CaCO_3$. Studies in USA, Canada and Japan indicate that the incidence of cardiovascular diseases is more in areas where drinking water is extremely soft (Hem, 1989). However, WHO (1984) showed no such definite relationship.

Chloride (Cl)

The limits of chloride have been put more from the point of view of taste rather than its adverse effect on human health. Water with a chloride concentration of more than 250 mg l^{-1} will be saline in taste. For persons suffering from hypertension, the chloride concentration should be less.

Iron and manganese

These may not have a direct effect on human health but they are of importance from aesthetic considerations. Certain types of bacteria, such as crenothrix, change the dissolved iron into an insoluble slime which causes plugging of pipes, valves and water meters. It also imparts a dirty reddish colour, stains clothes and fixtures and has a bad taste and odour. It also clogs the well screen and adjoining formation. These problems arise when the iron concentration approaches 0.3 mg l^{-1}. Therefore, the guideline value for iron is chosen as 0.3 mg l^{-1}. The use of chlorine is effective in stopping crenothrix growth; the common method of removal of iron from water is by aeration followed by sedimentation.

Fluoride (F)

This is beneficial when present in small concentrations (0.8 to 1.0 mg l^{-}) in drinking-water for calcification of dental enamel but it causes dental and skeletal fluorosis if present in higher amounts (Table 10.9). Higher concentrations of fluoride in drinking-water are also linked with cancer (Smedley, 1992). The permissible limit of F depends on temperature; a higher intake of fluoride can be permissible in colder climates (Hamill and Bell, 1986).

In areas of higher fluoride concentration, defluoridation can be adopted to supply safe drinking-water. The most common method, adopted in India, is the Nalgonda process developed by the National Environmental Engineering Research Institute (NEERI). The Nalgonda technique comprises the addition of lime and aluminum sulphate or aluminum chloride which helps in the removal of fluoride by flocculation. This method is quite simple and can be used in villages at a domestic level, and also for small communities.

Mercury (Hg)

Mercury is a toxic element and serves no beneficial physiological function in human beings. The main effect of mercury poisoning is in the form of neurological and renal disturbances; it also disturbs the cholesterol metabolism. There is no evidence that high doses of mercury are carcinogenic. The recommended guideline value of mercury in drinking water is 0.001 mg l^{-1}.

TABLE 10.9 Impact of fluoride in drinking water on health (Dissanayake, 1991)

Concentration of fluoride (mgl^{-1})	*Impact on health*
Nil	Limited growth and fertility
0.0–0.5	Dental caries
0.5–1.5	Promotes dental health
1.5–4.0	Dental fluorisis (mottling of teeth)
4.0–10.0	Dental and skeletal fluorosis (pain in back and neck bones)
>10.0	Crippling fluorosis

Lead (Pb)

The concentration in the drinking-water supply is kept low (10–20 $\mu g\ l^{-1}$) after treatment, but due to lead in water supply pipes and storage tanks, the lead in drinking water may be higher (> 300 $\mu g\ l^{-1}$). Lead is not essential for human beings, therefore the intake of lead should be kept to a minimum. Lead in high doses acts as a metabolic poison causing anaemia, behavioural and mental diseases. A guideline value of 0.05 mg l^{-1} of lead in drinking-water is recommended (WHO, 1984).

Arsenic (As)

Arsenic is toxic and carcinogenic (Smedley, 1992). Arsenic intake by humans is greater from food (e.g. seafood) and inhalation than from drinking-water. Earlier, the World Health Organisation recommended the limit for As in drinking-water as 50 $\mu g\ l^{-1}$, but this has recently been reduced to 10 $\mu g\ l^{-1}$ (Smedley and West, 1995).

Iodine (I)

The daily iodine requirement for the human diet is about 100-200 $\mu g\ l^{-1}$ (Smedley, 1992); deficiency of iodine causes goitre. As sea water is the main source of iodine, it is natural that it is deficient in areas away from the coast. This is the reason that goitre has been reported mainly from Alpine, Himalayan and Andean regions. An iodine concentration of less than 1 $\mu g\ l^{-1}$ is reported in goitrous areas in Nepal (Smedley, 1992). As high Ca in water suppresses the solubility of iodine, goitre is also observed in limestone areas.

Nitrate (NO_3)

The maximum prescribed limit of NO_3 in municipal water supplies is 50 mg l^{-1}. In many countries water supplies having high levels of nitrate have been responsible for blue baby disease in infants (infantile methaemoglobinaemia) and death. This problem does not arise in adults.

Pesticides

Among the pesticides, DDT and aldrin are important from the water quality point of view. The main effect of DDT and aldrin is on the nervous system and liver. There is no evidence of them being carcinogenic. The guideline values for DDT is 1 $\mu g\ l^{-1}$ and for aldrin it is 0.03 $\mu g\ l^{-1}$ (WHO, 1984).

Radionuclides

Radionuclides emit radiation in the form of alpha particles, beta particles, and gamma rays. Among the naturally occurring radionuclides ^{238}U is both an alpha and beta emitter and ^{226}Ra and ^{234}U are the typical alpha-emitting radionuclides while ^{90}Sr is among the man-made beta-emitters of interest.

Among the alpha-emitting radionuclides, ^{226}Ra, ^{224}Ra, ^{210}Po, ^{232}Th, ^{238}U and ^{238}U have high toxicity, and ^{90}Sr, ^{89}Sr, ^{134}Cs, ^{137}Cs, ^{131}I and ^{60}CO are the toxic beta-emitting radionuclides. From the water quality point of view, gross alpha and beta activity is of importance rather than that of individual radionuclides. WHO (1984) has recommended guideline values of 1 Bq l^{-1} (30 pci l^{-1})for gross beta activity and 0.1 Bq l^{-1} (3 pci l^{-1}) for gross alpha activity in drinking-water (Table 10.8). Guideline values represent a value below which water is considered potable without any further radiological examination.

10.7.2 Other uses

Water quality standards for use by livestock

Fundamentally, the same standards can be applied as for human beings. However, animals can have a greater tolerance of total dissolved solids (1000 mg l^{-1}). Standards also depend upon livestock type, age and food habits (Bouwer, 1978; Lloyd and Heathcote, 1985).

Industrial use

The quality criteria of water for industrial purposes depend on the type of industry, processes and products. The quality criteria of water for use in boilers is important as water used in high-pressure boilers should be free from suspended matter; it should have low total dissolved solids and no acid reaction. Low-pressure boilers can use water with total dissolved solids up to 5000 mg l^{-1} and $CaCO_3$ hardness up to 80 mg l^{-1}, while in high-pressure boilers total dissolved solids should be less than 50 mg l^{-1} and $CaCO_3$ hardness less than 1 mg l^{-1}. Therefore, water treatment is necessary before use in high-pressure boilers.

For the manufacture of pharmaceuticals and high-grade paper, water approaching or equalling the quality of distilled water is required. Very pure water is also desirable in nuclear reactors to keep the radioactivity caused by neutron activation of dissolved solids as low as possible (Hem, 1989). In the construction industry, the sulphate content of water is important to avoid deterioration of concrete. To take care of such problems, sulphate-resisting cements are developed to suit a range of sulphate concentrations in the soil and water (Lloyd and Heathcote, 1985).

Irrigation water

The quality standards for irrigation water are based on: (1) total dissolved solids which may affect the intake of water and other nutrients by plants through osmosis; (2) the relative concentration of alkalis and alkaline earths which affect the soil texture due to cation exchange, and thereby its permeability and drainage characteristics; and (3) the concentration of specific ions, viz. boron, selenium, cadmium etc., which are toxic to the growth of plants beyond certain levels. In addition to chemical characteristics of applied water, other factors which decide the suitability of a particular quality of water for use in irrigation are: (a) soil characteristics, (b) type of crop and its stage of growth, (c) climate, (d) depth to water-table, (e) method of irrigation, and (f) drainage characteristics of soil. The quality standards for irrigation water cannot be applied very rigorously as in the absence of water of the required quality, poor quality water can also be used by making necessary adjustments in farming and water management practices.

One of the earliest systems of classification of water for use in irrigation was given by Wilcox (1955), which is based on electrical conductivity (EC), percent sodium (%Na) and boron concentration. Percent sodium (%Na), is expressed as

$$\% \mathrm{Na} = \frac{\mathrm{Na}^+ + \mathrm{K}^+}{\mathrm{Ca}^{2+} + \mathrm{Mg}^{2+} + \mathrm{Na}^+ + \mathrm{K}^+} \times 100 \qquad (10.6)$$

where all the ionic concentrations are expressed in meq l^{-1}. Boron is useful in very small concentrations for the normal growth of plants, but it becomes toxic in greater concentrations depending upon the type of plants (Todd, 1980).

The US Salinity Laboratory of the US Department of Agriculture used **salinity hazard** and **sodium hazard** as the two important criteria for the classification of irrigation waters. Salinity hazard is a measure of EC and sodium hazard is expressed in terms of sodium adsorption ratio (SAR) which is defined as

$$\mathrm{SAR} = \frac{\mathrm{Na}^+}{[(\mathrm{Ca}^{2+} + \mathrm{Mg}^{2+})/2]^{\frac{1}{2}}} \qquad (10.7)$$

where all the ionic concentrations are expressed in meq l^{-1} **Adjusted SAR** has been suggested as another criterion to assess the suitability of water for irrigation (Ayers, 1975). Readers may refer to Todd (1980) and Lloyd and Heathcote (1985) for necessary details regarding various systems of classification for irrigation waters.

10.8 Groundwater contamination

During the last few decades, due to increased industrialization, urbanization and agricultural activities, the quality of groundwater as well as surface water has deteriorated considerably in some areas (Table 10.10). This has caused great concern both in developed and developing countries. It is comparatively easy to detect the contamination of surface-water sources like rivers and lakes, but as underground pollution cannot be observed directly, it is detected only after a long time has elapsed, and then it becomes a difficult task to rehabiliate the aquifer for safe water supplies. However, as the groundwater moves with a much lower velocity and so also does the plume of pollution, one can plan ahead and avoid the outcome of pollution by advance measures.

The term **contamination** is used for the addition of any solute into the hydrologic system as a result of man's activity, while the term **pollution** is restricted to a situation when the contamination attains levels that are considered to be objectionable (Freeze and Cherry, 1979). There could also be deterioration in water quality due to natural causes, namely dissolution of rock material. Contamination of groundwater can take place either from a wider source like percolation from agricultural fields on account of the application of fertilizers and pesticides, or from a point source like waste disposal sites. Over-exploitation of groundwater, especially in coastal areas, is also responsible for contaminating freshwater aquifers due to sea water intrusion. This aspect is discussed in section 17.6.

10.8.1 Movement of contaminants

The movement and interaction of contaminant with aquifer material and groundwater is a

TABLE 10.10 Major sources of groundwater contamination

Source of pollution	*Type*	*Principal contaminants causing concern*
Natural: Dissolution of minerals and rocks	Diffuse	Cl, F, Fe, As, Pb, Hg, Cd
Industrial	Point source	Hydrocarbons, chlorinated solvents, heavy metals, etc. depending on type of industry
Agricultural practices	Diffuse	Cl, No_3, pesticides and herbicides
Mining and other associated activities	Point source	Cl, SO_4, trace metals, e.g. Fe, As, Cu, Pb, Zn, Hg
Municipal wastes, e.g. landfills and other waste piles	Point source	TDS, CH_4, and various other organic and inorganic contaminants
Radioactive waste disposal	Point source	Radioisotopes
Miscellaneous, e.g. sea-water intrusion, deforestation, de-icing salts application, acid rain etc.	Diffuse	Cl, SO_4, trace metals

complex phenomenon. As the contaminant moves vertically downward towards the water-table, it passes through different zones. In the unsaturated zone, attenuation of the contaminant will take place as some chemicals are adsorbed on organic material and clay minerals, some are decomposed through oxidation and bacterial activity, and some are taken up by plants or released into the atmosphere.

After reaching the water-table, the dissolved contaminant will move along with the groundwater in the direction of its hydraulic gradient, mainly by advection (section 7.4.1). As the pore spaces in the zone of saturation are devoid of oxygen, there is no likelihood of oxidation. Further, due to very low groundwater velocity (<0.03 m d^{-1}), the contaminant will take several years to move from the source to the well point. The concentration of the contaminant in groundwater will also decrease with increasing distance of flow due to hydraulic dispersion and other attenuation effects.

As discussed in section 7.4.1, two types of hydrodynamic dispersion are prominent – the longitudinal dispersion which occurs in the direction of flow, and the transverse dispersion which occurs normal to the direction of flow. Due to hydrodynamic dispersion, the concentration of contaminant will decrease with distance from the source. The spread of the solute in the direction of flow will be greater than in the direction perpendicular to flow as longitudinal dispersivity is generally more than the transverse dispersion (Figure 10.12a). However, in fractured basalts at Idaho, USA, as the transverse dispersivity is more than the longitudinal dispersivity, the lateral spread of the chloride plume was longer than its spread in the direction of groundwater flow (Figure 10.12b). The effect of rock discontinuities on the spread of the contaminant from a uranium mill tailing pond is illustrated in Figure 10.13. This figure shows that the baseline natural uranium activity follows a linear pattern following the strike of the fractures. After 11 years of operation of the tailing pond, the uranium activity increased several-fold and the plume has spread to the north-west following the strike of fractures.

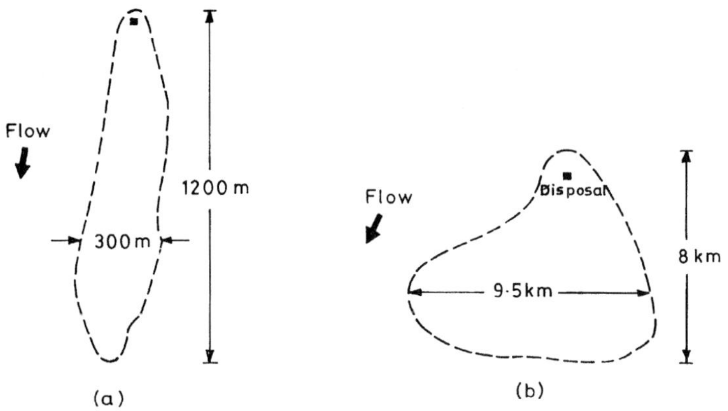

FIGURE 10.12 Effect of differences in longitudinal and transverse dispersivities on shapes of contaminant plumes: (a) chromium plume, Long Island, USA, after 13 years; (b) chloride plume in fractured basalt, Idaho, USA, after 16 years (after Miller, 1985).

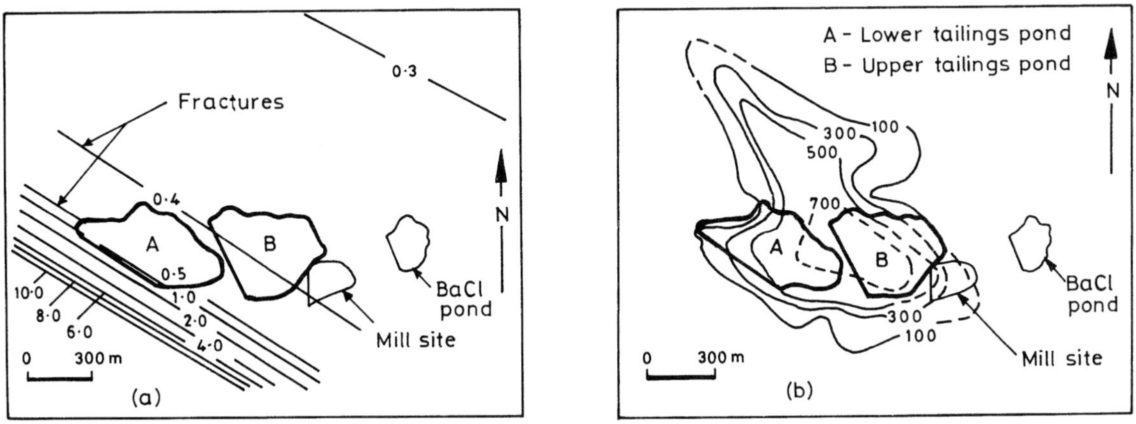

FIGURE 10.13 Effect of fractures on the spread of uranium activity from two unlined uranium tailing ponds in sandstone formation, Utah, USA. (a) Baseline natural uranium activity (μ Ci $ml^{-1} \times 10^7$) before the tailing ponds were put into service. This was due to natural leaching of uranium bearing rocks. (b) Elevated activity of natural uranium in the summer of 1981, i.e. after about 11 years of operation (after White and Gainer, 1985, reprinted with permission of Ground Water Publishing Co., *c*. 1985).

Another example of the importance of structural and also karst features in carbonate rock on contaminant migration from a chemical recycling facility in the US Midwest is described by Hasan (1996).

Adsorption and cation exchange reactions cause attenuation of solutes as discussed in section 10.5. Laboratory and field studies show that fractured shales in the unsaturated zone act as barriers for the migration of Cd^{2+} and Pb^{2+} obtained by leaching from sanitary land fills, fly-ash piles and other concentrated sources of these metals. These metals are retained by the shales through a combination of precipitation reactions, cation exchange and adsorption reactions (Angino and Wilbur, 1990).

10.8.2 Sources of groundwater contamination

Major sources of groundwater contamination are listed in Table 10.10.

Natural sources

The concentration of certain toxic or other substances beyond acceptable limits for human consumption can take place due to dissolution of aquifer materials. Increased concentration of fluoride, iron and arsenic in groundwater is usually due to this type of contamination.

The **fluoride** problem has received much attention in India and in several other countries (Handa, 1988). Fluoride-rich water in the East African Rift Valley, e.g. Ethiopia, is associated with calcium-deficient volcanic rocks and calcium-poor granitic rocks (Smedley, 1992). The hot springs in this area also have high fluoride levels (90 mg l^{-1}). In India, high fluoride (2 to 10 mg l^{-1}) water is reported mainly from granites and pegmatites which have F-bearing minerals, e. g. fluorite (CaF_2) and fluorapatite (Sinha and Kakar, 1974). Excessive concentration of fluoride in water is also reported from the eastern part of Sri Lanka, Thailand, China and several African countries.

Iron in groundwater is mainly due to the dissolution of iron oxides in laterites and other iron-bearing minerals especially in a high rainfall area, viz. coastal Orissa, Assam and Kerala in India. In these areas, Fe^{2+} is in the range of 0.4 mg l^{-1} to 46 mg l^{-1} and total iron is in the range of 6.83 to 55 mg l^{-1}. High concentrations of iron are also reported from the crystalline basement rocks in Malawi (Chilton and Foster, 1993) and other places.

Arsenic contamination in groundwater could be by the dissolution of minerals, like arsenopyrite usually associated with gold and tin deposits. In the past, arsenic was also used in the manufacture of insecticides which could also form a source of arsenic. In the western United States, high (> 50 μgl^{-1}) concentrations of arsenic are common in groundwater. These are associated with areas of sedimentary rocks derived from volcanic rocks, geothermal systems and gold deposits (Fetter, 1993). High concentrations of arsenic (0.3 to 1.1 mg l^{-1}), the highest value being 5 mg l^{-1}, have been reported from shallow alluvial aquifers in the depth range of 15 to 75 m from several districts of West Bengal in India and parts of Bangladesh, but the exact source of arsenic is still not well known. Irrigation in some areas in the USA has also liberated arsenic.

Industrial activities

The wastewater discharged from industries carries with it a variety of dissolved and suspended impurities, the composition of which depends on the type of industries and the processes used. At many places, effluents are discharged without proper treatment, thereby causing contamination of surface water and groundwater. For example, high concentrations of Cr, Ni, Cd, Pb and Zn in surface water and groundwater are reported from several places due to industries in north India (Kakar, 1988, 1990; Handa, 1994). Industrial pollution due to chlorinated hydrocarbons is reported from fractured limestones in the Copenhagen area, Denmark (Markussen and Moller, 1995) and from fractured sandstones in the Coventry region, UK (Lerner, 1995).

Hydrocarbons, viz. gasoline, kerosene, benzene/coal tar etc. are common examples of non-aqueous-phase liquids (NAPLs). These are usually immiscible in water and their transport is controlled by factors different from those that control aqueous flow. The reader may refer to Hasan (1996) regarding factors which control the subsurface transport of NAPLs with some case histories from the USA.

Mining industry

This is responsible for groundwater pollution mainly due to infiltration of leachate from mine waste and tailings. Dewatering of mines during operation causes entry of oxygen which results in the oxidation of sulphide ores resulting in strongly acidic (H_2SO_4) waters, known as acid mine-drainage; bacteria speed up the process. The acid mine-drainage water from coal mines is characterized by a high amount of suspended and dissolved solids, particularly iron and sulphate and low pH. Acidity (pH) depends on the amount of reactive pyrite present in the coal.

The problem of acid mine drainage from high organic sulphur Tertiary coals of Assam in north-eastern India is quite acute (Singh 1988). In this area, the problem is aggravated due to high rainfall of the order of 400 cm per yr^{-1}. The mine drainage water is highly acidic (pH = 2.3 to 4.0) and contains more than 3000 mg l^{-1} of sulphate and about 300 mg l^{-1} iron. It also has undesirable concentrations of heavy metals like As, Cd, Cr, Pb and Zn and therefore is a major source of pollution of surface water and groundwater. In the People's Republic of China, coal has a high content of sulphur and mine water usually has a pH of 2 to 3. The mine water contaminates surface water leading to a decline in fish and shrimp catches, and also the yield of farm crops irrigated with contaminated water (UNESCO-UNEP, 1989). Compared to colliery waters, drainage waters from metal mines have lower concentrations of sulphate but higher amounts of heavy metals (Hg, Pb, Cd, Cu etc.) due to lower pH.

Water from polymetallic (Zn–Pb–Cu) mines viz. Rangpo mine in Sikkim, India, has a low pH (5.0) and a higher amount of TDS (3616 mg l^{-1}), Pb (277 mg l^{-1}), Zn (35 mg l^{-1}) and Cu (1.80 mg l^{-1}) (Rai, 1994).

The electrical resistivity method is useful in determining both the extent and degree of acid mine drainage; and also for monitoring contamination due to the high conductivity of acidic water (Rogers and Kean, 1981).

Arsenic and mercury contamination of stream water, reported from the gold-mining belt of Ghana, appears to be a result of both mine effluent discharges and atmospheric emissions from the ore-roasting plant. Some of the local streams in the area have up to 170 $\mu g\ l^{-1}$ of arsenic (Smedley and West, 1995). Arsenic contamination (> 10 $\mu g\ l^{-1}$) of water is also reported from tin mines due to the presence of arsenopyrite from southern Thailand, Taiwan and the southern Andes (Smedley, 1992). Discharge of mercury from gold mining activities has contaminated streams and basement rock aquifers in the northern and central Precambrian shield in Brazil and Guiana, creating serious health risks for settlements (Reboucas, 1993).

Leachates from mine tailing dumps, which also contaminate surface water and groundwater, can be minimized by providing a cover of clay or some other impervious material which will reduce the entry of water and oxygen in the mine tailing dumps. Cover effectiveness is judged by the reduction of leachate production, its ability to remain water-saturated to check entry of oxygen, and avoid cracking (Herbert, 1992).

Municipal waste

Most of the municipal solid waste is disposed of in sanitary landfills. At some places, hazardous waste viz. organic chemicals and pesticides are also disposed in these landfills. The sanitary landfills are made in natural depressions or in man-made excavations such as abandoned quarries etc. The leachate from landfills may contain large numbers of contaminants; the total dissolved solids may be as much as 30 000 mg l^{-1}. Therefore, the leachate can cause severe damage to groundwater quality unless necessary preventative measures are adopted. The problem is more acute in a humid climate, especially in permeable and fractured formations where the water-table is at shallow depths. The leachate plume may extend to distances of several hundred metres in permeable formations which may last for many decades. In addition to the formation of leachate, a variety of gases such as CH_4 and CO_2 are also produced due to the biochemical decomposition of organic matter. CH_4 can be very dangerous as it is combustible and can cause severe damage. Therefore, landfill sites should be selected by necessary hydrogeological investigations to minimize the groundwater contamination by the leachate. In permeable formations, it is advisable to provide an impermeable lining of asphalt, clay or membrane over the bottom of the landfill to reduce infiltration of leachate. It is also necessary to construct cut-off drains around the landfill to prevent the flow of surface water into the landfill area. The reader may refer to Hamill and Bell (1986), Fetter (1988) and Mather (1995) for site selection and suitable design of various types of landfills.

Agricultural activities

One of the most important factors which cause contamination of groundwater is the greater use of inorganic fertilizers and pesticides for obtaining higher crop yields. This is particularly the case in the developing countries (Figure 10.14). In India, the consumption of fertilizers during 1996-97 is expected to be 16 million tonnes indicating an increase of about 10% from the previous year. Inorganic fertilizers are the source of nitrogen, phosphorus and potassium. Nitrogen, in the form of NO_3^-, is more soluble in groundwater than phosphorus while K^+ has low mobility due to cation exchange. A high amount of NO_3^- in groundwater can also be due to contamination from human and animal waste excreta. This is more common in fractured crystalline rocks with a thin soil cover. If the NO_3^- concentration exceeds 50 mg l^{-1}, it becomes undesirable for drinking purposes (Table 10.8).

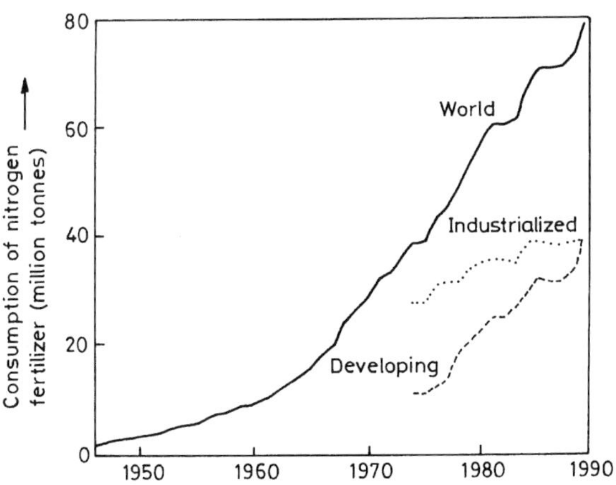

FIGURE 10.14 Consumption of nitrogen fertilizers (1946–89) (after Chilton *et al.*, 1995).

NO_3^- concentration is usually higher in shallow aquifers but with time and greater input, deeper aquifers are also likely to be contaminated. Studies in the UK show that the nitrate front moves in the saturated zone at 1 m yr^{-1} in chalk, to 2 m yr^{-1} in sandstones, with varying amounts of dispersion. A rise of NO_3 in groundwater from 14 mg l^{-1} in 1959 to between 26 and 52 mg l^- in 1970 is reported from East Yorkshire, UK, due to the application of fertilizers (Hamill and Bell, 1986). Peak values of nitrate concentration in British chalk aquifers are predicted early in the next century (Vrba, 1988).

In the Jaffna Peninsula of Sri Lanka, groundwater in Miocene limestones contains in excess of 90 mg l^{-1} of NO_3, which is mainly due to the increased use of inorganic fertilizers (Skinner, 1985). Higher concentrations of NO_3^- (> 100 mg l^{-1}) is reported from groundwater in the shallow unconfined aquifers of the semi-arid regions of western parts of India.

In addition to greater use of fertilizers, pesticides and herbicides are also being extensively used. The annual growth rate worldwide reached 12% in the 1960s, but now is around 3–4% (Chilton *et al.*, 1995). Increases in pesticide usage are now more rapid in many developing countries than in the developed world. In India, the average rate of application of pesticides to agricultural land is about 180 g ha^{-1} (Handa, 1994). Canter *et al.* (1987) gave a good account of the effect of pesticides on groundwater quality, based on field, laboratory and model studies. The contamination of groundwater by pesticides is dependent on the properties of the soil, and application factors. The chance for pesticide contamination is higher in surface water than groundwater due to the adsorption of pesticides by clay mineral grains, and organic matter and degradation by bacteria. However, some pesticides with higher solubility in water have significant mobility in some types of aquifer materials, such as clean sands and gravels (Freeze and Cherry, 1979). Preferential flow paths, e.g. macropores and fissures, in the unsaturated zone will also lead to fast movement of pesticides to the water-table.

The adverse effects of agriculture on groundwater quality can be reduced by sound management practices and optimum use of fertilizers, pesticides and irrigation water. LeGrand (1970) suggested that chances of groundwater contamination are reduced at those sites that have: (1) deep water-table, (2) high clay content, (3) water supply wells located on the upgradient side from the source of contamination, and (4) large distances between wells and contamination sources.

10.9 Radioactive waste disposal

Radioactive waste is generated in all the stages of the nuclear fuel cycle. This includes mining, milling and refining of uranium ore, fuel fabrication and fuel consumption in reactors, waste solidification and burial of solidified waste. The large amount of solid material obtained from mining and milling is a low-level radioactive waste. The uranium refining process also produces small quantities of solid or semi-solid low-level radioactive waste.

A large volume of high-level radioactive waste is produced from nuclear reactors for power production, weapons manufacturing and research. This waste is known as reactor waste. The volume of reactor waste is estimated to be more than 300 000 m^3 by the year 2000 in the USA (Freeze and Cherry, 1979). Radionuclides with short half-lives can be disposed of, on or in the ground, depending on their sorption by the mineral grains and organic matter till they decay to safe levels. Currently reactor waste is temporarily stored in iron, concrete or steel containers which are put underground, preferably in the unsaturated zone above the water-table. Ultimately these will be stored in the underground repositories.

A list of environmentally important radioactive isotopes produced at different stages from the nuclear fuel cycle, nuclear weapons test, research and medical waste is given in Table 10.11. Nuclear and health physics data for some selected radionuclides are listed in Table 10.12.

The wastes from mining, milling and refining processes usually contain ^{226}Ra, ^{230}Th and ^{238}U. ^{226}Ra is a common contaminant in surface and groundwaters associated with uraniferous rocks. It has a half-life of 1600 years and poses the biggest environmental problem as even small amounts of ^{226}Ra leached from waste rock or tailings can make groundwater unfit for drinking purposes. The maximum permissible limit of ^{226}Ra in drinking water is 3 pci l^{-1} = 10^{-9} mg l^{-1}.

Most of the reactor waste from nuclear power plants and nuclear weapons consists of ^{90}Sr, ^{137}Cs and ^{60}Co which have half-lives of 33, 28 and 6 years respectively and will decay to safe levels in about 1000 years. Longer containment is required for Pu and Am which, along with their radioactive daughters, may

TABLE 10.11 Sources of environmentally important radioactive isotopes (after Bedient *et al.*, 1993)

Source	*Radionuclides*
Mining waste, uranium, phosphate, coal	^{222}Rn, ^{226}Ra, ^{230}, ^{232}Th
Industrial wastes, e.g. nuclear power plant	^{59}, ^{63}Ni, ^{60}Co, ^{90}Sr, ^{93}, ^{99}Zr ^{99}Tc, ^{107}Pd, ^{129}I, ^{137}Cs
Nuclear weapons tests	^{3}H, ^{60}Co, ^{90}Sr, ^{137}Cs, ^{239}, ^{240}Pu
Weapons manufacturing, research and medical waste	^{134}, ^{137}Cs, ^{60}Co, ^{151}Sm ^{152}, ^{154}Eu, ^{237}Np, ^{239}, ^{240}, ^{242}Pu ^{241}, ^{243}Am

TABLE 10.12 Nuclear and health physics data for selected radionuclides (after UNESCO, 1980)

Radionuclide	*Half-life (yr)*	*Major (1) radiation*	*Critical organ*	*Biological half-life*	*MPC (2) (μCi ml⁻¹)*
^{3}H	12.3	β	Total body	12 days	3×10^{-3}
^{90}Sr	28.1	β	Bone	50 years	3×10^{-6}
^{129}I	1.7×10^{7}	β, γ	Thyroid	138 days	6×10^{-8}
^{137}Cs	30.2	β, γ	Total body	70 days	2×10^{-3}
^{226}Ra	1600	α, γ	Bone	45 years	3×10^{-8}
^{239}Pu	24 400	α	Bone	200 years	5×10^{-6}

(1) α = alpha particle; β = beta particle, i.e., electron; γ = gamma ray.
(2) MPC = maximum permissible concentration for water consumed by the general public without readily apparent ill effects.

require 500 000 years of isolation from the environment. The high-level radioactive waste containing various radionuclides may enter groundwater flow systems either by intentional release or accidental escape from radioactive waste management systems.

The migration of radionuclides from a repository will vary significantly, depending on the rock and site characteristics and the properties of the radionuclides. As most of the repositories are located in rocks of very low permeability, viz. crystalline rocks, shales and clays, the infiltration of water will be very slow, thereby there will be sufficient time for the substantial decay of radionuclides having a short half-life.

Fracture characteristics of the host rock (aperture, length and interconnections) and fracture infilling will have a considerable influence on the rate of transport (section 7.4). It is also important to visualize possible changes in the permeability of the fractures with time due to their opening or closure by processes of removal or deposition of clay material, and seismic activity, as we are thinking of safety measures for the disposal of radioactive waste in terms of thousands of years.

The movement of dissolved radionuclides in groundwater is retarded by several processes such as diffusion, adsorption, membrane filtration, chemical precipitation and radioactive decay (section 7.4). All nuclides have some degree of sorption except ^{14}C and ^{129}I. Sorption coefficients for different radionuclides vary significantly for different host materials such as granite, basalt, shale, clay and salt which form the common repositories for radioactive waste disposal (see Table 7.2). The degree of adsorption is also related to the pH of the solution. Readers may refer to Matthess (1982) for case studies of relative distribution of radioisotopes below some radioactive waste disposal sites.

The temporal and spatial distribution of concentration and flux through fractures for some typical high-level waste (HLW) radionuclides is given by Krishnamoorthy *et al.* (1992). It is shown that the concentration of ^{90}Sr and ^{239}Pu decreases significantly with distance along the fracture due to adsorption and decay, while ^{129}I does not show any marked decay. Further, although the peak flux of ^{129}I shows insignificant variation at different fracture lengths and aperture, that of ^{239}Pu is dependent on both these parameters.

As high-level nuclear waste has to be isolated from the accessible environment over a long period (>10^3 y), it has to be demonstrated that the disposal of high-level radioactive waste will not have any adverse effect on environment and human health, so that such programmes are acceptable to the public. These studies are quite challenging as they involve a precise understanding of the transport processes for predicting the possible migration of radionuclides with groundwater over long distances and time in spite of the fact that experiments can be made over much shorter times and distances. Therefore, laboratory studies followed by *in situ* field experiments in underground laboratories are necessary for what is known as 'safety synthesis' at potential repository sites. Underground laboratories for such studies are of two types; those associated with existing mines and other underground cavities, and those constructed specifically for testing purposes. Sub-seabed disposal as an alternative method is also suggested. Some of the well cited examples of field studies on land are from crystalline (granite) rock terrains, e.g. Stripa iron ore mine and the Aspo Hard Rock Laboratory (HRL) in Sweden, Underground Research Laboratory (URL) in Manitoba, Canada, the Grimsel Rock Laboratory in Switzerland and the Beishan area of Gansu province, China. Similar studies are also being done in Columbia basalts at Hanford, in Washington state and volcanic tuffs of Yucca Mountain, Nevada, USA. Other low permeability formations, viz. bedded and domal salt deposits in Germany and USA and clay formations in Belgium, Switzerland and Italy also form potential repositories for radioactive waste (Neerdael *et al.*, 1996; Put and Ortiz, 1996; Anon, 1996).

The study involves surface and borehole techniques for characterization of the subsurface geological, hydrogeological, and hydrochemical environment, and to assess the changes in the physical and chemical conditions in the rock mass and groundwater caused by excavation and disposal of nuclear waste. Subsurface geophysical methods are of much help in characterizing the fractures. Special hydraulic tests viz. cross-hole test and tracer injection test have contributed to a better understanding of the flow regime and interaction of solutes with the rock material (Simmons, 1985; Abelin and Birgersson, 1985; Put and Ortiz, 1996). The repository sites are provided with drifts for inspection purposes to monitor the migration of contaminants after waste emplacement. Readers may refer to Come *et al.* (1985) for a detailed account of the design and instrumentation of such experiments and interpretation of data.

At the Yucca Mountain, Nevada, USA, the repository for the disposal of high-level nuclear waste is proposed in the unsaturated (vadose) zone within bedded volcanic tuffs in a desert environment. The tuffs contain cooling fractures which are confined to individual beds; a number of large faults cut across these beds. The potential repository is located at a depth of approximately 350 m beneath the ground surface and 225 m above the water-table. Climate was an important reason, more than a decade ago, for choosing Yucca Mountain as the leading candidate for permanent nuclear waste storage in the USA. The area gets about 15 cm of rain a year. Faults and fractures will play an important role in the movement of contaminants. Larger fractures filled with gas or air, above the water-table, will provide an effective barrier to fluid flow. On the other hand, fracture zones and faults may serve as conduits for the transport of water from the surface to the water-table below the repository (Anon., 1996). In the unsaturated fractured rock, the contaminants would migrate both laterally and upward by capillary action in vapour form to the atmosphere and also downward to the zone of saturation (Evans and Nicholson, 1987). This would contaminate the biosphere as well as the groundwater. The rate of migration of the contaminants to the biosphere and to the zone of saturation would depend on fracture characteristics, degree of saturation, temperature conditions and air circulation in the vadose zone. The development of conceptual models for flow in variably saturated fractured rocks as at the Yucca Mountains is still incomplete. Recent studies have created doubts about the suitability of this repository as there are indications of fast seepage of rainwater through the vadose zone (Davis *et al.*, 1998). If all goes well, this repository is expected to accept waste in 2010.

10.10 Miscellaneous sources of groundwater contamination

In addition to the sources of contamination mentioned above, there are several other causes of deterioration of water quality: (1) deforestation, (2) over-irrigation especially in arid and semi-arid regions, (3) oil leaks and spills, (4) road salts, (5) sewage sludge, (6) urban run-off (7) thermal power plants, (8) deep well disposal of liquid waste, and (9) seawater intrusion, etc.

Deforestation may cause a rise in the water-table and a subsequent increase in the salinity of groundwater (Ruprecht and Schofield, 1991). In arid and semi-arid areas intensive irrigation leads to a rise in the water-table, thereby increasing the soil and groundwater salinity due to evapotranspiration losses. This problem is quite severe in flat alluvial plains, viz. Pakistan, Egypt, India and China (Greenman *et al.*, 1967; Shahin, 1985; Tyagi, 1994; Chen and Cai, 1995). Deterioration in groundwater quality due to sea-water intrusion is dealt with in

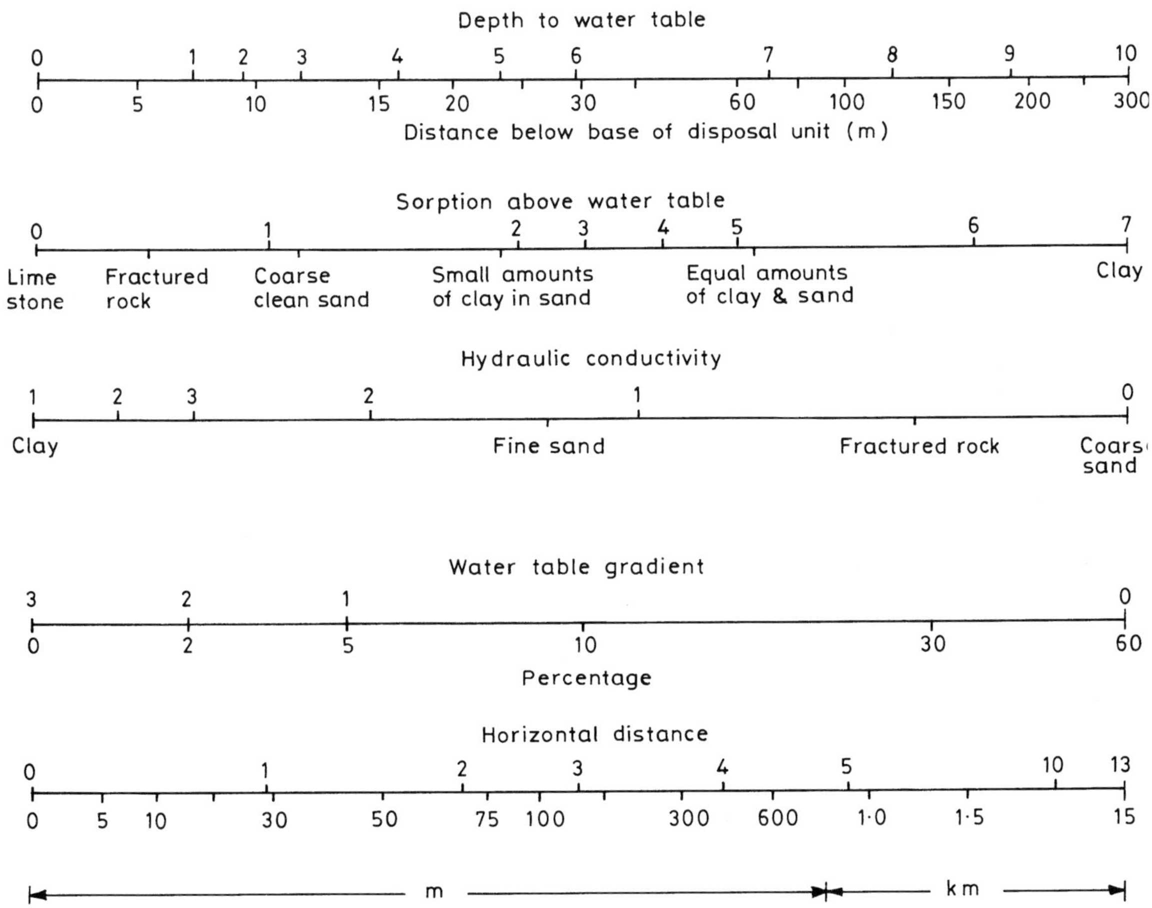

FIGURE 10.15 Rating chart for pollution assessment due to radioactive waste-disposal in both fractured rocks and unconsolidated sediments (after LeGrand, 1964; reprinted from Journal AWWA, Vol 56, No. 8, by permission, copyright 1964, American Water Works Association).

section 17.6. Readers may also refer to Bouwer (1978), Freeze and Cherry (1979), and Watson and Burnett (1993).

10.11 Evaluation of contamination potential and hazard

There is no precise method for evaluating the contamination potential and hazard from waste disposal sites. A precise knowledge of hydrogeological environment and chemical and biological aspects of the various types of wastes is necessary for such studies.

An empirical point count system for evaluating the contamination potential was suggested by LeGrand (1964). The system was designed for contaminants that attenuate or decrease in potency with time or by oxidation, chemical or physical sorption, and dilution through dispersion. Thus, it is suitable for a quick initial appraisal of sites, where hydrogeological data are scarce. Three rating charts were given for different geological environments; one such rating chart applicable for radioactive waste disposal in fractured rocks and unconsolidated alluvial formations is given in Figure 10.15. It uses parameters of depth to water-table, sorption above the water-table, hydraulic conductivity, hydraulic gradient and horizontal distance from the disposal site. A numerical value is read above the line from the chart for each of the above five factors, corresponding to the data below the line. The sum of the points for all the given factors is a measure of the pollution potential of the site (Table 10.13).

TABLE 10.13 Estimation of pollution potential by LeGrand method (after LeGrand, 1964)

Total points	*Probability of pollution*
0–4	Imminent
4–8	Probable or possible
8–12	Possible but not likely
12–25	Very improbable
25–35	Impossible

Based on the above method, the US Environmental Protection Agency (1978), developed the surface impoundment assessment (SIA) method for evaluating groundwater contamination potential. This method has been applied to a large number of wastewater ponds, septic tank systems and landfills (e.g. Canter, 1985).

Lately, a distinction is made between pollution potential and pollution hazard. Pollution potential is considered as a natural risk index, not involving the effect of man-made features, while groundwater pollution hazard is considered in terms of the probability of groundwater pollution in the area. Therefore, pollution hazard is based on pollution potential together with the society's pollution contribution in the area, given in terms of land use/cover, and hazardous material sites etc. The DRASTIC model developed by Aller *et al.* (1987) has been more frequently used in the GIS environment to assess groundwater pollution potential and hazard (for details see section 6.4.5). In one study, Kalinsky *et al.* (1994) found a positive correlation between the DRASTIC scores and contamination of well waters in Nebraska, USA, indicating the utility of the DRASTIC method for such studies.

Chapter eleven

Hydrogeology of crystalline rocks

11.1 Introduction

Crystalline rocks include plutonic igneous rocks (granites, diorites, etc.) and metamorphic rocks (gneisses, granulites, quartzites, marbles, schists and phyllites, etc). The plutonic igneous rocks, viz. granites, usually occur as large size intrusive bodies (plutons) while some other rocks, viz. dolerites and pegmatites, are of comparatively small size in the form of dykes and veins. The hydrogeological characters of dykes are decribed in Chapter Twelve as they are more commonly found in volcanic rock terrains. Like other hard rocks, the crystalline rocks are characterized by negligible primary porosity and permeability. However, weathering and fracturing can impart significant secondary porosity and permeability which is highly variable.

The crystalline rocks form important Precambrian shield areas in different parts of the world, mainly in Canada, Scandinavian countries, northeastern United States, India, Sri Lanka, China, Australia, Russia and many

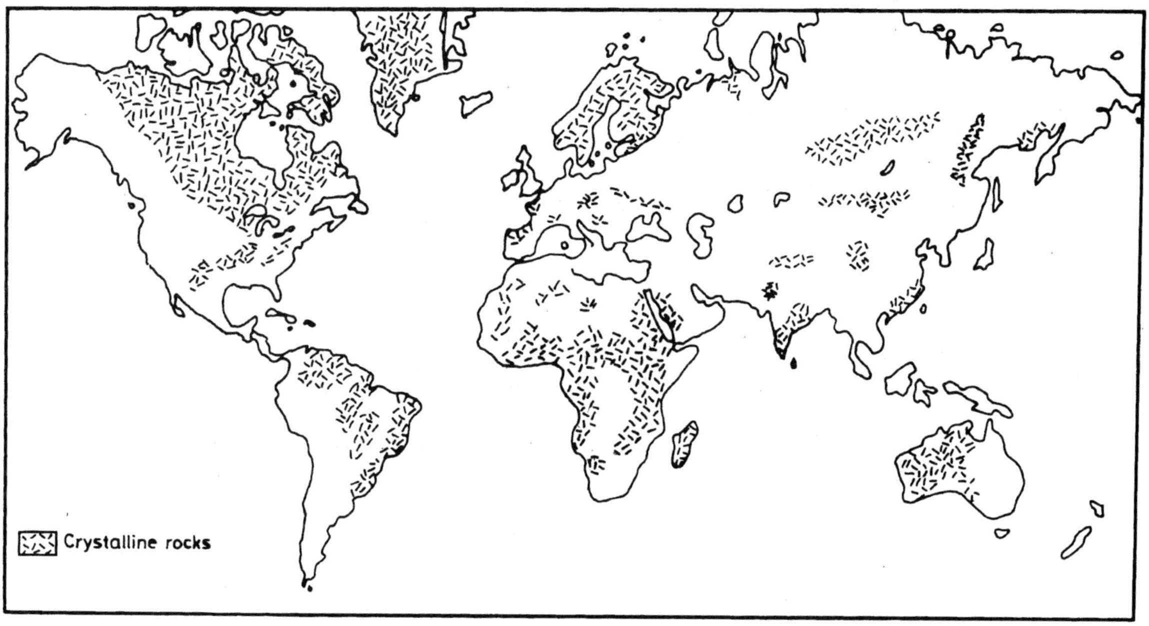

FIGURE 11.1 Major occurrences of the basement crystalline massifs on the continents.

African countries. Less extensive but important outcrops of crystalline rocks are also found outside the above shield areas, as in Spain, France and the Bohemian massif in Central Europe. The total extent of shield rock outcrops is estimated to be about 20% of the present land surface, i.e. about 30 million km^2 (Figure 11.1).

Earlier, the crystalline rocks, like other hard rocks, were not given due attention for groundwater development on account of low permeability and also difficulties in water-well drilling. However, in the last few decades, due to the need for safe drinking-water for vast rural populations, especially in the developing countries, crystalline rocks are being investigated in detail for groundwater development. Faster and more efficient methods of water-well drilling, like down-the-hole hammer (DTH), have also greatly promoted the construction of wells in crystalline rocks. Crystalline rocks also form a potential host rock for the safe disposal of high-level radioactive waste (section 10.9).

11.2 Landforms and drainage

The commonly developed landforms in crystalline rock terrains are structural hills, inselbergs, pediments, buried pediments, erosional valleys and valley fills (Figure 11.2). A representative aerial stereopair is given in Figure 11.3 and an IRS-image in Figure 4.10 (its interpretative map is Figure 11.4). The characteristic features including groundwater potential of these landforms are given in Table 11.1.

Structural hills
The morphology of structural or residual hills is controlled by large-scale rock structures and lithology. They evolve by the combined processes of tectonism and denudation. Since the rocks in this unit are hard and compact, they act as run-off zones; limited infiltration can take place along the weak planes like faults, fractures and joints. The groundwater may discharge as springs or seepages along the narrow valley portions, and the groundwater availability in general is very poor.

Inselbergs
These are small residual hills which stand in isolation above the general level of the surrounding erosional plains. Similar to the structural hills, they have negligible groundwater potential due to the small recharge area, steep slopes and low permeability of rocks.

Erosional valleys
These occur within the structural hills as narrow valleys, and the bedrock in the valleys is usually weak like phyllites. They have limited

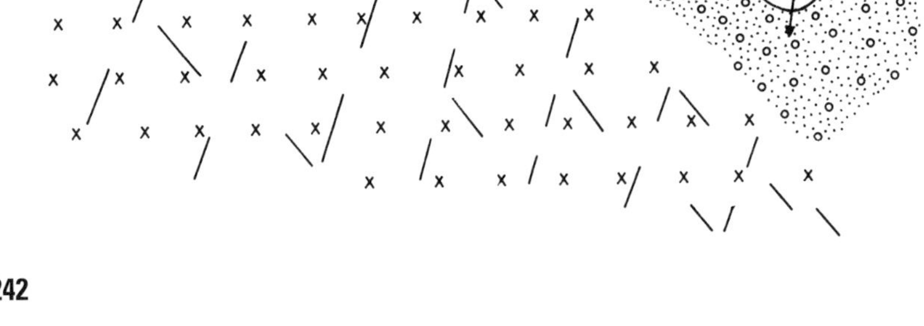

FIGURE 11.2 Schematic profile of landforms in a crystalline rock terrain.

(a)

(b)

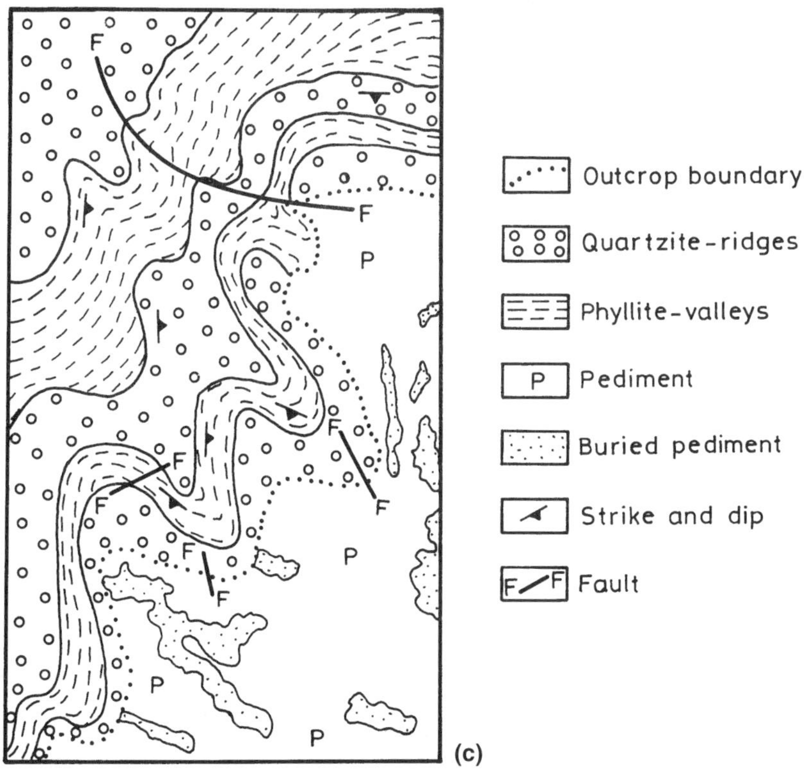

(c)

FIGURE 11.3 Figures (a) and (b) represent an aerial stereopair. The area comprises a typical metamorphic rock terrain exhibiting lithological bandings, folds, faults and fractures. Figure (c) is the interpretation map showing landforms, such as denudo-structural hills (ridges), erosional valleys, pediment, buried pediments and structural features, viz. faults (a & b, courtesy A. White).

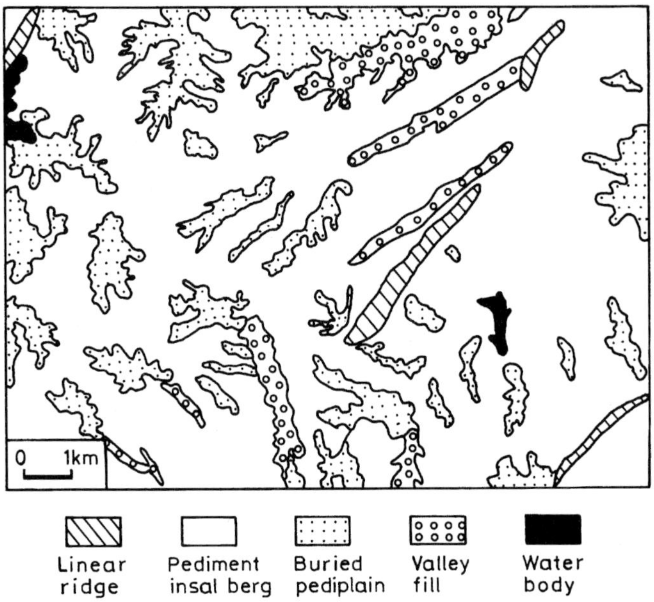

FIGURE 11.4 Landform interpretation map showing inselbergs/hills, pediments, buried pediments and valley fills in a granitic terrain. The corresonding remote sensing (IRS-LISS-II) FCC is shown in Fig. 4.10.

TABLE 11.1 Landforms in crystalline rock terrains (modified after Philip and Singhal, 1992)

Landform	*Description*	*Shape*	*Lithology*	*Groundwater potential*
Structural hills	Ridges and erosional valleys	Ridges and valleys	Competent rocks (quartzites and granite gneiss)	Poor
Inselbergs	Isolated residual hills in a vast plain	Island-like	Quartzites, granites etc.	Poor
Pediments	Low-lying rock outcrops, sometimes dissected by gullies	Irregular	Crystallines with thin veneer of soil cover	Moderate, suitable for bore-wells
Buried pediments	Low-lying bed rock covered with *in situ* material generally with high moisture	Irregular	*In situ* weathered material	Good, suitable for dug wells
Erosional valleys	Elongated depression occurring between two adjacent hills	Irregular generally elongated	Bedrock is of incompetent rocks viz. phyllites etc.	Poor
Valley fills	Marked by presence of palaeo-drainage pattern in pediments	–	Coarse sediments and rock debris	Good, suitable for dug and borewells

thickness of unconsolidated material and hence their groundwater potential is meagre.

Pediment
A broad, flat or gently sloping, rock-floored erosional surface or a plain of low relief. Such features are typically developed due to the processes of denudation by the subaerial agents including running water in an arid or semi-arid region at the base of an abrupt mountain front or plateau escarpment. At many places, the pediment is covered by a thin veneer of colluvial material; the pediment has low moisture content. Groundwater potential of this unit is limited due to the thin depth of weathered material and greater water-table fluctuation. Dug wells may not be very successful but borewells may yield small quantities of water. However, pediments developed along lineaments may form potential sources of groundwater.

Buried pediments
Formed when the sloping surface of the pediment gets gradually covered with a thick mantle of soil and colluvial material. The thickness of overburden may reach 20–100 m. The buried pediments have greater moisture and denser vegetation than pediments. The water-table fluctuation is relatively less and the recharge area is large. Hence, they form potential zones for groundwater development by dug wells and borewells.

Valley fills
A type of channel deposit developed by the process of deeper pedimentation, in an erosional environment in hard rock terrains. These are characterized by gentler slopes, greater moisture and denser vegetation. Therefore, the valley fills are the most important landforms for groundwater development in a crystalline rock terrain. It has been shown in Chapter Four that enhancement of multispectral data could lead to better discrimination of such landform features.

Besides the above, there could be special landform features like buried channels and palaeochannels which also form potential areas for groundwater development. As these are predominantly comprised of clastic fluvial sediments, they are discussed in Chapter Fourteen.

The drainage network shapes depend on surface geology and permeability characteristics of rocks. A dendritic drainage pattern is characteristic of massive crystalline rocks while rectangular/angular patterns are indicative of fractured rocks. Parallel patterns indicate steep slopes and often higher surface run-off.

11.3 Groundwater occurrence

The weathered layer (regolith) and fractures are the main sources of groundwater supply in crystalline rocks.

11.3.1 Weathered profile

The weathered layer, also called regolith, developed on crystalline basement rocks is an important source of groundwater, especially for the rural water supply in developing countries. Thick and areally extensive weathered layers can form a potential aquifer. However, if there is a good and perennial source of recharge, even a thin layer of 5–7 m thickness can be a good source of water supply.

A typical weathered profile consists of the following zones from top to bottom (Figure 11.5).

Zone A Sandy clay or clay-sands, often concretionary, usually a few metres thick.

Zone B **Saprolite** layer having mainly massive accumulation of secondary clay minerals, and characterized by high porosity and low permeability. Thickness may reach up to 30 m

Zone C **Saprock**, the parent rock progressively altered upwards with relict rock structures. Thickness may range from a few metres to about 30 m.

Zone D The parent rock with fractures, low porosity but moderate permeability.

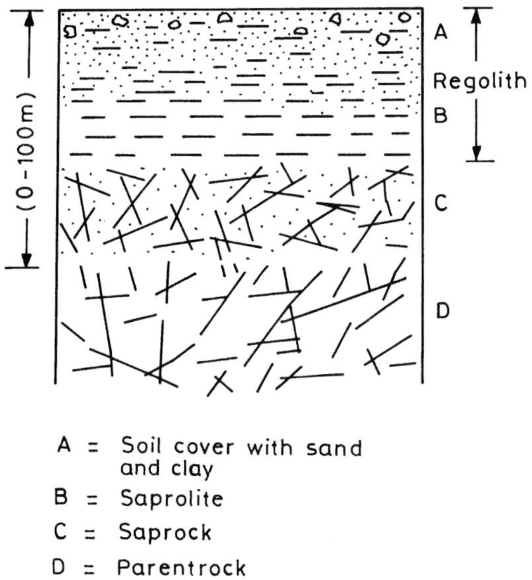

FIGURE 11.5 Idealized weathered profile in a crystalline rock terrain.

The thickness, areal extent and physical characteristics of the weathered layer vary from one region to another depending on climate, topography, lithology and vegetative cover.

(a) Climate

An important factor in the development of the regolith. In arid and semi-arid regions, the weathered layer is usually thin (< 1 m). In subhumid areas and humid tropic regions where annual rainfall exceeds 1000 mm, the weathered layer is very thick (65–130 m), e.g. in the upper regions of Ghana, the Ivory Coast and Togo (UNESCO, 1984a). Further, in some present-day arid areas, there may occur a thick weathered zone, for example in arid parts of Rajasthan (India), where although the present average annual rainfall is barely 380–460 mm, the weathered layer in places is found to be 25–30 m thick. Similarly, in Sudan and Nigeria, the weathered layer attains a thickness of 50 m (UNESCO, 1984a). Such thick regolith zones in arid regions must be related to a past pluvial climate.

(b) Topography

This also influences the development of a weathered layer. Extensive thick weathered layers are developed in erosional peneplain areas of low relief, as in Brazil, Sub-Saharan Africa, Sri Lanka, Peninsular India and parts of Australia (UNESCO, 1984a).

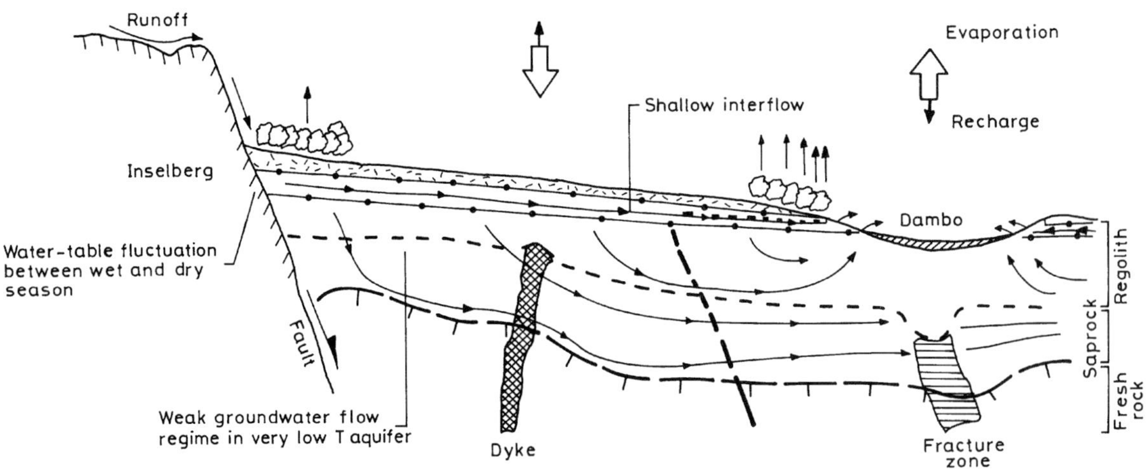

FIGURE 11.6 Generalized section of a weathered crystalline rock aquifer from Central Africa showing groundwater flow regime (modified after Chilton and Foster, 1993).

(c) Lithology and texture

Lithology and texture of the parent rock influence the thickness and permeability of the weathered layer. Coarse-grained salic rocks such as granite and orthogneiss give rise to thick and permeable weathered layers. In basic rocks, like gabbro and dolerite, although the weathered horizon may be equally thick, it is more clayey and therefore less permeable.

A generalized section of the weathered layer illustrating the groundwater flow regime in an African weathered basement aquifer is illustrated in Figure 11.6. Water-level in general represents and follows the ground topography in a subdued fashion. At shallow depths, in the interfluve areas, the water movement is vertically downward and at deeper levels there is a slower lateral movement towards the topographic depressions forming groundwater discharge areas (*dambo*). The flow pattern also influences water quality as water at shallow depths is of low mineralization but the deeper waters are strongly mineralized.

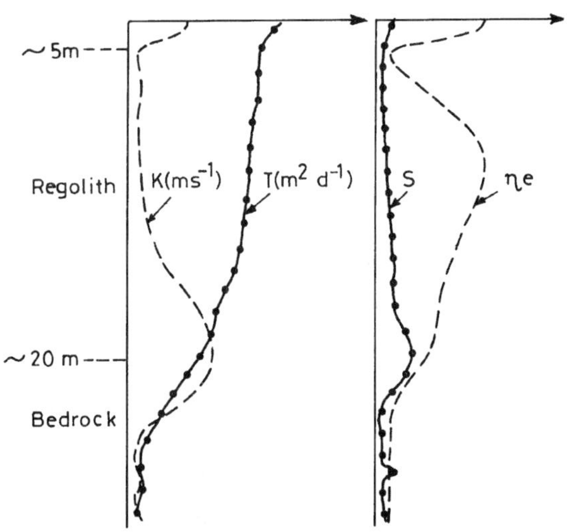

FIGURE 11.7 Conceptual hydrogeological model showing variation of hydraulic properties (K, T, S, ηe) in a typical African weathered basement rock aquifer (after Chilton and Foster, 1993).

Figure 11.7 shows possible depthwise variation in the hydraulic properties of regolith based on data from Zimbabwe and Malawi. The overall transmissivity is low (1–5 m^2d^{-1}) with occasional values of an order of several magnitudes lower or higher for the weathered material. Specific capacity also shows inconsistent values (Chilton and Foster, 1993).

11.3.2 Fractured media

The occurrence and movement of groundwater in massive crystalline rocks is mainly controlled by fractures and other discontinuities, which have been described in Chapter Two. Characterization of the fracture system is important in order to have a proper understanding of fluid flow and distribution of permeability. This involves identification of different sets of fractures based on their orientation (strike and dip), frequency, aperture, nature of filling, continuity and shape etc., which will define the geometry of the medium.

Open joints and fractures which are not filled with weathered or broken rock material form potential passages of groundwater movement but their permeability is greatly reduced when filled with clayey material such as smectite or montmorillonite. The extent of alteration and type of clay minerals which are formed depend on the original lithology and composition of the parent rock, tectonic history and composition of the circulating groundwater. The effect of fracture characteristics on rock permeability is discussed in section 8.2.1.

Alhough crystalline rocks extend over large areas and to greater depths, due to structural controls it is unlikely that the flow system extends over long distances. However, there are a few exceptions as observed at the Underground Research Laboratory (URL) in Manitoba, Canada, and in Stockholm, Sweden. At URL, sub-horizontal fractures with good permeability extend over a distance of about 1 km (Figure 11.8).

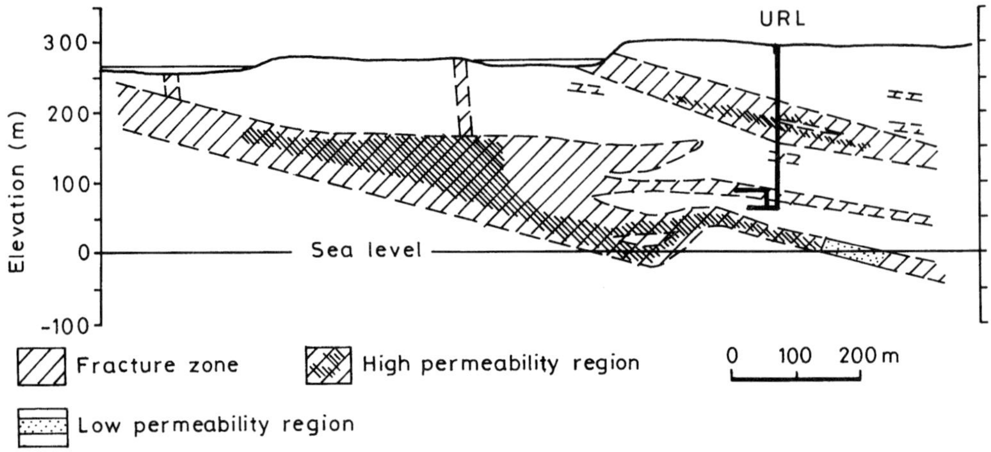

FIGURE 11.8 Geological cross-section across the Underground Research Laboratory (URL), Manitoba, Canada, showing the regional extension of sub-horizontal fracture zones (after Davison, 1985).

11.4 Hydraulic characteristics

As mentioned above, in crystalline rocks, the extent of weathering and fracture characteristics decide their hydraulic conductivity and other properties.

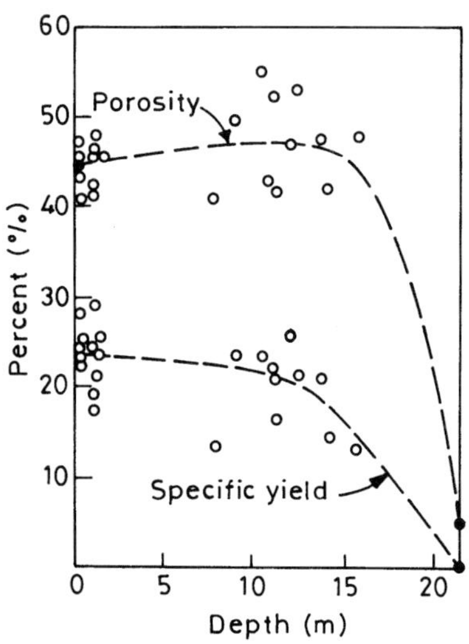

FIGURE 11.9 Range of porosity and specific yield values as estimated from laboratory tests on core samples in weathered metamorphic rocks from the Georgia Nuclear Laboratory area, Georgia, USA (after Davis and DeWiest, 1966).

The massive crystalline rocks have very low porosity and permeability. The porosity of unweathered crystalline rock mass usually varies between 0.1% and 1% while weathered rocks may have as much as 45% porosity. The range of porosity and specific yield in metamorphic rocks is shown in Figure 11.9. The hydraulic conductivity varies in the range of 10^{-6} to 10^{-3} ms^{-1} depending on the extent of weathering and fracturing. The range of hydraulic conductivity and permeability values for some sites in Europe and the USA are given in Figure 11.10.

As in other rocks types, hydraulic properties of crystalline rocks can be estimated from laboratory as well as *in situ* field tests (Chapter Nine). The *in situ* values of permeability determined from field tests in fractured rocks are significantly greater and more representative than laboratory values, because in the laboratory permeability measurements are usually made from compact and unfractured core samples while in the boreholes a greater thickness of rock with fractures is sampled. Permeability estimates from borehole tests may show greater variability depending on the aperture and frequency of fractures intercepted by the borehole. On the

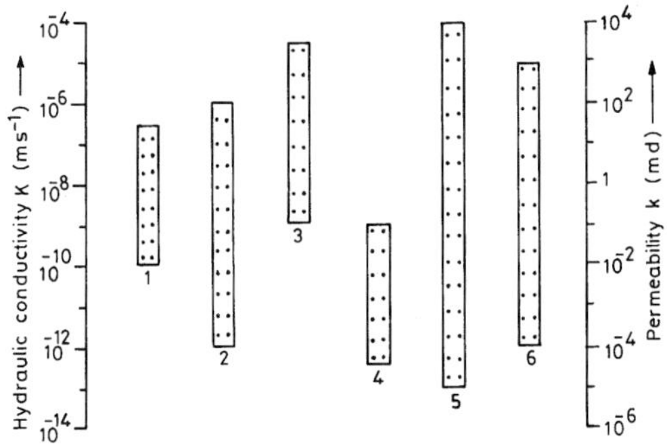

FIGURE 11.10 Range of hydraulic conductivity (K) and permeability (k) in some crystalline rocks estimated from *in situ* borehole tests (based on data from Brace, 1984, and Black, 1987). 1. Granite batholith, Monticello, SC, USA; 2. Altnabreac, Scotland; 3. Carynnen, Cornwall, UK; 4. HDR, Cornwall, UK; 5. Deep drilling in northern Switzerland; 6. Four sites in Sweden.

other hand, regional values show comparatively less variation due to proximity to REV (representative elementary volume) (Figure 11.11).

Pumping tests including slug tests are commonly used for estimating hydraulic characteristics of crystalline rocks (Chapter Nine) in areas where fractures are interconnected so that the medium is considered as a continuum. Slug tests

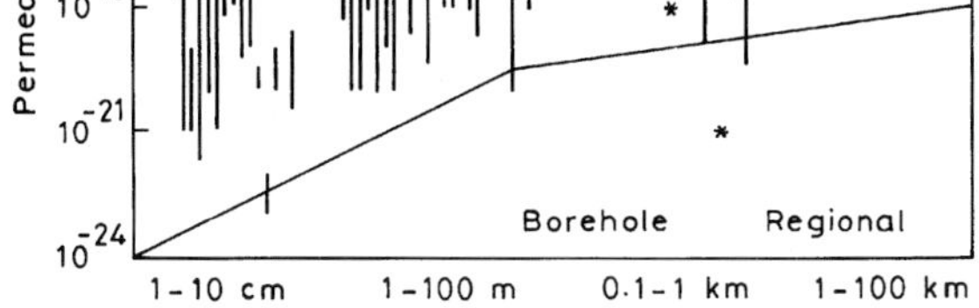

FIGURE 11.11 Variation in permeability values of crystalline rocks as a function of scale of measurement; bars mark the permeability range based on several reported values, and stars represent single values (after Clauser, 1992).

which involve injection of water may modify the hydraulic properties due to a decrease in the effective aperture of fractures. This is attributed to blocking of fractures by particulate material, and the precipitates formed due to chemical differences in the quality of naturally occurring water and the injected water. Therefore, necessary care is required to remove the suspended material from the injected water and its chemistry should also be known (Heath, 1985).

Pumping tests in hard rocks are performed either on large diameter dug wells which tap the regolith, or on borewells in underlying fractured rocks. In most fractured rock masses, local flow is controlled by the fracture system and is usually anisotropic. Therefore, configuration of multiple observation wells for determining anisotropic permeabilities of fractured rocks from the pumping test is necessary. In areas where weathered material (regolith) is underlain by fractured rocks, and groundwater development is from both these horizons, it is also necessary to have both shallow and deep observation wells to estimate the extent of interconnection between the two aquifer systems. The aquifer response could be similar to either a confined, leaky or unconfined type depending on the relative permeabilites of the constituent horizons. Various types of drawdown and build-up curves, depending on the nature of the fractures, boundary conditions and well characteristics are reported in literature (Davis and DeWiest, 1966; UNESCO, 1979; Kruseman and de Ridder, 1990; Boehmer, 1993). These aspects are discussed in Chapter Nine.

Pumping test data can be analysed by the conventional Theis and Jacob methods (Eagon and Johe, 1972; Wesslen *et al.*, 1977, among others), based on the assumption of continuum. Leaky aquifer and unconfined aquifer models are also used for aquifer parameter determination in crystalline rocks (CGWB, 1980; Sridharan *et al.*, 1990; Sekhar *et al.*, 1993; Boehmer, 1993; Levens *et al.*, 1994; Kaehler and Hsieh, 1994). Double porosity approach is used by Singhal and Singhal (1990) and Sidle and Lee (1995) to analyse pumping test data from fractured crystalline rocks. Methods for the analysis of test data from wells tapping a single vertical or horizontal fracture, and from fractured dyke systems (dykes and the adjoining fractured rocks) are described in Chapter Nine. Specific capacity values are also used to estimate transmissivity (section 9.2.4). A range of values of hydraulic properties of crystalline rocks are given in Table 11.2; values estimated at specific locations are shown in Table 11.3. These data may act as representative values and may be useful for groundwater resource esti-

TABLE 11.2 Range of hydraulic parameters in fractured crystalline rocks

Country	*Rock type*	*Borehole numbers*	*Mean transmissivity (m^2d^{-1})*	*Transmissivity (m^2d^{-1})*	*Hydraulic conductivity (md^{-1})*	*Testing method*
Zimbabwe	Mobile Belt gneiss	228	4.2	0.5–79	0.01–2.3	Pumping test[1]
	Younger granite	209	3.6	0.5–71	0.01–1.9	Pumping test[1]
	Older gneiss	392	–	0.5–101	0.01–2.8	Pumping test[1]
Malawi	Biotite Gneiss	2	–	–	0.1–0.2	Packer test[1]
United States	Granite	58	–	–	0.0–17	Injection test[1]
Europe, USA	Varied	8 site	–	–	1×10^{-8}–0.9	Packer test[1]
India	Granite & gneiss	–	–	5–50	–	Pumping test[2]

Source: [1] Wright (1992), [2] CGWB (1980)

TABLE 11.3 Hydraulic characteristics of crystalline rocks as estimated from pumping and packer tests

Rock type	*Country*	*T (m^2d^{-1})*	*K (ms^{-1})*	*Source*
Granite	Portugal	0.2–160	–	Carvalho (1993)
Granite-gneiss (fractured)	Zimbabwe	2–5	–	Chilton and Foster (1993)
Granite-gneiss (regolith)	Malawi	2–10	–	Chilton and Foster (1993)
Granite-gneiss	Sweden	1–2 $\times 10^{-1}$ to 10^{-2}	9×10^{-7} to 6×10^{-8}	Carlsson and Carlstedt (1977)
Granite	Sweden	50	2×10^{-4}	Wesslen *et al.* (1977)
Granite-gneiss	Norway	–	10^{-8} to 10^{-6}	Banks *et al.* (1992)
Granite	USA	–	10^{-13} to 10^{-9}	Sidle and Lee (1995)
Granite-gneiss and charnockites	India	3.65–7.2	–	Perumal (1990)
Granites & gneiss	India	5–50	10^{-4} 10^{-3} to	CGWB (1980)
Granodiorite	Canada	–	2×10^{-9} to 5×10^{-8}	Gale (1982b)
Chert	USA	–	1.84×10^{-12}	DeWiest (1969)
Chert	Australia	0.8–20	–	Pearce (1982)
Quartzite	India	54	–	Kittu and Mehta (1990)
Slates	USA	–	1×10^{-10}	UNESCO (1975)
Schist	USA	–	1×10^{-10} to 2×10^{-6}	UNESCO (1972)
Mica schist	Czech Republic	2×10^{-2} – 7×10^{-3}	–	Carlsson and Carlstedt (1977)

mation. A comparison of transmissivities of some crystalline rocks with those of alluvium is shown in Figure 11.12.

As pumping tests are expensive, spring discharge data can also be used to estimate hydraulic conductivity (section 13.6). In geotechnical and contaminant transport investigations, where permeability of discrete fractures and flow geometry is more important than the overall rock-mass permeability, borehole packer (Lugeon) tests are advantageous rather than conventional pumping tests. Special packer tests are designed to determine permeabilities of discrete fracture sets by keeping the orientation of the borehole perpendicular to the fracture set under consideration (section 9.2.1). The usefulness of injection packer tests in geotechnical investigations and pollution studies is described by Gale (1982a) and Banks *et al.* (1992). Levens *et al.* (1994) designed special packer tests for mine drainage studies.

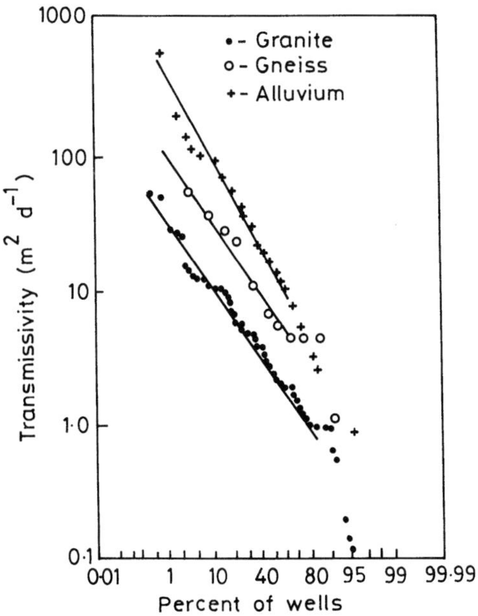

FIGURE 11.12 Log probability plot of transmissivity of some crystalline rocks and alluvium from Keonjhar district, Orissa, India (after Das *et al.*, 1996).

The tracer injection test in boreholes is used to study flow regime including the estimation of groundwater velocities and permeability of rock mass (section 9.3). Special tracer injection tests are also designed to estimate directional permeabilities and to study the mechanism and migration of solutes in fractures.

It may be emphasized that for a realistic appraisal of hydraulic characteristics in crystalline rocks, an integrated study involving structural mapping and controlled hydraulic *in situ* tests is necessary. The selection of a particular method would depend on the accuracy needed and purpose of study.

11.5 Water wells

The common groundwater structures for water supply in crystalline rocks are dug wells, borewells, dug-cum-borewells and collector wells. Infiltration galleries and horizontal boreholes are also constructed in suitable situations (Chapter Fifteen).

11.5.1 Well yields

Well yields in crystalline rocks are usually low being less than 20 m^3h^{-1}. Only a small percentage of wells have high yields (Table 11.4) and specific capacities of wells range from about 10 m^2d^{-1} in weathered crystalline rocks to about 18 m^2d^{-1} in fractured rocks (Houston, 1992). Data from India on well productivity from some crystalline rocks and alluvium are given in Figure 11.13.

Wells for a rural water supply are usually located in proximity to the demand site. A number of factors such as climate, lithology, topography, landform and depth, etc. influence the well yield.

Climate has an important bearing on well yields as it influences the depth of weathering, nature of weathered material and recharge potential.

TABLE 11.4 Yield of borewells in crystalline rocks

Rock type	*Region*	*Well yield (m^3h^{-1})*	*Well depth (m)*	*Source*
Granite-gneiss	Southern and eastern India	15–80	3–150	UNESCO (1984a), CGWB (1995b)
Granite-gneiss, Charnockite	Tamilnadu, South India	6–18	30–70	Perumal (1990)
Granite and gneiss	Kwara, Nigeria	9.5	50–75	Houston (1992)
Granite and gneiss	Victoria, Zimbabwe	4.1	20–30	Houston (1992)
Gneiss (fractured)	Maharashtra, India	48	164	CGWB (1995c)
Granodiorite and gabbro (fractured)	Lee Valley, USA	3.5–23	60–150	Kaehler and Hsieh (1994)
Pegmatite (fractured)	Karnataka, India	110	42	CGWB (1995c)
Quartz vein (fractured)	Bihar, India	55	137	CGWB (1995c)
Marble	Sri Lanka	2.4–24	–	Jayasena *et al.* (1986)
Quartzite	Sri Lanka	1.6–28.8	–	Jayasena *et al.* (1986)
Schist	Connecticut, USA	4.16	33.4	UNESCO (1972)
Slate	Maine, USA	3.42	–	UNESCO (1979)

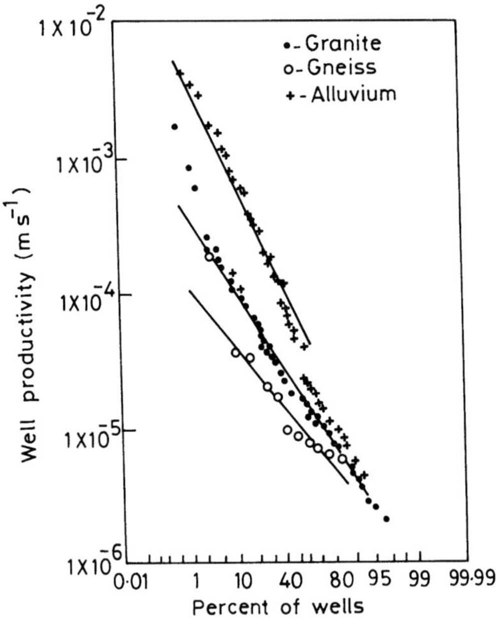

FIGURE 11.13 Frequency graphs of well productivity in fractured crystalline rocks and alluvium from Keonjhar district, Orissa, India (after Das *et al.*, 1996).

Lithology and rock texture also influence well yields. Generally, wells in quartz-rich and coarse-grained rocks like granites, pegmatites, quartz veins and quartzites have higher yields than other crystalline rocks as these rocks are more competent, and they develop and preserve good open joint systems. Higher yield from these rocks compared to slates and schists is due to their being more brittle, hence greater fracturing, and the coarse-grained character of their weathered product. High yields from wells tapping pegmatites and quartz veins are reported from several places, worldwide, for example from India (Karanth, 1992; CGWB, 1995c), USA (Florquist, 1973), and Zimbabwe (Owoade, 1993). Well yields in phyllites, schists and slates is usually low (Table 11.4) as these rocks have micaceous minerals which along with clay minerals, obtained as weathering products, tend to close the fractures. This is also the reason for their lower transmissivity as compared with granites and gneisses (Table 11.3). As compared to lithology, the climate, topography, landform and rock structures are more important to account for differences in well yields.

The **palaeoweathered** horizons can also form potential sources of water, if opportunities of recharge are available. For example, Henriksen

(1995) reported high yielding wells with median specific capacity of 5×10^{-6} m^2s^{-1}, terminating at the contact of tectonized phyllite and underlying weathered granite gneiss in Norway. Similarly, wells tapping weathered granite underlying Deccan basalt in central India have higher yields.

Topography and **landform** have a strong influence on well yield, especially of shallow wells, as they influence the thickness of the weathered zone (LeGrand, 1967; McFarlane *et al.*, 1992; Henriksen, 1995). Wells located in valleys have greater yields than those located on steep slopes, sharp ridges and interfluvial areas. Therefore, many drillers use only the topographic considerations in predicting the well yields.

LeGrand (1967) distinguished 18 types of topographic characteristics from the point of view of well yield. However, a recent reinterpretation of LeGrand's data shows that geomorphology alone is not the guiding factor in controlling well yields, but that fracture traces have a greater influence (Yin and Brook, 1992). As topographic depressions usually follow fracture traces/faults, and have a greater thickness of weathered material, the probability of well yields being greater in valleys is quite logical. Further, these depressions are also occupied by streams thereby providing a greater opportunity of recharge. The effect of distance from valleys on well yields is shown in Figure 11.14.

In Central Malawi, higher well yields are reported from areas of low relief as compared with areas with high relief adjacent to inselbergs (McFarlane *et al.*, 1992). Statistical analysis of well yield data from Norway also indicates the importance of geomorphological controls on well yield. Wells in valley bottoms and flat lands have the highest median yields, due to the greater thickness of superficial cover and source of recharge from surface water bodies (Henriksen, 1995). In Sri Lanka, the average well yield in the crystalline rocks from valleys is reported to be 3.6 m^3h^{-1}, but it is only 0.78 m^3h^{-1} from interfluvial areas (Hansmann *et al.*,

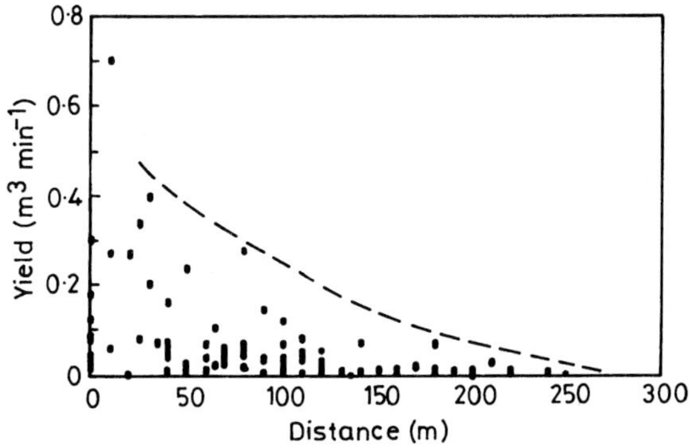

FIGURE 11.14 Well yields in relation to distance from valleys (after Hansmann *et al.*, 1992).

1992). In Uganda, out of 14 boreholes drilled in valley bottoms, 13 were productive with an average yield of 14.4 m^3h^{-1}, while those located on watersheds in hilly country had only 25% success, and the average yield was as low as 1.8 m^3h^{-1} (UNESCO, 1984a).

The influence of landforms on well yields is also demonstrated by Perumal (1990) from a study of granite–gneiss and charnockite formations in the Athur Valley of Tamil Nadu. Fault-controlled buried pediments, where the weathered horizon is thicker, give higher well yields as compared to other landforms (Table 11.5).

Rock structures and folding

Antiformal folds develop intense fracturing at crests which is responsible for higher secondary porosity and permeability, and hence greater well yields. However, in the case of open synformal folding, groundwater is more controlled by the dip of strata. In such open large-scale

TABLE 11.5 Yield of borewells in favourable landforms (after Perumal, 1990)

Landform	*Yield (m^3s^{-1})*
Valley	0.003–0.006
Shallow buried pediment	0.0006
Buried pediment (fault controlled)	0.020–0.026

structures, doubly plunging synclinal folds (i.e. tectonic basins) may coincide with groundwater basins. For example, in the Precambrian basement in southeastern Botswana, successful boreholes appear to be preferentially situated in areas where large scale synforms intersect (Dietvorst *et al.*, 1991).

Fractures and lineaments

Fractures and lineaments form important loci for groundwater. Lineaments generally represent vertical or near-vertical fractures along which the rocks, in certain situations, may be also deeply weathered (Figure 4.16). This may be revealed by a resistivity survey (section 5.3). Lineaments are also characterized by a higher degree of saturation and thick vegetative cover. The importance of lineaments for locating successful wells has been reported from several crystalline rock terrains, e.g. Kaehler and Hsieh (1994). The yield of wells located along lineaments in phyllites and quartzites in Andhra

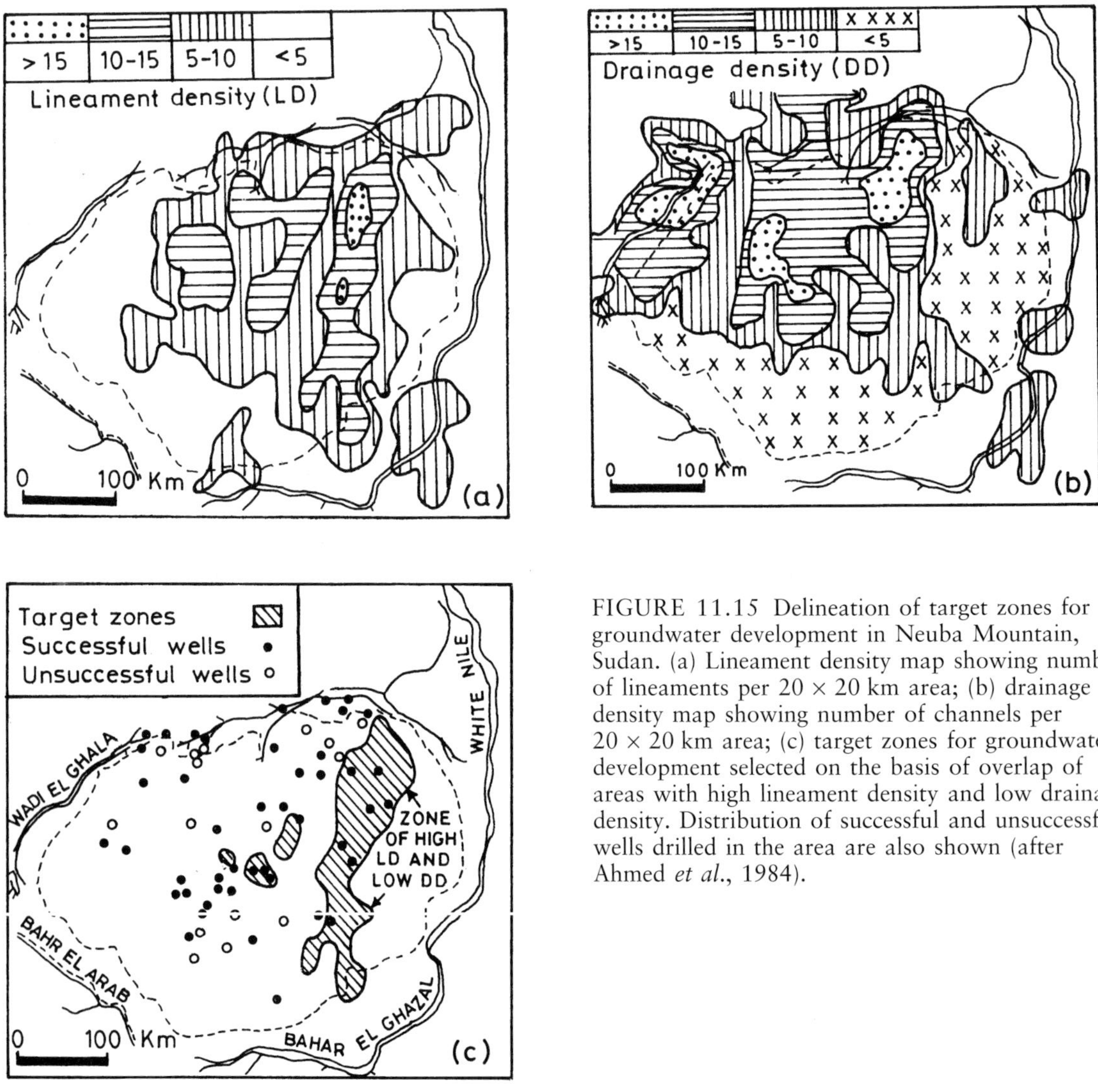

FIGURE 11.15 Delineation of target zones for groundwater development in Neuba Mountain, Sudan. (a) Lineament density map showing number of lineaments per 20 × 20 km area; (b) drainage density map showing number of channels per 20 × 20 km area; (c) target zones for groundwater development selected on the basis of overlap of areas with high lineament density and low drainage density. Distribution of successful and unsuccessful wells drilled in the area are also shown (after Ahmed *et al.*, 1984).

Pradesh, India, was four-orders of magnitude higher than elsewhere in similar rock types (Waters *et al.*, 1990). Tensile fractures are more productive than shear fractures as the latter have smaller apertures and are often filled with mylonite. High well yields (50 to 80 m^3h^{-1}) are reported from several sites located at the intersection of major tensional fractures (Waters *et al.*, 1990).

Near-horizontal sheet joints when interconnected by sub-vertical fractures form an important source of water for shallow borewells (Carruthers *et al.*, 1993). Such sheet joints are more open and closely spaced near the ground surface but decrease in frequency and aperture lower down.

Lineaments are best mapped as aerial photographs, and satellite images (section 4.8.10). Digital image processing of remote sensing data can be carried out to enhance lineaments and also landforms. A collective study of lineaments and landforms can be extremely helpful in locating target zones (Figure 4.20). Further, GIS-based modelling relating yields of existing wells with various spatial characteristics like distance from valleys, lineaments, and surface reservoirs may be used for potential groundwater zoning (Hansmann *et al.*, 1992). Lineament density maps can be combined with drainage density maps to define target zones (Figure 11.15). An example of the importance of integrated study involving geophysical, structural and geomorphological investigations for locating high-yielding wells is given in section 15.2. However, in places there may be limitations to using lineaments for locating successful wells because of their variation in dip, width and possible tightening at depth due to shearing (Greenbaum, 1992; Carruthers *et al.*, 1993).

11.5.2 Optimum depth of wells

The optimum depth of well is the depth at which an unsuccessful well should be abandoned, so as to give the lowest average cost per successful well. The concept of optimum depth of wells holds good in crystalline fractured rocks as the permeability usually decreases with depth (section 8.2.3). While deciding the optimum depth of wells, it is necessary to consider what makes a well successful or unsuccessful in a region. The cost of well construction, water use, well yield and water quality together decide the success or failure of a well. Even a relatively small well yield (about 1×10^{-4} m^3s^{-1}) of good quality water may make a well, fitted with a hand pump, successful for a rural domestic water supply (Houston, 1992). On the other hand, for a piped water supply to small communities and for minor irrigation, the yield requirement from a well with an electric submersible pump should be higher (>0.002 m^3s^{-1}) (section 15.1.2).

Read (1982) suggested that a curve between the function–ln [(1 – S(*d*)], where S(*d*) is the estimated cumulative success rate to depth (*d*) when plotted against depth (*d*) gives a better idea of the optimum depth of a well for a given expectd discharge. A smooth curve is drawn through the resulting step-graph. The optimum depth is indicated by the point where a straight line through the origin is tangential to the curve (Figure 11.16).

An overall decrease in well yield with depth is reported from various crystalline rock terrains in different parts of the world (Davis and Turk, 1964; UNESCO, 1979; Woolley, 1982; Henriksen, 1995). Based on well productivity data, Davis and Turk (1964) suggested the optimum depth of wells in crystalline rocks to be between 50 and 60 m (Fig. 11.17). Similar conclusions are drawn by several other workers (Table 11.6).

There are also cases where well yields do not necessarily decline with depth. For example, in a recent study, Loiselle and Evans (1995) did not find any evidence of either decrease in fracture density or fracture yield up to a depth of about 180 m in the crystalline rocks of Maine, USA. High yields from deep fracture horizons

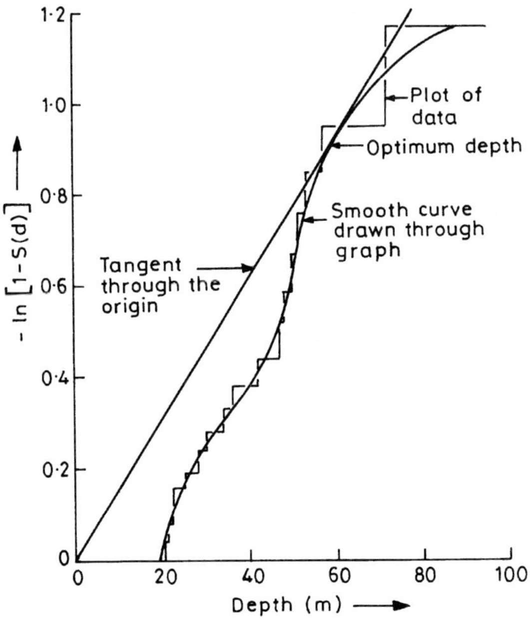

FIGURE 11.16 Plot of -ln [1-S(d)] against depth of 27 wells drilled in gneisses and schists of Haast Bluff area, NT, Australia. S(d) is cumulative success rate to depth 'd' where 5×10^{-4} m^3s^{-1} is classified as success. The straight line drawn through the origin, tangential to the curve, indicates an optimum depth of about 60 m (after Read, 1982).

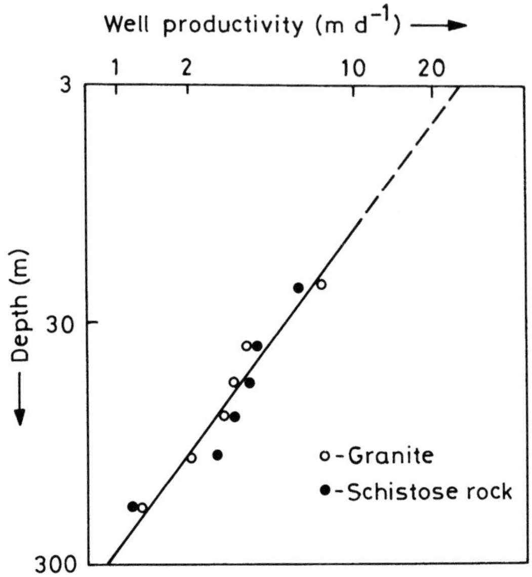

FIGURE 11.17 Plot showing decrease in well productivity with depth in crystalline rocks of Eastern United States (after Davis and Turk,1964).

(>100 m) are also reported from Sri Lanka and India. In Sri Lanka, deep fresh-water flow is reported along the Trincomalee submarine canyon indicating a considerable loss of fresh groundwater into the sea. Recent data on well drilling in parts of the Indian shield also show a considerable improvement in well productivity with an increase in depth up to 300 m (Jagannathan, 1993; CGWB, 1995c).

From the above it can be summarized that although there is a general decrease of well yield with depth in crystalline rocks, the variation is case specific and may depend upon the type of formations, type of fractures and other variables. In the case of sheet joints, formed due to near surface processes, one would expect a decrease of well yield with depth. However, lineaments and other deep-seated fractures may be in an open state even at greater depth.

TABLE 11.6 Data on optimum depth of wells in different regions and rock types

Region	*Rock type*	*Optimum depth (m)*	*Source*
Cyprus	Gabbro	170–200	Read (1982)
Satpura hills, India	Granite gneiss & schist	45–60 45–75	Uhl (1979) Read (1982)
North Carolina, USA	Granite	75–90	Read (1982)
Karamoja, Uganda	Gneiss & schist	30–92	Faillace (1973)
Zimbabwe	Greenstone and gneiss	40–80	Wright (1992)
Norway	Granite and gneiss	40–60	Henriksen (1995)

11.6 Assessment of groundwater recharge

Estimation of recharge is of importance for the assessment of groundwater resources of an area. In temperate and tropical regions, rainfall is the main source of groundwater recharge. In addition to this, other important sources of recharge are influent seepage from rivers and canals, inter-basin groundwater flow and return flow from irrigation. Various methods of estimation of groundwater recharge are described in section 17.1.3. The estimation of groundwater recharge in crystalline and other hard rocks is faced with greater problems as compared with sedimentary rock aquifers, on account of their heterogeneity. Estimates of recharge from rainfall by using different methods in some crystalline rock terrains is given in Table 11.7; there is a wide variation in recharge depending upon local geology and climatic conditions. In most cases the recharge is around 5% to 15% of rainfall.

The groundwater residence time in the crystalline rocks may vary from only a few years in shallow weathered horizons to as much as several thousand years in deep fractures, as revealed from ^{226}Ra, ^{3}H and ^{14}C data (Gale *et al.*, 1982; Sukhija and Rao, 1983; Silar, 1996). This indicates fast movement of groundwater in the weathered horizon but very retarded movement and lack of interconnection in deep fractures.

In areas where sources of natural recharge are limited, artificial recharge may be practiced, depending on the availability of water. An account of various methods of artificial recharge under different hydrogeological conditions is given in section 17.8.2. In crystalline rocks, on account of their typical geomorphic setting and hydrogeological characteristics, the common methods of artificial recharge are: (a) percolation tanks (infiltration ponds), and (b) subsurface dykes. Contour bunding and other soil conservation methods also promote natural groundwater recharge.

11.7 Groundwater quality

In crystalline rocks, the groundwater flow mainly takes place through fractures. Therefore, the contact area between water and the rock matrix is small, as compared to rocks possessing primary porosity. Further, the crystalline rocks contain mainly silicate minerals which weather very slowly. Therefore, in general, groundwater in metamorphic and plutonic igneous rocks has relatively small amounts of dissolved solids (TDS) and the water is of

TABLE 11.7 Estimated values of groundwater recharge (mm) from rainfall

Area/Country	*Annual rainfall (mm)*	*Method of estimation*					*Reference*
		Base flow measurement	*Water-table fluctuation*	*Chloride balance*	*Soil moisture*	*Isotope method*	
Malawi	750–900	30–200		95–234	–	–	Wright (1992)
Zimbabwe	750–1400	50–200 (5% to 20%)		80–115	–	22–33	Chilton and Foster (1993)
Kwara, Nigeria	1700–2000	–	–	–	(15%)	–	Houston (1992)
Vedavati Basin, India	590–620		17–90 (3%–15%)			(6.2%)	Duba (1973) Athavale (1985)
Ponnani and Rao Noyil basin, India	800–1300				(13%–21%)		Sukhija and (1983)
			(5.7% to 12.20%)				UNESCO (1984a)
					(4.5% to 9.8%)		Athavale (1985)

NB: Values in brackets give recharge as percentage of rainfall.

$Ca-HCO_3$ type. However, in arid and semi-arid regions, due to low rainfall and excessive evaporation, groundwater may have high salinity.

11.7.1 Effect of lithology

Igneous rocks are classified as acidic, basic and ultramafic types. Acidic igneous rocks (e.g. granites) and their metamorphic equivalent (gneisses) have appreciable amounts of quartz and other aluminosilicate minerals such as feldspars and micas. Quartz has very low solubility under existing pH and temperature conditions. Other silicate minerals, e.g. feldspars and micas, dissolve incongruently in water having dissolved CO_2. This causes release of common cations like Ca^{2+}, Mg^{2+}, Na^+ and K^+ into water, leaving behind a clayey residue which is rich in aluminosilicates (clay). It is observed that nearly all groundwaters from igneous rocks plot in the kaolinite field of stability diagrams, suggesting that alteration of feldspars and micas to clay (kaolinite) is quite common (Freeze and Cherry, 1979). This is accompanied by a rise in pH and HCO_3 concentration of water.

The amount of total dissolved solids in water from crystalline rocks is usually less than 300 mg l^{-1}. The silica content is low (10–30 mg l^{-1}), its solubility increases with temperature. There is dominance of Na^+ over K^+, as sodium is more soluble than potassium, and the latter is more easily fixed on clay minerals in the rock matrix. Ca^{2+} is also dominant over Mg^{2+} and HCO_3^- is more than Cl^- and SO_4^{2-} (Table 11.8). The ratio of Na^+/Ca^{2+} in aqueous extracts is generally much lower as compared to the mole ratio in the parent rocks, indicating greater solubility of Ca^{2+} to that of Na^+. Granitic rocks usually have moderately acidic waters (pH = 6.0 to 6.8).

At places, the fluoride content of groundwater in acidic igneous rocks (granites and pegmatites), and mica schists may be greater than the drinking-water permissible limits (<1.5 mg l^{-1}), due to the presence of fluoride-bearing minerals like apatite, fluorite and biotite in the rocks (Chapter Ten). Among other trace elements, groundwater in the granitic and metamorphic rocks may have a high concentration of Pb, Cr, Fe, Mn and low I. For example, Smedley (1992) reported high Pb (around 0.15 mg 1^{-1}) and Cr concentrations in acidic waters from granitic terrains in Ghana.

The basic igneous rocks (e.g. gabbros) are rich in Ca and Mg minerals, e.g. plagioclase feldspars and pyroxenes. The groundwater in these rocks, therefore, has high ratios of Ca/Na and Mg/Ca. The total dissolved solids are usually less than 400 mg l^{-1} (Table 11.8). The total iron content is moderate but silica is high (> 20 mg l^{-1}).

The ultramafic igneous rocks (e.g. dunites and peridotites) are made up mainly of magnesium and iron silicate minerals (olivines and pyroxenes). Although mafic and ultramafic

TABLE 11.8 Chemical characteristics of groundwater from different types of crystalline rocks (mgl^{-1})

Rock type	*TDS*	*Ca*	*Mg*	*Na*	*K*	*Cl*	*SO_4*	*HCO_3*	*Fe*	*SiO_2*	*Location*
Granite	223	27	6.2	9.5	1.4	5.2	32	93	1.6	39	Maryland, USA[1]
Granite gneiss	137	28	1.9	6.8	4.2	1.0	1.4	121	2.7	31	Baltimore, USA[2]
Diorite	347	72	4.1	10	2.8	6.5	115	114	0.04	22	N. Carolina, USA[1]
Gabbro	359	32	16	25	1.1	13	10	203	0.06	56	N. Carolina, USA[1]
Olivine-tuff (ultramafic)	281	20	42	19	–	7	22	279	–	31	Arizona, USA[2]
Quartzite	52	1.6	5.8	2.8	–	9.9	2.0	18	–	8	Transvaal, S. Africa[1]
Schist	221	27	5.7	16	0.7	2.5	9.6	138	0.11	21	Georgia, USA[1]

Source: [1] = Hamill and Bell (1996); [2] = Hem (1989)

rocks contain little or no quartz and have lower amounts of total silica than acidic rocks (granites), the amount of dissolved silica in groundwater may be high, indicating chemical instability of associated minerals. The Mg content in these rocks is quite high; therefore, the spring water from these rocks is reported to be of Mg-HCO_3 type (Table 11.8).

Low-grade metamorphic rocks (phyllites and slates) have comparatively high porosity but low permeability. Therefore, on account of greater contact of groundwater with rock matrix, the total dissolved solids and concentration of cations and anions is greater than in high-grade metamorphic rocks like granulites and gneisses. Groundwater in pyrite bearing slates has higher SO_4^{2-} and lower pH. In mica-schists, which are equivalent to granite in chemical composition, the groundwater is alkali-rich but in amphibolites, which are equivalent to basic igneous rocks, groundwater is rich in alkaline earths (Matthess, 1982).

11.7.2 Hydrochemical zoning

In Precambrian shields and massifs of crystalline rocks, vertical zoning in the chemical composition of groundwater is reported from several areas (Figure 11.18). Near-surface groundwaters are dilute and are of Ca-Na-HCO_3 type. With increased depth and length of flow path, at greater depths (150–500 m), the groundwater becomes brackish and tends towards Na-Ca-HCO_3-Cl type. At increasingly greater depths (>500 m), the water is usually saline having 5000–50 000 mg l^{-1} TDS and the composition is dominated by Na-Ca-Cl and Ca-Cl types. Such a general pattern has been reported from the Canadian shield (Gascoyne and Kamineni, 1993; Gascoyne *et al.*, 1996) and the granitic suite of rocks in Sweden (Smellie *et al.*, 1995). The pore fluids at depths >400 m in the granitic batholith of Manitoba, Canada, have TDS > g 100 l^{-1}. Groundwater from very deep levels (>700 m) in granites and other crystalline rocks from many other parts of the world is also reported to be highly saline (TDS >10 000 mgl^{-1}) (Couture *et al.*, 1983; Edmunds and Savage, 1991; Kessels and Kuck, 1995). Edmunds and Savage (1991) reported high salinities (up to 31 600 mg l^{-1}) from boreholes at a depth of 2400 m in granites of the UK.

The saline waters in crystalline rocks are typically depleted in Na^+ relative to Cl^-, which makes them different to the saline waters formed either due to entrapment of sea water or dissolution of evaporite deposits. The change from Ca-Na-HCO_3 type to Na-Ca-HCO_3 type with increased depth has been explained by the hydrolysis of plagioclase feldspars which liberates Na and Ca coupled with loss of Ca by precipitation and/or ion exchange (for Na) on clay minerals in the fractures (Gascoyne and

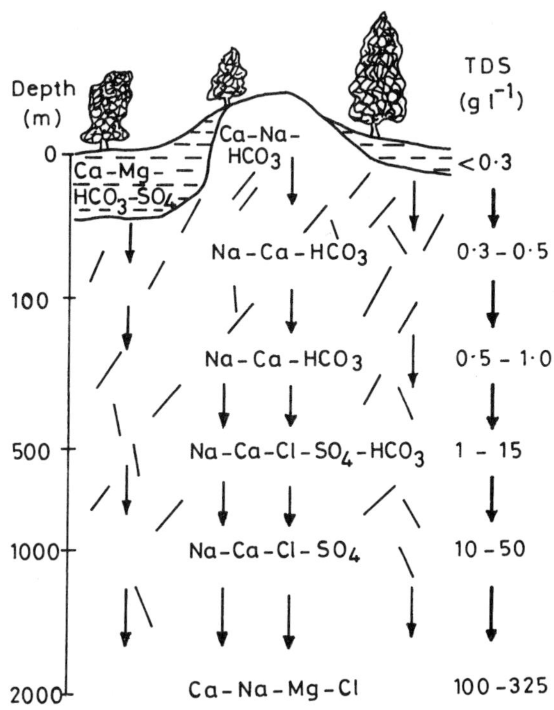

FIGURE 11.18 Schematic representation of the chemical evolution of groundwater with depth in plutonic igneous rocks (after Gascoyne and Kamineni, 1993).

Kamineni, 1993). Several other explanations have been given for the origin of deep thermal saline waters including migration of sedimentary formation brines (Couture *et al.*, 1983; Kessels and Kuck, 1995), marine transgressions, residual hydrothermal fluids, breakdown of fluid inclusions, radiolytic decomposition of water and water–rock interaction (Edmunds and Savage, 1991). The $CaCl_2$ pore fluids in the Canadian shield are believed to be millions of years old, formed due to the interaction of rock and ancient sea water or brines (Gascoyne *et al.*, 1996).

The composition of groundwater in deep-seated crystalline rocks is not only of geochemical interest but is also important for the disposal of radioactive waste and tapping of geothermal power.

11.7.3 Groundwater contamination

Crystalline rocks are more vulnerable to contamination from anthropogenic activities, as the vadose zone is usually thin and rapid flow takes place through fissures and joints. In arid regions, although the water-table is usually deep as the weathered layer is coarse-grained and devoid of clayey material, the contaminants may move faster. High nitrate in groundwater is reported from several crystalline rock aquifers (Reboucas, 1993). Contamination from pesticides is likely to be present in fractured rock aquifers and coarse-grained regolith with a shallow water table. Movement of leachates from landfills along fractures also contaminates groundwater in crystalline rocks. Faecal contamination of shallow wells and boreholes in the weathered basement aquifer of Malawi and Botswana is widespread. In such cases borewells with casing pipes should be preferred to conventional dug wells in order to reduce chances of pollution.

Acid rains affect groundwater quality especially in those hard rock terrains which consist of non-carbonate rocks, as reported from the granitic rocks of Scotland and metamorphic and granitic rocks of the Bohemian Massif in the western part of the Czech Republic (Mather, 1993). Crystalline rocks, due to their low permeability, form potential repositories for high-level radioactive waste. These aspects are discussed in section 10.9.

Chapter twelve

Hydrogeology of volcanic rocks

Volcanic rocks are formed by the solidification of magma at or near the ground surface. The most common volcanic rocks are basalts, which are of basic composition. Acidic and intermediate rock types such as rhyolites and andesites have comparatively very limited occurrences. Basalts are formed due to the eruption of lava either on the ground surface (subaerial eruption) or on the sea floor (submarine eruption). Eruption on the land surface could be of fissure type (**plateau basalts**), covering large areas on the continents or of **central type**, which is of limited distribution mostly forming volcanic cones (Figure 12.1).

12.1 Weathering, landform and drainage

Volcanic rocks possess characteristic landforms, weathering and drainage patterns. Generally, they are highly susceptible to weathering. Older extrusive rocks exhibit deeper weathering implying a thicker soil and vegetation cover; younger volcanic rocks are generally barren of vegetation. In the tropical climate, basalt weathers into 'chernozem', also known as 'black cotton soil' which being rich in clay (montmorillonite) is quite impervious but has high fertility. Under good leaching conditions, the ultimate weathering products of basalt are laterite and bauxite.

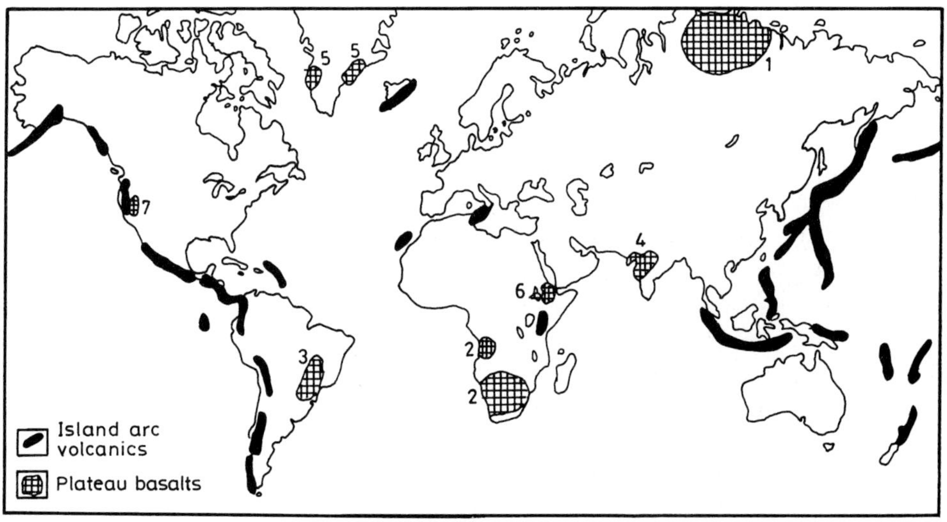

FIGURE 12.1 World distribution of major Mesozoic and younger plateau basalts and active volcanic arcs. 1=Siberian Traps; 2=Karoo Province; 3=Parana Volcanics; 4=Deccan Traps; 5=Thulean Province; 6=Ethiopian Basalts; 7=Columbia River Basalts.

A variety of landforms develop in volcanic rock terrains which are linked to the composition of lava and type of eruption. They also govern the hydrogeological characteristics of the terrain. Acidic lava is viscous, and therefore restricted in extent, often forming steep-sided bulbous domes. Basic lava has a relatively lower viscosity and is commonly of a ropy type. It exhibits flow structures and possesses an oblate outline with flat topography.

The central type of eruption is marked by sloping cones, calderas, vents and conical landforms. Eruptions of the fissure type are flatter near the centre and become serrated along the periphery, the flows generally have a rough surface topography and discordant contacts with the bedrock. The lava flows are often interbedded with weathered and pyroclastic material and other neo-volcanic sediments. From a distance, on scarp faces, a number of flows may collectively impart the impression of rough subhorizontal bedding. The landscape looks like a series of giant steps, each flow producing a steep escarpment and a flat top (mesa) (Figure 12.2). This step-like topography is known as *trappen* in German from which the word trap is derived. Due to such a landscape, the basaltic flows which cover large parts in Central India are known as Deccan Traps (Figure 12.3). These lava flows are associated with dykes which acted as feeding channels and now occur as ridges, forming barriers for surface water and groundwater flow (Figure 12.4).

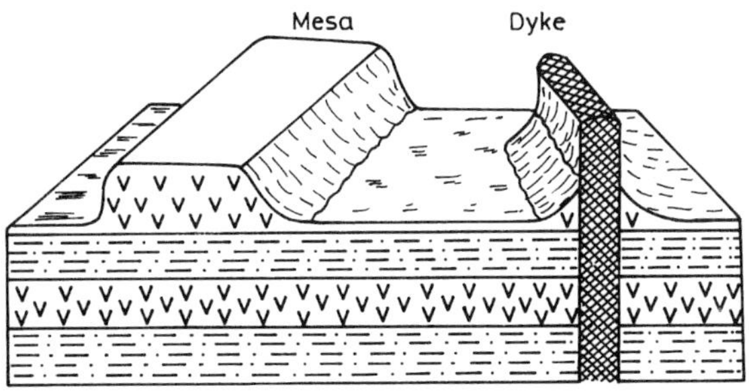

FIGURE 12.2 Formation of 'mesa' and dykes in volcanic terrain.

Lava tubes and tunnels are locally important subsurface features. A lava tube or tunnel is formed when the surface of a lava flow has cooled and hardened, but the interior remains more fluid and happens to drain out from beneath the solidified crust, leaving behind a tunnel-like void.

Basalts typically exhibit columnar jointing

FIGURE 12.3 Extensive layering produced by successive eruptions in Deccan Traps, Western India, (courtesy H. Kulkarni).

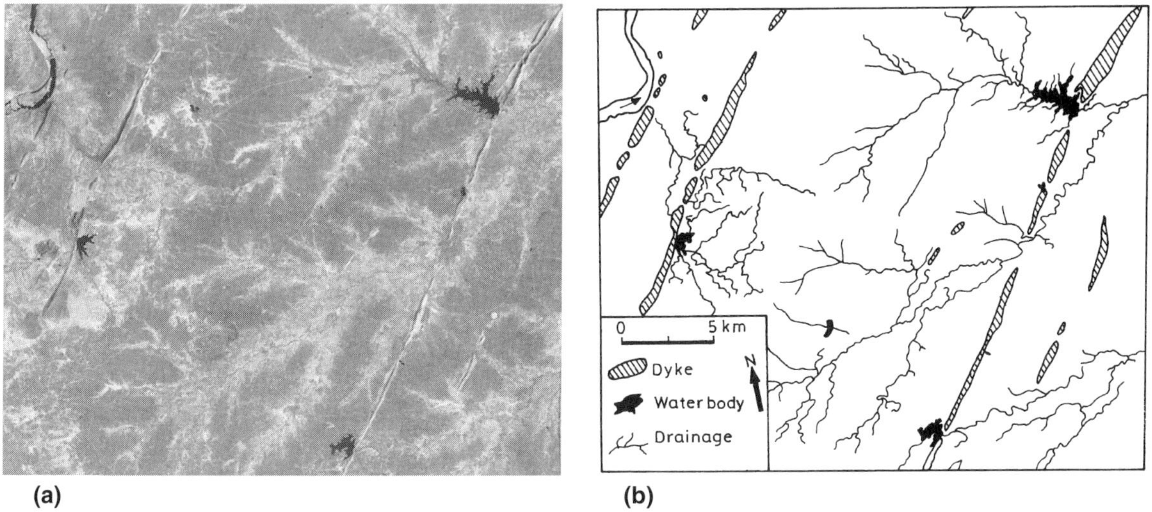

FIGURE 12.4 (a) LANDSAT TM image of a part of central India showing damming of surface streams by dykes; (b) interpretation map.

FIGURE 12.5 Columnar joints in basalts, N. Ireland (after Reader's Digest Association, 1980).

(Figure 12.5). Some of the columns have 4, 5, 8 or 10 sides, but the majority of them are near-perfect hexagons, about 30–50 cm in diameter. They are formed due to contraction of the cooling lava, forming prismatic patterns in the solidifying rock. As shrinkage continues, the joints extend through the rock mass resulting in the network of vertical joints forming columns. The columnar joints cause anisotropy and strong vertical hydraulic conductivity in basalts.

The lava flows may obliterate the pre-existing water divides, resulting in changes in surface and groundwater flow regimes, as is exhibited in the Snake River Plain in Idaho, USA (Stearns, 1942). Lakes may be formed either due to closed depressions, viz. crater lakes, and also due to damming of river courses by lava flows. In areas of tectonic activity, volcano-tectonic lakes are formed as in New Zealand and Indonesia (Ollier, 1969). The drainage pattern is influenced by the type of eruption and rock characteristics; radial and annular drainage patterns are associated with a central type of eruption.

Surface drainage, in volcanic rocks, increases with weathering and age; older lavas display high drainage density and a dendritic pattern. Surface drainage in young volcanic rocks, however, may be almost absent owing to high porosity and permeability. For example, an area of about 25 000 km^2 in southern Idaho, USA, underlain by basalt, has no surface run-off (Stearns, 1942). Also in tropical volcanic islands such as Hawaii, despite heavy rainfall, which may exceed 750 cm per year, most of the streams are flashy and are seasonal. This is primarily due to a high rainfall and peculiar watershed characteristics which favour a high rate of flashy surface run-off; and also the high permeability of the basalts and soil cover which favour a high infiltration rate. Here, drainage basins are of a small size, and watersheds are characterized by steep slopes and steep valley walls with little channel storage. Therefore, in such areas surface storage of water is difficult due to the high rate of infiltration (Peterson, 1972). Stearns (1942) has highlighted the problems of construction of reservoirs and dams in basaltic terrains.

12.2 Hydrogeology

12.2.1 Plateau basalts

Plateau basalts, also known as continental flood basalts, are widespread in various parts of the world. They are of different geological ages, but the most common ones on the continents are of late Mesozoic and younger ages. Some of the important occurrences of plateau basalts from the hydrogeological point of view are listed in Table 12.1.

The plateau basalts usually consist of a number of flows of varying thicknesses superimposed over each other. The thickness of the individual layer ranges from less than 1 m to more than 30 m, most being between 10–30 m. The individual layer is much thicker in continental flood basalts as compared to basalts of oceanic islands (Figure 12.6). This could be due to greater original surface gradients of oceanic islands, as compared to that of continental plateau areas (Davis, 1974).

The lateral extent of individual flow units in plateau basalts depends upon the composition and viscosity of lava, rate of supply and loss of volatiles during cooling. Basaltic lava with low silica and low viscosity spreads over larger areas, forming thin and extensive sheets as compared to acidic lavas such as rhyolite and andesite which are more viscous (Figure 12.7). The lateral extent of individual flow may vary from a few tens of metres to as much as hundreds of kilometres. Many basalt flows occupy ancient river valleys and therefore locally enclose alluvial sediments of varying lithologies and thicknesses, which are important sources of water supply.

TABLE 12.1 Major plateau basalt provinces on the continents

Country	*Basalt Province*	*Age*	*Area (km^2)*	*Maximum thickness (m)*
India	Deccan Traps	Upper Cretaceous to Eocene (65–60 Ma)	500 000	1500
Brazil	Parana Volcanics	Lower Cretaceous (140–120 Ma)	900 000	2000
USA	Columbia and Snake River basalt	Miocene to Quaternary (17.5–6 Ma)	200 000	1500
Russia	Siberian Traps	Middle Paleozoic to early Mesozoic	1.5×10^6	700
South Africa	Karoo Province	Early Jurassic (190 Ma)	3×10^6	

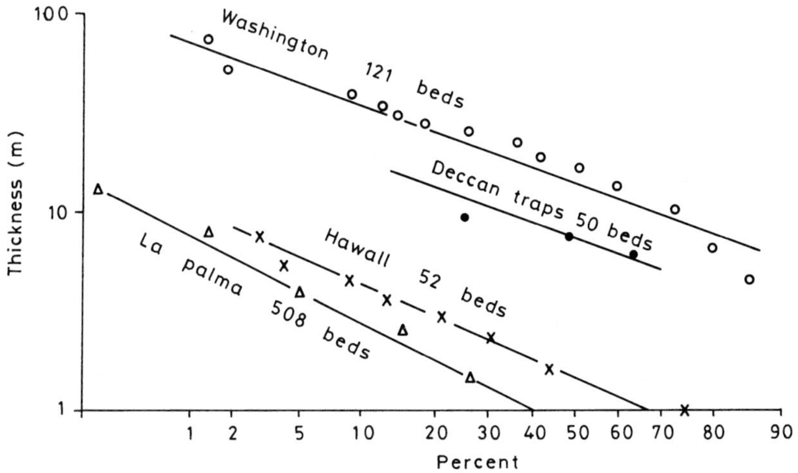

FIGURE 12.6 Log-probability plot of bed-thickness of basalt flows. Data for Washington, Hawaii and La Palma are from Davis (1974) and data for Deccan Traps are from Krishnan (1949).

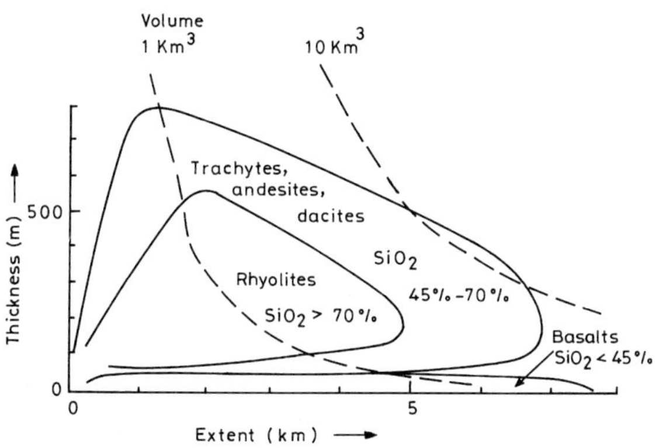

FIGURE 12.7 Variation in thickness and lateral extent of lava flows depending on chemical composition (SiO_2%) (after Walker, 1973).

Many continental lava flows develop an autobrecciated top surface which is also progressively pushed beneath the flow as it advances (Figure 12.8). This tends to produce a relatively permeable blocky top and base to many lava flow units (Mathers and Zalasiewicz, 1994). Description of some important plateau basalt provinces is given below.

Columbia–Snake River basalts

These are of Miocene to Quaternary age and occupy an area of more than 200 000 km^2 covering the states of Washington, Oregon and Idaho in the USA. The total thickness is more than 1500 m. The individual flows are very thick (50–150 m), which can be traced for as much as 250 km indicating high fluidity of lava. The basalt flows are either flat or slightly dipping with an angle of 1–2 deg to the southeast (Fetter, 1988). The interflow surfaces and sedimentary layers are highly permeable. Shrinkage cracks create vertical permeability. Deeper aquifers are more productive than shallower ones (Fetter, 1988).

Deccan Traps

They cover an area of about 500 000 km^2 in the western and central parts of India (see Figure 17.2). Originally they would have

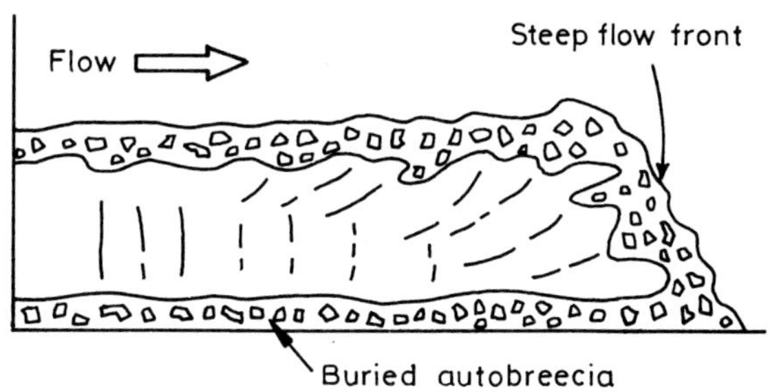

FIGURE 12.8 Development of autobrecciated material on the top surface of lava flow and its burial due to downward sliding along the steep-moving front.

covered an area of more than 1×10^6 km^2. The age of the Deccan Traps, based on argon data, is estimated to be 65 to 60 Ma (Duncan and Pyle, 1988). Based on deep seismic-sounding surveys, the greatest thickness of basalt is reported to be about 1500 m near the western coast (Kaila, 1988). The thickness decreases towards the east, and it becomes only a few tens of metres in the eastern and southeastern parts. The eruption took place through wide fissures as is evidenced by the presence of dolerite dykes. These dykes extend in length from a few hundred metres to 70 km or more with an average width of 1–10 m. The dykes have a NW–SE trend on the western coast and an E–W trend in the central region.

In the Deccan Traps as many as 29 lava flows are reported from a borehole at Bhusawal in Maharashtra (Krishnan, 1949). The thickness of individual flow units varies from a few metres to 50 m. The individual flows are separated from each other by residual soil locally called **bole** and other interflow sedimentary deposits (intertrappeans). The lower part of an individual flow is usually massive, and it becomes vesicular and amygdaloidal towards the top (Figure 12.9). The **bole** is usually of a red colour but occasionally it may be green coloured also, the thickness being usually less than 1 m. It is rich in clay minerals and is believed to be a product of atmospheric weathering and/or hydrothermal alteration of amyg-

(a)

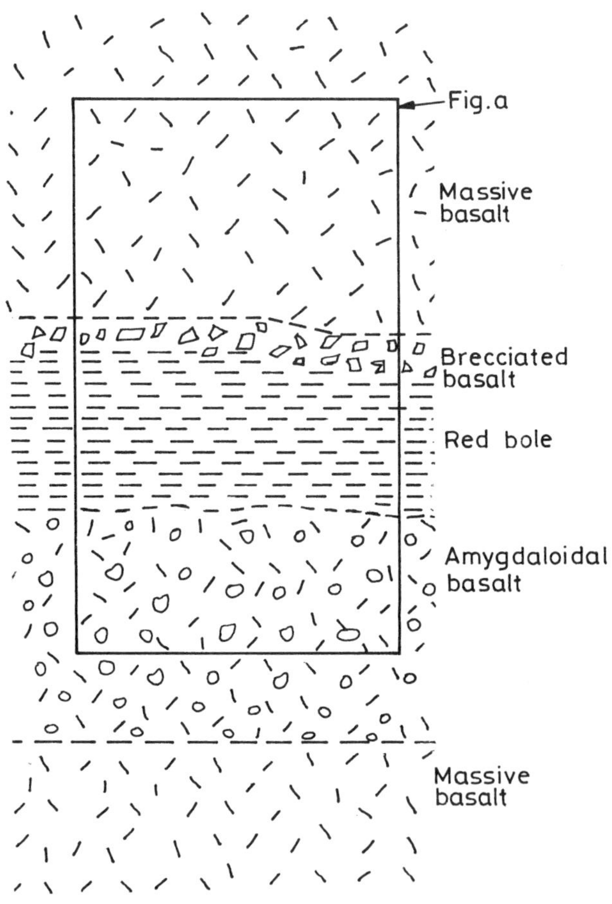

(b)

FIGURE 12.9 (a) Field photograph showing layering within a flow in Deccan Traps (courtesy H. Kulkarni); (b) interpreted profile.

daloidal basalt or of pyroclastic material (Inamdar and Kumar, 1994; Wilkins *et al.*, 1994). Being rich in clay, the bole layer usually serves as a confining or semi-confining layer. However, when fractured it forms an aquifer as is seen in many dug well sections. Natural gamma ray logging is useful in the detection of red boles (Versey and Singh, 1982).

Rajmahal Traps

They are of Jurassic age and at present cover an area of about 400 km^2 in the eastern part of India. A large part of the traps is covered by alluvium as revealed from boreholes. The basalts have developed good secondary permeability due to fracturing. The Rajmahal Traps were regarded as an early manifestation of Deccan Trap volcanism, but recent geochemical data indicate gross dissimilarities between the two.

12.2.2 Volcanic islands – submarine eruptions

Submarine basalts show distinct hydrogeological characteristics as compared to subaerial continental types, due to differences in the physical conditions of solidification of lava. The submarine basalts are characterized by a pillow structure in which the intervening spaces between the pillows form easy channels for the movement of water. Pillow structures can also develop under continental conditions when lava is erupted into lakes or in water-saturated sediments. In comparison to plateau basalts, the submarine basaltic flows show greater heterogeneity and are commonly associated with pyroclastic materials with low porosity and permeability.

Volcanic islands are of two types: (1) oceanic islands, viz. Hawaii and French Polynesia in the Pacific Ocean and Mauritius in the Indian Ocean, which are of basaltic composition and therefore belong to the Basaltic Province, and (2) island-arc islands, viz. Indonesia, New Guinea and the Philippines, which are formed of mainly andesitic lavas and hence belong to the Andesitic Province. The basaltic oceanic islands can further be classified into two types – high islands and low islands. Some of the high islands, e.g. Hawaii islands rise to a height of more than 5000 m above the floor of the Pacific Ocean. They are occupied mainly by young basalts and may be partly flanked by marine and terrestrial sediments. Low islands e.g. the Cook islands in the Pacific and Laccadive (Laksha Dweep) Islands in the Arabian Sea, are only a few metres above sea level which may be partly or fully covered with coralline limestone reefs forming atolls underlain by volcanic rocks (CSC, 1984).

The basaltic flows on volcanic islands are usually thin (6 m or less) but form main aquifers due to their high permeability. The high permeability is primarily due to clinker zones in the **aa** type flows, lava tubes and gas vesicles in the **pahoehoe** flows, columnar vertical joints and the irregular openings between the lava flows (Peterson, 1984). In places, the flows are interbedded with ash beds which form confining layers.

The central part of the Hawaii islands is characterized by a swarm of closely spaced vertical or steeply inclined dykes which cut across the gently dipping lava flows. The dykes, being more or less impermeable, serve as barriers against lateral movement of groundwater and thereby form water compartments in which groundwater occurs at different heights (Figure 12.10).

In the Indian Ocean, Mauritius is an important volcanic island. The island is totally composed of basalts varying in age from 7 Ma to Recent. Younger flows are vesicular and scoriaceous with high permeability (Rogbeer, 1984).

12.2.3 Dykes and sills

Dykes are vertical or steeply inclined intrusive igneous bodies which cut across the pre-existing

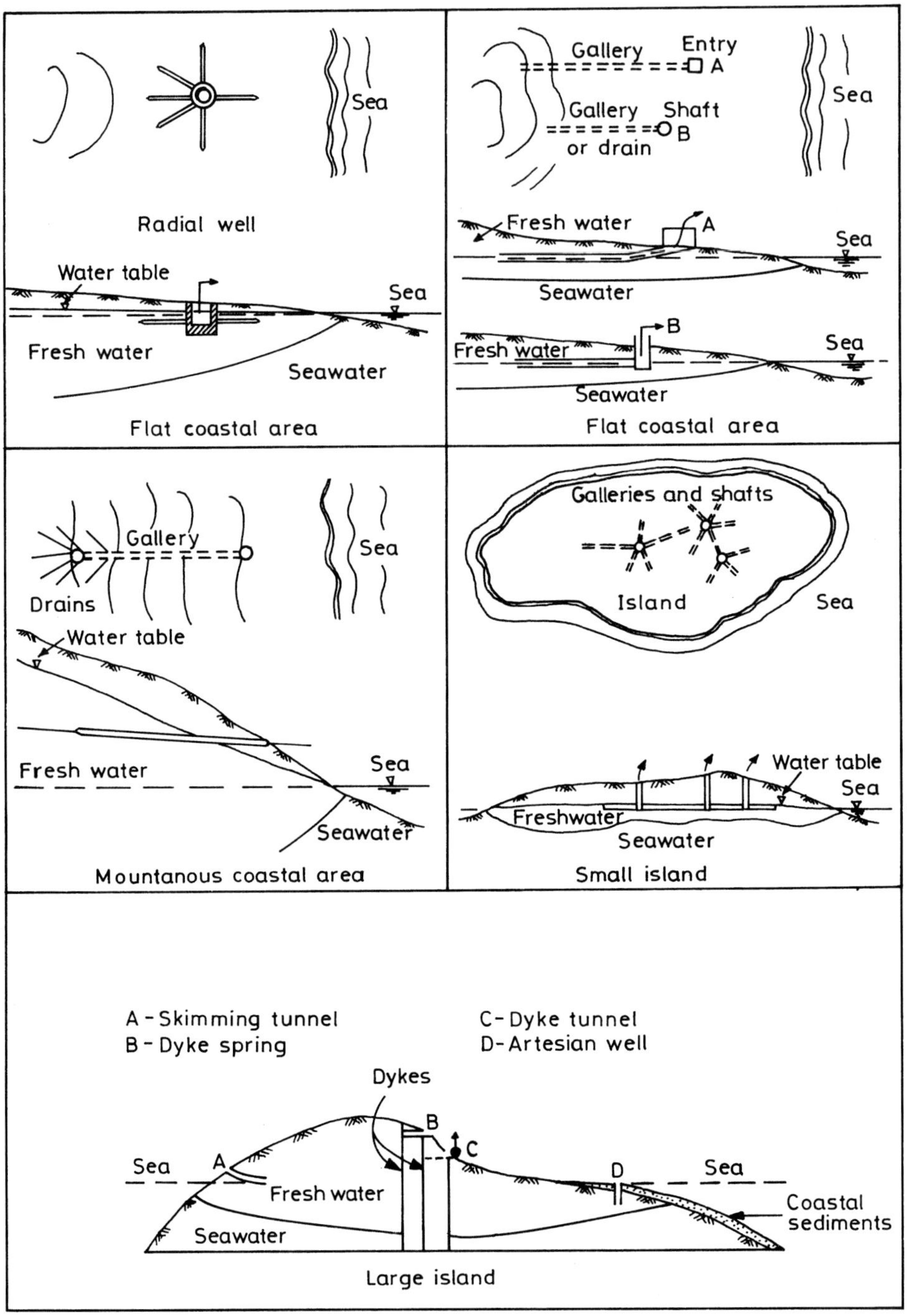

FIGURE 12.10 Different types of groundwater structures used in volcanic islands and coastal areas (modified after UNESCO, 1987).

rocks. They are usually of basic composition. Dykes vary in thickness from a few decimetres to hundreds of metres, but widths of 1–10 m are most common. In length they may be from a few metres to several kilometres long. They represent feeders for the lava flows. As dykes are usually more resistant to erosion than the country rock they stand out prominently as wall-like ridges. Massive and unweathered dykes form barriers for lateral movement of groundwater, thereby confining large volumes of groundwater. On the other hand, fractured dykes may form good aquifers. One of the dykes in the Palaghat Gap (western coast of India) extends for a strike length of about 14 km and is highly fractured, forming a potential source of groundwater. The discharge from some of the wells in this dyke is of the order of 240–840 $m^3 d^{-1}$ (Kukillaya *et al.*, 1992). Sometimes dykes also cause fracturing of the adjacent country rock due to thermal effects. Drilling into the country rock close to the dyke is recommended in such cases, as in the Precambrian crystalline basement rocks of South Africa, Western Australia and north-east Brazil (Boehmer and Boonstra, 1987). Pumping tests in dyke areas from Botswana indicate that dykes which are thicker than 10 m serve as groundwater barriers, but those of smaller width are permeable as they develop hydraulic continuity with the country rock through cooling joints and fractures (Bromley *et al.*, 1994).

Sills are nearly horizontal tabular bodies which are commonly concordant and follow the bedding of enclosing sedimentary rocks or the older lava flows. Some of the sills are very thick and extend over large areas, as in the Karoo System of South Africa. Due to their low permeability, except when fractured, sills form perched water bodies, as in the Hawaiian islands.

12.3 Groundwater occurrence

Groundwater in volcanic rocks occurs under perched, unconfined and confined conditions. Perched water occurs above the regional water-table due to the presence of impervious formations, viz. sills, ash beds and dense basalt flows. Water under confined conditions is reported where the pervious lava beds are confined between impervious sedimentary beds (Stearns, 1942; Fetter, 1988). Confined conditions are also created when vesicular or fractured basalt is sandwiched between massive units (Singhal, 1973). Free-flowing artesian conditions are reported from some places, viz. Honolulu (Stearns, 1942) and Java island in Indonesia (Krampe, 1983). Water under artesian conditions is also reported from the Deccan Traps (Kittu, 1990; Rana and Vishwakarma, 1990). Dykes also cause lateral confinement of water forming compartments as described earlier.

The water-table gradients are low (<1:1000) in areas where volcanic rocks have high permeabilities, as in Columbia–Snake River basalts. The depth to the water-table varies depending upon the recharge and permeability of volcanic rocks, and the depth to the water-table is considerable (>150 m) in areas where the basalts are highly permeable (Ollier, 1969). In the Deccan Traps, which are less permeable than Columbia River basalts, the water-table is generally shallow (<10 m) and it follows the ground topography.

The general velocity of groundwater in basalts is high, but not as high as in carbonate aquifers. The velocities are reported in the range of about 3.6–255 $m d^{-1}$ (Shelton, 1982).

12.4 Hydraulic characteristics

Primary porosity and permeability in volcanic rocks depend on the rate of cooling, viscosity of magma and degassing during cooling. Acidic volcanic rocks, like rhyolites and trachytes, are usually more massive than basalts and therefore have lower porosity and permeability, with some exceptions.

The various openings which impart porosity

and permeability to basaltic rocks are (a) scoariae, (b) breccia zones between flows, (c) cavities between pahoehoe lava flows, (d) shrinkage cracks, parallel to the flow surfaces or columnar joints, (e) gas vesicles, (f) lava tubes, and (g) fractures and lineaments (Stearns, 1942; UNESCO, 1975). The order of importance may, however, vary in different areas. **Pahoehoe** flows and pillow lavas are more permeable than the thick dense **aa** flows. The lower and upper brecciated parts of flows develop good permeability (Figure 12.8). Other interflow spaces may also impart permeability. Sheet and columnar joints formed due to cooling, and other fractures and lineaments produced as a result of later tectonic activity, sometimes impart high permeability. Table 12.2 gives the range of values of porosity and hydraulic conductivity of volcanic rocks, indicating that the variation in permeability is almost nine orders of magnitude. Such a variation indicates their importance both from the point of view of water supply as well as for the selection of suitable sites for the disposal of radioactive waste (Dietz, 1985; Evans and Nicholson, 1987).

The hydraulic conductivity of pyroclastic deposits depends on the degree of consolidation and welding. The unconsolidated pyroclastic deposits, especially those which are reworked, show higher conductivities, as in Indonesia, Japan and in central California, USA (Stearns, 1942; Davis and DeWiest, 1966). Welded tuffs which are formed at high temperatures by the fusion of rock fragments have low porosity in the range of 5–20% and very low conductivity (10^{-9}–10^{-7} m s^{-1}) depending on the degree of welding (Sharp *et al.*, 1993).

TABLE 12.2 Porosity and hydraulic conductivity values in volcanic rocks. (Based on data after Davis, 1969; Freeze and Cherry, 1979; UNESCO, 1975 and Sharp *et al.*, 1993)

Rock type	*Porosity (%)*	*Hydraulic conductivity (m s^{-1})*
Basalt		
Dense	0.1–1	10^{-11}–10^{-8}
Vesicular	5–11	10^{-9}–10^{-8}
Fractured, weathered	10–17	10^{-9}–10^{-2}
Pumice		
Pyroclastics, (tuffs)	87	
Zeolitized	39	4×10^{-9}
Welded	5–20	10^{-9}–10^{-7}

Volcanic rocks also show anisotropy. Horizontal hydraulic conductivity is several times greater than vertical hydraulic conductivity due to the presence of interflow spaces and horizontal fractures. In Gran Canaria, Spain, the ratio of K_h/K_v is reported to be between 20 and 100 (Custodio, 1985). Therefore, the rate of horizontal flow may be several orders of magnitude greater than the vertical flow. However, at places, columnar jointing may impart high vertical conductivity which may be of the order of 10^{-1} m s^{-1}.

A decreasing trend in porosity and hydraulic conductivity of volcanic rocks with increasing geological age is reported from several places. This is attributed to the filling of vesicles and fractures by secondary minerals such as zeolites, calcite and secondary silica. For this reason pre-Tertiary basaltic rocks of Brazil, Deccan Traps of India and basalts of eastern United States form generally poor aquifers, as compared to the Tertiary basalts of Columbia–Snake River area and Hawaiian islands (USA), and Canary Islands (Spain). Pillow lavas of Triassic age in Connecticut (USA) and the island of Guam also show lower conductivity due to the infilling of interstices with secondary minerals, as compared with the younger Tertiary pillow lavas from Snake River Canyon, USA (Stearns, 1942) and Pleistocene and Holocene basalts of Mexico and Mauritius (Niedzielski, 1993; Rogbeer, 1984) (Table 12.3).

The above generalized relationship of hydraulic conductivity with age holds good mainly for primary porosity and permeability. However, at times the older basalts can acquire even higher secondary porosity and permeability due to frac-

TABLE 12.3 Hydraulic properties of volcanic rocks

Rock type	*Age*	*Location*	*T* (m^2d^{-1})	*S*	*Source*
Dacite (weathered)	–	Red Hill, Australian Capital Territory	0.5–1.2	5×10^{-5}	Jacobson (1982)
Andesite (vesicular & fractured)	Lower Permian	Banana area, Queensland	1.5–200 average 65	–	Pearce (1982)
Basalt (fractured)	Jurassic, Cretaceous	Rajmahal, India	90–330	1.4×10^{-4}	Khan and Raja (1989)
Basalt	Cretaceous–Eocene	Deccan Traps	10–700	10^{-3}–10^{-1}	Adyalkar and Mani (1974); Versey and Singh (1982); Kittu and Mehta (1990)
Basalt	Miocene	Columbia Snake River area, USA	202–22 780 average 5310	2×10^{-2}–6×10^{-2}	Walton and Stewart (1961)
Basalt	Miocene Quaternary	Gran Canaria Spain	50–300	–	Fernandopulle *et al.* (1974); Custodio (1985)
Basalt (fractured)	Pleistocene–Holocene	Mexico	605–865	–	Niedzielski (1993)
Basalt	Pliocene	Republic of Djibouti	34–5100	10^{-3}–10^{-4}	Jalludin and Razack (1993)

turing, as is reported from some places in the Deccan Traps, and even older Rajmahal Traps of Jurassic age, from India (Khan and Raja, 1989).

The groundwater flow characteristics in basalts are similar to those in other hard rocks. Accordingly, basalts may have discrete fracture flow, or may form a continuum medium when fractures, lava tubes and vesicles are interconnected. Therefore, estimation of aquifer parameters in volcanic rocks is besieged with similar problems to those in other fractured rocks. In literature there are many examples where the conventional methods of Theis, Jacob and Thiem were used to determine the aquifer properties from pumping test data (viz. Walton and Stewart, 1961; Davis, 1969; Adyalkar and Mani, 1974; Jalludin and Razack, 1993). Further, packer tests have been used for estimating permeability of discrete zones. The Papadopulos model has been considered for analysing pumping test data from large diameter dug wells (Rao, 1975). The double-porosity method has also been used for estimating transmissivity of matrix blocks and fractures (Singhal and Singhal, 1990).

Step-drawdown test data have been analysed by Uhl *et al.* (1975), Rao (1975), and Mishra *et al.* (1989) using Jacob's and Rorabough methods for estimating well and formation loss coefficients as well as transmissivities. Methods of analysis of pumping test data from dykes and adjacent country rocks are described in section 9.2.4. Acidic rocks, e.g. dacites and andesites, usually have low transmissivities as compared with basalts. Hydraulic characteristics of some volcanic rocks from different parts of the world, as estimated from pumping tests, are given in Table 12.3.

In recent years, a greater interest has developed in the hydrogeological characteristics of basalts and tuffs as they could be potential host rocks for the disposal of radioactive waste, due to their low permeability. Therefore, detailed *in situ* hydraulic tests have been conducted in Columbia River basalts at Hanford, USA (Dietz, 1985) and in fractured tuffs at the Nevada test site in the USA (Russell *et al.*, 1987), to assess their hydraulic parameters.

12.5 Groundwater development

Springs are an important source of water in volcanic terrains. They usually show high discharge, quite comparable to those in carbonate rocks (Table 12.4). In the USA, out of the 66 first magnitude springs (discharge 2.83 m^3s^{-1} or more), 36 issue from basalts (Stearns, 1942). In eastern Australia, some springs show a discharge as much as 5 m^3s^{-1} (UNESCO, 1975). In volcanic islands, dykes and ground topography control the location of springs (Figure 12.10). The spring discharge depends on the size and interconnection of fractures and other passageways like volcanic tubes etc. Relatively constant discharge is expected where the fractures are narrow and the groundwater body is large, while greater variability is observed when flow is through wide fractures and other openings which are well-developed and abundant.

TABLE 12.4 Discharge values of some major springs in basalts

Location	*Discharge (m^3s^{-1})*	*Source*
Datta Spring, USA	1.4–3	UNESCO (1975)
Thousand Springs, Idaho, USA	15–20	UNESCO (1975)
Kalauaa Spring, Pearl Harbour, Oahau Island, USA	0.4–1	UNESCO (1975)
Oaxtepec and Aqua Hedrionda Springs, Mexico	0.7	Niedzielski (1993)

Dug wells and borewells are constructed to tap groundwater. Horizontal boreholes, tunnels and galleries are also suitable to tap perched water on hills or on impermeable ash or tuff layers between flows and intertrappeans in plateau basalts and also in volcanic islands (Figure 12.10).

Yield of large diameter dug wells in the Deccan Traps of India is reported to be 0.01–0.05 m^3s^{-1} for moderate drawdown (Adyalkar and Mani, 1974). Specific capacity of these wells is in the range of 1×10^{-4} to 5×10^{-3} m^2s^{-1} depending on the permeability of basalts (Deolankar, 1980). Borewells in the Deccan Traps are usually of 10–20 cm diameter and 40–50 m depth having a yield of 1 to 2×10^{-3} m^3 s^{-1}. The well productivity generally decreases beyond a depth of 70 m. Hydro-fracturing helps to increase well yield (Madhure and Sakare, 1990). Borewells located on intersections of lineaments have higher yields (0.02–0.04 m^3s^{-1}) for a drawdown of 3–5 m (Kittu, 1990; Subramanian, 1993). Borewells in older Rajmahal traps of Jurassic age in eastern India have an yield of 2×10^{-3} for a drawdown of less than 1 m from fractured horizons (Khan and Raja, 1989). In Columbia River basalts of the USA, typical well yields from confined aquifers located at depths between 150 and 300 m are 0.06–0.12 m^3s^{-1} (Fetter, 1988). Specific capacity of wells in Columbia basalts is reported to vary from 2×10^{-6} to 8×10^{-2} m^2 s^{-1} with an average value of 2×10^{-3} m^2 s^{-1} (Maxey, 1964).

In coastal areas and volcanic islands, basal groundwater which floats over sea water, is mostly developed by radial wells (Maui type), tunnels, galleries, shafts and a combination of these to minimize the chances of sea water intrusion (Figure 12.10). The type of structure depends upon topography, rock structure and their permeabilities. Confined aquifers in coastal sediments are also an important source of water supply tapped by screened wells. Such wells in Honolulu, Hawaii, produce at the rate of 0.025 m^3s^{-1} (UNESCO, 1975).

The **Maui** type of tunnels may produce up to 2 m^3s^{-1} depending on their length and nature of rocks (Fetter, 1988). In places the tunnels dip slightly towards the entrance so that water can flow out by gravity. Sometimes they are also provided with vertical bores (well shafts) to tap simultaneously deeper aquifers, as in the Canary Islands. These tunnels or galleries are of a few kilometres in length and are located at a depth of 1000 m below ground level (Custodio, 1985). Water confined in the dyke compartments is either discharged as springs or is tapped by dyke tunnels.

12.6 Groundwater recharge

The recharge in volcanic rocks can be estimated by the same techniques as in other rocks (section 17.1.3). The main source of recharge is the direct infiltration from precipitation, seepage from streams and return flow from irrigation. Fog drip could also be an important source of recharge as reported from the Maui island in Hawaii (Stearns, 1942) and La Palma island of Spain (Veeger *et al.*, 1989).

Recharge in old volcanics is generally low which is attributed to weathering effects and low permeability of rocks resulting in high runoff. In the Deccan Traps, based on water balance and water-table fluctuation methods, groundwater recharge is estimated to vary from 10% to 20% of the rainfall (Adyalkar and Rao, 1979). Tritium data also gave similar values of recharge. Even in high rainfall areas, recharge is low due to the high relief and impervious nature of the Deccan Traps. In arid regions of different parts of the world, the groundwater recharge in basaltic terrains is reported to be about 10% of the annual rainfall (UNESCO, 1975). Recharge in the fractured tuffs at the Nevada nuclear test site, USA, located in an arid climate, was estimated to be approximately 8% of all precipitation. At this site a comparison of the travel time of water with the pre-nuclear testing period indicated increased permeability due to induced fracturing as a result of nuclear testing (Russell *et al.*, 1987). In the Hawaiian islands, where annual rainfall varies from 500 to 6300 mm, the recharge in older basalts is 6% to 10% and in younger volcanics 30% to 36% of the annual rainfall (Wright, 1984). In Mauritius, where basalts are of Quaternary to Recent age, and average annual rainfall is 1200 mm, the recharge is computed to be 24% of the rainfall (Rogbeer, 1984).

Deeper confined and semi-confined aquifers may be recharged by leakage from shallow horizons through vertical fractures, and partly by lateral subsurface flow from outcrop areas. In Columbia River basalts, good recovery of water-levels after pumping from deeper aquifers, and other hydraulic data indicate high vertical hydraulic conductivity and interconnection between aquifers. On the other hand, the age of groundwater as determined by the ^{14}C method ranges from modern to as old as 32 000 yr BP; age increasing with increasing depth. This indicates lack of interconnection between shallow and deeper aquifers (Fetter, 1988). Low tritium content and greater salinity of water in the deeper aquifers of the Deccan Traps is also regarded as indicating very limited or negligible vertical movement of water (Versey and Singh, 1982).

As in carbonate rock aquifers, spring discharge data are also used to determine the recharge pattern and the groundwater potentiality of volcanic rocks. Analysis of spring flow data from fissured basalts of Mexico shows a double recharge mechanism as in karstic terrains (section 13.6). In this case, the main recharge takes place immediately within a period of 1–3 days after rainfall through open fractures, but the other recharge is slow due to diffused flow through fine fissures and is delayed by 1–3 months with respect to a rainy period. This diffused flow is mainly responsible for dry weather discharge (Niedzielski, 1993).

TABLE 12.5 Chemical composition of groundwater (in mg l^{-1}) from volcanic rocks

Rock type	*TDS*	*Ca*	*Mg*	*Na*	*K*	*Cl*	*SO_4*	*HCO_3*	*Fe*	*SiO_2*	*Location*
Rhyolite	234	14	5.8	20	5.2	4.0	7.7	112	–	62	Oregon, USA[1]
Columbia River Basalt	280	24	15	12	5.3	15	1.6	156	0.43	50	Oregon, USA[1]
Deccan Trap	505	62	28	24	–	37	30	294	–	30	Hyderabad, India[1]
Basalt	–	24	15	12	5.3	15	1.6	156	0.43	50	Oahau & Hawaii, USA[2]

Source: [1] White *et al.* (1963)
[2] Morrison *et al.* (1984)

Large-scale groundwater development has resulted in the lowering of the water-table at several places in the Deccan Traps area. Therefore, various methods of artificial recharge are used; recharge by percolation (infiltration) tanks is quite effective. We return to this subject in section 17.8.2.

12.7 Groundwater quality

Basalts mainly consist of minerals rich in Ca and Mg. Therefore, groundwater in basalts has a higher amount of alkaline earths (Ca^{2+} and Mg^{2+}) and a lower amount of alkalis (Na^+ and K^+). The TDS is usually less than 500 mgl^{-1} (Table 12.5). Among the anions, HCO_3 is dominant over Cl^- and SO_4^{2-}. The pH is usually in the range of 6.7 to 8.5 and the SiO_2 content is high (> 30 mgl^{-1}). In general, the groundwater in basalts is of the Ca-Mg-HCO_3 type. In the case of acidic volcanic rocks (rhyolites), groundwater has a higher concentration of SiO_2 and Na^+ and the waters are of the Na-HCO_3 type (Table 12.5).

In areas where volcanic rocks have high permeability, there are greater chances of groundwater contamination. Pollution of groundwater in volcanic islands, due to sewage and increased use of pesticides, has been reported from some places (CSC, 1984). However, in areas where permeable basalts are covered with some impervious material, the chances of downward migration of contaminants are minimized, as in the Columbia–Snake River area in the USA (Davis and DeWiest, 1966). Overexploitation of groundwater in oceanic islands can lead to sea-water intrusion. A case study of such a possibility from a small island in the Arabian sea is described in section 17.7.6.

Chapter thirteen

Hydrogeology of carbonate rocks

13.1 Introduction

Carbonate rocks are sedimentary type of rocks containing more than 50% carbonate minerals which are mainly calcite, $CaCO_3$, and dolomite, $CaMg(CO_3)_2$. The carbonate minerals could be a result of chemical precipitation, organic processes or may occur as detrital material. Some of the dolomites may be a result of diagenesis known as dolomitization which involves replacement of calcite by dolomite to a varying extent. The term limestone is used for those rocks which contain more than 90% carbonates. If the rock contains more than 50% but less than 90% carbonates, it is termed as arenaceous limestone or an argillaceous limestone, depending upon the relative amounts of quartz and clay minerals. Chalk is a type of limestone which is soft and white in colour and is rich in shell fragments.

The carbonate rocks occupy more than a quarter of the continental areas of the world. Locally, their cumulative thickness may be as much as 10 000 m (UNESCO, 1984b). In the geological past, most limestones and dolomites were deposited during the intermediate phases of Caledonian, Hercynian and Alpine tectonic cycles and their ages are mainly Palaeozoic and Mesozoic.

Carbonate rocks have attracted greater attention of hydrogeologists in the past few decades, both from the point of view of water supply as well as potential sites for waste disposal. About 25% of the world's population, both in rural and urban areas, depends on water supply from these rocks. Carbonate rocks are also of great economic value having large deposits of oil, gas and ore minerals.

13.2 Weathering and landforms – the karst

Carbonate rocks (limestones and dolomites) are soluble to a large degree in water rich in carbonic acid giving rise to a characteristic landscape known as **karst**. The term karst is of Yugoslav origin as karst features were first described from the Dalmatian coast of the Mediterranean area in the former Yugoslavia.

The development of karst landforms is governed by a number of factors such as palaeoclimates, tectonic movements and eustatic changes, mainly during Pleistocene and Holocene times. The direction of groundwater movement is also an important factor in the development of karst. The karst development depends upon two main factors: (a) availability of water rich in CO_2, and (b) sufficient permeability of the rock in the form of fractures and other discontinuities (Figure 13.1). Higher rainfall and humidity along with a higher concentration of CO_2 promote karstification (equation (13.1)).

$$CaCO_3 + CO_2 + H_2O \rightleftarrows Ca\,(HCO_3)_2 \qquad (13.1)$$

The presence of soil cover will promote dissolution, as partial pressure of CO_2 in the soil zone is higher than in the atmosphere due to the

FIGURE 13.1 Solution cavities (karren) along discontinuities in Miliolite limestone of Saurashtra, India (after Nair, 1981).

weight of the overburden, and additional CO_2 is also derived from the decay of organic material. Conversely, a decrease in CO_2 pressure will cause precipitation of calcite where the groundwater either discharges on the surface through spring outlets or when the CO_2 escapes through the cave openings. The carbon cycle, therefore, plays an important role in karst development, as the carbonate rocks, being a huge storehouse of carbon, form an important sink or source of atmospheric CO_2 due to the processes of dissolution and deposition (Yuan, 1996).

There is a distinct relationship between the degree of karstification and the lithology and the thickness of the stratigraphic unit. A rule of thumb is: the thicker the limestone sequence, the more important and dominant will be the effect of karstification on the hydrological regime of the area. If the carbonate rocks are thin, then only a superficial karst will develop (UNESCO, 1984b).

Dissolution of limestones and development of cavities become more pronounced due to sea-level changes and the mixing of waters of different compositions, viz. fresh and sea water in coastal aquifers. This situation can also occur anthropogenically, due to the disposal of waste water of a composition different to that of resident water in the carbonate aquifer (Deike, 1990).

Karst is well developed in all geographic latitudes, viz. karst of the Urals is in part at latitude 70°N and that of South Australia at latitude 35°S (Komatina, 1975). Karst covers large areas in different parts of the world, namely CIS (former USSR), USA, Central America, South China, North Africa (Atlas Mountains), and southern Iran. The karst of Central America, namely Jamaica, Cuba, Mexico and Guatemala together with that of Florida and Texas in the USA constitutes one of the largest karst belts in the world, covering an area of about 500 000 km^2. In China, karst landforms are well developed in the carbonate rocks ranging in age from Proterozoic to Quaternary. The bare karst terrain of south China is one of the largest and most densely populated tropical karsts in the world covering an area of about 9×10^5 km^2. It is developed in Palaeozoic carbonate rocks characterized by low porosity and thin soil cover. The total karst water resources in China are estimated to be 2029×10^3 m^3 out of which 1847×10^3 m^3 is in south China and only 182×10^3 m^3 in north China (Chen and Cai, 1995). In England and France, chalk of Cretaceous age forms highly productive aquifers. The karst processes have been widespread not only in the Recent but also in the past geological periods.

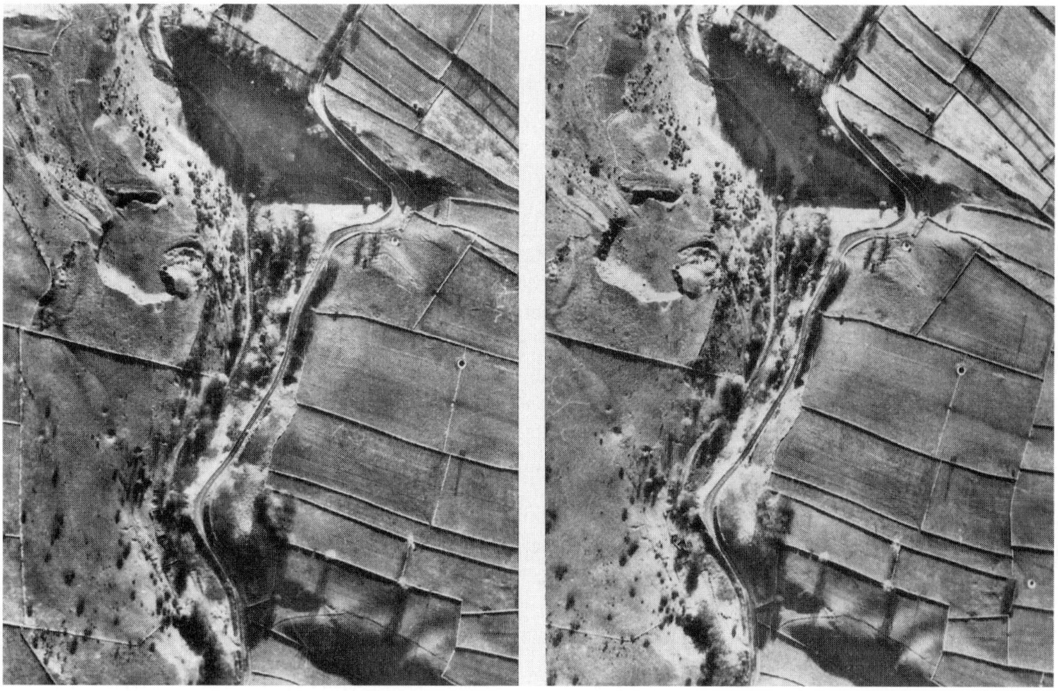

FIGURE 13.2 Stereo-photographic pair showing limestone terrain characterized by depressions, dry valleys and mottled surface due to variation in surface moisture (courtesy Aerofilms Ltd).

Karst characteristics and types

The karst is characterized by a number of features, viz. sinkholes and other depressions, long dry valleys, sparse streams, internal drainage, springs and bare rocks. Sometimes due to weathering, a red-coloured iron-rich material known as **terra-rossa** covers the land surface. The various karst features can be identified on aerial photographs, multispectral images and SLAR images (Figures 13.2 and 13.3). On panchromatic stereo aerial photographs, fresh limestone surfaces appear light toned, whereas terra-rossa is dark-toned; variation in moisture usually leads to a mottled texture. Sudden or gradual disappearance of streams (sinking creeks) or dry river beds may be observed. Similarly, springs and a sudden increase in stream discharge is also characteristic of karst terrains. On SLAR images, the radar return is related to the variation in surface moisture and higher surface roughness (due to subsidences, vegetation etc.). Figure 13.3 shows a SLAR image of karst features in Indonesia.

The characteristic subsurface features are the caverns and solution channels; the solution activity in carbonate rocks results in the lowering of the ground surface. In zones of intense fracturing, this effect is more pronounced giving rise to depressions known as **dolines, uvalas** and **poljes**. The surface drainage is directed underground. In the old stage of landform development, due to the collapse of the land surface, the underground caverns which were earlier filled with water and air are destroyed, and the water-table comes closer to the ground surface.

Although karst phenomenon is not restricted to any particular climatic zone, there is no doubt that the degree of solution is related to climate, as it depends on the temperature regime and

FIGURE 13.3 SLAR image of karst terrain in Indonesia characterized by numerous pits and depressions (after Sabins, 1983).

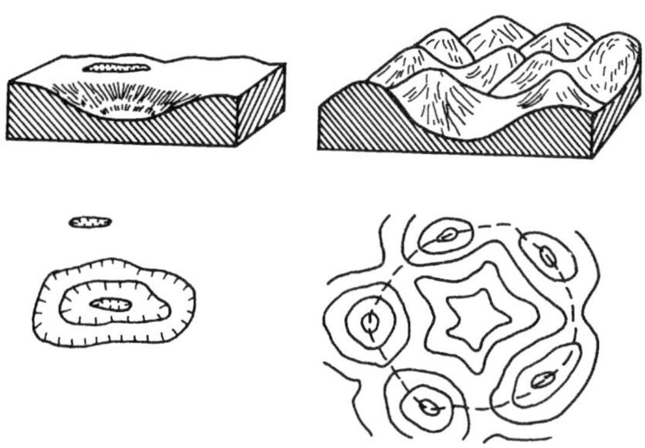

FIGURE 13.4 Comparison of karst terrains in temperate and tropical climates. Doline are characteristic of temperate climate and cone karst of tropical areas.

atmospheric precipitation. Therefore, attempts have been made to identify different types of karst landforms that develop in temperate and tropical regions. Temperate karst regions are usually characterized by broad depressions, e.g. dolines, uvalas and poljes with subordinate development of positive forms such as pinnacles and cones. Tropical karst, on the other hand, is dominated by positive forms (Figure 13.4).

The common landforms of karst areas developed in humid climates are briefly described below. For more detailed descriptions, the reader may refer to Burger and Dubertret (1984) and Ford and Williams (1989), among others.

Doline is a Yugoslav term for a depression or valley in karstic terrain. Dolines are circular in plan and vary from a few metres to about one kilometre in diameter. The mechanism for the development of dolines is shown schematically in Figure 13.5. Dolines are called 'sinkholes' in North America and 'swallow holes' in England. In the limestone area of Florida, USA, due to solution and subsidence, lakes are formed which are also called 'sinkholes'. Several dolines coalesce to form **uvalas.** A **karst cave** is a solution opening in carbonate rocks that is large enough for human entry. The formation of caves and cavern passages are facilitated by the dissolution of carbonate rocks along the various systems of joints (Figure 13.6). The caves may be formed in the vadose zone or in the deep phreatic region (Ford and Williams, 1989).

Tower karst, also known as **stone forest,** is one of the most spectacular features of the karst terrains in South China which is characterized by the development of residual hills or peaks separated by large depressions (Figure 13.7). It covers an area of 500 000 km^2 in South China. The formation of tower karst within Upper Devonian limestones and dolomitic limestones is related to intense fracturing of these rocks during Mesozoic and Tertiary episodes. Stone forest aquifers have very large lateral permeabilities but very small storage capacities as the aquifers are thin. This is responsible for large fluctuations in karst water discharge and alternating drought and flood hazards which are characteristics of these naked karst areas. Therefore, tower karst has rather poor development potential, though it is an important source of local water supply.

Palaeokarst refers to karst developed largely or entirely in ancient geological times during the periods of emergence above sea-level. As palaeokarsts are decoupled from the present

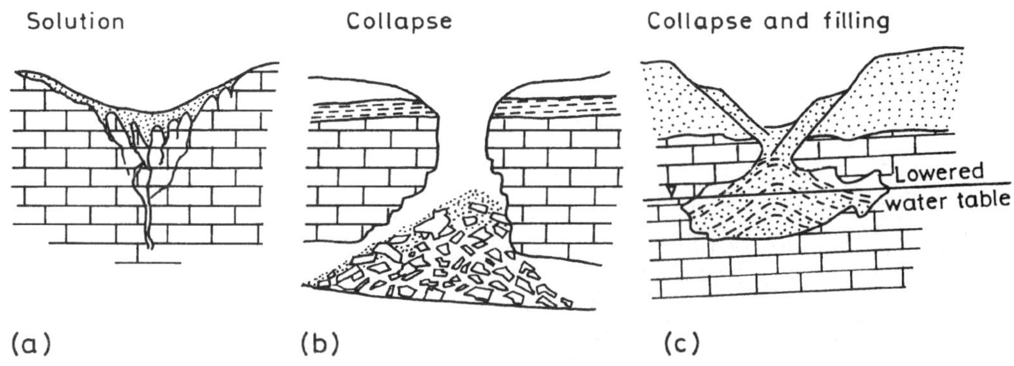

FIGURE 13.5 Formation of different types of doline by: (a) solution, (b) collapse, and (c) collapse and filling.

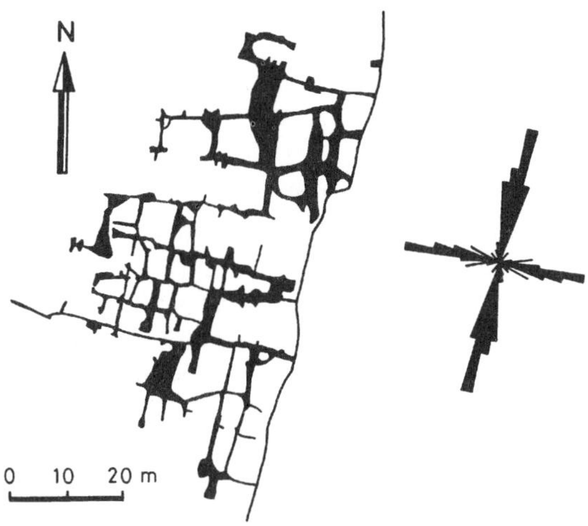

FIGURE 13.6 Map of Herron Cave, Indiana, USA, showing the pattern of cavern passageways in relation to joint orientations. The rose diagram includes 136 joint measurements (after Powell, 1977).

hydrogeochemical system, they are also termed as **fossil karst**.

Palaeokarsts are of different geological ages and have been described from various parts of the world (Bosak *et al.*, 1989). The earliest palaeokarsts are of Proterozoic age as reported from China and Canada. Palaeokarsts are of importance both from the point of view of supply of good quality water, as well as for temporary subsurface storage of imported water (Burdon and Al-Sharhan, 1968). Palaeokarsts are also of great economic importance as they contain at places economic deposits of oil and gas, silver, lead and zinc, bauxite, phosphate and other minerals, such as lead-zinc deposits in Morocco, Algeria and Tunisia, and oil and gas in the USA (Bosak *et al.*, 1989).

13.3 Springs

Springs are characteristic features of karst terrains, some of which have high discharge (Table 13.1). They are an important source of water supply especially in hilly terrains. Springs with high discharge can also be a potential source for the generation of hydropower. Karst springs can be classified based on: (a) morphology, (b) variability of discharge, and (c) hydrodynamic conditions.

Springs may emerge on the land surface and also on the bottom of a lake or sea. On the land surface, springs are formed by stream erosion or other denudational processes. Spring locations are usually controlled by geological discontinuities like faults and impervious beds (Figures 13.8, 13.9). Springs occurring at different altitudes can be tapped by tunnels; one such tunnel was constructed for water supply to the city of Innsbruck, Austria, having a discharge of 0.6–1.7 m^3s^{-1} (Graziadei and Zotl, 1984).

Submarine springs are formed when the dis-

FIGURE 13.7 Tower karst of Guilin area, China (courtesy Yuan Daoxian).

TABLE 13.1 Discharge values of some of the major karst springs (after Ford and Williams, 1989; Korkmaz, 1990; Bonacci, 1995)

Spring location	*Discharge (m^3s^{-1})*			
	Mean	*Maximum*	*Minimum*	*Basin areas (km^2)*
Buna, Yugoslavia	40	440	2	–
Ljubljanica, Yugoslavia	39	132	4	1100
Chingshuli, China	33	390	4	1040
Vaucluse, France	39	260	4.5	2100
Ombla, Croatia	25	106	4.0	800–900
Silver, USA	23	36	15.3	1900
Kirkgoz, Turkey	16	19	12	1800

charging water in the bottom of the lake or sea is under sufficient hydrostatic pressure to overcome the weight of the overlying surface water (Figure 13.8). Sometimes, due to differences in hydrostatic pressure, reversal of flow may take place when sea water moves landwards through the solution cavities, making fresh groundwater brackish (Figure 13.8); the proportion in which fresh water and sea water get mixed will vary seasonally. Submarine karst springs are most common along the eastern coast of the Adriatic Sea in the former Yugoslavia.

Submarine springs are important for groundwater supply to meet local needs. The springs result in local eddies and currents in sea water and hence may be detected by photography (Figure 13.10). Further, they frequently cause thermal anomalies in surface water due to mixing. Hence, thermal IR sensing can enable detection of such freshwater discharge sites (Figure 13.11) and this can be corroborated by field measurements of density, salinity and temperature distribution in the sea water. The remote sensing surveys for delineation of submarine freshwater springs are timed to enable detection of perennial freshwater discharges. The thermal IR data (pre-dawn) provides zones of thermal anomalies. The land in the night is cooler than the sea and therefore, freshwater which has a temperature corresponding to that of the land is cooler than the sea water. After eliminating thermal patterns due to surface drainage, the remaining sites of thermal plumes can be attributed to subsurface discharge. Lee (1969) has given nomograms for estimating spring discharge

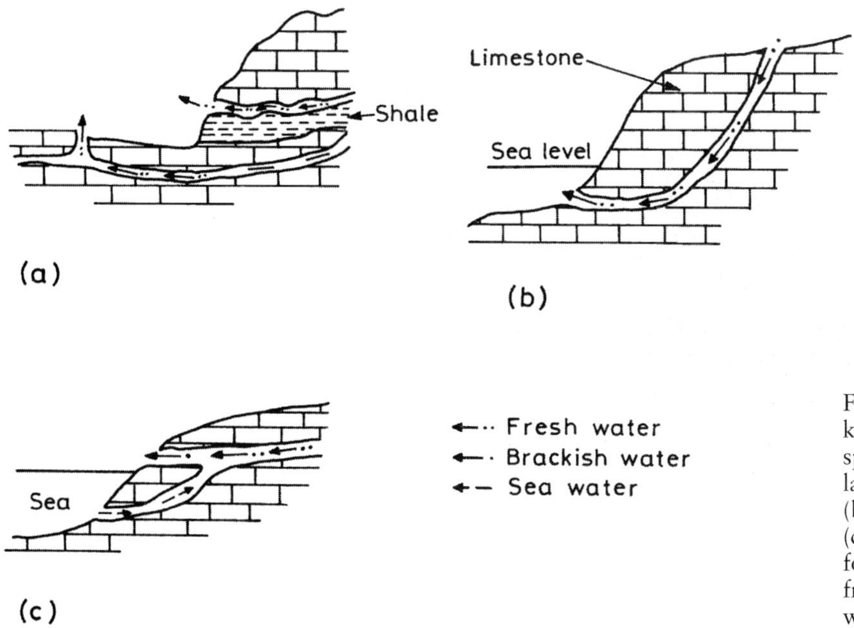

FIGURE 13.8 Types of karst springs: (a) freshwater spring controlled by lithologic layering and fissures; (b) submarine spring; (c) brackish water spring formed due to mixing of fresh groundwater and sea water.

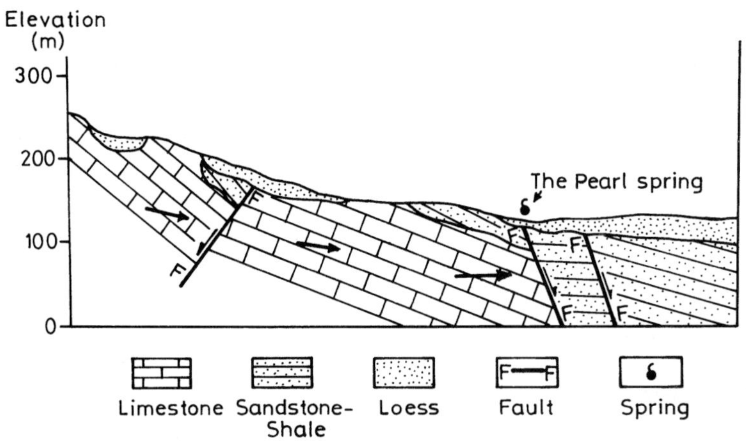

FIGURE 13.9 Hydrogeological profile of the Pearl spring, Henan province, PR China. High discharge from the spring is controlled by a fault in karstified limestone (after Zhengzhou *et al.*, 1996).

based on data about the area of thermal plume and temperatures of air, sea water and groundwater. The photogrammetric photographs may also be interpreted for lineament–geological structures, and multiband data provide areas of higher surface moisture and healthy vegetation. Superimposition of all the above data on to a common base map may yield valuable information about the fracture zones which control the groundwater movement (Figure 13.11).

Considering the variability of spring discharge, the following types can be recognized:

(a) Fluctuating springs. These show changes in discharge related to precipitation. The degree of fluctuation also depends upon the size of the groundwater reservoir; if the storage is large, fluctuations may be less, otherwise large fluctuations may take place. For example, the Sreid spring, emerging

FIGURE 13.10 Freshwater submarine spring in karst formations in the Adriatic Sea. The spring discharges under high hydrostatic pressure caused by the high elevation of the recharge area in the hinterland (after Alfirevic, 1966).

from Jurassic and Cenomanian dolomitic limestones in Lebanon, shows a time lag of three months between maximum precipitation in the catchment area during January, and maximum discharge (3 m^3s^{-1}) which is in April (Khair and Haddad, 1993). Springs in the stone forest region of the Guilin area in South China also show great fluctuation in discharge from 0.001 m^3s^{-1} to 7 m^3s^{-1} which indicates both uneven precipitation and also the heterogeneity of the system (Yuan *et al.*, 1990).

(b) Rhythmic springs. These show periodic (regular) variation in discharge, unconnected with seasonal or tidal influence. These springs are also referred to as periodic or ebb- and flowing springs. They are usually dammed springs

(a)

(b)

(c)

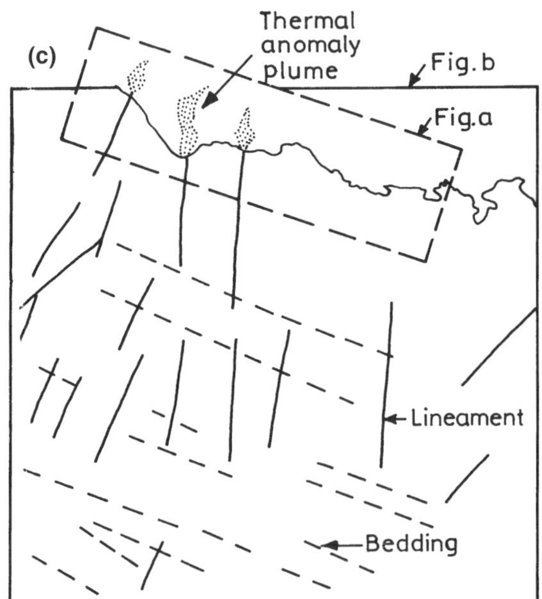

FIGURE 13.11 Submarine coastal springs in limestone: (a) thermal IR aerial scanner image (predawn) showing discharge of cooler (darker) groundwater into warmer (lighter) sea water; (b) NIR-band aerial photograph of the same area acquired at noon-hours; (c) interpretation map. A number of fractures can be delineated on the aerial photograph apparently controlling freshwater discharge into the sea (courtesy J. Bodechtel).

with a siphoning reservoir system, controlling their discharge (Ford and Williams, 1989).
(c) Springs of changing outlet where the discharge point shifts from time to time.
(d) Springs of constant discharge which have more or less constant discharge.

Hydrodynamically, karst springs can be classified as descending, ascending, or combined type. **Descending springs** drain water from an unconfined aquifer and show seasonal variation in discharge. An analysis of their recession curve can yield data about the size of their catchment and volume of water in storage. **Ascending springs** are fed by a confined aquifer. They show less variation in their discharge, chemical composition and temperature. Truly ascending springs are rare. Most of the springs are of **combined** type being fed by aquifers with varying hydrodynamic conditions.

13.4 Hydrogeological characteristics

The geomorphological features and rock permeabilities control the position of the water-table to a large extent. A deep water-table is observed under high relief and in thick permeable formations, while a shallow water-table develops in low topographic conditions and in areas where impervious beds underlie permeable karstic horizons at shallow depth. Groundwater movement is fast through bigger cavities and is slow and of diffuse type through small fractures and intergranular spaces. This is reflected in the trend of water-level fluctuations and spring discharge. Seasonal water-level fluctuations are usually high, being of the order of 20–80 m. This is due to the large infiltration capacity of karstic regions leading to the quick inflow of water into solution cavities.

On account of fracturing and development of solution cavities, tracers give a better idea of groundwater flow than the mapping of a water-table or potentiometric surface (Davis and DeWiest, 1966; Thrailkill, 1985). Principles of groundwater flow in fractured rocks as discussed in Chapter Seven also hold good in carbonate rocks. Therefore, the flow characteristics in these rocks can be explained by discrete fracture flow, equivalent porous medium and double porosity models. For the purpose of regional groundwater resource assessment and management, the continuum character of the aquifer (equivalent porous medium) can be an acceptable generalization, especially when fractures and solution cavities are closely spaced and well interconnected (Maslia and Prowell, 1990). The reader may refer to LaMoreaux *et al.* (1975), LaMoreaux (1986), Bogli (1980), UNESCO (1984b), and Paloc and Back (1992) for a more detailed account of the hydrogeology of carbonate and karstic formations.

Hydraulic properties

Karst terrains have their own characteristic distribution of porosity and permeability, both in the unsaturated and saturated zones. Porosity in carbonate rocks is both of primary and secondary types, the latter being more important. Primary or intergranular porosity is usually very small. Variation in total bulk porosity is reported as being from near zero to as high as 45% (Table 13.2). Chalk and calcareous tuff have high porosity; in chalk, in spite of high porosity, since the pores are small (1.0 to 0.012 μm), primary permeability is low and specific retention is high. However, the rock may acquire high secondary porosity and permeability, to varying extents, depending on fracturing and dissolution of mineral matter. Dolomites are usually more porous than limestones due to reduction in volume as a result of dolomitization. On the other hand, diagenetic processes namely cementation etc. tend to reduce the porosity and permeability.

Hydraulic conductivity, transmissivity and storativity

K,T and S of carbonate rocks depend on lithology, intensity of fracturing and solution activity.

TABLE 13.2 Total porosity values of some carbonate rocks (after UNESCO, 1984b)

Lithology	*Location*	*Total porosity (%)*
Compact limestone	Italy	0.2–14
Fissured limestone	Buxton, UK	14
Poorly fissured limestone	Dundee, USA	2.2–9.4
Poorly fissured limestone	UK	1.4–1.6
Oolitic limestone	Monk's Park, USA	20
Chalk	France	30–45
Dolomite	Micheldeau, UK	9 to 22
Marble	–	0.4 to 2.1
Calcareous tuff	–	20 to 32

The lithology controls the mechanical and dissolution properties of rocks and hence the development of fractures and solution cavities. Limestone and dolomite usually have well-developed fractures as compared with clayey marl and chalk formations. A greater amount of matrix material and clay will also lower rock permeability. The primary permeability of unfractured limestones and dolomites is commonly less than 10^{-7} ms^{-1} at near surface temperatures (Freeze and Cherry, 1979). Fracturing and solution impart higher permeabilities to these rocks. The hydraulic conductivity of the fractured rock mass is considerably greater than the massive part as the main movement of groundwater is along fractures and joints (Table 13.3). For this reason, the transmissivities are largely anisotropic, as is evidenced from the drawdown ellipse obtained from the pumping test and cave passageway orientation (Greene and Rahn, 1995). In chalk, the vertical hydraulic conductivity of 5×10^{-8} ms^{-1} is about 500 times greater than that attributed to intergranular conductivity (Hamill and Bell, 1986).

Generally, the hydraulic conductivity of carbonate rocks shows a decreasing trend with increasing geological age. Often higher values of hydraulic conductivity are observed in the carbonate rocks of Quaternary age than in older carbonate rocks (Figure 13.12). This figure also shows that although older limestones have high aggregate fissure density, hydaulic conductivity is lower, possibly due to the healing and infilling of fractures by calcite with time. Therefore, K is directly related to the density of open frac-

TABLE 13.3 Comparison of hydraulic conductivities (K) of unfractured and fractured carbonate rocks (after UNESCO, 1984b)

Rock type	*K values of unfractured rocks* (ms^{-1})	*K values of fractured rocks with one fracture per metre*	
		Opening (mm)	*K along the fractures* (ms^{-1})
Limestone	10^{-13}–10^{-12}	0.1	0.7×10^{-6}
	10^{-11}–10^{-9}	4.0	0.5×10^{-1}
Dolomite	10^{-10}	6	1.6×10^{-1}
Chalk	10^{-7}–10^{-6}	–	10^{-5}–10^{-4}

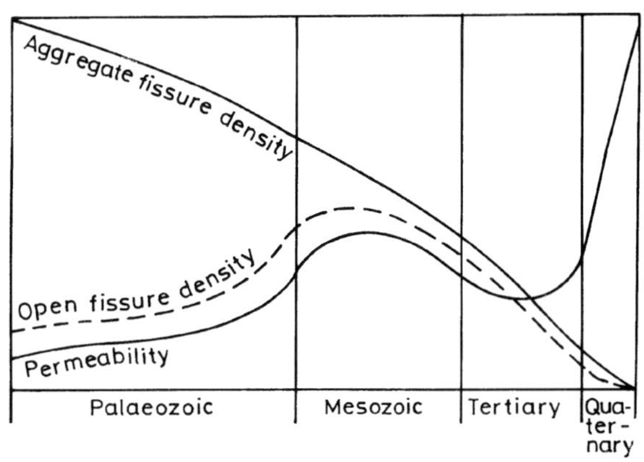

FIGURE 13.12 Variation in hydraulic conductivity and fissure density with geological age in carbonate rocks (after UNESCO, 1984b).

tures rather than the total number of fractures. In some cases, however, higher hydraulic conductivity is also observed in Precambrian limestones and dolomites (section 13.5).

The karst areas generally show a variation with depth in hydrogeological characteristics. As the porosity, hydraulic conductivity and storativity depend on fracturing and dissolution, they generally reach a maximum value close to the lower position of water-table and then decrease with depth (Figure 13.13). A decreasing trend in transmissivity (T) and storativity (S) with depth can also be due to decreasing intensity of fracturing, as is demonstrated from a study in the Cretaceous chalk aquifer from England (Figures 13.14 and 13.15). Giusti (1977) also reported a decrease in K with depth from Puerto Rican karst ranging from about 1 ms^{-1} in the upper parts to 10^{-4} ms^{-1} in the basal parts. Such a trend is also reported from many other carbonate rock terrains

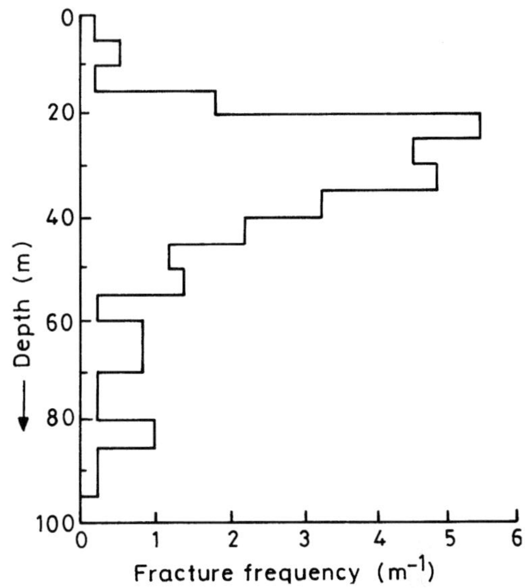

FIGURE 13.14 Fracture distribution with depth in the Cretaceous Chalk aquifer of London Basin, UK (after Owen, 1981).

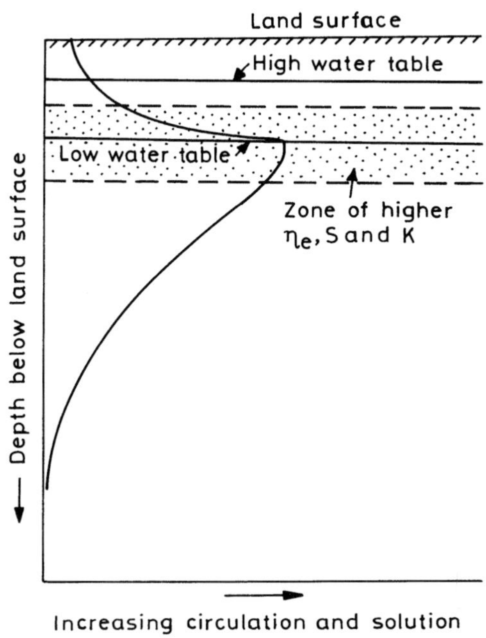

FIGURE 13.13 Relationship between hydraulic properties and depth in carbonate rocks. Note that the maximum values of η_e, S and K are close to the water table (after UNESCO, 1984b).

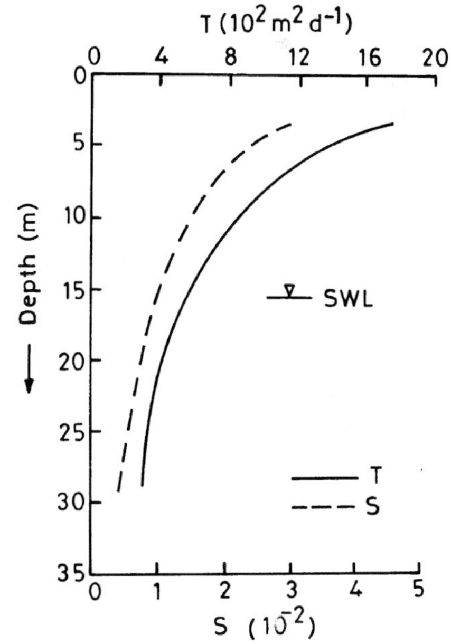

FIGURE 13.15 Variation in transmissivity (T) and storativity (S) with depth in the Cretaceous Chalk aquifer of London Basin, UK (after Owen, 1981).

TABLE 13.4 Transmissivity and storativity values of some carbonate rock aquifers

Location	*Rock type (age)*	*T* (m^2d^{-1})	*S*	*Source*
Ohio, USA	Limestone & dolomite (Silurian–Devonian)	4–128	2×10^{-4}	Eagon and Johe (1972)
Georgia, USA	Dolomites (Upper Floridan)	5907	3.5×10^{-4}	Maslia and Prowell (1990)
Israel	Fractured limestone (Jurassic)	86–11 000	0.01	UNESCO, (1984b)
Lebanon	Fractured limestone	86–5000	0.001–0.024	UNESCO (1984b)
Witwaterstand, S. Africa	Karstified dolomite	3300–4600	–	Boehmer (1993)
Mt Larcom, Queensland, Australia	Limestone (Middle Devonian)	10–3000, average 500	–	Maslia and Prowell (1990)
Damman, Saudi Arabia	Limestone (Palaeocene)	55 720	–	Otkun (1977)
London Basin, UK	Chalk (Lias)	15–370	–	Water Res. Board (1972)
Sanibel Island, Florida, USA	Limestone & dolomite (Miocene)	119	2.7×10^{-5}	Motz (1982)
Coastal Saurashtra, India	Miliolite limestone (Pleistocene)	200–900 175–950	– 0.12	Nair (1981) Rushton and Raghava Rao (1988)
Ghataprabha, India	Kaladgi limestone (Precamabrian)	150–500	–	Angadi (1986)
Chattisgarh, India	Charmuria Limestone (Precambrian)	400	–	Khare (1981)

(Otkun, 1977; Rushton and Raghava Rao, 1988; Arihood, 1994). The infilling of solution cavities by clastic sediments will also result in the lowering of hydraulic conductivity at deeper levels (Huntoon, 1992a).

The hydraulic properties of carbonate rocks can be estimated in the laboratory and in the field from slug tests, pressure injection (Lugeon) tests, pumping tests and tracer injection tests, as described in Chapter Nine. Spring discharge and groundwater-level data are also used to estimate hydraulic conductivity (section 13.6). The conventional Theis–Jacob drawdown method has often been used for estimating aquifer characteristics from pumping test data in carbonate rocks (Zeizel *et al.*, 1962; Csallany, 1965; Eagon and Johe, 1972). Leaky aquifer and double-porosity models have also been used to estimate aquifer properties in carbonate rocks, e.g. Motz (1982); McConnell (1993). Values of T and S as estimated from borehole tests in some carbonate rocks are given in Table 13.4.

As in other fractured rocks, due to anisotropy and heterogeneity, the measured value of hydraulic conductivity is a function of the reference volume of the rock mass, which may vary by more than three orders of magnitude, depending on the scale of measurement (Figure 13.16). The greatest difference is in karstified rocks (Rovey and Cherkauer, 1995). In chalk from the London Basin, laboratory values of K

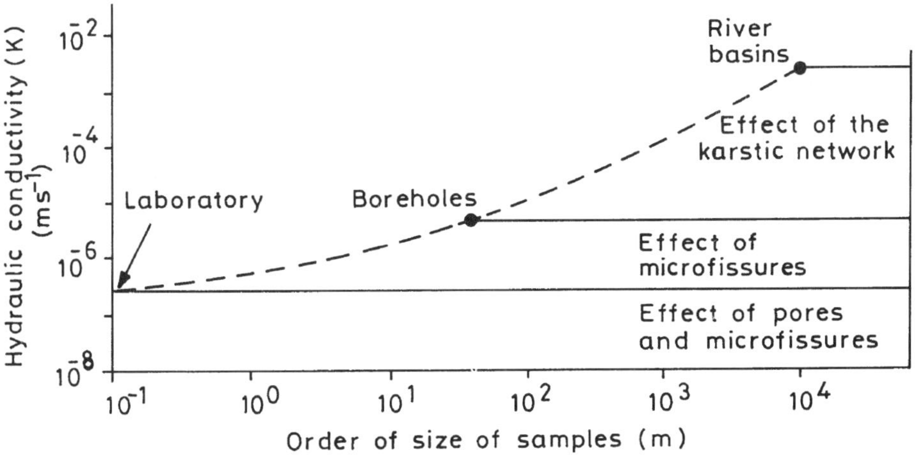

FIGURE 13.16 Schematic representation of the effect of scale on the estimated hydraulic conductivity of karst. Laboratory values obtained from core samples are several orders of magnitude lower than those obtained from borehole tests and regional assessment (after Kiraly, 1975).

using permeameters were 5×10^{-9} ms^{-1} as compared with the values of K (2×10^{-6} to 7×10^{-5} ms^{-1}) estimated from pumping tests (Water Res. Board, 1972).

13.5 Groundwater development

Springs are the main source of water supply in karst areas, especially in hilly regions. As in other rock formations, dug wells and boreholes are common structures constructed for water supply.

Dug wells of 5–10 m diameter are common in the stone forest aquifers of China (Huntoon, 1992b). The yield of such wells varies from 2×10^{-4} m^{3}s^{-1} to 2×10^{-3} m^{3}s^{-1} with small drawdown. Very high discharges of the order of 0.13 m^{3}s^{-1} with practically no drawdown is reported from the Precambrian (Vindhyan) limestones of the Borunda-Bilara area in western India. In the chalk of South-east England, a common practice is to sink a well and provide adits to intercept the maximum number of vertical fractures. The yield of such wells is in excess of 0.15 m^{3}s^{-1}. However, suitably located smaller diameter wells in chalk aquifers are more successful (Hamill and Bell, 1986).

The yield of borewells from rocks with good secondary porosity and permeability can be very high. For example, the usual discharge from wells in the Ocala limestone of Eocene age in Florida and Georgia, USA, is 0.063 m^{3}s^{-1} and one well gave a yield of 0.475 m^{3}s^{-1} with a drawdown of 3 m only (Maslia and Prowell, 1990). The sustained yield of borewells from fissured chalk aquifers of unconfined and confined zones in England is reported to be 0.05–0.10 and 0.002–0.05 m^{3}s^{-1} respectively (Owen, 1981). The yield of borewells from partly karstified Pakhal and Cuddaph limestones of Precambrian age in central India ranges from 0.01–0.02 m^{3}s^{-1} for 3 m drawdown.

Specific capacity and well productivity data are useful in comparing the relative performance of wells in different hydrogeological environments. **Well productivity** value is the specific capacity per unit saturated thickness of the aquifer adjusted to a common radius and pumping period (section 9.7.3).

The well productivity in carbonate rocks depends on: (a) fractures, lineaments and other structural features, (b) solution features, (c) topography, (d) depth, and (e) palaeoweathered horizon. The importance of fracture traces, lineaments and solution features has been demonstrated by several workers (for example, Lattman and Parizek, 1964). Figure 13.17 indicates that wells located on fracture traces have higher yields than nonfracture trace wells. A

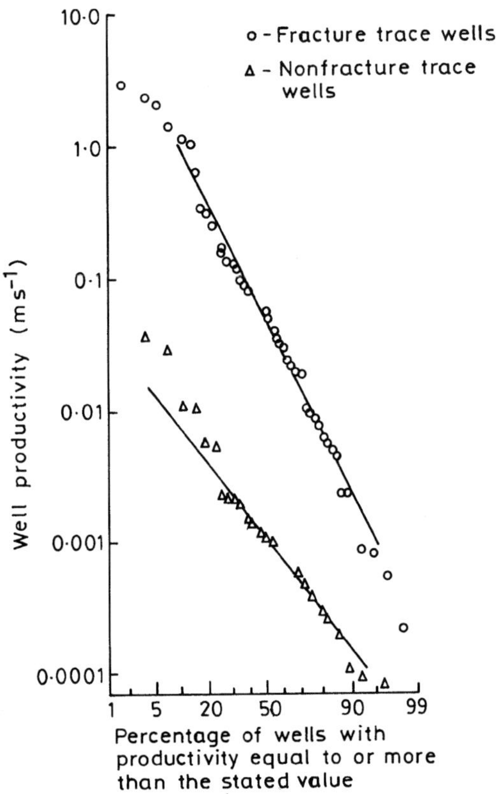

FIGURE 13.17 Production–frequency data for water wells in relation to their location with respect to fracture traces (lineaments) in carbonate rocks (after Siddiqui and Parizek, 1971).

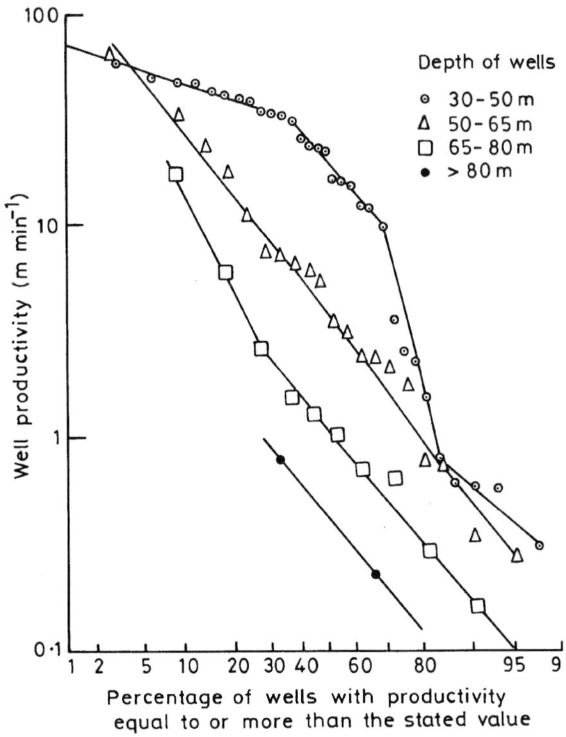

FIGURE 13.18 Production–frequency plots in relation to depth of water wells in Charmuria limestone, Raipur Basin, India. Note the decreasing trend of well productivity with depth (after Khare, 1981).

study of well productivities in limestones from Maryland, USA, suggested that specific capacity of wells located within a distance of 30 m from the lineaments was about 15 times greater than those located at longer distances, indicating the importance of lineaments for locating high-yielding wells (LaRiccia and Rauch, 1977).

Solution features, which are mainly controlled by rock structures, also influence well productivity e.g. solution openings developed along horizontal bedding planes are important for the water supply from wells, as there is a greater probability of vertical wells striking horizontal openings rather than vertical joints. In folded carbonate rocks, wells drilled in the crest of the anticlines are likely to be more productive than those in synclinal areas as the intensity of fracturing and solution will be more pronounced in the former case. The well productivity usually decreases with depth due to decrease in fracturing and solution (Walton, 1970). A similar trend is observed from Charmuria limestone in the Raipur basin, India (Figure 13.18).

Topography influences transmissivity and thereby well productivity. For example, in the Upper Cretaceous chalk aquifer of the Seine valley in France, transmissivity was estimated to be 16 000 m^2d^{-1} along the valleys and less than 20 m^2d^{-1} in the interfluves (UNESCO, 1984b).

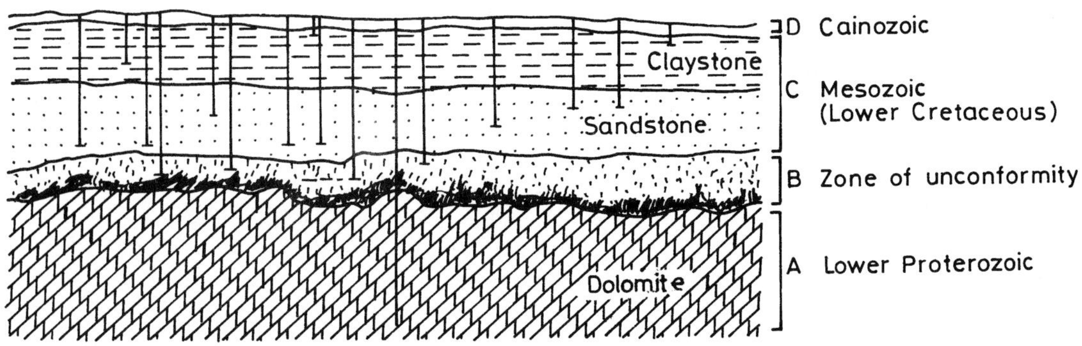

FIGURE 13.19 Hydrogeological section in Darwin Rural Area, NT, Australia. B is the hydro-stratigraphic unit formed due to unconformity. This horizon consisting of chert, dolomite and other weathered materials forms the most productive horizon with well yields upto 0.7 m^3s^{-1}. (after Verma and Qureshi, 1982).

Palaeoweathered horizons and unconformities form a good source of water to wells as is illustrated from a study in the Darwin Rural Area, NT, Australia (Figure 13.19). In addition to the above factors, methods of well construction also influence well yields (Knopman and Hollyday, 1993). Acid treatment is very effective in improving well discharge. This aspect is discussed in Chapter Fifteen.

13.6 Assessment of groundwater recharge

The following four methods have been used for assessing groundwater recharge in karst regions.

(a) Conventional water balance method.
(b) Chloride method. This is suitable for the estimation of groundwater recharge in carbonate rocks because run-off is very low, and the rocks contribute very little to the chloride content of groundwater (UNESCO, 1975).
(c) Isotope techniques using tritium (^3H) and stable isotopes (^2H and ^{18}O) are well-suited for hydrogeological studies in karstic aquifers. Measurement of ^3H, ^2H and ^{18}O helped in determining the relative recharge from rainfall and snowmelt in the Alpine karstic aquifers (Long, 1995). Nair *et al.* (1993) used ^3H and ^{14}C for studying the recharge pattern in the Vindhyan limestone of the Jodhpur district in India.
(d) Spring hydrograph analysis.

The first three methods are described in section 17.1.3 while the spring hydrograph analysis is given below. Some values of recharge based on spring flow data are given in Table 13.5.

Usually, karst rocks have a high rate of infiltration which at places may be even 80% of precipitation (Boni *et al.*, 1984). Khair and Haddad (1993) have reported infiltration rates to be 40% to 44% of the annual rainfall (1600 mm) in fractured carbonate rocks of Jurassic and Cenomanian age in the Lebanon. Even in arid areas, recharge rates can be high

TABLE 13.5 Recharge rate as estimated from spring discharge data in some Mediterranean countries

Country	*P (mm)*	Q_w *(mm)*	$R_p = Q_w/P \times 100$
Tunisia*	463–633	108–264	23–42
Greece*	1150–1400	583–723	50.7–51.6
Israel*	600–750	229–335	30.5–53.0
Croatia**	–	270	–

P = rainfall (mm)
Q_w = recharge; spring discharge divided by the intake area
R_p = rate of recharge in terms of percentage of rainfall
Source: * UNESCO (1975); ** Korkmaz (1990)

due to high permeability of the surface. For example, in the karst areas of Saudi Arabia, recharge is about 45% of the yearly rainfall (145 mm) (Al-Saafin *et al.*, 1990).

Spring hydrograph analysis

The hydrograph of a spring discharge shows a shape similar to that of a surface water stream, i.e. a rising limb, peak and a recession limb (section 1.2). It also shows a time-lag between precipitation and peak of the hydrograph. In permeable formations, like karstic limestones, the peak of the spring hydrograph will be broader and flatter with a greater time-lag. In contrast to this, in less permeable formations, the hydrograph has a sharper peak due to small aquifer storage. In karstic aquifers, the peak and the recession limb of the spring hydrograph may also have several bumps indicating complex recharge and flow mechanism (Figure 13.20).

A quantitative analysis of spring hydrograph recession is generally based on Maillet's formula (equation (13.2))

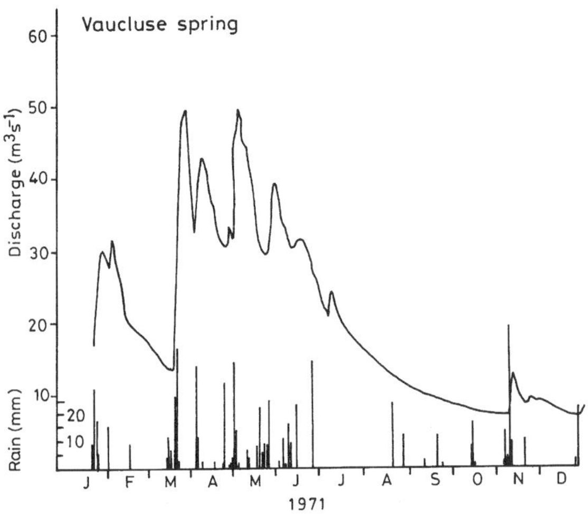

FIGURE 13.20 Hydrograph of Vaucluse spring in karst formation, France. Rainfall data is also shown (after Ford and Williams, 1989).

$$Q_t = Q_o e^{-\alpha t} \tag{13.2}$$

where Q_t is the discharge rate at time t, Q_o is discharge rate at any previous time during recession, t is the elapsed time between Q_t and Q_o, e is the base to the Napierian logarithm, and α is the recession (discharge) coefficient of dimension (T^{-1}). The coefficient, α will be a function of the aquifer transmissivity, storage coefficient or specific yield and the catchment geometry. In logarithmic form to the base 10, equation (13.2) is written as

$$\log Q_t = \log Q_o - 0.4343\, t\alpha \tag{13.3}$$

The recession coefficient, α may be obtained directly from equation (13.3) as

$$\alpha = \frac{\log Q_o - \log Q_t}{0.4343\,(t_2 - t_1)} \tag{13.4}$$

The recession curve usually plots as a straight line on a semi-log paper with time on the arithmetic scale. Sometimes these semi-logarithmic plots consist of two or more segments (Figure 13.21) with different values of α indicating complex hydrogeological characteristics (permeability and storativity) of the aquifer. Higher

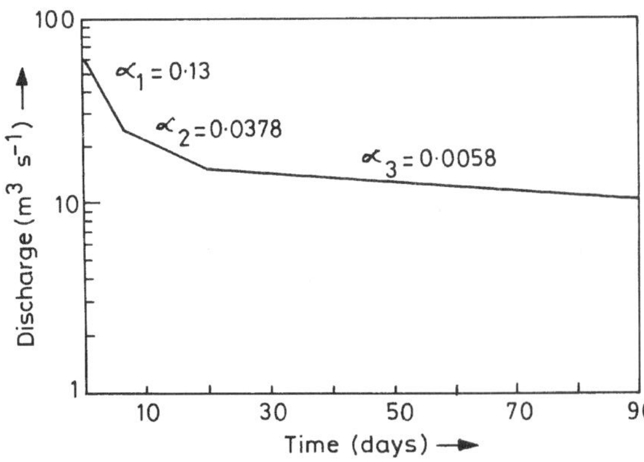

FIGURE 13.21 Composite hydrograph recession of Ombla spring, Croatia. Note that the recession coefficient (α) decreases with increase in time of recession (after Milanovic, 1981).

values of α, at the earlier time of recession indicate fast drainage from large solution cavities and fractures, while the smaller values toward the later part of recession point towards diffuse flow through smaller fractures and pore spaces. In such cases, spring discharge can be expressed as

$$Q_t = Q_{o_1} e^{-\alpha_1 t} + Q_{o_2} e^{-\alpha_2 t} + \cdots Q_{o_n} e^{-\alpha_n t} \qquad (13.5)$$

The concept of a half-flow period, $t_{0.5}$, has also been used to evaluate the recession constant value. The concept of a half-flow period is similar to the half-life concept in nuclear physics, which is defined as the time required for the base flow of the stream or spring to halve i.e. $2Q_{t0.5} = Q_o$. By substituting this into equation (13.2)

$$Q_{t_{0.5}} = 2Q_{t_{0.5}}\ \beta^{t_{0.5}} \qquad (13.6)$$

hence

$$\frac{1}{2} = \beta^{t_{0.5}} \qquad (13.7)$$

and

$$t_{0.5} = \text{constant}/\log\beta \qquad (13.8)$$

Like recession coefficient, α, half-flow period, $t_{0.5}$ is a direct measure of the rate of β recession and can be used as a means of characterizing exponential base-flow recessions of a spring. It is sensitive to changes in the spring flow domain and can have values in the range of zero to infinity.

With large values of β and small values of $t_{0.5}$, the recession curve will be steep, indicating fast drainage of the aquifer and little groundwater storage, i.e. the conduit function is prominent. Small values of β and large values of $t_{0.5}$, are characteristic of slow drainage from an extensively fractured or porous aquifer, with large storage capacity and limited recharge (Ford and Williams, 1989).

Estimation of groundwater recharge

Groundwater recharge can be estimated from spring flow data using equation (13.9) based on the principle of continuity.

$$AR = \frac{Q_2}{\alpha} - \frac{Q_1}{\alpha} + \int_{t_1}^{t_2} Q\,dt \qquad (13.9)$$

where A is the recharge area (L^2), R is the recharge rate (LT^{-1}), t_1 and t_2 are the instance of time at the end of one dry season and the beginning of the next one, Q_1 and Q_2 are the spring discharge rate (L^3T^{-1}) at time t_1 and t_2 respectively, and α is the recession coefficient, as defined earlier. Based on this approach, Korkmaz (1990) analysed the spring discharge data of the Kirkgoz spring in the Mediterranean karst of Turkey. The average annual recharge was estimated to be 493×10^6 m^3, i.e. 0.27 m yr^{-1} (Table 13.5). Bhar (1996) used monthly discharge data from the Kirkgoz spring to estimate recharge on a monthly basis. Annual values of recharge obtained by Bhar (1996) are in close proximity to those obtained by Korkmaz (1990).

Evaluation of hydraulic parameters

Considering the relationship between the recession coefficient, α and the aquifer characteristics, Domenico (1972) provided equation (13.10) for estimating the transmissivity, T

$$T = \frac{4\alpha L^2 S}{\pi^2} \qquad (13.10)$$

where L is the groundwater path length, i.e. the average distance from a catchment divide to the spring, and S is the coefficient of storage or specific yield. Equation (13.10) is valid when the thickness of the saturated zone is high enough so that the change in the saturated thickness during the drainage period can be neglected. Further, in fractured and karstic formations, variation in the value of α due to rock heterogeneity and inaccuracy in the estimation of L in the field, may not provide realistic estimates of T. However, in some cases the values of T estimated from equation (13.10), and from pumping tests, are quite comparable (Stasko and Tarka, 1996).

Hydraulic conductivity, K can also be estimated from spring discharge and groundwater-level data using equation (13.11) based on Dupuit's assumption for steady-state flow in an unconfined aquifer.

$$K = \frac{2.3Q \log r_2/r_1}{\pi (h_2^2 - h_2^2)} \qquad (13.11)$$

where, h_1 and h_2 are the groundwater levels in two observation wells at distances r_1 and r_2, from the spring's outlet, respectively. Equation (13.11) is the well-known Thiem's equilibrium equation which is used for estimating K from pumping tests in unconfined aquifers under steady-state conditions.

Hydraulic conductivity, K of the karstic limestone at the Ombla spring, Croatia, was estimated using equation (13.11) by Bonacci (1995). The layout of the observation wells is shown in Figure 13.22. K was computed to be in the range of 2×10^{-3} to 5×10^{-3} ms^{-1} for the two sets of water-level observations. Bhar (1996) used a search technique, for estimating transmissivity and storativity of the confined karstic aquifer feeding the Kirkgoz spring, Turkey, based on monthly spring discharge data. T was estimated to be 686 m^2d^{-1} and S was 0.0013.

13.7 Groundwater quality

Extensive work has been done on the thermodynamic equilibrium for studying the dissolution–precipitation process for $CaCO_3$ and $MgCO_3$ species (Garrels and Christ, 1965; Back and Hanshaw, 1965; Hem, 1989). Such studies help in predicting whether a given water is unsaturated and can dissolve more calcite at a given temperature, supersaturated and can precipitate calcite, or is at equilibrium. The measurement of pH, temperature, calcium and carbonate concentrations, and ionic strength will provide this information.

Groundwater in limestones contains more alkaline earths and carbonate ions, relative to alkalis, chloride and sulphate. The pH of the groundwater is generally above 7. As the

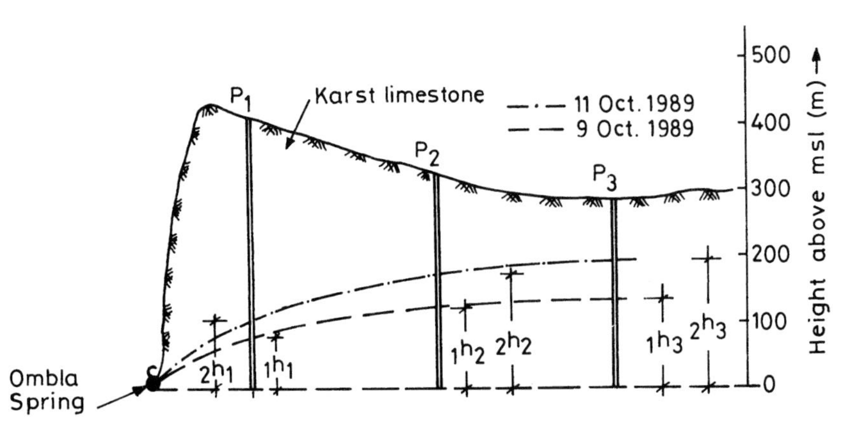

FIGURE 13.22 Section across the karst massif at Ombla spring, Croatia, showing the position of three piezometers (P_1, P_2 and P_3) at distances of 1.62, 4.66 and 7.40 km respectively from the spring outlet. Groundwater levels in the three piezometers on 9 October 1989 and 11 October 1989 are also shown (after Bonacci, 1995, reprinted from *Journal of Hydrology*, with kind permission of Elsevier Science-NL, Amsterdam, the Netherlands).

contact area between groundwater and rock matrix is limited due to fissure flow, groundwater in carbonate rocks usually has low total dissolved solids. In dolomites, the ratio of γMg/ γCa in the groundwater is rather less than in the rock due to preferential solution of calcium carbonate.

13.8 Concluding remarks

Although karst areas are an important source of water supply, they may also pose several hydrological and environmental problems as listed below.

(a) **Scarcity and poor predictability** of groundwater supplies due to great variation in permeability, both laterally as well as vertically. Further, zones of high permeability need not form potential sources of groundwater supply as the water may rapidly drain out from these places and the water-table may be very deep.
(b) **Rapid drainage of groundwater** causing alternate flooding and drought conditions. For example, in the karst area of South China, due to high infiltration, the water-table comes close to the ground surface in the rainy season but in the dry period the water-table recedes considerably resulting in drought conditions (Huntoon, 1992b).
(c) **Scarcity of surface perennial streams** and variability of flow due to large infiltration.
(d) **Poor soil cover** which is infertile and easily eroded, resulting in poor agricultural production and reduction in spring discharge.
(e) **Leakage of surface reservoirs.**
(f) **Unsatisfactory waste disposal** environment due to high permeability and thin soil cover, causing erratic and fast movement of the pollutant to the zone of saturation without getting purified. It is also difficult to predict the movement of contaminants due to rock heterogeneity. Therefore, the sanitary landfill and other waste disposal sites in such areas should be located with great care and regular monitoring of surface and groundwater quality is essential.
(g) **Instability of the ground surface** due to the development of subsurface solution cavities which may cause land subsidence. This also creates land use problems such as construction of roads, mining, agricultural and other civilian activities.

Chapter fourteen

Hydrogeology of clastic formations

Under clastic formations, both unconsolidated and consolidated sediments have been included. Unconsolidated sediments include various admixtures of boulder, sand, silt and clay deposits. These on consolidation form clastic sedimentary rocks, e.g. sandstone, siltstone and shale.

14.1 Unconsolidated sediments

Most of the unconsolidated sediments were deposited during the last few million years, and are of Quaternary–Recent age. They have been formed under different sedimentary environments, namely fluvial, glacial, eolian and marine. The coarse-grained sediments (gravel and sand) form potential aquifers due to their high hydraulic conductivity and storativity. Being incoherent in nature, unconsolidated or semi-consolidated sediments are largely unfractured, except some glacial deposits like clay tills.

14.1.1 Fluvial deposits

The fluvial deposits are characterized by typical landforms, such as alluvial fans, flood plains, and terraces etc. which can be identified on aerial photographs and space images. Fluvial deposits form ideal aquifers as they occur along river valleys and in areas of even topography with adequate recharge. Some of the river valleys are deep and narrow, e.g. the Jordan river valley along the border of Jordan and Israel, and the Owens river valley in California, USA. Others form broad plains, e.g. the Indo-Gangetic basin and the North Plain in China.

Buried valleys and **palaeochannels** are of special interest for groundwater development in hard rock terrains. They contain thick deposits of gravel and sand and are a result of fluvial and fluvio-glacial processes involving shifting of river courses due to either tectonic or climatic causes. Palaeochannels are identified by their typical sinous serpentine shape and form. In an arid or semi-arid terrain, the palaeochannels may posses anomalous moisture and vegetation, in comparison to the adjoining areas. Table 14.1 gives a summary of physical conditions of palaeochannels and the corresponding spectral characters on remote sensing data.

An interesting example of palaeochannel is furnished by the 'lost' Saraswati river in India, which is said to have been a mighty river in the olden **Vedic or pre-vedic** times (about 5000 years BC). It used to flow in the area now occupied by the Thar desert in western India. Here, the palaeochannels can be identified in most cases on the basis of higher moisture content and vegetation patterns (Figure 14.1). Using Landsat images, the ancient course of the river Saraswati has been delineated for a distance of about 400 km (Yashpal *et al.*, 1980; Bakliwal and Grover, 1988). Some of the buried

TABLE 14.1 Physical conditions of palaeochannels and their spectral characters on remote sensing data

Physical conditions associated with the palaeochannel	*Spectral character*
Higher surface mositure	Medium to dark in visible bands; dark on NIR and SWIR images; dark on SLAR images
Preferential denser vegetation cover	Dark on visible bands; light toned in NIR; light on SLAR images
Dry sand cover; no surface anomaly in terms of vegetation; bedrock occurs at a greater depth	Darker on longer wave-length SLAR images
Alluvial cover, thick gravelly river bed resulting in relatively less surface mositure and poor vegetation	Very light on visible and NIR bands

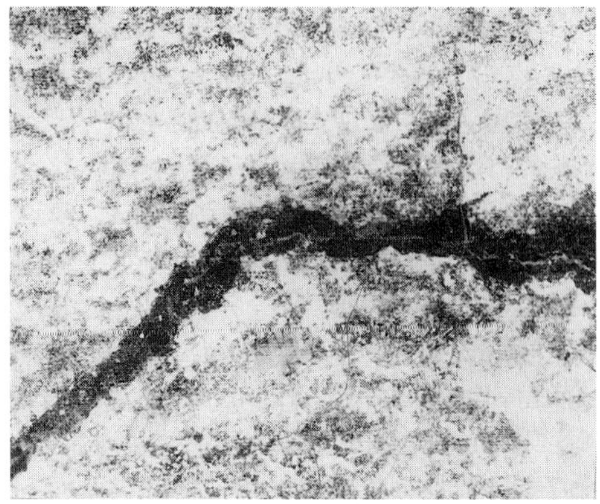

FIGURE 14.1 Landsat MSS band-2 image showing the presence of palaeochannel of the 'lost' Saraswati river in northwestern India. The bed is about 6–8 km. wide and is marked by flush vegetation.

segments of the river Saraswati are a potential source of groundwater (Venkateswarlu *et al.*, 1990). In arid and hyper-arid regions, the buried channels are of added importance for water supply, owing to acute water scarcity conditions. The SIR-A data in the eastern Sahara has demonstrated the applicability of SLAR data for delineation of palaeochannels in hyper-arid regions (McCauley *et al.*, 1982; Elachi *et al.*, 1984). In the sand-covered limestone terrain of the Rajasthan desert, India, Mehta *et al.* (1994) interpreted the presence of relict rivers on the basis of ERS-SAR images (Figure 14.2).

14.1.2 Glacial deposits

Unconsolidated glacial deposits occur mainly in North America, northern Europe and Asia. Isolated occurrences are also observed on major mountain chains such as the Alps, the Andes, and the Himalayas, while close to the equator, glaciation was restricted to Mt Kenya and Kilimanjaro in East Africa. Glacial deposits in North America and elsewhere have been studied in detail both as a source of water supply and also as host rock for the disposal of waste (Cherry, 1989; Nilsson and Sidle, 1996).

Glacial tills are unstratified and typically poorly sorted ice-transported sediments in which the grain size varies from clay fraction to boulder-size material. In some areas tills are thick and unfractured but wherever fractured, they provide active hydraulic connections and potential contaminant pathways (Hendry, 1982; Ruland *et al.*, 1991). Fracture spacing usually increases with depth thereby affecting vertical distribution of permeability (Figure 14.3).

Fractures are often filled with calcite and gypsum. In some areas, thin rootlets are observed along the fractures to depths of 5–10 m below the ground surface (Freeze and Cherry, 1979). Shallow fractures are regarded to be a result of alternate cycles of wetting and drying, freezing and thawing. The origin of deep open fractures in the unweathered clays is more problematic. These are attributed to stress changes caused by post-Pleistocene crustal rebound or to stress changes caused by the advance and retreat of the last Pleistocene glacier (Cherry, 1989).

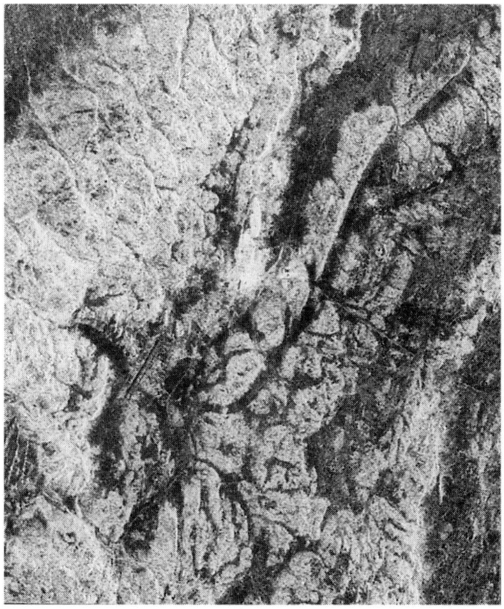

FIGURE 14.2a Palaeochannel as delineated on the ERS SAR image of Rajasthan, Thar desert, India. The darker tones in the palaeochannel area are due to local thicker sand cover, as compared to the adjacent area. (Image courtesy: ERS-1 AO IND1.3 Project, 'Geological and geomorphological mapping for mineral exploration'.)

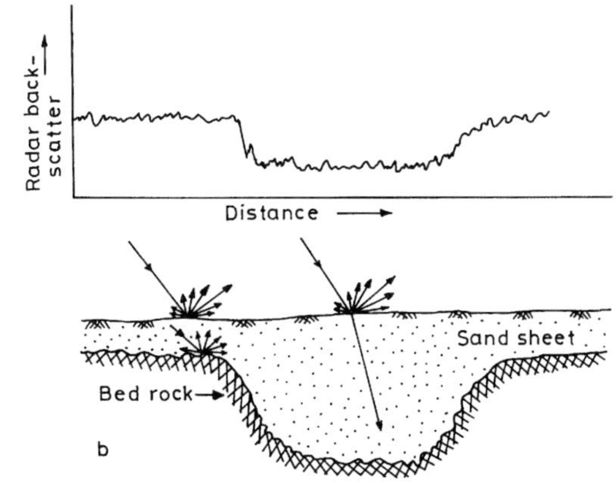

FIGURE 14.2b Schematic interpretation. The lower radar backscatter is attributed to the thicker sand cover in the palaeochannel area.

Volume changes due to geochemical processes, such as cation exchange, also have been suggested for the development of fractures (Freeze and Cherry, 1979).

Glacial tills have porosities in the range of 25 to 45%; clay tills have higher porosities. Due to their unsorted character, the matrix (intergranular) permeabilities of glacial tills are low (10^{-10}–10^{-9} ms^{-1}). However, due to weathering and fracturing, they acquire higher hydraulic conductivities (K) of the order of 10^{-9}–10^{-6} ms^{-1}. The upper parts of the till deposits usually have higher hydraulic conductivity due to greater intensity of fracturing and weathering (Ruland *et al.*, 1991; Jones *et al.*, 1992) (Figure 14.3). The higher values of hydraulic conductivities in weathered and fractured tills will facilitate efficient drainage of agricultural lands

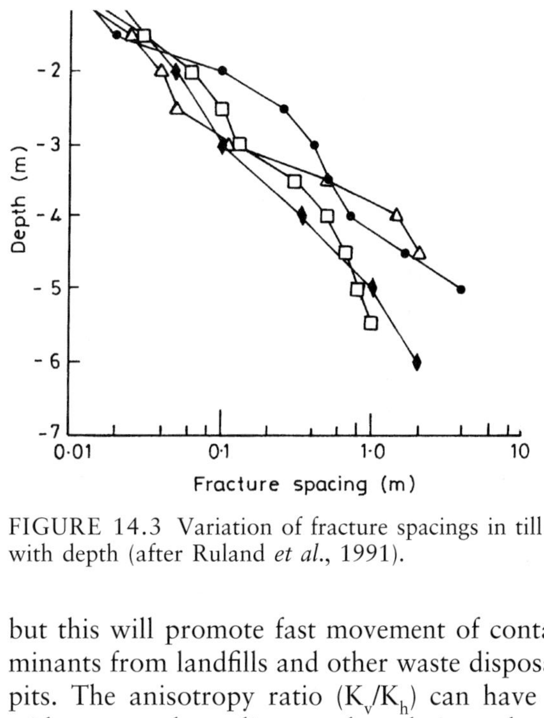

FIGURE 14.3 Variation of fracture spacings in till with depth (after Ruland *et al.*, 1991).

but this will promote fast movement of contaminants from landfills and other waste disposal pits. The anisotropy ratio (K_v/K_h) can have a wider range depending on the relative role of horizontal layering and vertical fracturing. Laboratory values of hydraulic conductivity measured from core samples are of a two to four order of magnitude smaller than that determined from field tests (Jones *et al.*, 1992).

Detailed pumping tests with an array of observation wells, distributed radially, can provide a better idea of aquifer parameters including anisotropy (Edwards and Jones, 1993). Tracer injection tests are also carried out for the estimation of hydraulic parameters. Conservative tracers like chloride were used to estimate the potential of groundwater contamination in fractured tills in Denmark (Nilsson and Sidle, 1996). Tritium profiles indicate active groundwater circulation in fractured tills up to a depth of 5–10 m (Ruland *et al.*, 1991).

Water in glacial aquifers is generally very hard and has a high concentration of totally dissolved solids and sulphate. In areas of active groundwater circulation, water quality is usually good. A high concentration of SO_4^{2-} (1000–10 000 mg l^{-1}), reported from weathered clays, is attributed to oxidation of organic sulphur or pyrite (Cherry, 1989). In such cases, the amount of SO_4^{2-} can be used as a natural tracer to study the downward solute migration through fractures in the unweathered clays.

The chemical evolution of groundwater in the fractures is influenced by the composition of the matrix and of pore water as well as the diffusion gradients. Studies in fractured glacial till indicate that different ions, e.g. Ca^{2+} and Cl^- behave differently due to variation in their diffusion coefficients and adsorption by the matrix. Laboratory studies show that Ca^{2+} passes through the fractures more rapidly than Cl^- due to the smaller diffusion coefficient of Ca^{2+} and adsorption of Ca^{2+} within the porous matrix (Grisak *et al.*, 1980).

14.2 Consolidated sediments

The common clastic sedimentary rocks are sandstone, siltstone and shale. These are formed in almost all environments including marine, fluvial, deltaic, lacustrine and eolian. Usually, sandstones and fine-grained clastic rocks, e.g. shales and siltstones, occur as alternating beds with varying thicknesses in most sedimentary sequences. Although sandstones, being more permeable, are the main interest as a source of water supply, their potentiality as well as water quality is greatly influenced by the composition of intergranular cementing material and properties of the interbedded shales and siltstones.

Sandstones being more resistant to erosion usually form hills, ridges and scarps while shales erode easily forming hill slopes and valleys. For the same reason, sandstones have low to medium drainage density but in shales drainage density is high. In sandstone, the drainage pattern is rectangular or angular due to rock discontinuities; shales show typical dendritic pattern. A comparison of landform and drainage characteristics in these two rock types is illustrated in Figure 14.4.

14.2.1 Sandstones

The hydrogeological properties (porosity and permeability) of sandstones depend on their textural characteristics, which in turn are influenced by the depositional environments and subsequent changes due to cementation, consolidation and fracturing. The porosity, hydraulic conductivity and specific yield of sandstones is lower than sands due to compaction and cementation. Common cementing materials are clays, carbonates, silica and iron oxides. As the degree of compaction depends on depth of burial, the porosity of sandstones decreases systematically with depth and age; poorly cemented sandstones may have porosity of about 35%. Hydraulic conductivity of sandstones may vary from 1×10^{-6} to 1×10^{-4} ms^{-1} (Table 14.3). Geologically, older sandstones have lower porosity and hydraulic conductivity due to greater compaction and cementation, but there are also several exceptions. Stratification imparts anisotropy; hydraulic conductivity along the bedding plane is usually higher than across it. In Berea Sandstone, USA, permeability parallel to the bedding was found approximately four times higher than that normal to

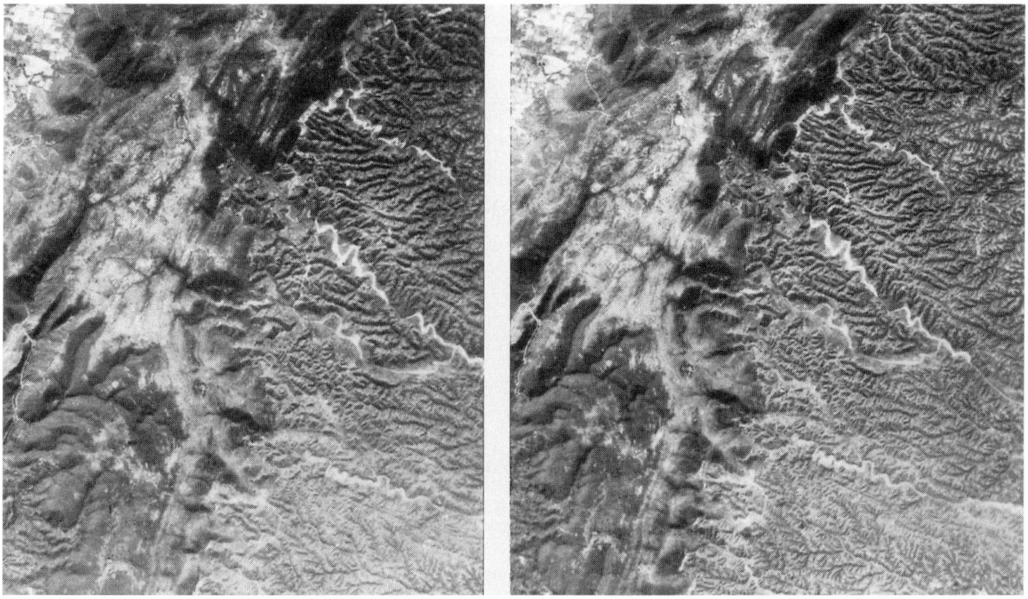

FIGURE 14.4 Stereo aerial photos showing sandstones and shales marked by differences in landform and drainage (courtesy A. White).

the bedding (Lee and Farmer, 1993). At other places, fracturing may impart higher vertical conductivities.

Younger sandstones of Mesozoic and Tertiary ages form the most productive aquifers covering large areas, e.g. Nubian sandstone in North Africa, Dakota sandstone in the USA, Ranmark and Murray Group of sandstone in Australia, the Jurassic sandstone in the Great Artesian Basin (GAB) in eastern Australia and Cuddalore sandstone (Miocene–Pliocene) in India. They form some of the best known confined aquifer systems extending over several thousand square kilometres having high transmissivities (Table 14.3). A case study of numerical modelling to estimate the effect of pumping of groundwater to facilitate opencast mining of lignite from Cuddalore sandstone in India is described in section 17.7.6.

Groundwater in the Nubian sandstone is dated as being old (about 35 000 years) and was probably recharged during the pluvial climate in Pleistocene time. Therefore, groundwater extraction in such areas should be planned carefully to avoid groundwater mining. Similar situations also exist in several other arid areas (section 17.8.1).

Fractured sandstones

In firmly cemented sandstone, intergranular permeability is negligible but secondary permeability due to fracturing is of significance. Fracturing in sandstones could develop due to unloading, tectonic movements and at their contact with dolerite dykes. Fractures developed due to expansion, as a result of unloading, will be more prominent only up to a shallow depth (50 m or so), but tectonic fractures related with folding and faulting can be deep-seated. Fracturing due to both tectonic movements and intrusion of dolerite dykes is reported from the Permo-Triassic sandstones of the Karoo Supergroup in Botswana and other places in South Africa. As a result of fracturing, sandstones have acquired good transmissivity and specific yield

TABLE 14.2 Hydraulic properties of glacial deposits

Type of deposit	*Location*	*T* (m^2d^{-1})	*K* (ms^{-1})	*Sy*	*Source*
Till (fractured clayey)	Manitoba and Montreal, Canada	–	10^{-10}–10^{-9}	–	Cherry (1989)
Till (fractured)	Alberta, Canada	–	5×10^{-9}–2×10^{-7}	–	Jones *et al.* (1992)
Till (weathered)	Iowa, USA	–	4×10^{-4} (av.)	0.04 (av.)	Edwards and Jones (1993)
Glacial outwash	Central Illinois, USA	43–1664 88 (av.)	2×10^{-6}–7×10^{-6}	0.5	Walker and Walton (1961)
Glacio-fluvial	Michigan, USA	1440–1920	10^{-5}–10^{-4}	0.04–0.35	Kehew *et al.* (1996)

(Table 14.3) (Bromley *et al.*, 1994; Sami, 1996). Triassic sandstones in the UK also show much lower integranular hydraulic conductivity than the bulk aquifer conductivity as determined by pumping tests, which is explained due to dominant fracture flow component (Hamill and Bell, 1986). Similarly, other fractured sandstones also show moderate to high transmissivities (Table 14.3).

14.2.2 Shales and siltstones

Fine-grained argillaceous clastic rocks, e.g. shales and siltstones, are formed by compaction and lithification of clay and mud deposits. Porosity of freshly deposited clays and muds is high (50% to 80%), but due to compaction on burial, porosities are reduced to less than 30%. Shales usually have porosities in the range of 1.5% to 2.5%. Intergranular permeabilities in shales are low (10^{-13} to 10^{-9} m s^{-1}) so that groundwater cannot move faster than a few centimetres per century through intact shales (Table 14.3).

Even if the primary permeabilities of shales and siltstones are low, they are capable of transmitting large quantities of water and solutes over large contact areas by leakage across litholigic boundaries. For example, a 30 m thick siltstone bed with a hydraulic conductivity of 10^{-9} ms^{-1} having a hydraulic head difference of 3 m perpendicular to the bedding, will transmit about 3153 m^3 of water each year through each square kilometre of its surface area.

Low permeability argillaceous rocks (clays and shales) have attracted greater attention from hydrogeologists in recent years from the point of view of disposal of high-level radioactive waste. With this in view, detailed investigations are in progress in the Underground Research Laboratories in the Boom Clay, Belgium, and opaline clay/shale at Mt Terri in Switzerland (Put and Ortiz, 1996; Neerdael *et al.*, 1996). Special *in situ* interference tests were designed in these underground laboratories to assess the hydraulic conductivity and storativity (Table 14.3).

Fractured shales and siltstones

Fracturing can impart good hydraulic conductivity (10^{-7} to 10^{-4} ms^{-1}) to shales, siltstones and other fine-grained clastic rocks which are otherwise impervious (Table 14.3). In the USA, notable examples include the Brunswick shale of Triassic age in New Jersey and adjoining states of New York and Pennsylvania, shales of Pennsylvanian age in the Central States and fractured siltstones of Devonian age in the Appalachian basin of New York state. It is opined that fractured shales represent a double porosity system, in which the fractures control the major amount of flow while the intervening porous blocks contribute slowly by transient flow (Neuzil, 1986). In a heterogeneous sedimentary sequence, thin incompetent beds, such as shales, will be more intensely fractured as compared with thicker units of strong and resistant formations, such as mudstones. This is evidenced from the Brunswick Formation in the USA, where mudstone horizons at shallow depths have lower values of K (10^{-6} m s^{-1}) while shales at deeper levels (20–45 m) are fractured forming aquifer horizons with hydraulic conductivity (K) varying from 2×10^{-5} to 5×10^{-4} m s^{-1} (Michalski, 1990). Devonian siltstones of the Appalachian plateau, USA, exhibit two sets of fractures – subhorizontal bedding plane fractures and vertical fractures. Unlike vertical fractures, the spacing of bedding fractures increases with depth due to greater confining stress at deeper levels. Slug tests in these siltstones indi-

TABLE 14.3 Hydraulic properties of sandstones and argillaceous rocks

Rock type	*Age*	*Location*	*T* (m^2d^{-1})	*S*	*K* (ms^{-1})	*Source*
Charmuria sandstone (fractured)	Algonkian	Raipur, India	3–13	–	3×10^{-7}–3×10^{-6}	Khare (1981)
Berea sandstone	Mississipian	USA	–	–	3×10^{-6}	Read *et al.* (1989)
Sherwood sandstone (fractured)	Triassic	UK	600 (av.)	0.05–0.25	–	Kimblin (1995)
Karoo sandstone (fractured)	Permo-Triassic	Botswana, South Africa	200	10^{-4}–10^{-2}	–	Bromley *et al.* (1994)
Nubian sandstone	Cretaceous	Egypt	100–15 000	3.5×10^{-4}	–	Shata (1982)
Renmark Group sandstone	Lower Tertiary	Victoria, Australia	800–900	1.8×10^{-4}	–	Lawrence (1975)
Cuddalore sandstone	Miocene	Tamilnadu, India	1000–5000	2×10^{-4}–5×10^{-4}	–	Gupta *et al.* (1989)
Gunderdehi shale (fractured)	Algonkian	Raipur, India	2–18	–	10^{-8}–10^{-7}	Khare (1981)
Shales and siltstones (fractured)	Upper Permian	Queensland, Australia	1–21 12 (av.)	–	–	Pearce (1982)
Siltstones (fractured)	Devonian	Appalachian Plateau, New York, USA	–	–	10^{-8}–10^{-6}	Merin (1992)
Brunswick shale (fractured)	Triassic	New Jersey, USA	708–933	10^{-4}	10^{-5}–10^{-4}	UNESCO (1972) Michalski (1990)
Pierre shale	Cretaceous	USA	–	–	10^{-14} 10^{-10}	Neuzil (1986)
Opaline clay	Mesozoic	Mt Terri, Switzerland	–	–	10^{-11}	Neerdael *et al.* (1996)
Boom clay	Neogene	Belgium		$S_5 = 1.3 \times 10^{-5}$ m^{-1}	10^{-12}	Put and Ortiz (1996)

cate that closer spacing of bedding plane fractures at shallow depths (<7 m) impart hydraulic conductivities (2×10^{-8} to 2×10^{-6} m s^{-1}) which are 10 to 100 times greater than those obtained from deeper wells (50 m) (Merin, 1992).

14.2.3 Groundwater development

Groundwater from clastic consolidated rocks is tapped both by dug wells and drilled wells, the latter being more common. The common method of water-well drilling in open textured sandstones is the hydraulic rotary method. In hard and well cemented sandstones, percussion and DTH methods are used (section 15.3). Geologically younger sandstones are less cemented and therefore are more productive. The yield of 100–3000 m deep wells, in the Jurassic sandstones of the Great Artesian Basin in eastern Australia, is 0.05–0.1 m^3s^{-1}. Here water from deeper wells has a high temperature (100°C). Well yields vary from less than 3×10^{-4} m^3s^{-1} in compact sandstones to 3×10^{-2} m^3s^{-1} in open textured rocks. Specific capacity of wells in Cambrian–Ordovician sandstones of northern Illinois, USA, is in the range of 53–70 m^2d^{-1} indicating the hydraulic conductivity to be in the range of 1×10^{-5} to 1.8×10^{-5} m s^{-1} (Walton and Csallany, 1962). Well yields from fractured Charmuria sandstone of Precambrian age and Athgarh sandstone of Jurassic ages in India are of the order of 1.5×10^{-3} m^3s^{-1} and 10^{-3} to 10^{-2} m^3s^{-1} respectively depending upon the degree of compaction and fracturing (Khare, 1981; CGWB, 1995b). In compact sandstones, well yields can be increased by shooting using explosives and to a lesser extent by acid treatment, viz. shooting increased the specific capacity of sandstone wells in northern Illinois by an average of 22% to 38% (Walton and Csallany, 1962).

Integrated geophysical and geological studies have proved very valuable in the location of high-yielding fracture zones in the sandstone formation of Karoo Supergroup, especially in areas with complex geology obscured by overburden (Bromley *et al.*, 1994). Figure 14.5 gives a comparison of the specific capacity of wells sited on fractures mapped by the above approach as compared with those in unfractured aquifers. This figure also indicates that major subvertical fractures identified by VLF survey have the highest well yield. Studies by Sami (1996) in the Karoo sandstone of South Africa also show that wells located in the fracture zone in the vicinity of dolerite dykes together with those located along stream channels of higher orders have higher yields. Well yield in fractured shales will be low to moderate. However, in the absence of other good aquifers, wells in fractured shales may yield some water. Well yield from Lower Proterozoic fractured shales in Darwin Rural Area, NT, Australia, is reported to be 1×10^{-3} to 3×10^{-3} m^3s^{-1}.

14.3 Water quality

The chemical composition of groundwater in clastic rocks, as in other rock types, depends on their lithology and texture. In purely siliceous sandstones, the rocks consist almost wholly of quartz and therefore there are very few elements that can be dissolved in water even in arid regions. The groundwater in such rock types has low pH (about 6 or 5) and the HCO_3 content is also low; Ca, Mg, Cl and SO_4 are also in small concentrations; Na may exceed Ca. In rocks having lime or gypsum, the water has a greater concentration of Ca and HCO_3. In arid and water-logged regions, shallow groundwaters may have higher contents of salts due to evaporation (Greenman *et al.*, 1967; Tanwar, 1983; Kulkarni *et al.*, 1989). In coastal areas, groundwater may show complex chemical characteristics due to mixing of fresh groundwater and sea water as well as cation exchange.

Sandstone aquifers, in large sedimentary basins show both horizontal (lateral) and vertical hydrochemical zoning (section 10.6). Groundwater in the recharge area is mainly of HCO_3 type which

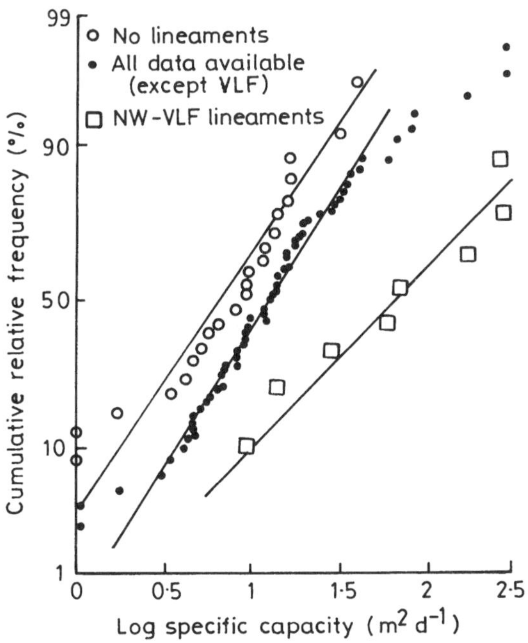

FIGURE 14.5 Cumulative frequency of specific capacity of wells in fractured and non-fractured locations in Karoo sandstones, Botswana. Note that NW–VLF lineaments are most productive (after Bromley *et al.*, 1994, reprinted by permission of journal of *Ground Water*).

changes to SO_4-Cl type in the direction of flow. A decrease in SO_4^{2-} in the direction of flow is also reported from some places, e.g. groundwater in the Fox Hills sandstones in Dakota, USA, shows an increase in salinity but decrease in SO_4^{2-} from recharge to discharge areas. The source of SO_4^{2-} in the recharge area is presumably due to pyrite and/or gypsum dissolution. The decrease in SO_4^{2-} appears to be a result of sulphate reduction in the presence of reduced carbon (lignite) while the increase in chloride is attributed to the cross-flow of saline water from underlying Pierre shale (Thorstenson *et al.*, 1979). Leakage from interbedded aquitards (clay, siltstone deposits) causing an increase in Cl^- content of groundwater is also reported from the Triassic Sherwood sandstone aquifer in northwest England (Kimblin, 1995).

Shales and siltstones, being very fine and porous, provide an enormous contact area with groundwater. The permeabilities are also very low, thereby increasing the residence time and greater opportunity for dissolution of rock material. Therefore, groundwater in these formations has high total dissolved solids (>1000 mg l^{-1}) and the amount of SO_4^{2-} and Cl^- is more than HCO_3^-; both Ca-Mg and Na rich waters are reported. In shales, SO_4^{2-} is the dominant anion while Cl^- is dominant in both shales and sandstones. Cation exchange is a characteristic phenomenon in modifying the chemical characteristics of groundwater due to the presence of clay minerals. Clay and shales also play an important role in acting as semi-permeable membrane (osmotic filters) in spite of their low permeability as described in section 10.5. Such a process is believed to be responsible for the occurrence of saline waters in non-marine sediments (Back and Hanshaw, 1965; Neuzil, 1986).

Movement of tracers and contaminants in fractured sandstone and shale will behave as in other dual-porosity aquifers with rapid movement in fractures, and diffusion into low permeability but porous matrix. Fracture skins formed due to the chemical alteration of sandstones into oxides of iron and manganese considerably lower the rate of diffusion into the matrix. These aspects have been discussed in section 7.4.

Chapter fifteen

Water wells

Water wells are vertical shafts or holes used by mankind since time immemorial for obtaining groundwater for drinking purposes and other uses. The type of well to be constructed depends on the nature of the geological formation, depth to the water-table, yield requirement and economic consideration. Sometimes horizontal or sub-horizontal wells or galleries are also constructed under favourable hydrogeological conditions to tap subsurface water. Wells serve other purposes also, such as subsurface exploration, artificial recharge and waste disposal.

15.1 Types of wells

For large-scale groundwater supplies, especially in unconsolidated and semi-consolidated formations of high permeability, tube wells are recommended. These wells are usually 100–300 m deep, or even more, depending on the depth of aquifer horizons, and water quality. Tube wells are provided with screens and are often gravel packed, creating a zone of high permeability around the well which also increases its effective diameter.

In hard rocks, due to their low permeability and depth limitations, the common structures for groundwater withdrawal are dug wells, borewells and a combination of these types. In the present text, only those types of wells are described which are suitable in hard rock formations. The technology of well drilling, design and construction is, however, more developed for unconsolidated formations (Campbell and Lehr, 1973; Driscoll, 1986).

15.1.1 Dug wells

Traditionally in most of the developing countries of Asia and Africa, large diameter dug wells are popular in low permeability hard rock aquifers. They can be constructed by local labour and are ideally suited when the weathered layer is thick and the water-table is shallow. There are hundreds of thousands of such wells in tropical, sub-tropical and temperate regions of the world.

Dug wells are about 3–10 m in diameter and have a depth of 10–15 m (Figure 15.1a). Sometimes they extend in depth to 40 m or even more but the diameter is reduced telescopically with depth (Figure 15.1b). In weathered rock mass, these wells are circular in cross-section, but in fractured rock, they are made rectangular so that the strike of joints, foliation and other discontinuity planes in the country rock are at right angles to the longer axis of the rectangle (Figure 15.2). Such a well gives a higher

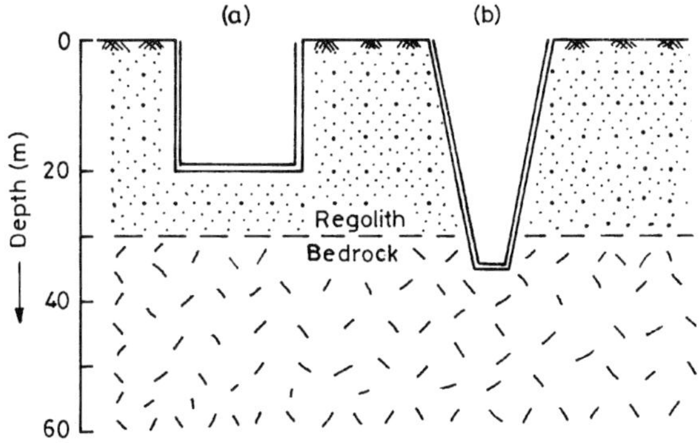

FIGURE 15.1 Types of dug wells: (a) shallow dug well, and (b) deeper dug well with gradual reduction in diameter.

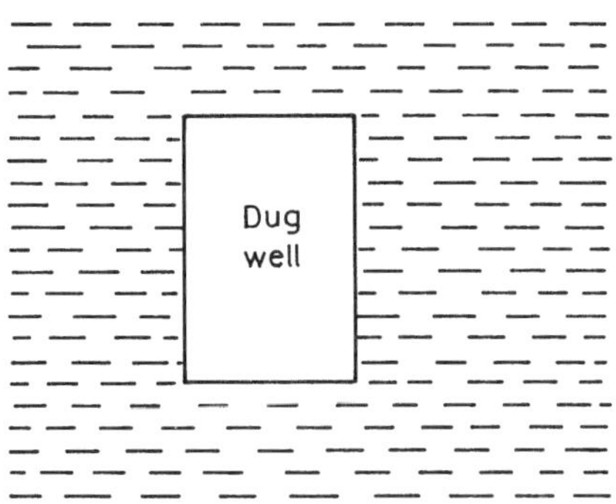

FIGURE 15.2 Schematic plan of a rectangular dug well. The longer axis of the rectangle is kept at right angles to the strike of foliation or fracture planes to intercept maximum number of discontinuities.

yield as compared with a circular well, as it is likely to intersect a larger number of fractures.

Dug wells are usually unlined, except with a masonry curbing in the upper 5 m or so. Sometimes, they are also provided with perforated concrete rings in the water-bearing zone. Dug wells of large diameter in low permeability rocks, as in crystalline formations, also serve as storage reservoirs, into which water seeps slowly during the recuperation period after pumping. These wells are pumped for 3–4 h and are left idle, preferably overnight, for full recuperation which may take about 6–8 h. Shallow dug wells have the risk of getting dry in the summer season or during a drought. To ensure an adequate supply, dug wells must extend at least 2–3 m below the water-table.

TABLE 15.1 Yield of dug wells in crystalline rocks of India (after Subramanian, 1992)

Rock type	*Well yield (m^3d^{-1})*		
	Recharge area	*Discharge area*	*Canal command area*
Granite	29–78	48–128	300–772
Charnockite	9–53	42–96	60–360
Schist	9–32	18–54	–

In earlier years, large size dug wells were at places provided with steps for direct accessibility into the well. Another ancient mode of abstraction of water has been through '*mhot*' (leather bag) or persian wheel driven by animals. In recent years, centrifugal pumps, which are coupled to a diesel engine or electric motor, have become more popular.

The yield characteristics of dug wells depend upon several factors, namely:

(a) Landform – whether located in pediment, buried pediment or valley fill areas.
(b) Regolith – its thickness and permeability.
(c) Fracture characteristics of bedrock.
(d) Local groundwater regime: whether the well is located in groundwater recharge or discharge area (Table 15.1).
(e) Depth of water-table and its fluctuation.

A dug well in an unconsolidated formation can yield 10–30 m^3h^{-1} but in hard rocks, the yield is usually less than 5 m^3h^{-1}. The yield may change with the season, e.g. in West Africa yield is 5 m^3h^{-1} in the rainy season and only 1 m^3h^{-1} in dry periods (UNESCO, 1979).

Dug wells are used for irrigation, livestock and domestic water supplies. When used for drinking purposes, the well should be protected by providing a parapet wall and cement sealing around the well to protect against surface water contamination, and a tin shade to inhibit contamination from airborne material.

15.1.2 Borewells

Small diameter drilled wells (borewells) are increasingly being used in several countries in Asia, Africa and South America for domestic water supply and minor irrigation. These wells in hard-rock terrains were earlier drilled by conventional percussion (cable tool) or rotary methods. However, in recent years, due to the introduction of the down-the-hole air hammer (DTH) drill, the construction of such wells in hard rocks has become faster and economical.

By this method, in hard rocks a well, say, of 155 mm diameter and of 60 m depth can be drilled in one or two days.

Borewells have the advantage of tapping a greater thickness of aquifer in the weathered layer and also the underlying fractured rocks (Figure 15.3). Sometimes, saturated clay horizons within the weathered layer create problems in well construction due to caving. This problem is usually taken care of by providing a blank casing.

Depth of borewells is usually 30–70 m while the diameter depends on the type of pump. Wells which are provided with hand pumps are of 100 mm diameter but for electric submersible pumps the minimum diameter should be 150 mm. Mark II and Mark III handpumps developed in India are being used successfully for rural water supply in several countries.

Borewells are either of the open-end type in massive hard rock, or they are provided with a screen and packed with gravel when they tap the weathered layer (Figure 15.3). A steel casing or PVC pipe is provided to protect the well against caving, and cement grout is used to protect from pollution.

The yield from a borewell depends mainly on fracture characteristics, and varies considerably from one region to another and within short distances. The various factors which affect yields of borewells in crystalline rocks are: (a) climate, (b) lithology, (c) geomorphologic setting, and (d) rock structures. These are discussed in section 11.5. Criteria for siting borewells are given in section 15.2. A minimum yield of 1.5 m^3h^{-1} is usually required for wells fitted with hand pumps and 7.5 m^3h^{-1} for wells fitted with submersible pumps. However, the yield criteria will vary from one country to another and also from one region to another depending on economics. In several countries of Africa, e.g. Malawi, Zimbabwe and Nigeria, a successful well fitted with a hand pump may have a yield of 0.72 m^3h^{-1} (Wright and Burgess, 1992). Typical well yields of borewells in different rock types are given in respective chapters on hydrogeology (Chapters Eleven to Fourteen).

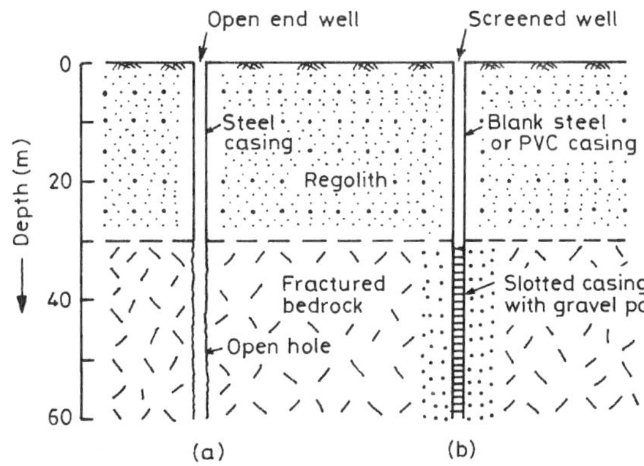

FIGURE 15.3 Types of borewells: (a) open-end type, and (b) screened well.

Borewells are preferred over dug wells under the following hydrogeological conditions.

(a) Insufficient thickness of weathered horizon.
(b) Deep water-table.
(c) Larger seasonal fluctuations of water-table.
(d) For supplying safe drinking water.

Relative merits of dug wells and borewells are listed in Table 15.2.

15.1.3 Horizontal borewells

Borewells may also be oriented horizontally or sub-horizontally in areas of high relief where fractures and joints are vertical, or where groundwater occurs as perched water above an impermeable outcropping bed, or is trapped behind dykes or other vertical barriers. The horizontal borewells are preferably aligned at right angles to the most transmissive fractures for intercepting the maximum number of such fractures for a given length of borehole (Figure 15.4). These wells are free flowing, and are provided with valves or other flow control devices at the outlet. The horizontal or inclined borewells may have problems of stability when

TABLE 15.2 Relative merits of dug wells and borewells

Dug wells	*Borewells*
1. Difficult to protect against pollution	1. Easy to protect
2. May require about six months for construction	2. A well of 150 mm diameter and 60 m depth can be constructed in 1 or 2 days by DTH machine
3. Wastage of agricultural land as each well occupies quite a large area	3. Well occupies very small space
4. May go dry during prolonged drought	4. Less affected by seasonal fluctuation; therefore more dependable for water supply in drought period
5. Pumped by centrifugal pumps which are cheap and easy to maintain	5. Submersible pumps are required which are expensive and more difficult to maintain
6. Can be pumped only intermittently, at higher rates of about 5 to 10 m^3h^{-1}	6. Can be pumped constantly with an average rate of about 5 m^3h^{-1}
7. The unit cost of water, considering all the items of input and benefit, is higher	7. The unit cost of water is lower considering all the items of input and benefit.

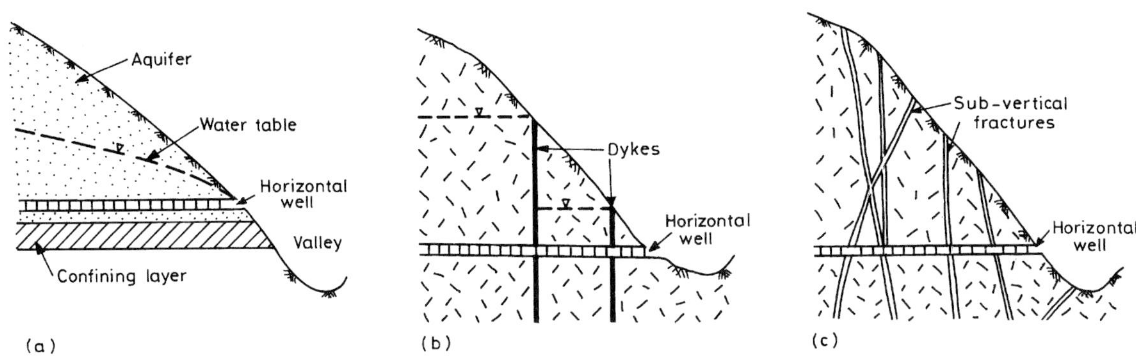

FIGURE 15.4 Horizontal wells tapping: (a) perched water from above an impervious layer, (b) confined water from dyke compartments, and (c) confined water from steeply inclined fractures.

the rocks are highly fractured (Banks *et al.*, 1992). Horizontal borewells generally have a yield of 10–50 m^3d^{-1}; high yields (300–1000 m^3d^{-1}) are reported from weathered Tertiary andesites in the Lake Tahoe Basin, California, USA (Figure 15.5).

15.1.4 Dug-cum-borewell and collector well

Dug-cum-borewells and collector wells are more or less similar in principle as both of them have a large size open well with extension bores for tapping a greater thickness or surface area of the aquifer (Figure 15.6).

The term dug-cum-borewell is used in India. In such a well one or two small diameter (50–90 mm) vertical bores are drilled at the bottom of the existing dug well, to a depth of about 30 m, for tapping a deeper fractured aquifer (Figure 15.6a). Horizontal or inclined boreholes may also be drilled to tap vertical or steeply inclined joints or permeable layers, e.g. inter-trappean formations in basalts (Singhal, 1973). The yield of dug-cum-borewells in crystalline

FIGURE 15.5 Example of horizontal wells tapping weathered andesite at a height of 2400 m above m.s.l. in the Lake Tahoe Basin, California, USA (after Matthews, 1985).

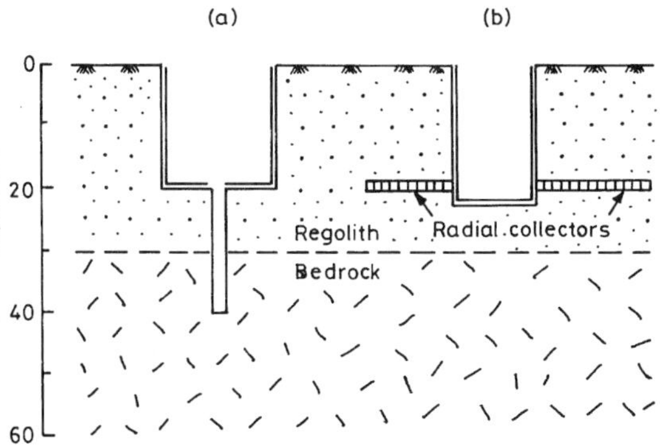

FIGURE 15.6 (a) Dug-cum-bore well in crystalline rocks. (b) Collector well in weathered crystalline rock.

rocks of south India is usually in the range of 5–15 m^3h^{-1} (Singhal *et al.*, 1987), i.e. an increase of about 50% to 100% compared to that of a dug well. Dug-cum-borewells have another advantage over dug wells as the recuperation period is relatively rapid and the well may provide a perennial source of water round the year.

A typical collector well, also known as **Ranney well**, is constructed in the river bed/or terrace deposits (predominantly sand and gravel). It consists of a central caisson of about 4 m diameter and 20–40 m depth, and at the bottom is provided with radiating screen pipes to facilitate recharge from the river (Figure 15.7). The yield of such wells may range between 4000 and 100 000 m^3d^{-1} with an average of about 25 000 m^3d^{-1} (Hamill and Bell, 1986). In India, collector wells have been recently constructed at a number of places in the flood plain of the river Yamuna for water supply to the cities of Delhi, Agra and Mathura.

Collector wells, although traditionally constructed in alluvial formations, have also been found to be quite successful in the weathered crystalline rock terrain of Zimbabwe, Malawi and Sri Lanka (Wright, 1985, 1992). These wells are usually of 2–3 m diameter and 20 m depth. They are provided with radial collectors of 90–150 mm diameter up to 40 m length at the base of the well (Figure 15.7). They show less drawdown and fast recovery in comparison to borewells and dug wells. The yield from

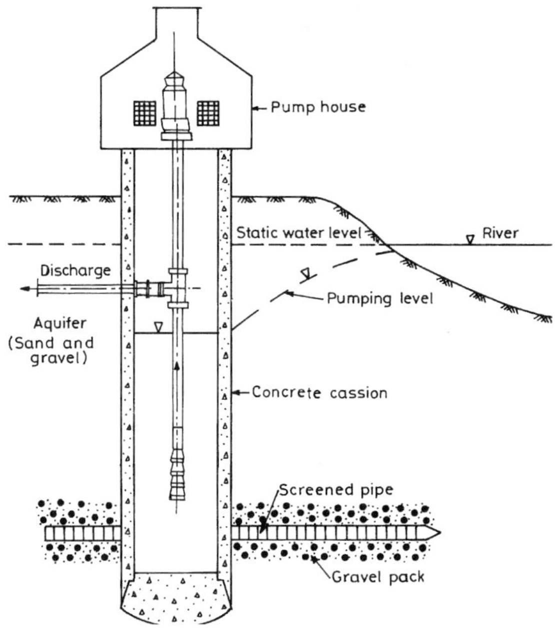

FIGURE 15.7 A Ranney type collector well in alluvium adjacent to the river bed.

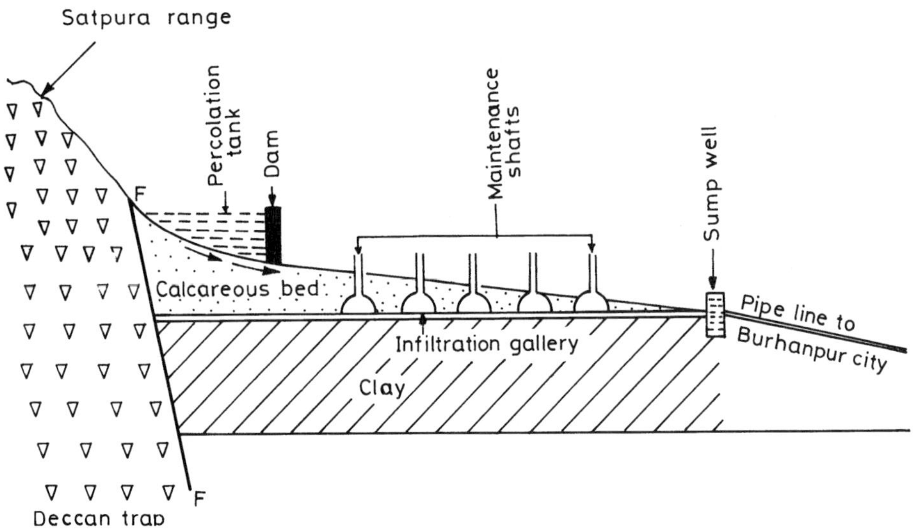

FIGURE 15.8 The hydrogeological setting of an infiltration gallery (qanat) for supplying water to Burhanpur City, Madhya Pradesh, India (courtesy S. Romani).

collector wells in the basement rocks is reported to be of the order of 1×10^{-3} to 8×10^{-3} m^3s^{-1} with drawdowns of 2–3 m. In comparison, the typical well yield from slim boreholes is 1×10^{-4} to 1×10^{-3} m^3s^{-1} with a drawdown of 30 m or even more (Wright, 1992).

15.1.5 Infiltration galleries (*qanat*)

Infiltration galleries are constructed in thin aquifers which are in hydraulic continuity with some surface water body, as in such cases vertical wells may not provide sufficient quantities of water. '*Qanat*' is an Arabic term for a canal. The *qanat* principle of intercepting a gently sloping water-table was conceived in ancient Persia about 3000 years ago and is still practiced in many countries in the Middle East. It was adopted and perpetuated by Spanish colonialists in South America. The *qanat* system consists of a number of vertical shafts linked by a tunnel which intercepts the water-table; the slope of the tunnel is less than that of the water-table. Figure 15.8 gives an outline of an infiltration gallery built in the river alluvium in central India during the Moghul period in the 15th century. Recently a percolation tank was added to augment recharge to the alluvial aquifer. *Qanat* is regarded as an elegant solution of water management as it makes over-exploitation of water resources virtually impossible and thereby avoids mining of water. Although infiltration galleries are usually constructed in permeable unconsolidated formations, their use for groundwater extraction in weathered and fractured crystalline rocks has also been advocated (Ruden, 1993). They could form a potential method of tapping the thin weathered rock mantle.

Galleries and horizontal tunnels are well known from basaltic terrains in the USA and Spain. These are a common mode of groundwater abstraction in oceanic islands for skimming freshwater floating over sea water (section 12.5). Many such structures are built on the basaltic formations in the Canary Islands in the Atlantic and the Hawaii islands in the Pacific, and also in islands made of coral limestones (Falkland, 1984).

15.2 Criteria for siting wells

Wells for domestic and minor irrigation are to be located within a reasonable distance from the place of use. This puts a severe constraint on the site selection of such wells.

Based on the experience of early high-yielding wells, target areas may be defined followed by mapping of significant lineaments using satellite images and aerial photographs and ground reconnaissance. This should be followed by geophysical surveys at identified sites, and finally drilling. Criteria for the site selection of common types of wells in hard rocks are briefly described below.

Dug wells

These are constructed in areas with a thick (>10 m) and permeable weathered layer. They are more successful in groundwater discharge areas like valleys and buried pediments where the depth and seasonal fluctuation of the water-table is limited. Canal command areas where perennial irrigation is practised and sites in proximity to surface water tanks, are also ideal for dug wells (Table 15.1). Remote sensing data can help to delineate areas of relatively deeper weathering, buried pediments and valleys. Vertical electrical sounding is useful in ascertaining the saturated thickness of the weathered layer. Usually seismic refraction technique is also very appropriate to assess the nature of the weathered material and identify permeable zones within a practical range of 10–20 m (section 5.4).

Borewells

Fractures and lineaments are important for locating borewells. The relation between fracture systems and streams is significant as the streams may recharge the aquifer due to flow along fractures. Therefore, drilling may be done on such fractures that are hydraulically connected to the streams (UNESCO, 1972).

LeGrand (1967) has emphasized the importance of sheet joints for groundwater supplies from shallow wells in granitic and gneissic rocks. In Australia, such types of joints are often very open near the ground surface and up to a depth of about 100 m. Borewells tapping these fractures have a yield of about 0.01 m^3s^{-1} (UNESCO, 1972). Similar inference is also drawn from studies in the Archean granite of southeast Zimbabwe (Carruthers *et al.*, 1993).

Studies in the shield rocks of a number of regions (India, Africa, Scandinavia, etc.) show that tensile fractures are more productive than shear fractures. Broad criteria for making a distinction between the two are given in section 2.2.3. It may be emphasized that the mere presence of fractures is not sufficient for locating borewells in crystalline rocks, as fractures when filled with clay minerals will not serve any useful purpose. It is also difficult to distinguish between clay-filled and water-saturated fractures by any geophysical or remote sensing method.

In addition to the study of fractures and lineaments by remote sensing, the resistivity, seismic refraction and EM techniques are regarded as potential exploration tools for siting of wells. Resistivity and seismic surveys of the azimuth type can be helpful in ascertaining the fracture orientation and their relative saturation and hydraulic conductivities in different directions (sections 5.3.2 and 5.4.2).

EM methods are growing popular in groundwater investigations as they are fairly rapid because they do not require electrode contact with the ground (section 5.2). However, EM methods are usually too expensive to be applied routinely for a rural water supply. The data from African countries indicate that the success rate in water well construction increases by using both EM and resistivity techniques (Table 15.3). Radon gas survey has also proved useful in locating water-bearing fractures in some areas (Pointet, 1989; Wright, 1992). Buried fracture zones which are targets for drilling of borewells can be identified by a careful interpretation of EM, resistivity and radon surveys (Wright, 1992). Figure 15.9 shows the case of a possible erroneous siting of a borewell which could be avoided by a close and judicious interpretation of both EM (horizontal loop) anomaly and radon anomaly data. In the example shown, the

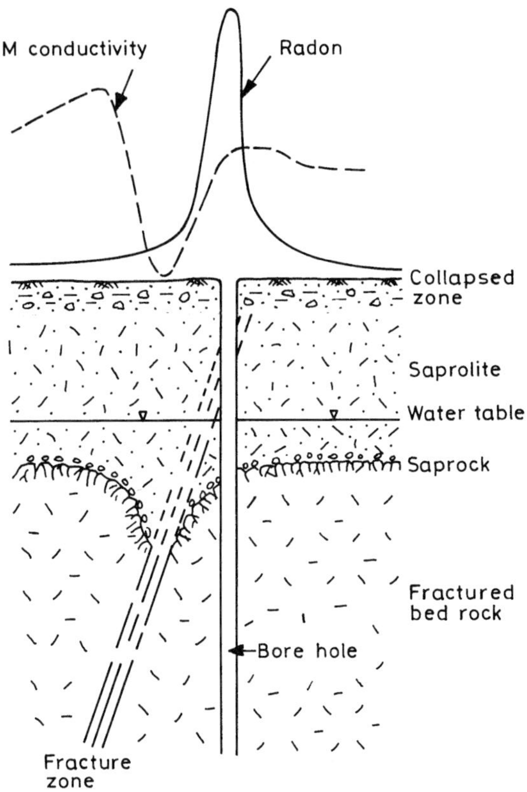

FIGURE 15.9 Example of the radon anomaly and EM (horizontal) loop anomaly across a dipping fracture zone and a possible erroneous siting of borewell which could be avoided by careful interpretation of data. For details see text (modified after Wright, 1992).

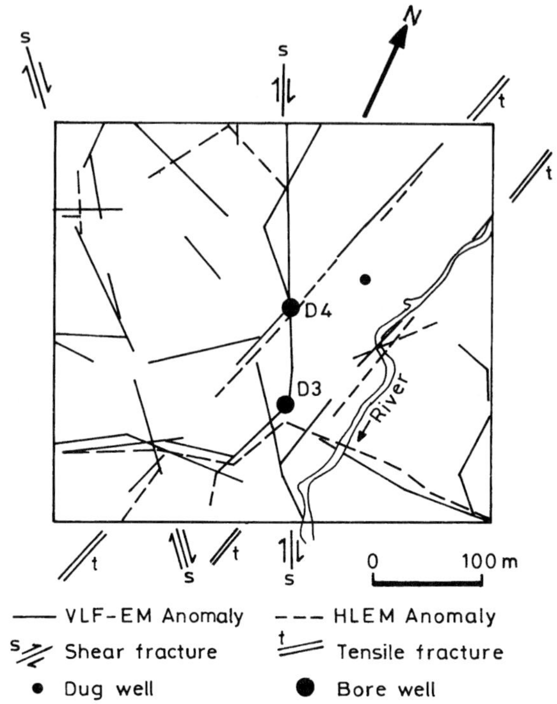

FIGURE 15.10 Map of a part of crystalline basement of West Africa, showing EM anomaly axes and traces of tensile and shear fractures. The borewells sited at the intersection of important tensile and shear fractures coincident with EM anomalies were found to be highly successful. The borewell D4 gave an exceptionally high yield of 7 m^3h^{-1} due to its proximity to the flood plain (after Boeckh, 1992).

positive peak of the radon anomaly, indicating higher moisture circulation, is in broad agreement with the negative peak of the EM anomaly. This would normally tempt one to drill a vertical borewell close to the radon peak, which would miss the dipping fracture zone at depth. A closer look reveals the asymmetry in the EM anomaly on either side of the minimum value and also a slight shift of the EM minimum with respect to the radon peak. These features indicate the dipping nature of the fracture, which ought to be considered for the correct siting of borewells. Further, the importance of integrated study using satellite images, aerial photographs and geophysical data for the siting of high yielding wells in the crystalline rocks is also demonstrated from another study in West Africa (Figure 15.10).

In some areas, lineaments of a particular orientation may be more potential than another. For example, Bromley *et al.* (1994) observed that W–NW trending lineaments deciphered from EM–VLF surveys gave the best results in identifying such potential lineaments in the buried Karoo sandstones in Zimbabwe (section 14.2.3).

Horizontal borewells and infiltration galleries

Horizontal borewells are effective in areas of

TABLE 15.3 Success rate for various siting techniques (after Barker *et al.*, 1992)

Technique	*Percentage of successful boreholes*
EM + resistivity	90
Resistivity	85
EM	82
Hydrogeology	66
Air photo	61
Logistical	50

high relief where rocks are steeply dipping or have other vertical discontinuities in the form of dykes or fractures. Infiltration galleries are preferred to tap thin pervious horizons like intertrappean sediments and vesicular horizons. They are of additional advantage in volcanic islands to check deterioration of water quality due to sea water encroachment. For successful siting of different type of wells, the various factors, viz. topography, structure, landform, lithology, lineaments, source of recharge etc. need to be collectively considered, and this can be accomplished in a GIS environment (section 6.4.3).

Dug-cum-borewells

These are preferred in areas with comparatively lesser thickness of overburden and greater water-table fluctuation. The orientation of boreholes in such wells is controlled by the fracture pattern so that boreholes are aligned perpendicular to the fracture orientation. Instead of constructing a new dug-cum-borewell, an existing dug well is usually revitalized by extension drilling.

Collector wells with radial strainers are preferred in local depressions in bedrock surfaces with an adequate thickness of regolith. In volcanic islands such types of wells (also known as '*Maui*' wells) also check deterioration in water quality due to sea-water encroachment.

15.3 Methods of well construction

The method of well construction depends on the type of well and nature of geological formation. Dug wells in weathered rock are excavated by hand using pick and shovel. In hard rocks, power-driven equipment and explosives are used to facilitate excavation. Tube wells and borewells in unconsolidated sand and gravel formations are drilled by rotary and reverse rotary circulation method. In bouldery formations, the cable tool method is more effective. In crystalline rocks, air rotary and down-the-hole (DTH) hammer methods are recommended for drilling small-size boreholes of 30–100 m depth.

In volcanic rocks, e.g. basalts, the suitability of a particular method depends on the texture and permeability of the various flow units. In the Deccan Traps, where basalts are generally massive, the down-the-hole hammer method is widely practised for drilling 30–50 m deep and 100–150 mm diameter borewells. On the other hand, on some volcanic islands, viz. Hawaii and French Polynesia, where there are thick permeable horizons, deep (up to 400 m) and 200 mm diameter high yielding (0.1 m^3s^{-1}) wells are drilled by a combined air hammer and rotary machine (Wright, 1984).

In fractured and karstified limestones, the cable tool method is more advantageous as it will avoid problems of lost circulation. If lost circulation is not a major problem, the choice of the rig will depend on the number of holes to be drilled and the urgency of work. The cable tool method has an additional advantage as it facilitates interpretation of lithological and water quality changes during drilling.

Table 15.4 gives the comparative merits of various drilling methods. The suitability of these methods in the common geological formations is indicated in Table 15.5; readers may refer to Driscoll (1986) for the details of various drilling methods. A brief description of some common methods of drilling is given below.

15.3.1 Cable tool method

In the cable tool method, also known as the percussion method, drilling is done by alternate

TABLE 15.4 Comparison of common water well drilling techniques

Drilling method	*Formations for which best suited*	*Usual range of possible depth and diameter*	*Typical penetration rates* ($m \cdot h^{-1}$)	*Advantages*	*Disadvantages*
Cable tool (or percussion)	Unconsolidated clays and, gravel, and boulders; consolidated and hard rocks, viz. sandstones, limestones, crystalline rocks	Depth from 100 to 600 m; diameter from 80 to 600 mm	Clay and shale (0.2–0.6); sandstones (0.6–1.25); hard rocks (0.04–0.12)	Low cost, simple operation; good sampling of rock cuttings & water; low water requirement	Slow penetration, casing is required in loose material
Direct circulation hydraulic rotary drilling	Unconsolidated sand, silt, soft consolidated rock	Depth from several hundreds to few thousand metres; diametre 80 to 300 mm	Unconsolidated material (6.5); consolidated rocks (0.4–0.6)	Fast method, capable of drilling deep wells; casing usually not required	High cost of rig, requires large quantities of water, samples contaminated with mud
Reverse circulation hydraualic rotary drilling	Unconsolidated sand, silt, clay and other soft rocks	Depth from 80 to 200 m, diameter 400 to 1500 mm	About 10	Capable of drilling large diameter wells; suitable for gravel packed wells; no contamination with mud	High cost of rig, requires large quantities of water; not effective in large-size gravel and boulder formation
Air rotary and down-the-hole hammer drilling	Very hard dry rock, viz. crystalline rocks, basalts and massive limestones	Depth up to 200 m – diameter 50 to 400 mm	3 or more	Fast rate of drilling and most effective in hard rock; water or mud not required; good sampling of rock cuttings	Not effective below water-table, casing required in unconsolidated overburden

TABLE 15.5 Relative performance of drilling methods in some common types of geological formations

Type of formation	*Drilling method*		
	Cable tool	*Hydraulic rotary*	*Air rotary & DTH hammer*
Loose sand and gravel	Difficult	Rapid	NR
Loose boulders	Suitable casing required	NR	NR
Firm shale	Rapid	Rapid	NR
Sandstone	Slow	Slow	NR
Limestone and dolomite	Rapid	Rapid	Very rapid
Limestone, fractured	Rapid	Difficult	Rapid
Limestone, karstified	Rapid	Difficult	NR
Massive igneous and metamorphic rocks	Slow	NR	Rapid

NR: Not Recommended

lifting and dropping of a heavy string of tools. The drill bit is massive and heavy with a relatively sharp chisel edge which crushes the rock by impact. In the zone above the water-table, water is added to the hole to make a slurry of the crushed rock so that it can be removed easily by a bailer. There is usually no need to provide a casing in a consolidated formation but in unconsolidated and loose rock materials casing is necessary to check the caving of the hole.

The cable tool method is a more versatile method for drilling in a variety of rock formations, like boulders and fissured and cavernous rocks. It is capable of drilling holes of 80–600 mm diameter to depths of 100–600 m. If a hole is of a small diameter, it can be drilled to a relatively greater depth.

The main advantages of the cable tool method are that representative data about the rock formations and water quality as well as static water-levels in different formations can be obtained. Further, it also requires a lower quantity of water for drilling which is advantageous in arid and semi-arid regions. The main drawbacks are the slow rate of drilling and limitations about the diameter and depth of drilling. Driving of casing pipe during drilling in unconsolidated rocks as well as maintaining the plumbness of the hole, is often problematical.

15.3.2 Direct circulation hydraulic drilling

Direct circulation hydraulic rotary method is suitable for drilling deep wells of up to 30 cm diameter, or even more, in unconsolidated formations. The hole is drilled by rotating a hollow bit attached to the lower end of a drill collar and a string of drill pipe. Drilling mud is circulated into the hole through the drill pipe, which brings out the rock cuttings.

The direct rotary method is used widely for drilling oil wells and water wells. This technique is advantageous due to the rapid rate of drilling in a variety of geological formations. Two types of bits are generally used – the drag (fish tail) and the roller cone (rock) bit. It is difficult to use this method in highly jointed and cavernous formations due to a heavy loss of drilling fluid, and also in loose bouldery formations as the bit may be unable to crush such material. In some areas, the availability of an adequate quantity of water for drilling may also be problematical.

15.3.3 Reverse circulation rotary method

As the name indicates, the reverse circulation method has a reversed circulation of drilling fluid so that the drilling fluid, along with rock cuttings, is pumped out through the drilling pipe. This method is advantageous in drilling comparatively larger diameter (400–1500 mm) high-capacity wells to depths of 200 m, especially in situations where gravel-packed wells are to be constructed. This method is more suitable for soft unconsolidated clay and sand formations. If the rock consists of cobbles and boulders of a size larger than the drill pipe, or openings in the bit, then drilling by this method is not possible as such a material can not be brought out through the drill pipe.

15.3.4 Air rotary method

In air drilling methods, instead of water, air is used as the drilling fluid. There are two such methods – direct rotary air and down-the-hole hammer methods. In both these methods, a large compressor is used to force the air through the drill pipe and bit into the hole which brings the cuttings to the ground surface. The air also cools the bit.

Air drilling methods are most effective for drilling small diameter (100–400 mm) holes to depths of 200 m in consolidated sedimentary formations and other hard rocks where chances of caving of the hole are rare. These methods are also advantageous for drilling in fissured rocks in areas where the availability of water for drilling is limited. Large diameter holes can also be drilled by using foams and special additives to the air to reduce loss of air to the formation (Driscoll, 1986). Use of foams also helps in speedy removal of cuttings.

Down-the-hole hammer method

This is one of the recent methods of air drilling; the equipment consists of a tungsten-carbide bit attached to the hammer and drill pipe (Figure 15.11). The hammer is operated by compressed-air, and is capable of delivering 500 to 1000 blows per minute. A rotation speed of about 40 rev/min is usually satisfactory, although in abrasive rocks, such as quartzites, the rotation may be at about 10 rev/min. These machines are capable of drilling 4 in. (102 mm) and 6 in. (152 mm) diameter wells up to a depth of about 50 m. Large diameter (380–760 mm) holes can also be drilled by a set of six or eight down-the-hole hammers, arranged concentrically around a leading (pilot) hammer, which are rotated as one unit to give a large-size hole. The down-the-hole hammer method is more efficient than the cable tool method as in the former the rock cuttings are continuously removed by air.

The compressor forms an important component of air drilling as it has to provide adequate air volume at a certain operating pressure in order to reduce drilling time. Compressed air is supplied to the hammer at a pressure of 100–110 psi (690–758 kPa). The upward velocity of air through the annular space between the drill pipe and the sides of the hole should be about 90 m s^{-1}, or more for the effective removal of rock cuttings (Driscoll, 1986).

The advantages of the down-the-hole method are: (a) rapid removal of rock cuttings; therefore the rate of drilling is fast, (b) the method is especially suited for fractured hard rocks such as basalts, granites, quartzites, limestones and dolomites, and (c) water requirement for drilling is negligible. Further, yield characteristics from a particular horizon can be estimated during drilling. The main disadvantage of the down-the-hole hammer machine is the damage caused to the bits by abrasion with continued use. Furthermore, the down-the-hole hammer method is also unsuitable for unconsolidated formations like loose sand, clay and cavernous limestones.

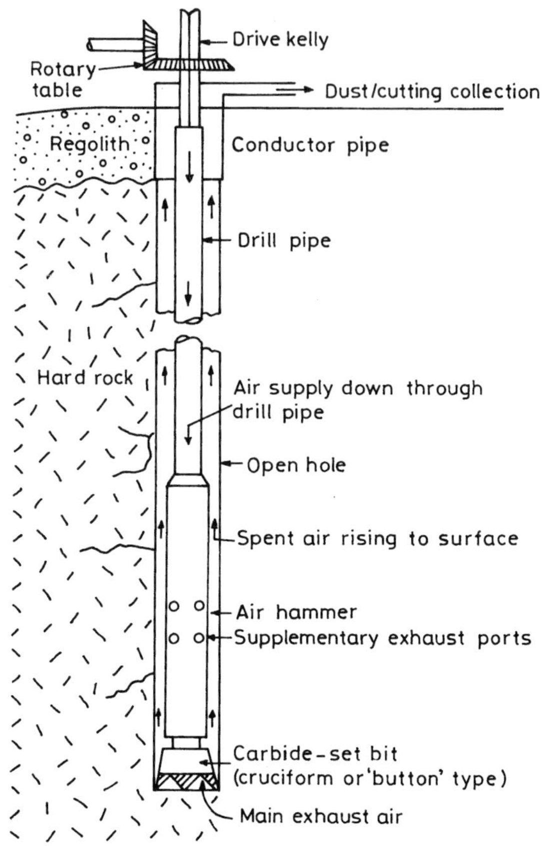

FIGURE 15.11 Down-the-hole hammer drill (after Cruse, 1979, reprinted from *Journal of Hydrology*, with kind permission of Elsevier Science-NL, Amsterdam, the Netherlands).

15.4 Methods of well development

After a well is completed, it is developed to increase its yield and efficiency. In unconsolidated formations, this is achieved by removing finer material from around the well screen thereby leaving an envelope of coarse-grained material. In gravel-packed wells, well development is also recommended to increase their efficiency.

15.4.1 Development techniques in unconsolidated formations

In unconsolidated rocks, well development methods include pumping, surging and hydraulic jetting (Todd, 1980; Driscoll, 1986).

Pumping is the simplest method where the well is pumped at rates higher than its design capacity. This causes higher velocities which help in removing finer material. The **over-pumping** method of well development is more effective in non-stratified formations where gravel-packed wells are constructed.

Well development by **surging** is achieved by either using a plunger or compressed air. The downward motion of a plunger (surge block) forces the water from the well into the aquifer through the screen while its upward motion causes back-washing. This results in the removal of fine material into the well which should be periodically extracted with the help of bailers. Surging is more effective in aquifers containing clay or mica as these minerals are likely to clog the screen. In hard rocks, well development by surging can be done in the casing above the open hole. In air surging, compressed air is injected into the well through an air pipe which causes surging and also lifts water with washout material to the surface.

High velocity jetting by water is an effective method of well development, both for open-end wells in hard rocks and also in screened wells. Water with high velocity jet action enters the formation through the screen openings agitating the aquifer material adjacent to the screen. The fine-grained aquifer material is dislodged and enters into the well by the turbulent flow of water.

Deflocculants like polyphosphates are also used for well development as they disperse the clay particles. This method is effective both in unconsolidated and consolidated formations.

15.4.2 Development of wells in hard rocks

In hard rocks, wells are either provided with screens or they are the open-end type without screens. The well face and the fractures are often clogged by the fine rock material produced during drilling. Removal of this material improves the well yield. In addition to the methods described above, the other techniques which are used for increasing well yields, especially in hard rocks, are actually directed towards increasing the rock permeability rather than well development. These are therefore more appropriately called 'aquifer development methods' or 'aquifer stimulation methods'. The common aquifer development techniques are acidizing, blasting or shooting, and hydraulic fracturing.

Acidizing

In this method, hydrochloric acid is used in carbonate aquifers and other calcareous formations to dissolve carbonate minerals deposited on the wall of the well screen as encrustation, or those within the rock fractures or voids, to improve the well yield. In siliceous rocks, hydrofluoric acid can be used for a similar purpose. The amount of hydrochloric acid required to treat an encrustated screen depends on the screen diameter and its length. Other acids which are used to remove carbonate encrustations are sulphuric acid and hydroxyacetic acid (Driscoll, 1986).

In an advanced method of pressure acidizing, a larger volume of muriatic acid (commercial grade of HCl), is pumped into the aquifer at high pressure. This will result in developing a larger volume of aquifer due to deep radial penetration of the acid (Hamill and Bell, 1986). Experience has shown that acidizing is more effective in fractured carbonate rocks (Figure 15.12).

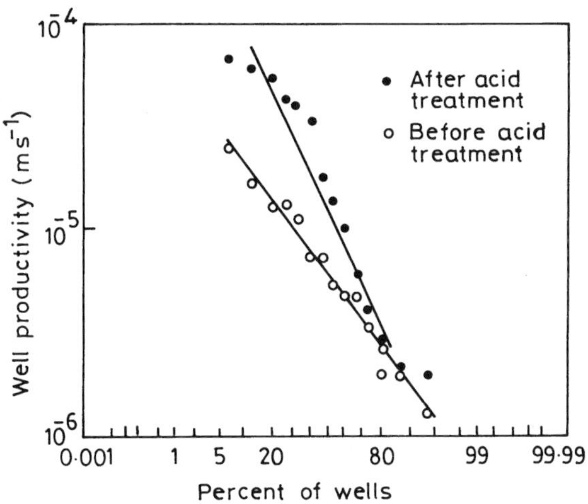

FIGURE 15.12 Effect of acid treatment on well productivity in carbonate rocks (after Csallany and Walton, 1963).

Blasting or shooting

Explosives (viz. nitroglycerine–liquid or solid) are used to shoot rock wells to enhance well yields, this being accomplished by an increase in well diameter, development and enlargement of fractures and removal of fine-grained material and encrusting deposits from the well face. It is necessary to use proper blasting procedure and explosives in a given situation depending on the rock type, size of well and depth of water. Care should also be taken when blasting is done in a populated area to avoid any damage to life and property. Earlier, small explosive charges, usually 14–45 kg, were used but more recently larger charges of 400–900 kg have been used in igneous rocks for better results (Driscoll, 1986).

Excellent results in improving well yields by blasting in hard rocks is reported from several places. Explosives have also been used to rehabilitate old wells, as specific capacities of wells may sometimes decrease with time due to clogging. Shooting of wells in sandstone aquifers of Cambrian–Ordovician age in northern Illinois, USA, improved the specific capacities from a small percentage to 71% (Walton and Csallany, 1962). Shooting of wells by using mild explosives in limestone rocks of the Raipur basin, India, improved the well discharge by more than ten times (Adyalkar, 1983).

Hydraulic fracturing – or hydrofracturing

This is an advanced technique to enhance the well yields in hard rocks by opening existing fractures and providing better interconnection between them in the vicinity of the well. This method owes its origin to oil reservoir engineering where it is employed to create reservoir fractures for increasing oil productivity during secondary recovery. More recently, this method is used to create new fractures in HDR systems.

For hydrofracturing, an inflatable packer is lowered into the well and is inflated a little above the production zone, thereby isolating this zone. Sometimes the production zone is isolated by two packers (Figure 15.13). Water is pumped

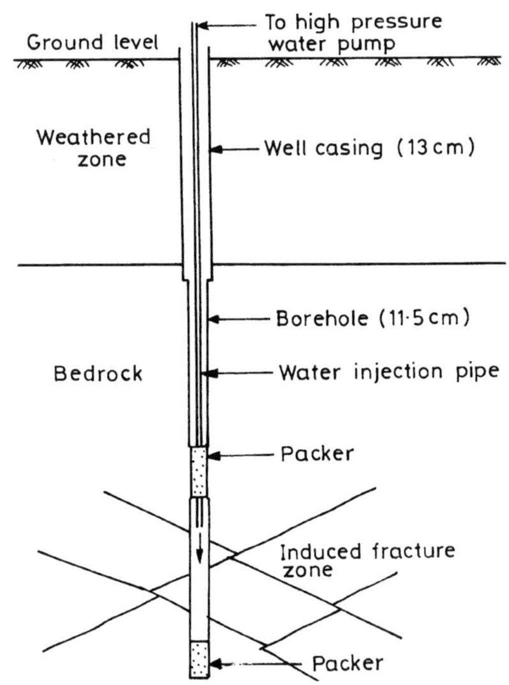

FIGURE 15.13 Schematic set-up of hydrofracturing in a hard rock aquifer (after Klitten, 1991).

into the isolated zone under pressure exceeding the pressure exerted by the overburden.

The overburden pressure for every 1 ft (0.3 m) depth will be about 1 psi (6.9 kPa). Therefore at a depth of about 100 ft (30 m) the overburden pressure can be overcome if water is injected under a pressure of more than 100 psi (690 kPa). Usually water is pumped into the isolated zone at pressures 5000–7000 kPa depending on the depth of this zone and the nature of the rock (Driscoll, 1986). Higher pressure is required in massive rocks with few fractures. Hydraulic fracturing is caused due to rock failure when fluid pressure increases to the point that effective stress becomes negative and exceeds the tensile strength of the rock. It is preferred that the wells for hydrofracturing are drilled normal to the existing fractures so that the induced fractures are developed at right angles to the existing ones to provide interconnectivity.

Before hydrofracturing, it is necessary to clean the borehole walls of any clay coating or drill cuttings, either mechanically or using chemical additives such as diluted hydrochloric acid to remove carbonate deposits. Experience has shown that the productive zone to be subjected to hydrofracturing should be identified beforehand perhaps by electrical logging, and also the diameter of the hole by calliper logging. This will help in placing the packer of the right size at the proper depth. It is also necessary to determine safe injection pressure from step-rate tests. If the temperature of the injected water is different from the formation water, the pressure required to extend the hydrofracture will change. The geometry of hydrofractures is not well understood as it depends on existing fractures and matrix permeability. Pressure and temperature changes during hydrofracturing may also cause precipitation of minerals in fractures and thereby may reduce the permeability.

As the newly created fractures are likely to heal with time, sometimes propping material such as sand, special beads or small particles of high-strength plastic are introduced along with the injected water to keep the cracks open, after the water pressure is removed. This is called 'sand fracing' (Hamill and Bell, 1986).

Hydraulic fracturing may increase the permeability of the aquifer for about a 100 m radius or so around the well, by about 200%. A distinct improvement in the yield of wells by hydraulic fracturing in a variety of rocks is reported from many countries (Williamson, 1982; Sheila and Banks, 1993). Experience shows that it is more economical to improve the yield of failed wells by hydrofracturing rather than to drill new wells (Talbot *et al.*, 1993).

The effect of hydrofracturing on improving the transmissivity is greater in hard crystalline rocks (gneisses, granites and migmatites) than compared with soft rocks like schists and low-grade amphibolites. Klitten (1991), based on studies in Tamilnadu, India, reported an increase of 100% in the transmissivity of hard crystalline rocks while in softer rocks the increase was only 20% to 30%. Results from Australia show that the specific capacity of shallow low-yielding wells improved by about six times in granites and schists but in phyllites the ratio was 0.95, indicating deterioration instead of improvement (Williamson, 1982).

Chapter sixteen

Geothermal reservoirs and hot dry rock systems

16.1 Introduction

Most geothermal reservoirs and hot dry rock (HDR) systems are located in fractured low porosity geological formations. In such settings, fractures provide conduits for fluid flow through such rocks. Therefore fractures, either natural or those artificially created, are important for the successful operation of geothermal reservoirs and HDR systems.

Geothermal reservoirs are located in areas of youthful mountain building, tectonic plate boundaries and volcanic activity, where subsurface temperatures are higher. They form potential reservoirs for the generation of electrical power. Geothermal fluids have also been used for centuries for domestic and greenhouse heating and for therapeutic and recreational purposes.

There are three types of geothermal energy systems: low enthalpy or hot-water dominated systems (50–150°C), high enthalpy or vapour–dominated systems (150–300°C) and hot dry rock (HDR) systems (50–300°C). In low enthalpy systems, the hot groundwater is used as a source of heat. In high enthalpy systems, steam is the extracted fluid which is used to drive turbines for generating electricity. In hot dry rock (HDR) systems, water is circulated to deeper levels in artificially created fractures where it is heated up and hot water and steam flow back to the ground surface to be used as a source of geothermal energy.

Some of the earlier notable power plants generating energy from geothermal resources are in Italy, New Zealand, Japan, western United States, Iceland and Chile. Several new geothermal fields have been discovered in recent years. For example, in the last 20 years, three large geothermal plants have been established in southeast Asia – Bulalo (426 MW) and Tiwi (330 MW) in the Philippines, and Awibengkok (330 MW) in Indonesia. Each of these geothermal fields is a water-dominated reservoir associated with an andesitic volcano in an island arc setting (Williamson *et al.*, 1996). In Russia, the Northern Caucasus, the Volga river area, West Siberia and the Far East are identified as the most promising areas for development of geothermal energy. In India, the main geothermal fields are located in the northwestern Himalaya, Narmada valley in central India and along the west coast. However, as they are mostly low temperature (<150°C) systems, their potential for power generation is limited (GSI, 1991).

Some of the areas, e.g. Narmada geothermal zone in central India, are characterized by high heat flow (≈300 mW m^{-2}) and high thermal gradients (≈120°C km^{-1}) which can be a potential area for geothermal power development (GSI, 1991; Chandrasekharam, 1996). Feasibility of binary cycle power plants at some sites in central India and at Puga in the Ladakh area in Himalaya has also been established (Pitale and Padhi, 1996).

16.2 Hot dry rock (HDR) systems

In some areas with steep geothermal gradients, at depths greater than 3000 m, temperatures are in the range of 200° to 500°C and rocks are dry. In such areas the attempt is to extract heat from the hot dry rock by circulating water through drill holes. As the geological formations at these places are mostly granites which are impermeable, it is necessary to create fractures for facilitating the movement of fluids by either a suitable array of multiple large-yield nuclear explosions or by hydraulic thermal stress fracturing. In the latter case, cold surface water is used to pressurize natural or stimulated fractures. This causes rock shrinkage and development of additional fractures. A second well is drilled later to intersect the hydraulically fractured region, thereby a closed loop circulation is established. At Los Almos, New Mexico, USA, two wells were drilled into hot but impermeable granite which was hydraulically fractured to create a connection between the wells. Water is pumped through one of the wells and flows back through the other (production) well (Figure 16.1). In this process, the circulating water is heated due to hot rocks at depth. Several other areas, viz. Carnmenellis granite in Cornwall, UK, and Soultz-Sous-Forets in the Rhine Graben in France have also been investigated. The data obtained from these studies provide information about the properties of fractures and their effect on flow characteristics with time.

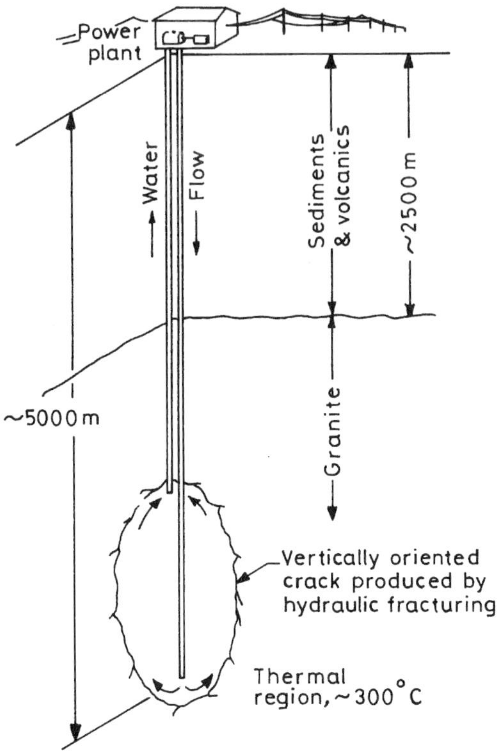

FIGURE 16.1 Hot Dry Rock (HDR) geothermal system produced by drilling and hydraulic fracturing (after Collie, 1978).

Studies in Carnmenellis granite, UK, have been carried out by the researchers of the Camborne School of Mines (McCann *et al.*, 1986; Baria, 1990). Four boreholes were drilled to a depth of approximately 300 m. Reservoir development required a reduction in the hydraulic impedance of the borehole walls, and an increase in the permeability of the rock mass between the boreholes, for efficient extraction of heat. The hydraulic impedance of borehole walls was achieved by explosive stimulation and the bulk permeability of rock mass between the boreholes was increased by hydraulic stimulation. Cross-hole seismic experiments were carried out to delineate the cavity created by explosive stimulation.

An HDR experiment is in progress at Soultz-Sous-Forets in the Rhine Graben in France, where two deep wells of 3590 m and 3876 m with a separation of 450 m have been drilled in the granitic basement which is under a 1400 m thick cover of Triassic sedimentary rocks (Gerard *et al.*, 1996). The granite at that depth is highly fractured. Hydraulic performance of wells was improved by hydraulic stimulation, by injecting several thousand cubic metres of water at high flow rates (up to 0.06 m^3s^{-1}). The hydraulic performance of one of the deep wells was increased by hydraulic stimulation from 5×10^{-4} m^3s^{-1} to 2×10^{-2} m^3s^{-1}, the water temperature being around 135°C. The pore water in granite was highly saline which is most probably derived from the sedimentary cover.

16.3 Flow characteristics

During exploitation of geothermal reservoirs, fractures and joints influence the movement of injected water as well as transport of heat from the rock to the circulating fluid. In a closely fractured homogeneous medium, the injected water, which is at a lower temperature, will sweep the heat out of the rock more effectively during its movement. On the other hand, if fractures are isolated and widely spaced, injected water may flow directly through the fractures towards the production well without extracting sufficient heat from the reservoir rock. Therefore, the cold water front will reach the production well in advance.

In a double-porosity reservoir, the matrix blocks will not only influence the transfer of fluids but also the heat from the blocks to the fracture. The hydraulic response of a double-porosity geothermal reservoir is expected to be the same as of a double-porosity aquifer; a double-porosity framework has been applied to the Geysers Geothermal Reservoir in California.

In geothermal reservoirs, fracture properties and thereby permeability may change with time during the operation of the system. Decrease in pressure due to fluid extraction from the reservoir can cause closure of fractures. On the other hand, fluid injection into the reservoir will cause opening of fractures. Thermal changes during extraction will also affect the fracture permeability with time. Injection of water, which is at a lower temperature, causes contraction resulting in increase in permeability. This is manifested as decrease in flow impedance, i.e. an increase in the rate of injection with time (Figure 16.2). Injection of water may also lead to changes in permeability due to precipitation and dissolution of minerals.

The change in permeability can be monitored by repeat well discharge tests (section 16.4). As in other fractured aquifers and petroleum reservoirs, the permeability of geothermal reservoirs can also be improved by acid treatment and more often by hydraulic fracturing.

16.4 Reservoir properties

In geothermal and petroleum reservoirs, the usual practice is to express the transmissivity in terms of intrinsic transmissivity or permeability-thickness (kb) where k is the intrinsic permeability in darcy and b is the saturated thickness of reservoir in metres; kb is expressed in darcy-metre (d-m) (10^2 d-m = 1 m^3 or 1 d-m = 0.01 m^3). Productive geothermal reservoirs usually have a kb value of 10 d-m or more, and in some cases it is 100 d-m. Values

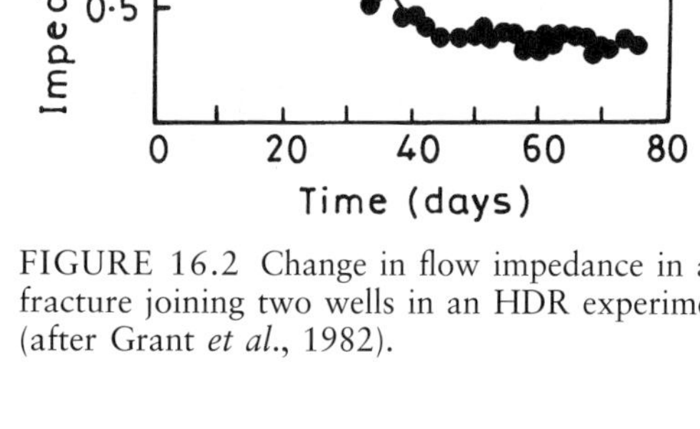

FIGURE 16.2 Change in flow impedance in a fracture joining two wells in an HDR experiment (after Grant *et al.*, 1982).

less than 10 d-m indicate poor potentiality of a geothermal reservoir.

In addition to permeability-thickness, storativity, S*, is also important to know the fluid reserves. Storativity, in the case of geothermal reservoir, is expressed as

$$s = \eta \beta b \qquad (16.1)$$

where η is the porosity, β is fluid compressibility and b is the aquifer thickness; S* is expressed in m Pa^{-1}.

It has been discussed in Chapter Eight that in a confined aquifer, storativity depends on the compressibility of the rock matrix and of the fluid. In geothermal reservoirs there are two fluid phases – liquid and vapour, both of which have different compressibilities; therefore, the storativity value of the reservoir will also depend on the relative proportion of the two phases. The compressibility of liquid water is very small (4.7×10^{-10} Pa^{-1}), as compared to that of vapour (1.4×10^{-6} Pa^{-1}). Therefore, even a small amount of vapour in the reservoir will result in a much greater storativity. This implies that a large storativity value in a high temperature reservoir does not always mean a thick high-porosity aquifer. In a poor liquid phase system, storativity can be converted to porosity-thickness (ηb) by simply dividing S* with liquid water compressibility. However, if reservoir temperatures are high, it is better not to convert the storativity into a porosity-thickness value (Grant *et al.*, 1982).

The presence of the vapour phase also influences the transmission of pressure changes in the geothermal reservoirs. In a low-temperature liquid dominated system, the pressure changes are transmitted over long distances in short times, but in a high temperature two phase (liquid and vapour) reservoir, the pressure change will move out slowly. Therefore, the effect of boundary, if any, will be reflected only if the test is of a longer duration. The fractured nature of the geothermal reservoir also makes the flow mechanism and interpretation of test data more complex. In spite of these limitations, the hydraulic parameters of geothermal reservoirs can be estimated by well discharge tests, similar to those described in Chapter Nine.

Single-well tests

In the single–well test, the effect of discharge (constant or variable) or injection from a single well is observed in the same well. In a drawdown-test, the shut-in is opened to flow. It is assumed that the flow rate is constant. The drop in pressure with time is noted. In a pressure build-up test, commonly used in a petroleum reservoir, discharge is produced at a constant rate for a certain period of time and then shut-in. Pressures are measured during the shut-in period as a function of shut-in time. The data are plotted on semi-logarithmic paper and permeability can be estimated by the Horner, and Miller–Dyes–Hutchinson methods (Ramey, 1977).

Interference test

In a well interference test, one or more wells are discharged simultaneously and the effect is observed in one or several adjacent wells. This is the common method of estimating the hydraulic properties of aquifers (see Chapter Nine). Interference tests can also be used when the well production is achieved by injection, as is commonly the case in the petroleum industry. The interference tests are of greater value as compared to the single-well tests because they scan the reservoir properties over a greater area. Interference tests are also useful in identifying boundary conditions but the presence of a vapour phase in a geothermal reservoir creates problems for such an analysis. The effect of wellbore storage and skin can also be identified from this type of test data.

The test data can be interpreted by using an appropriate flow model of homogeneous, double-porosity and discrete fracture flow, as in the case of aquifer tests (Chapter Nine). Both type curve and straight line methods of solution are used by plotting pressure change against time, on double-logarithmic and semi-logarithmic paper respectively (Grant *et al.*, 1982). McEdwards and Tsang (1977) argued that the type-curve solutions used widely in the field of hydrogeology and petroleum engineering are not applicable to geothermal reservoirs testing where there are two or more producing wells, each of which have a unique flow rate history. For such conditions, they suggested a computer-based least square fitting technique which is of advantage for accounting effects of two or more flowing wells, unique flow rate histories, presence of recharging or discharging boundaries, wellbore storage and skin effects.

As in the case of aquifers, the pressure conditions and levels in wells tapping geothermal reservoirs are also affected by changes in barometric pressure, tides, rainfall and other sources. Therefore, well test data should be corrected for these changes induced by external effects. It is also necessary to have reliable measuring equipment which can stand the high temperature and salinity of geothermal fluids. Injected water during the interference test may cause changes in reservoir permeability due to induced pressure, temperature variation and deposition of precipitates on the well face. The reservoir permeability will also change as it is temperature dependent; the decrease in temperature causes thermal contraction and thereby additional fractures and higher permeability.

An interference test at Wairakei, where the rocks are acid volcanics (rhyolitic tuffs, dacites and andesites) of Pliocene to Recent age, gave permeabilities of the order of 0.05–0.5 darcy (Elder, 1981). At Raft River in Idaho, USA, geothermal wells penetrate sedimentary formations and igneous rock (quartz monzonite). Here interference tests gave kb = 70 d-m whereas single-well tests gave kb values of 5–20 d-m, indicating varying permeabilities. Other borehole tests and analysis of barometric and tidal effects in this area also indicate varying porosities; the monzonites have low (≤ 5%) while sedimentary formations have high (14–24%) porosities (Grant *et al.*, 1982). A long duration (33 days) well discharge test for geothermal exploration from a 2615 m deep well penetrating Triassic sandstones at Southampton in the UK gave a reservoir transmissivity of 6 m^2d^{-1} (3.5 d-m). The drawdown pattern indicated the bounded nature of the reservoir (Downing *et al.*, 1984).

Tracer tests

Tracer injection tests are useful for the estimation of reservoir properties, and also determination of preferential flow paths and hydraulic interconnectivity of fractures. Both radioactive and non-radioactive tracers are used. In Broadlands, New Zealand, interference well test data indicated the homogeneous reservoir character, but tracer tests indicated preferential flow (Grant *et al.*, 1982). Tritium tracer studies at Manikaran geothermal field in India gave velocities of 72 m d^{-1} and direction of flow from north to south for geothermal waters in fractured quartzites (Athavale *et al.*, 1992). Tracer tests are also used for siting injection and extraction wells.

16.5 Ground subsidence

Ground subsidence due to withdrawal of geothermal fluids has also been reported from several areas, e.g. Wairakei, New Zealand; Imperial valley and the Geysers, California, USA. Subsidence in geothermal fields poses problems similar to other areas where subsidence is a result of withdrawal of water or oil. However, the monitoring of ground deformation in geothermal areas becomes more difficult as these are generally located in active seismic zones. All control points move horizontally as well as vertically, and establishing baseline data is a more difficult problem (Grant *et al.*, 1982).

In geothermal fields, the elastic stresses

change due to withdrawal of heat and fluids. Hence, in these areas the subsidence depends both upon fluid pressure and temperature changes, unlike in groundwater and petroleum reservoirs where the changes are largely isothermal. Therefore, in groundwater and petroleum reservoirs, the subsidence can be checked by re-injection of water leading to maintenance of pressure. However, in geothermal reservoirs, the re-injection of fluids will not cure temperature changes. In some cases, the thermal effect may outweigh the pressure effect, e.g. at the Geysers, California, the temperature effects causing subsidence were four times greater than the pressure effects (Grant *et al.*, 1982). Ground subsidence at Wairakei (New Zealand) is well documented where the ground surface has subsided by as much as 5–10 m (Elder, 1981). The greatest rate of subsidence is at a distance of 1 km east from the main production area (Figure 16.3), which is difficult to explain (Grant *et al.*, 1982). Re-injection of thermal fluids may also reduce the permeability of the formation, due to the precipitation of salts. The injection of fluids in seismically active areas may also induce seismic activity. The reservoir temperature and thereby its energy potential will also be reduced by injection.

16.6 Water quality

The chemical composition of geothermal waters varies widely (Table 16.1). The composition of waters is mainly controlled by the composition and content of gas, e.g. a large concentration of H_2S will give rise to acid-sulphate waters, while systems rich in CO_2 will form Na-HCO_3 or Ca-HCO_3 waters. Chloride and chloride-sulphate waters are more common. The SiO_2 content in geothermal waters is typically high due to greater solubility of quartz at higher temperatures. Chloride-rich waters have a higher concentration of silica. SiO_2 and Na-K geothermometry has been used to estimate the base temperature of geothermal reservoirs and also the extent of mixing with meteoric water (Balmes, 1994; Raymahashay, 1996).

At places, for example in Puga Valley, India, economic deposits of sulphur and borax are associated with hot-water springs. Here the B content is in the range of 116–140 mg l^{-1} (GSI, 1991). Large deposits of calcite, aragonite and travertine are also associated with some thermal springs, e.g. Mammoth Hot Springs, Yellowstone National Park, USA. Thermal waters are also well known for their medicinal and curative properties.

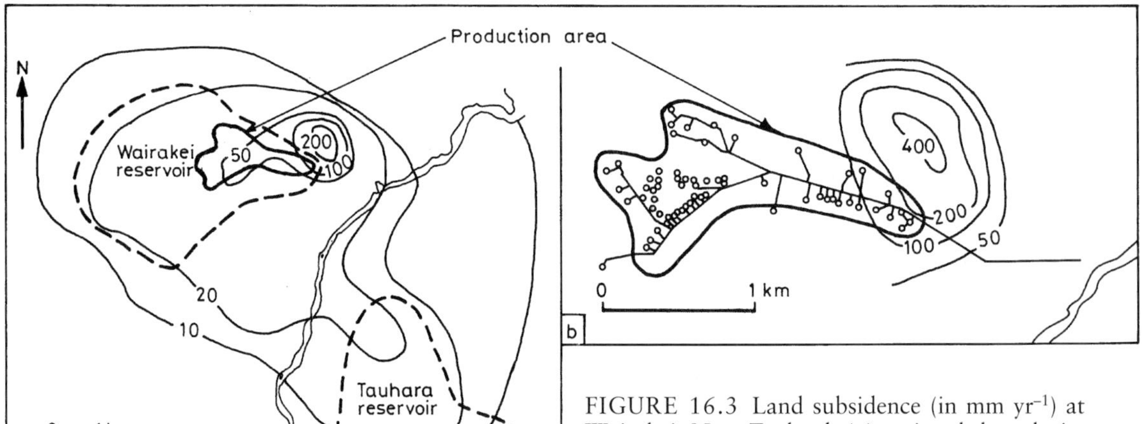

FIGURE 16.3 Land subsidence (in mm yr^{-1}) at Wairakei, New Zealand: (a) regional data during 1956–61; (b) detailed data for production area during 1964–74 (after Elder, 1981).

TABLE 16.1 Chemical composition of water from some thermal springs (in mg l^{-1})

Location	*Yellowstone National Park, USA*	*Puga, India*	*Manikaran, India*	*Surdulica, Serbia*	*Tongonian Geothermal Field, Philippines*
Rock type	Volcanic tuff	Ophiolite	Quartzite	Granaodiorite	Andesite
Temperature (°C)	94	55–81	93	78–126	55–81
pH	9.6	7.4–8.1	8.0	7.5–9.0	7.3–8.2
TDS	1310	2020–2278	546	1024–1302	–
Na	352	530–640	83	18–287	2020–2208
K	24	68–80	25	8.6–18.8	172–201
Ca	0.8	2–17	52	12–32	56–111
Mg	0.0	1–5	4	1–7	0.1–0.3
HCO_3	–	–	195	354–732	–
SO_4	23	123–149	50	260–480	81–107
Cl	405	370–424	145	43–60	3383–3863
SiO_2	363	120–160	90	40–90	245–289
Source	Hem (1989)	GSI (1991)	Romani & Singhal (1970)	Hadzisehovic *et al.* (1995)	Balmes (1994)

16.7 Origin of geothermal waters

The origin of geothermal waters has been a matter of great interest. Both geochemical and stable and radioactive isotope data indicate that geothermal waters are mainly of meteoric origin (Elder, 1981; Minzi *et al.*, 1988). Plots of δD versus δ^{18}O values of thermal fluids from some geothermal fields is given in Figure 16.4 which show that the deuterium (D) content is the same as that of local precipitation but the δ^{18}O is higher. This oxygen isotope shift is attributed to progressive equilibration of

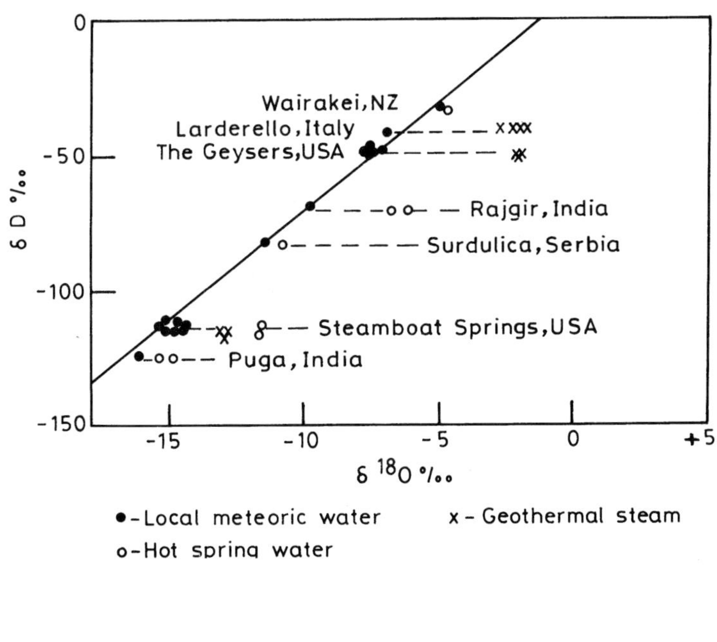

FIGURE 16.4 δ^{18}0 and δD variation in geothermal fluids (based on data from Craig, 1963; GSI, 1991; Hadzisehovic *et al.*, 1995; Raymahashay, 1996).

oxygen in the water with silicates and carbonate rocks.

A comparison of tritium (^3H) in thermal waters with local meteoric water from several geothermal fields also shows large-scale mixing with meteoric water (Grant *et al.*, 1982; Raymahashay, 1996; Zhiming *et al.*, 1996). Some highly saline thermal waters may also originate due to the mixing of meteoric waters and deep-seated brines as is evidenced from the concentration of boron and its isotope, ^{11}B (Balmes, 1994; Vengosh *et al.*, 1994).

The age of the geothermal waters at various sites is estimated to be in the range of less than 100 to 1500 years by tritium and ^{14}C dating (Hadzisehovic *et al.*, 1995; Zhiming *et al.*, 1996). Such variable ages in the same geothermal system in fractured rocks point towards both fissure flow, which is fast, and the delayed diffused flow through rock matrix.

Chapter seventeen

Assessment and management of groundwater resources

Assessment and management of groundwater resources is important for their optimum utilization and to avoid any adverse effects. First of all, it is necessary to have a prior knowledge of the distribution as well as hydraulic and geochemical characteristics of aquifers, as discussed in earlier chapters. It is also necessary to know the amount of natural recharge and groundwater abstraction. Management of groundwater resources is simple when natural recharge is more than the abstraction. However, when the aquifer approaches full development, the management problems increase significantly. Abstraction more than natural recharge leads to decline in groundwater-levels which causes uneconomic pumping lifts, land subsidence, seawater encroachment in coastal aquifers and so on. On the other hand, extensive use of surface water for irrigation may cause water logging and salinization of water and soil. These problems may arise both in granular rock formations and fractured rocks. Groundwater modelling is a useful tool to study problems related to the effect of various stresses on groundwater regime. The various technical aspects of groundwater assessment and management are outlined in this chapter. Most of the examples cited are from granular formations as such studies in fractured rocks are limited, but the general approach is the same.

17.1 Water budget and groundwater balance

Water budget is a quantitative assessment of the total water resources of a basin over a specific period of time. It helps in evaluating the availability of surface and subsurface water for planning the various utilization patterns and practices, and thereby efficient management of water resources. This involves the preparation of: (1) hydrological budget which considers both surface and groundwater resources, and (2) groundwater budget which is more specific for groundwater assessment and management. The various components of the groundwater budget are discussed in detail by Sokolov and Chapman (1974).

17.1.1. Hydrological balance

The hydrological balance equation can be written as

$$S_m + S_g + S_S = P + I_s + I_g - R - ET - O_S - O_g \pm n \quad (17.1)$$

where S_m = change in soil moisture; S_g = change in groundwater storage; S_S = change in surface water storage; P = precipitation; I_s = surface inflow into the basin from outside through canals, rivers etc.; I_g = subsurface inflow from adjoining basins; R = run-off; ET = evaporation and evapotranspiration; O_S = surface outflow to other basins; O_g = subsurface outflow to the adjacent basins, and n = errors in estimation or undetermined elements, viz. losses through deep percolation etc.

Changes in soil moisture storage, S_m, are negligible if balance is prepared on an annual basis. However, for a seasonal balance, S_m will be significant. It can be determined from the soil moisture profile at the beginning and end of the period. Change in groundwater storage, S_g, can be estimated from groundwater level fluctuation at the beginning and end of the period and the specific yield of the aquifer. The change in surface water storage, S_S, can be estimated from reservoir/lake data. The weighted average depth of precipitation is estimated from rain-gauge data usually by constructing Theissen polygons or from isohyetal maps. Surface inflow and outflow, I_s and O_S, are estimated from river and canal discharge data. Subsurface inflow and outflow, I_g and O_g, can be computed from Darcy's law (Q = TIL) based on data of aquifer transmissivity (T), hydraulic gradient (I) and width of the aquifer (L). Evaporation and transpiration losses (ET) include evaporation from surface water and evapotranspiration from soil, natural vegetation and cropped areas. An idea of evaporation from surface water is obtained from evaporation pan data. ET losses from soil and vegetative cover are more complex and can be estimated using lysimeter and various empirical equations based on types of vegetation and other parameters. In case field hydrological data are not available in an area, remote sensing–GIS techniques can be used to generate surrogate hydrological data (e.g. Meijerink *et al.*, 1994).

17.1.2 Groundwater balance

The groundwater balance equation for a given time period can be written as

$$R_p + R_c + R_t + I_r + R_b - ET - T_p - S_e - O_g \pm \Delta S \pm n = 0 \quad (17.2)$$

where R_p = recharge from precipitation; R_c = recharge from canal seepage and field irrigation; R_t = recharge from tanks; I_r = influent seepage from rivers; R_b = subsurface inflow from adjacent basins; ET = evapotranspiration; T_p = draft from groundwater; S_e = effluent seepage to rivers; O_g = outflow to other basins; ΔS = change in groundwater storage, and n represents undetermined elements of the balance and errors in the estimation of various balance elements. S and n can be either positive or negative.

Each item in the groundwater balance equation (17.2) is estimated independently. The various items can be expressed either in terms of volume or depth of water using consistent units. The duration of the water balance could be a water year or a shorter period. The span of the water year may vary from one country to another based on the annual precipitation pattern. In India, the water year is taken from 1st November to 31st October of the next year. The water year can be subdivided into monsoon period – 1st June to 30th November, and non-monsoon period – 1st December to 31st May. In the USA, the water year starts on 1st October and ends on 30th September and is designated by the calender year in which it ends. The groundwater balance equation can also be used to estimate any one component provided the other items are determined by other methods.

For groundwater budgetting, first, the area of study has to be delineated. For this purpose, the best approach is to select a groundwater basin as a unit. In the past, the practice in India and in many other countries has been to take the interfluve '*doab*' area or an administrative zone ('*taluka*' or district) as a unit for such a study. However, such boundaries may not coincide with the groundwater divides. Ideally, groundwater balance should be prepared separately for each aquifer unit of a groundwater basin. The reliability of the groundwater balance will depend on the accuracy of the estimation of different parameters.

17.1.3 Estimation of various components of groundwater recharge

In granular aquifers, recharge is usually areally distributed. The water-table fluctuations in granular aquifers are also small, and rarely range more than a few metres. By contrast, in fractured aquifers, recharge is rapid and water-table fluctuations may range up to a few tens of metres. This is more true when fracturing is sparse and the rocks are either bare of soil cover or have only a thin overburden.

Recharge from precipitation (R_p)

The following methods can be used from estimating recharge from precipitation (R_p).

Water-table fluctuations method
R_p can be estimated from equation (17.3)

$$R_p = S_y h \qquad (17.3)$$

where S_y is the specific yield of the zone of water-table fluctuation and h is the weighted average rise in water-table. Although h can be known quite accurately, large errors in R_p may arise due to unrepresentative value of S_y.

Spring discharge method
The springs discharge recession data are also used to estimate recharge as discussed in section 13.6.

Chloride balance method
In this method, relative concentration of a stable ion like chloride present in precipitation, groundwater or steady-state portion of soil profile and in the dry fallout is measured. Mandel and Shiftan (1981) gave equation (17.4) for estimating recharge by the salt balance method in well flushed regions where airborne salts are the only source of chloride in groundwater.

$$R_p = P\left(\frac{C_p}{C_g}\right) + \left(\frac{F_d}{C_g}\right) \qquad (17.4)$$

where R_p = average annual groundwater recharge in m yr^{-1}; P = average annual rainfall in m yr^{-1}; C_p = average annual chloride content in rainwater, in mg l^{-1}; C_g = average chloride (Cl^-) content in groundwater in mg l^{-1}; F_d = average annual dry fallout of chloride, in g $m^{-2}yr^{-1}$.

The salt balance method has limitations in the following respect (Mandel and Shiftan, 1981): (i) the assumption that the study area is well flushed may not be valid everywhere; (ii) the airborne chloride data based on limited measurements may not be representative; (iii) estimation of dry fallout component is less satisfactory, and (iv) there may be an additional input of chloride from rock weathering and use of fertilizers. In spite of these limitations, chloride balance method has provided reliable results from several countries in a variety of hydrogeological environments (Vacher and Ayers, 1980; Sharma and Hughes, 1985; Sukhija *et al.*, 1988; Gardner *et al.*, 1991).

Isotope techniques
Isotope techniques are being increasingly used for estimating groundwater recharge as well as for understanding the mechanism of recharge. Tritium (^3H) and other environmental isotopes, viz. ^{18}O, ^2H, ^{14}C and ^{32}Si have been commonly used as tracers and for dating of groundwater (section 3.5.2). These studies are also important for management of water resources.

Tritium (^3H), produced in the upper atmosphere, enters the hydrological cycle in precipitation. The concentration of ^3H has varied from time to time, depending upon the detonation of the nuclear devices by different countries since 1952. The tritium concentration reached peak values in 1963. Studies show that ^3H isotope infiltrates in the same manner as water. Therefore, ^3H concentration in the soil profile provides useful data regarding groundwater recharge.

There are two methods of computing recharge by using tritium: (a) environmental tritium method (peak and integral methods), and (b) injected tritium method. In the environmental tritium peak method, recharge is estimated from equation (17.5)

$$R_p = \frac{100S_m}{P} \quad (17.5)$$

where R_p is as defined earlier, S_m is the total soil moisture in the soil column, from the surface up to the depth where the 1963 peak is located, and P is the total rainfall since 1963 to the time of investigation. In the integral method, the total amount of tritium in rain since 1952 to the time of investigation is determined and also the amount of tritium and moisture content present in the soil profile. Recharge as a percentage of rainfall R_p, is calculated from equation (17.6)

$$R_p = \frac{\Sigma a_x m_x}{\Sigma A_i P_i} \times 100 \quad (17.6)$$

where a_x is tritium concentration (TU) of soil at depth x; m_x is moisture content of soil at depth x (in cm); A_i is tritium concentration (TU) in precipitation; and P_i is precipitation (in cm) in different years since 1952.

Groundwater recharge can also be estimated by monitoring the vertical movement of the injected tritium. This is known as the **injected tritium** or 'tritium tagging' method. It has been used quite widely in India for estimating groundwater recharge in sedimentary as well as hard (fractured) rock formations (Table 17.1). This method is based on the assumption that the soil water in the unsaturated zone moves downward in discrete layers. Any freshwater added near the soil surface due to precipitation or irrigation will move downwards by pushing the older water beneath and this in turn will push still older water further below, thereby ultimately the water from the unsaturated zone is added to the groundwater reservoir. This flow mechanism is known as the piston flow model. Therefore, if tritium (^3H) is injected at any particular level, the vertical movement of this tagged layer can be monitored by measuring the concentration of ^3H at different depths. The position of the tracer will be indicated by a peak or maximum in the tritium activity versus depth plot. Molecular diffusion, dispersion, and other aquifer heterogeneities may cause broadening of the peak.

Hydrological balance method

Equation (17.2) as described earlier, has commonly been used for estimating groundwater recharge (R_p) when all other remaining parameters in the equation are known or computed. For example, this method was used for computing groundwater recharge in the weathered and fractured granodiorite and diorite in the Lee valley of California, USA (Kaehler and Hsieh, 1994).

Empirical method

Several empirical equations based on local rainfall characteristics, water-level fluctuation and hydrogeological parameters have also been proposed to obtain an approximate idea of groundwater recharge (R_p) from precipitation (Karanth, 1987). However, these are area specific and cannot be extrapolated to regions of differing hydrogeolgoical setups.

Recharge from canal seepage

The seepage losses from canals in alluvial areas can be computed from equation (17.7)

TABLE 17.1 Average annual groundwater recharge from rainfall as estimated by tritium injection method in some rock types in India.

Rock type and age	*Region*	*Av. annual rainfall (mm)*	*Recharge (R_p) as % of rainfall*	*Source*
Alluvium (Recent)	U.P.	990	22	Goel *et al.* (1975)
	Haryana	460	18	Goel *et al.* (1975)
Kamthi sandstone (Permian)	Andhra Pradesh	1430	15	Athavale (1985)
Cuddalore sandstone (Miocene)	Neyveli, Tamil Nadu	1200	12–13	Sukhija *et al.* (1996)
Basalts (Deccan Traps) (Cretaceous)	Godavari Basin, Andhra Pradesh	650	9.2	Athavale (1985)
Granite-gneiss and schist (Precambrian)	Vedavati Basin, Andhra Pradesh	560	6–2	Athavale (1985)
Granite gneiss (Precambrian)	Tamil Nadu	500–600	5–10	Athavale and Rangarajan (1988)

$$\text{Canal losses } (m^3 s^{-1} km^{-1}) = \frac{C}{200}(B+D)^{\frac{2}{3}} \qquad (17.7)$$

where B and D are the bed width and depth of the channel respectively in metres, and C is a constant, being 1.0 for intermittent running channels and 0.75 for continuously running channels.

Recharge from irrigated fields

The magnitude of recharge from irrigated fields depends on the source of the water (surface or groundwater), method of irrigation, type of crop and nature of the soil. When irrigation is from surface water, 35% to 40% of the applied water recharges the groundwater body while in the case of irrigation from groundwater sources, the return flow will be generally between 30% and 35%.

Other components

Other components of groundwater balance equation, viz. influent and effluent seepage from rivers and flow from/to the adjacent basins can be computed from Darcy's law as mentioned earlier.

Leakage across aquifer–aquitard boundaries can be estimated from equation (17.8) based on Darcy's law.

$$Q_v = \frac{K' \Delta h A}{b'} \qquad (17.8)$$

where Q_v is vertical leakage; K′v is the vertical hydraulic conductivity of the aquitard; Δh is the head difference between the two aquifers separated by aquitard of thickness *b′*, and A is the plan area of the aquifer through which the vertical leakage occurs. In leaky aquifers, Q_v forms an important component in recharge computations.

Recharge from tanks is estimated by the water balance equation considering surface inflow into the tank as input and outflow through sluices, evaporation and infiltration as output. ET losses are estimated as given in section 1.2.

Groundwater draft (T_p)

This occurs mainly through wells. Annual groundwater draft of a well is computed by multiplying its annual working hours and

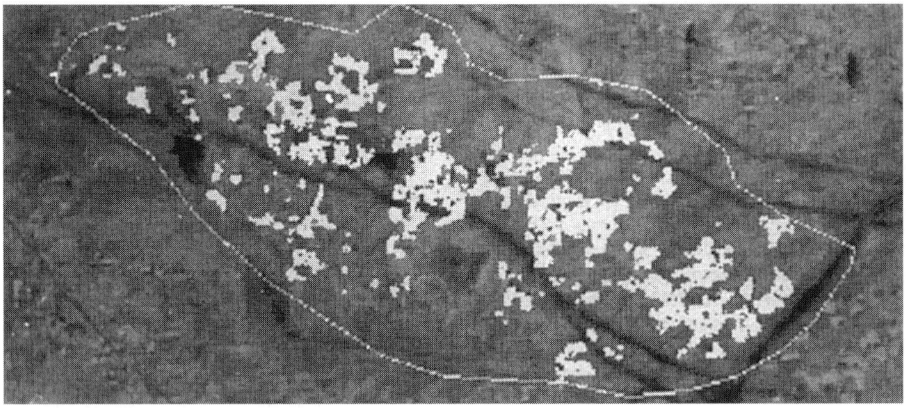

FIGURE 17.1 Processed IRS-LISS-II data (NDVI image superimposed over NIR band) showing areas (in white) irrigated by large diameter wells in Anantpur district, Central India. Such remote sensing data products along with crop inventories are useful for estimating groundwater draft from wells (after Meijerink *et al.*, 1994).

average discharge. The number of working hours can be calculated by the hourly consumption of electrical or diesel energy. The groundwater draft can also be estimated from data on irrigation water requirements of crops in the command area of the well. Remote sensing data can also be used in mapping the irrigated fields (Figure 17.1), where irrigation is only from groundwater. This information along with the cropping calendar and crop-water requirement can be used to estimate the groundwater draft.

Adhoc norms

In the absence of required data for computations of various components of groundwater balance, certain adhoc norms as given in Table 17.2 have been used for Indian conditions.

17.2 World's water resources

Several researchers have attempted assessment of global water resources, e.g. Nace (1971), Shiklomanov (1990) among others. As per the estimates of Shiklomanov (1990), about 96.5% of the total water resources, i.e. 1338 million km^3 occur in the oceans, and only 3.5% i.e. 47 million km^3 occur in the continents (Table 17.3). However, only 5 million km^3, which is about 10% of the total water on the continents, can be utilized as freshwater. The remaining freshwater is stored in glaciers and polar ice caps and as deep groundwater which is difficult to extract. Rivers and freshwater lakes, which are among the main sources of surface water supply, contain only 93×10^3 km^3 of water, which is only 1% of the total amount of fresh groundwater on the earth.

The global estimates do not show regional imbalances in the availability and requirement of water. Due to variations in climatic and hydrogeological conditions, water resources are very irregularly distributed throughout the world (Table 17.4). It is estimated that a few 'rich' countries enjoy more than 50% of the total world water resources. The water demand also varies due to diverse hydro-economic conditions and population density.

The overall availability of water resources of a country can be expressed by the annual or seasonal run-off per unit area or per unit habitant (Table 17.5). Therefore, per capita availability of water is decreasingly proportional to the increase in population. For example, while in 1950, the per capita availability of water for the whole world was about 33 300 m^3yr^{-1}, in 1983 it was 9500 m^3yr^{-1} and in 1990 it was only 8500 m^3yr^{-1}. Further, these values may not be very helpful in the assessment and management of water resources on a local or even

TABLE 17.2 Adhoc norms for recharge estimation (after CGWB, 1995a)

Recharge components	*Norms (%)*	*Recharge components*	*Norms*
Rainfall infiltration (% of normal annual rainfall)		**Canal seepage**	
a) Sandy	20–25	Unlined canals, normal type of soil	0.15 to 0.21 m^3d^{-1} per m^2 of wetted perimeter
b) Areas with higher clay content	10–20	Unlined canal, sandy soil	0.26 to 0.31
Friable, highly porous sandstones	10–15	Line canal	20% of the above values
Weathered and fractured granites	10–15	**Return seepage from irrigation fields**	
Unweathered granites	5–10	Surface water	35% of water delivered at the outlet. For paddy irrigation 40%
Vesicular and jointed basalt	10–15	Groundwater	30% of water delivered at the outlet. For paddy irrigation 35%
Weathered basalt	4–10	**Seepage from tanks**	44 to 60 m per year over the total water spread
Phyllites, limestones, quartzites, shales, etc.	3–10	**Influent seepage**	To be computed by Darcy's law Q = TIL

TABLE 17.3 Water storage on earth (after Shiklomanov, 1990)

Type of water	*Reference area ($km^2 \times 10^3$)*	*Volume ($km^3 \times 10^6$)*	*Share of world reserves (in %)*	
			Total water	*Fresh water*
World oceans	361 300	1338	96.5	–
Total groundwater	134 800	23	1.7	–
Fresh groundwater	134 800	10	0.76	30.1
Soil moisture	82 000	0.017	0.001	0.05
Glaciers and permanent snow pack	16 232	24.0	1.74	68.7
Underground ice in the permafrost zone	21 000	0.300	0.022	0.86
Water reserves in lakes	2058	0.176	0.013	–
Freshwater lakes	1236	0.091	0.007	0.260
Salt-water lakes	822	0.085	0.006	–
Swamps	2683	0.011	0.0008	0.03
Rivers	148 800	0.002	0.0002	0.006
Water in the biosphere	510 000	0.001	0.00007	0.003
Water in atmosphere	510 000	0.012	0.0009	0.04
Total water reserves	510 000	1385	100	–
Fresh water	148 800	35	2.53	100

TABLE 17.4 Water balance of the continents (mm yr^{-1}) (after Zektser and Loaiciga, 1993)

Water balance component	*Eurasia*	*Africa*	*North America*	*South America*	*Australia (without islands)*	*Earth's land surface*
Precipitation	728	686	670	1650	440	834
Evapotranspiration	430	547	383	1065	393	540
Total river run-off	298	139	287	585	47	294
Surface flow	298	91	203	375	40	204
Groundwater contribution	82	48	84	210	7	90

TABLE 17.5 Water availability in some countries in the world for 1983 (after Shiklomanov, 1990)

Country	*Area* ($km^2 \times 10^3$)	*Population* (10^6)	*Long-term mean annual streamflow* (km^3)	*Per unit area* ($\times 10^6\ m^3 m^{-2}$)	*Per capita* ($m^3 \times 10^3$)	*Percentage of world streamflow*
Brazil	8512	130	9230	1084	71	20.7
USSR	22274	275	4740	213	17	11
PR China	9561	1024	2550	267	2.5	5.7
Canada	9976	25	2470	248	99	5.6
India	3288	718	1680	511	2.3	3.8
USA	9363	234	1940	207	8.3	4.4
Norway	324	4	405	1250	99	0.9
Yugoslavia	256	23	256	1000	11	0.6
France	544	55	183	336	3.4	0.4
Finland	337	5	110	326	22	0.2
Total world*	134 800	4665	44 500	330	9.5	–

* Without Antarctica

regional scale as they do not account properly for: (a) seasonal and yearly variability of stream flow; (b) the interaction between surface water and groundwater; (c) water quality, and (d) per capita water use depending on social and economic factors (Dingman, 1994).

17.3 Groundwater resources of India

India has prominently an agricultural economy which makes a greater demand on water for irrigation with diverse topography, climate, geology and hydrological conditions. The distribution of available water resources (both surface and groundwater) in the country is highly uneven both in space and time. Rainfall is the main source of groundwater recharge which, being seasonal, is termed as **monsoon.** The distribution of the average annual rainfall on the Indian sub-continent is greatly variable, being maximum ($\simeq$ 250 cm) along the western '*ghats*' (coast) and the NE Himalaya while lowest (<30 cm) is in the western part of the country where arid conditions prevail (Figure 17.2).

The availability of groundwater is mainly controlled by rainfall distribution and the hydrogeological characteristics of the terrain. The distribution of various groundwater provinces is also shown in Figure 17.2 and their main hydrogeological characteristics, including yield potential, are outlined in Table 17.6. The vast alluvial plain in the northern part of India forms the most productive region where groundwater has been extensively used for irrigation. About two-

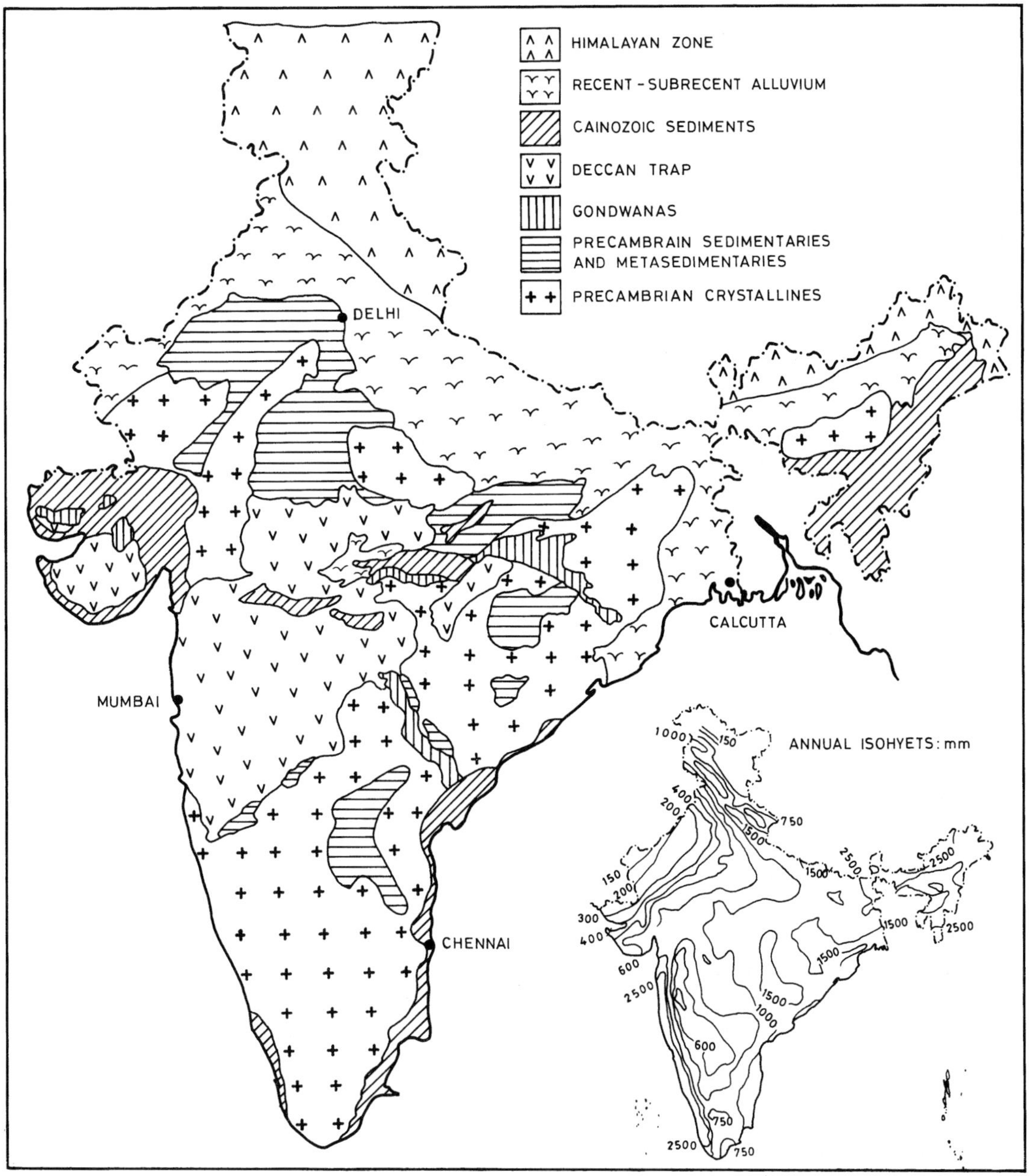

FIGURE 17.2 Groundwater provinces of India (modified after CGWB, 1989).

TABLE 17.6 Hydrogeological characteristics of major groundwater provinces of India

Groundwater Provinces/ Regions	*Location*	*Lithology (age)*	*Main features of aquifers*	*Hydraulaic properties* $T(m^2d^{-1})$	*Yield potential* (m^3s^{-1})
Indo-Gangetic alluvium	Northern India covering an area of more than 1 million km^2	Alluvium with beds of sand, silt and clay with occasional beds of gravel, max. thickness more than 2000 m (Sub-Recent to Recent)	Shalalow aquifers unconfined, deeper ones confined or of leaky type	T = 1000–5000 S = 10^{-4}–10^{-3}	0.04–0.11
Cainozoic sedimentary basins	Eastern coast, NE India, and western parts	Unconsolidated to semi-consolidated sandstones, shales and limestones	Deeper aquifer under confined condition; at places flowing wells	T = 500–5000 S = 10^{-5}–10^{-3}	0.01–0.04
Deccan Traps	Central and western India covering an area of about 500 000 km^2.	Basaltic lava flows generally flat lying, maximum thickness about 1500 m in the western coast (Upper Cretaceous to Lower Eocene)	Main source of groundwater are (a) weathered and fractured horizon (b) interflow spaces, (c) intertrappeans (d) vesicular horizons	T = 10–700 S = 10^{-3}–10^{-1}	0.001–0.03
Gondwana Province	Structurally controlled basins mainly in central India	Semi-consolidated sandstones and shales with coal seams (Carboniferous to Lower Cretaceous)	Shallow aquifers unconfined, deeper ones confined	T = 50–500 S = 10^{-3}–10^{-1}	0.01–0.10
Precambrian sedimentary basins	Four discrete structural basins (a) Cuddapah (b) Raipur (c) Vindhyan (d) Western Rajasthan	Consolidated sandstones, shales and limestones (Proterozoic)	Intergranular porosity low; fractures are the main source of water in sandstone and shales; solution cavities in limestones	T = 5–500 S = 10^{-3}–10^{-2}	0.01–0.04
Precambrian crystalline province	It occupies nearly half of the country in Central and South India	Crystalline rocks, viz. granites, gneisses and schists (Precambrian)	Weathered mantle (regolith) is the main source of water supply; fractures and lineaments also facilitate groundwater movement	T = 5=50 S = 10^{-3}–10^{-2}	0.001–0.005

thirds of the surface area of the country is covered with hard (crystalline and basaltic) rocks where the availability of groundwater is limited.

In the past, several attempts have been made to assess the total utilizable water resources of India. Out of a total of 1140 km^3 of annual utilizable water, the utilization in 1990 was 572 km^3 which comprised 382 km^3 from surface water and 190 km^3 from groundwater. With an increasing population, which may cross the one billion mark by the end of the present century, and pace of development, there will be a considerable increase in water requirements for various basic amenities and developmental purposes like drinking and municipal water supplies, irrigation, thermal and hydropower generation, industrial requirement, navigation and maintenance of ecology. Accordingly, the projected demand of water for various purposes by 2000AD is estimated at 750 km^3 comprising 500 km^3 of surface water and 250 km^3 of groundwater (Table 17.7). Water scarcity conditions are predicted in the next 25–30 years due to massive demands for irrigation and drinking purposes. Table 17.7 indicates that the entire utilizable water resources of the country would be required to be put to use by the year 2025AD.

Even at present there are indications of over-exploitation of groundwater as manifested by the lowering of the water-table both in the alluvial aquifers of north India and in the crystalline rocks of south India. In some areas the situation is alarming as dug wells have dried up causing great misery to farmers (Figure 17.3). This needs necessary water conservation methods by adopting efficient methods of irrigation and avoiding wasteful use of water.

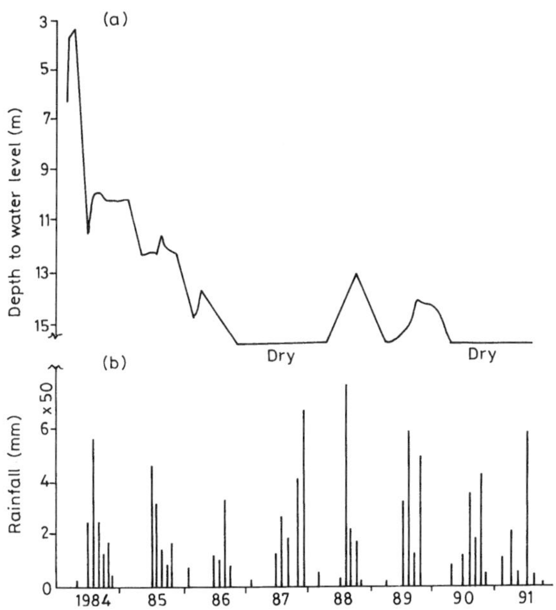

FIGURE 17.3 (a) Hydrograph of an observation well in the granite aquifer in Nalgonda district, South India. (b) Rainfall records note that although rainfall pattern is unchanged, there is a considerable decline in water-level. (After Reddy and Raj, 1997.)

TABLE 17.7 Estimates of water resources of India (after IWRS, 1996)

(a)	**For the year 1990**	
1.	Annual precipitation	= 4000 km^3
2.	Average available surface water	= 1870 km^3
3.	Utilizable surface water	= 690 km^3
4.	Replenishable groundwater	= 450 km^3
	Total Utilizable water	= 1140 km^3
(b)	**Water Utilization in 1990**	
1.	Irrigation	= 460 km^3
2.	Domestic	= 25 km^3
3.	Industries	= 15 km^3
4.	Energy	= 19 km^3
5.	Other	= 53 km^3
	Total	= 572 km^3
	Out of this 572 km^3, 382 km^3 is from surface water and 190 km^3 from groundwater	
(c)	**Projected demands for the year AD 2000**	
	Surface water	= 500 km^3
	Groundwater	= 250 km^3
	Total	= 750 km^3
(d)	**Projections of water utilization for the year AD 2025**	
1.	From surface water	= 700 km^3
2.	From groundwater	= 350 km^3
	This includes	
(a)	Irrigation	= 770 km^3
(b)	Domestic	= 52 km^3
(c)	Industries	= 120 km^3
(d)	Energy	= 71 km^3
(e)	Other uses	= 37 km^3
	Total (a) to (e)	= 1050 km^3

TABLE 17.8 Common causes of groundwater level fluctuations (modified after Freeze and Cherry, 1979)

Causes of fluctuations	*Unconfined*	*Confined*	*Natural*	*Man-induced*	*Short-lived*	*Diurnal*	*Seasonal/ long-term*
Groundwater recharge	✔	✔	✔				✔
Evapotranspiration	✔		✔			✔	
Tidal effects near oceans	✔	✔	✔			✔	
Atmospheric pressure effects	✔	✔	✔			✔	
External loading		✔		✔	✔		
Earthquakes		✔	✔		✔		
Nuclear explosion		✔		✔	✔		
Groundwater pumpage	✔	✔		✔			✔
Irrigation and drainage	✔			✔			✔
Urbanization	✔	✔		✔			✔

17.4 Groundwater level fluctuations

Fluctuations in groundwater levels indicate changes in groundwater storage due to meteorological factors and other environmental impacts (Table 17.8). A continuous decline of water-levels is observed in areas where there is a greater withdrawal of groundwater than recharge, or continuous drought for several years. On the other hand, a rising trend in water-levels points towards excessive recharge and less withdrawal. A decreasing trend in water-level can cause ground subsidence, especially in unconsolidated formations, while a rising trend is responsible for water logging, and water and soil salinization. The main causes of groundwater level fluctuation are as follows.

Recharge from rainfall and snowmelt

Water-level in wells, both in unconfined and confined aquifers, responds to infiltration due to rainfall and snowmelt. Fluctuations due to snowmelt and rainfall are of a seasonal nature. In north Europe and America, the highest water-levels are observed in late spring and lowest in winter. In India where the main source of recharge is the monsoon rainfall, the highest water-levels are observed in September–October and lowest in May–June.

Losses due to evapotranspiration

In shallow unconfined aquifers, where the water-table is close to the ground surface, the water-level in wells fluctuates in response to losses due to evaporation and transpiration. The loss due to evaporation from the water-table depends on its depth below the ground surface and the soil structure which controls the capillary rise. For the same depth of water table, losses due to evaporation will be greater in fine-grained soils due to a higher capillary rise, as compared with coarse-grained material. The rate of transpiration also depends on the type of vegetation. In the shallow water-table condition where the plant roots draw water directly from the zone of saturation, the fluctuation is due to direct losses from the water-table due to evapotranspiration. Lowering of the water-table during the day is a result of evapotranspiration, and recovery during the night is due to the absence of such losses. These aspects have been discussed in Chapter One.

Atmospheric pressure

Changes in atmospheric pressure can produce sizeable fluctuations of water-levels in wells tapping confined aquifers. The relationship is inverse, i.e. an increase in atmospheric pressure causes lowering of water-levels. This is explained due to the elastic behaviour of the confined aquifer (Todd, 1980). The barometric efficiency of the aquifer, B is expressed as

$$B = \frac{s_w}{s_p} \tag{17.9}$$

where s_w is the change in the water level and s_p is the change in barometric pressure expressed in terms of a column of water. B is usually in the range of 0.20–0.75.

Wind

Wind and cyclones cause reduction in pressure which have a similar effect as a vacuum pump causing a sudden reduction of air pressure within the well; this results in a quick rise of water-level. After the storm passes, the air pressure in the well will rise and the water-level will accordingly be lowered.

Ocean tides

In coastal aquifers and near some lakes and streams, water-levels in wells show semidiurnal fluctuations in response to tides. In an unconfined aquifer, the response is due to direct entry of sea water into the aquifer. However, in confined aquifers, which are separated from the ocean water by an impervious confining layer, the response in groundwater levels is due to the change in the load on the top of the aquifer due to the weight of the column of ocean water. A rise in the tide will cause additional loading resulting in the compression of the aquifer, and thereby rise in the piezometric surface. This is contrary to the effect and changes in barometric pressure. The ratio of the change in piezometric level, s_w to change in height of tide, s_t is known as the tidal efficiency (C).

$$C = \frac{s_w}{s_t} \tag{17.10}$$

Equations (17.11) and (17.12) relate the tidal efficiency (C) and barometric efficiency (B) to the elasticity of a confined aquifer respectively (Ferris *et al.*, 1962).

$$C = \frac{\alpha/\eta\beta}{1 + \alpha/\eta\beta} \tag{17.11}$$

and

$$B = \frac{1}{1 + \alpha/\eta\beta} \tag{17.12}$$

where α is the bulk modulus of compression of the solid skeleton of the aquifer, β is the bulk modulus of compression of water and η is the porosity of the aquifer. The sum of the barometric and tidal efficiencies equals unity (i.e. $B + C = 1$). The tidal efficiency is a measure of the incompetence of overlying confining layers to resist pressure changes due to loading.

Earthquakes

Earthquakes can lead to small short-lived fluctuations of water-levels in wells penetrating confined aquifers. These fluctuations are caused by the compression and expansion (dilation) of elastic confined aquifers due to the passage of seismic waves. In order to adjust to the pressure changes, the water level in the well tapping a confined aquifer first rises and then falls. The amounts of rise and fall of the water-level, with respect to the initial position, are approximately the same. However, in cases where the earthquake has caused deformation of the aquifer, water-levels may not return to the original position. Wells in limestone aquifers show greater fluctuations than wells in unconsolidated granular material (Ferris *et al.*, 1962).

As the average velocity of seismic (Rayleigh) waves is approximately 4000 m s^{-1}, the effect of the earthquake on groundwater levels will be manifested with some time lag at places away from earthquake centres. The large magnitude earthquake can cause widespread fluctuations in water-levels. The Alaskan earthquake of March

27, 1964, (magnitude 8.4 to 8.8 on the Richter scale) produced water-level fluctuations all over North America, ranging from about 2.0 to 7.5 m in wells tapping confined aquifers in Pleistocene and late Tertiary strata. None of these wells showed full recovery, suggesting changes in the physical structure of the aquifer and an increase in pore space due to rearrangement of grains or fractures as a result of horizontal extension and/or elastic dilation (Plafker, 1969).

Underground nuclear testing

Underground nuclear tests are carried out in either vertical shafts or horizontal tunnels usually in hard rocks (igneous and metamorphic) in the zone of aeration. In areas where the water-table is shallow, the testing is done below the water-table. Abrupt rise in groundwater levels are reported due to detonations of nuclear, and also high energy chemical explosions at several sites in China, the USA, Canada, Russia (Kazakhstan) and several places in Europe (Matzko, 1994). This is explained as being a result of increase in pore water pressure on account of closure of fractures, as the shock waves pass through the dense and low porosity rocks. As the pressure in the explosion cavity is disseminated quickly (in less than 1 min), the rise is followed by a quick drop in water-level, probably due to the filling of the collapsed cavity. It may take about 10 months for the re-establishment of the original water-level due to regional groundwater recharge (Matzko, 1994).

Charlie *et al.* (1994) reported maximum rise of groundwater levels to be 4.8, 0.8, 0.04 and 0.02 m at distances of 650, 2170, 2550 and 3400 m from the detonation site. A 17 m rise in water-level was recorded at a distance of 307 m from a nuclear test in Nevada, USA, which had an explosive yield of 30 kt. Therefore, the yield of the nuclear test will govern the magnitude of water-level fluctuation and the maximum distance up to which such changes will be noticed (Figure 17.4). This relationship has a potential to determine the yield of underground nuclear detonation from water-level data (Charlie *et al.*, 1994).

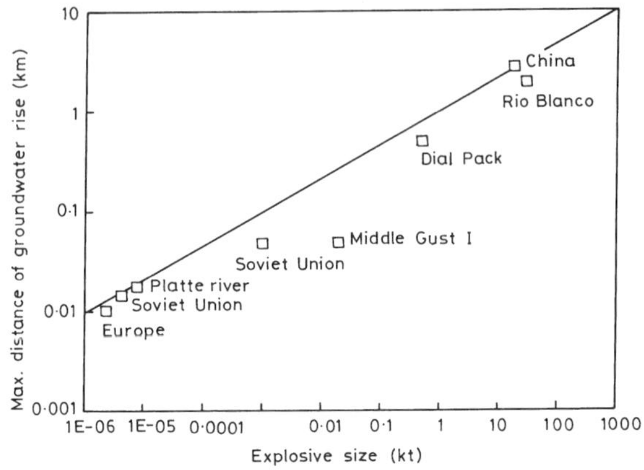

FIGURE 17.4 Relation between nuclear explosion size and distance of maximum groundwater rise (after Charlie *et al.*, 1994). (Reprinted from *Engineering Geology*, Vol. 38, 1994, with kind permission of Elsevier Science-NL, Amsterdam, the Netherlands).

Urbanization

Urbanization reduces the infiltration due to the construction of buildings, impervious pavements and storm sewers. This, coupled with increased withdrawal of groundwater for various uses, causes a water-level decline. For example, in Calcutta city, water-levels have declined by 4–8 m between 1958 and 1992 due to excessive pumpage (Figure 17.5). In several other metropolitan cities worldwide, decline of groundwater levels has resulted in land subsidence (section 17.5). In coastal areas, decline of groundwater level due to excessive pumpage has also led to seawater intrusion as described in section 17.6.

In the UK, during the late 1800s and early 1900s, due to the heavy withdrawal of groundwater from the Cretaceous chalk formations in the London Basin and Permo-Triassic sandstones in Birmingham, Liverpool, Manchester and Nottingham areas, the groundwater levels declined by some tens of metres below ground

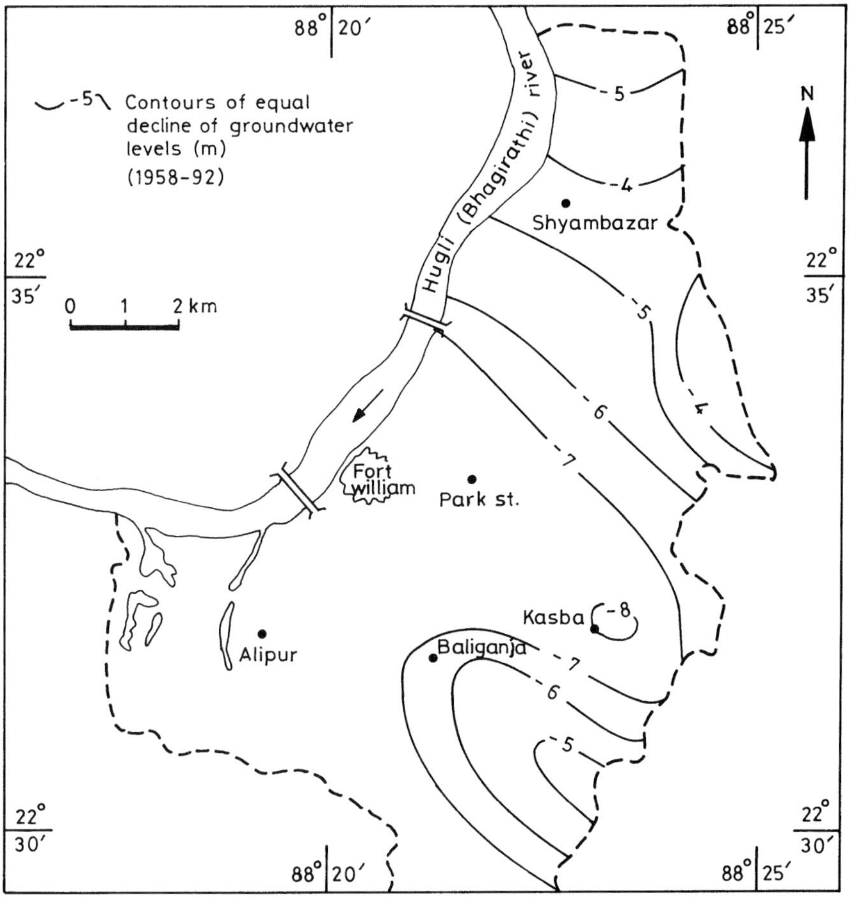

FIGURE 17.5 Decline of groundwater levels between 1958–92 in the alluvial aquifer below the city of Calcutta, India (redrawn after Goswami, 1995).

level over large areas. This resulted in the decline of water-levels in earlier flowing wells which then ceased to flow. However, in the last 20–25 yr, a reverse trend resulting in a rise of water levels at the rate of about 1 m yr^{-1} is reported from these cities in the UK (Figure 17.6). This is attributed to the reduction in groundwater abstraction and additional recharge from sewers, lawns and leaking pipelines. The rise in groundwater levels has posed problems of flooding of basements, rail tunnels and stability of other underground structures (Wilkinson and Brassington, 1991). Similarly, in Louisville, Kentucky, USA, increasing groundwater levels have caused problems of structural settlement, drainage to basement floors and disruption of underground tunnels (Hamill and Bell, 1986). Preventive measures would include control of groundwater levels by pumping or carrying out remedial engineering works. The disposal of pumped water is likely to pose a problem in heavily urbanized areas like London. It would also be necessary to take necessary care to avoid over-pumping, as this may then lead to problems of subsidence and deterioration in water quality. Therefore, long-term management of the groundwater resources is necessary.

Perennial irrigation

Excessive seepage losses from unlined irrigation channels and irrigated fields have caused a

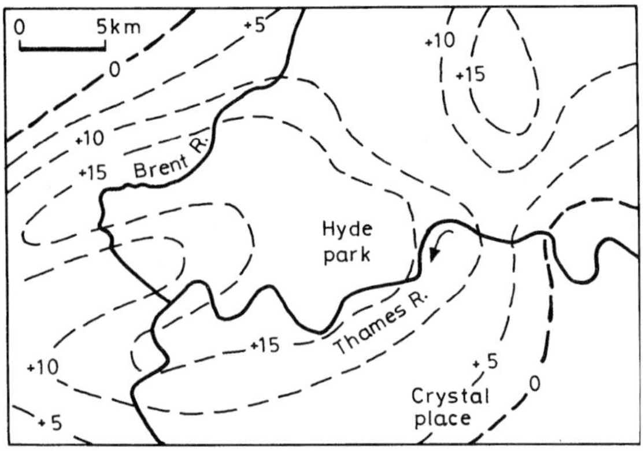

FIGURE 17.6 Rise of groundwater levels in the Chalk aquifer below London (1985–90) (all contours in metres) (after Marsh and Davies, 1983).

conspicuous rise in groundwater levels, particularly in semi-arid parts of the world, viz. in Pakistan, India, China, and Egypt. This has resulted in waterlogging and soil salinization thereby reducing the soil fertility. In Pakistan, the rise in the water-table in Rechna '*doab*' (interfluve), in between rivers Ravi and Chenab, was as much as 28 m from the pre-irrigation period (1900) to the post-irrigation period (1960) (Figure 17.7). In India, the total extent of the waterlogged area is estimated to be about 2.5 million ha which is about 10% of the total irrigated area (Tyagi, 1994). In China, waterlogging is prevalent in the alluvial plain of the Yellow River in the northern part of the country, which has a semi-arid climate (Chen and Cai, 1995). In Egypt, about 8400 km^2 or 30% of the total arable land is salt-affected or waterlogged (Shata, 1982). Reclamation of waterlogged areas by surface and subsurface drainage is one of the top priorities in several countries, worldwide, for augmenting agricultural production.

17.5 Land subsidence

Over-abstraction of groundwater is a common cause of land subsidence. Land subsidence is also a result of excessive withdrawal of other fluids, e.g. oil, gas and geothermal fluids. Heavy

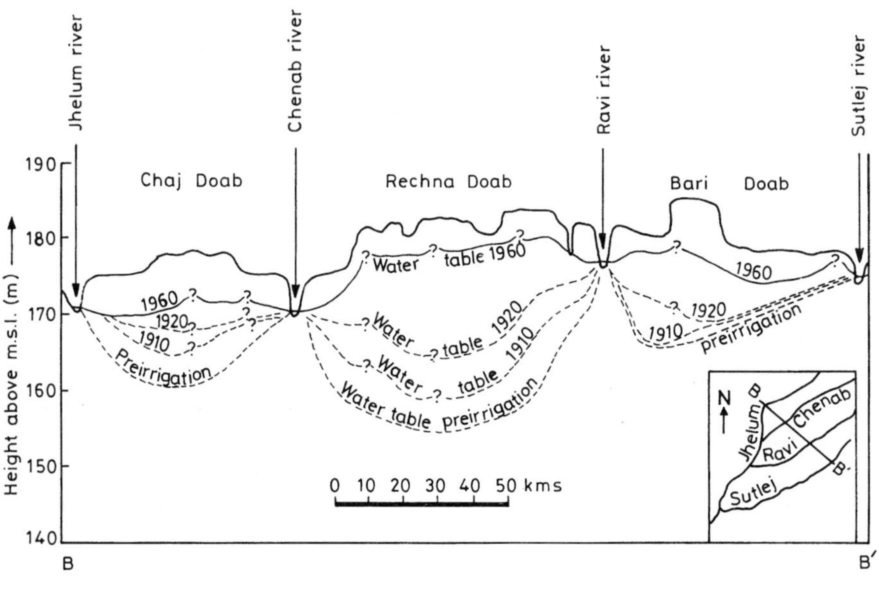

FIGURE 17.7 Profile across the northern part of Pakistan showing rise in groundwater levels due to irrigation in various interfluves ('doabs') of the Punjab alluvial plain (after Bennett *et al.*, 1967).

withdrawal of groundwater and other fluids causes reduction in pore-water pressure, and thereby an increase in the effective weight of the overburden. The relationship is given by the equation (17.13)

$$\sigma' = \sigma - p_w \qquad (17.13)$$

where σ is the total stress, σ' is the effective stress and p_w is the pore-water pressure. Therefore, a reduction in pore-water pressure results in the compaction of both the aquifer and the interbedded clay layers. The total potential compaction of the confining layer (clay) is much greater compared with that of sand, as clay has greater compressibility. However, due to the lower hydraulic conductivity of clay, the rate of compaction of clay is slower than that of a sand aquifer. Therefore, although the compaction of a sand aquifer is immediate and elastic, that of clay is much greater but inelastic. The amount of compaction also depends on the clay mineralogy, being about three times greater in montmorillonite as compared with kaolinite, while illite has intermediate values (Waltham, 1989). There is also a time lag between the withdrawal of fluid and the start of subsidence.

The main effects of land subsidence are: (a) inundation of coastal areas; (b) modification of surface drainage gradients; (c) damage to well casings; (d) activation of faults and opening of fissures; and (e) damage to buildings and other surface structures.

There are several examples of land subsidence due to heavy withdrawal of groundwater, worldwide. One of the worst examples is from Santa Clara Valley in California, USA. Other well-known examples are from several mega cities, viz. Tokyo and Osaka in Japan, Bangkok in Thailand, Venice in Italy and Mexico City in Mexico. In Mexico City, which is located on alluvial fan and lacustrine deposits, heavy withdrawal of groundwater caused land subsidence of about 9 m. In China, overpumping of groundwater in Beijing, Tianjin and several other cities on the North China Plain has caused significant drawdown which at places has exceeded 50 m resulting in land subsidence. In Beijing, the pumping of groundwater has increased by more than 50% during the last 10 years causing a head decline of 20 m and land subsidence of up to 59 cm. In Shanghai, the sinking rate is reported to be as much as 24 cm annually during 1949 to 1965, the maximum subsidence having been 2.63 m (Xuanjiang, 1994). Large-scale extraction of groundwater in the Xian city area, which was the ancient capital of China, has caused the lowering of the groundwater level resulting in ground subsidence (Figure 17.8). Subsidence has also caused surface rupturing and development of ground fissures damaging buildings and other man-made structures (Lee *et al.*, 1996). In Venice, Italy, subsidence has resulted in an inundation of large areas by sea water.

In some areas, large-scale pumping of groundwater is also required to enable mining by opencast method, e.g. of lignite. This has also resulted in land subsidence, as reported from some areas in Germany and Hungary (Briechie and Voigt, 1980; Alfoldi, 1980). Extraction of oil and gas can also cause surface subsidence. One of the earliest examples is from the sinking of Gaillard Peninsula in Texas, USA, where the ground surface sank below the surface of the bay in 1920 as a result of extraction of oil. In Japan, heavy withdrawal of gas and brine from poorly consolidated sediments in the coastal city of Niigata caused sinking of a part of the city below the sea level in 1961 (Marsden and Davis, 1967). Subsidence from geothermal fields is described in section 16.5.

As the main cause of land subsidence is the overexploitation of groundwater, the best remedial measures will be to reduce groundwater withdrawal and also adopt some methods of artificial recharge. These measures have proved effective in many areas. For example, in Mexico City the rate of subsidence was reduced from

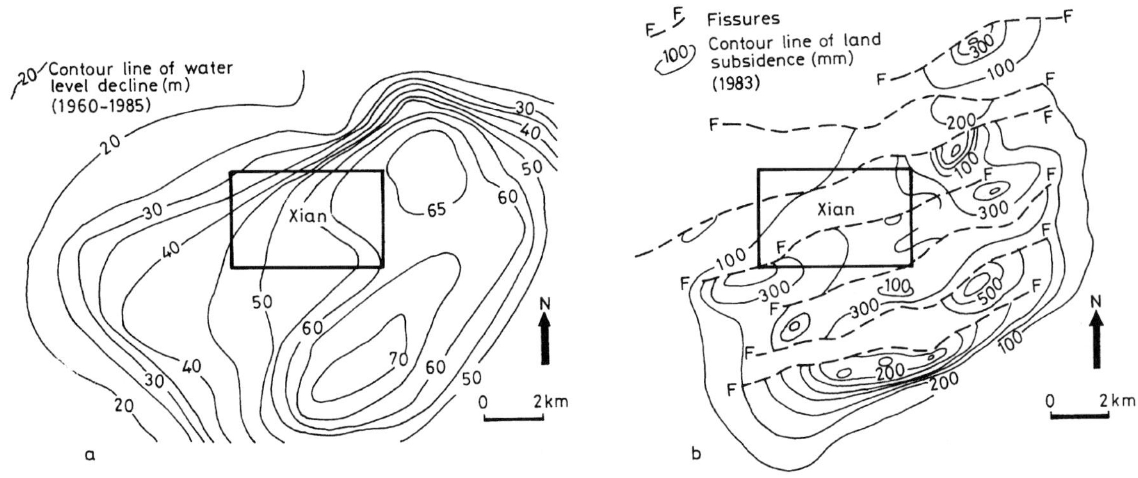

FIGURE 17.8 (a) Decline of groundwater levels (1960–85) in the area around the city of Xian, PR China. (b) Subsidence of the land surface in the same area (after Lee *et al.*, 1996, reprinted from *Engineering Geology* with kind permission of Elsevier Science-NL, Amsterdam, the Netherlands).

900 mm yr^{-1} to around 30 mm yr^{-1} (Waltham, 1989). In China, reduction of groundwater abstraction and artificial recharge reduced annual subsidence in Shanghai from 110 mm in 1961 to 1.5 mm in 1992. Similarly, it was reduced in Tianjin from 86 mm in 1985 to 17 mm in 1991 (Chen and Cai, 1995). Imported water supply will also reduce groundwater demand and thereby reduce subsidence. In view of this, interbasin water transfer can be beneficial (section 17.8.3).

17.6 Fresh–sea-water relationship in coastal areas

The coastal aquifers in different parts of the world form important sources of water supply. Some of the coastal areas are heavily urbanized leading to a greater demand for groundwater.

17.6.1 Freshwater discharge to the sea

Usually, the groundwater gradient is seaward and so a large amount of fresh groundwater is discharged into the sea as (1) submarine springs, and (2) leakage through the aquifers and semi-confining layers exposed in the sea bottom (Figure 17.9). Submarine springs are common in the karst formations e.g., bordering the Adriatic Sea as discussed in section 13.3. Springs are also widely distributed on the submarine slope of volcanic islands with mountainous topography, e.g. Hawaii and the Philippines. The interaction of groundwater seepage with the ocean water and sediments is also of importance for the formation of nodules of iron, manganese and other elements.

The freshwater outflow to the sea is an important resource which should be assessed and exploited judiciously. Attempts are therefore made to estimate groundwater discharge in different parts of the world. Based on hydrogeological data, Zektser and Dzhamalov (1988) and Zektser and Loaiciga (1993) computed the total discharge of groundwater to oceans to be 2400 km^3yr^{-1} which includes 1485 km^3yr^{-1} from the continents and 915 km^3yr^{-1} from islands. In a recent study, Moore (1996) – based on ^{226}Ra content of coastal waters and sediments in the southeastern part of the Atlantic

FIGURE 17.9 Freshwater flux (surface and subsurface) into seawater along the interface of the Po river plains and the Adriatic sea. Note that the interface extends for a width of about 100 km. (HCMM night thermal IR image dated 25 November 1978; Courtesy NASA).

coast – estimated groundwater flux to oceans to be about 40% of river water flux. Other studies also show that groundwater flux to oceans is in the range of 0.01–0.1 md^{-1} which is about 10% of the surface water flux. Therefore, in order to avoid wasteful discharge of freshwater into the sea, it is necessary to identify such areas for optimum groundwater development without disturbing the fresh–sea-water balance adversely as discussed below.

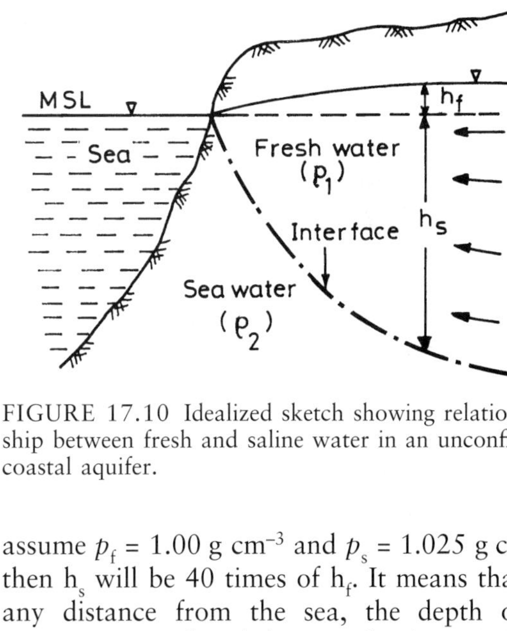

FIGURE 17.10 Idealized sketch showing relationship between fresh and saline water in an unconfined coastal aquifer.

17.6.2 Sea-water intrusion

The relation between freshwater and sea water in coastal aquifers was first given independently by B. Ghyben and B. Herzberg (in Todd, 1980) (equation 17.14)

$$h_s = \frac{\rho_f}{\rho_s - \rho_f} h_f \qquad (17.14)$$

where h_f = height of water-table above m.s.1; h_s = depth of the fresh–sea-water interface below the m.s.1; ρ_s = density of sea water, and ρ_f = density of freshwater (Figure 17.10). If we assume p_f = 1.00 g cm^{-3} and p_s = 1.025 g cm^{-3}, then h_s will be 40 times of h_f. It means that at any distance from the sea, the depth of a stationary interface below sea level is 40 times the height of the freshwater table above sea level.

As the sea water is not stationary and is moving inland, the head in the sea-water body in the aquifer will be less than the mean sea level.

For such a non-equilibrium condition, where both the saline and freshwater are in motion, the depth of interface is given by Lusczynski (1961) as

$$h_s = \frac{\rho_f}{\rho_s - \rho_f} h_f \frac{\rho_f}{\rho_s - \rho_f} H_s \qquad (17.15)$$

where H_s is the head difference in the sea-water wedge and m.s.l. (Figure 17.11). If H_s is zero, then equation (17.15) reduces to equation (17.14). As per this dynamic concept, which is actually the case in nature, the depth of fresh–sea-water interface will be greater than that given by the Ghyben–Herzberg relationship. In a confined aquifer (Figure 17.12), the equilibrium position of the intruded sea-water wedge is given by

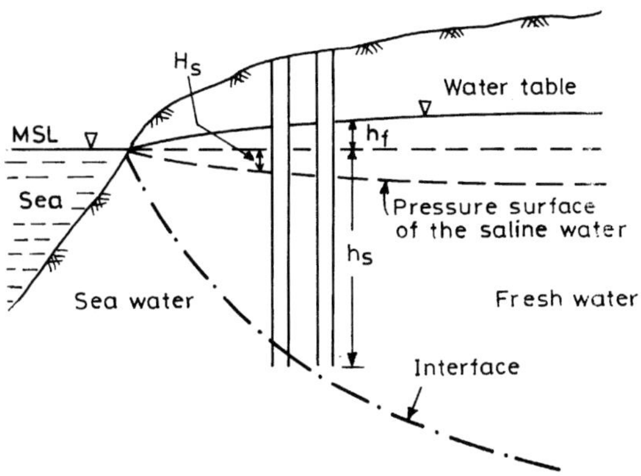

FIGURE 17.11 Nonequilibrium conditions in fresh and saline water in an unconfined coastal aquifer.

$$L = \frac{1}{2} \frac{\rho_s - \rho_f}{\rho_f} \frac{Kb^2}{q} \qquad (17.16)$$

where L is the length of the sea-water wedge, K is the hydraulic conductivity, b is the saturated thickness of the aquifer and q is the seaward freshwater flow per unit width of the ocean front. Equation (17.16) indicates that the length of the sea-water wedge (L) has the following characteristics: (a) it is directly proportional to the hydraulic conductivity of the aquifer; therefore the most permeable formations, like some of the volcanic rocks, karstic formations and fractured horizons, will be more affected by sea-water intrusion; (b) it is directly related to the square of the aquifer thickness, b, i.e. in thick aquifers and old river channels, filled with gravel and other coarse material, the salt-water wedge will penetrate to longer distances; (c) the wedge length, L is inversely proportional to the freshwater flow to the sea, q, therefore greater penetration of the wedge is expected in areas with less recharge, such as in arid and semi-arid climates, or where groundwater basins are of a smaller size; (d) heavy withdrawal of groundwater will also reduce freshwater outflow, q, to the sea, thereby increasing the length of the sea-water wedge. Thus, by pumping from a coastal confined

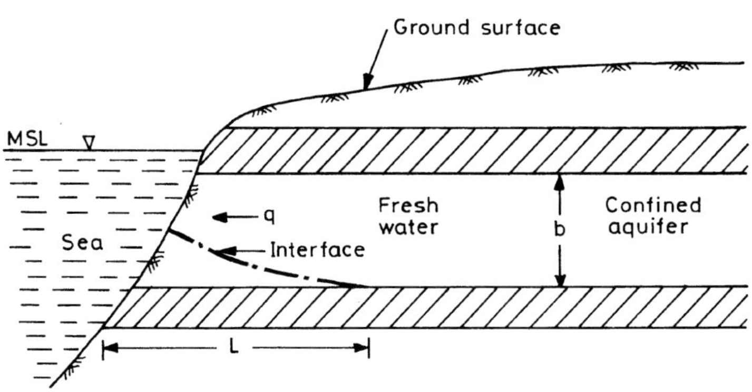

FIGURE 17.12 Seawater-wedge in a confined coastal aquifer.

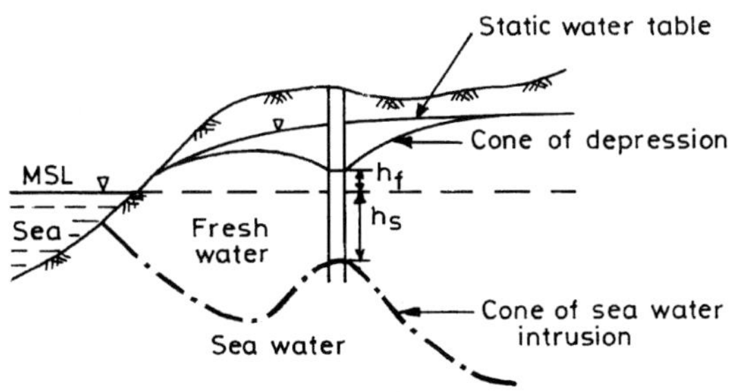

FIGURE 17.13 Formation of cone of sea-water intrusion due to pumping in an unconfined coastal aquifer.

aquifer, the freshwater flow, q will decline and the sea-water wedge will advance inland. This phenomenon is known as sea-water encroachment or intrusion. In an unconfined aquifer, due to the lowering of the water-table as a result of overpumping, sea-water intrusion can also take place (Figure 17.13). It is therefore necessary to properly plan the groundwater exploitation in coastal areas to avoid sea-water encroachment.

In multilayered coastal aquifers exposed on the ocean floor, the fresh–sea-water wedge will develop in each aquifer separately depending on the piezometric head in each aquifer, its thickness, hydraulic conductivity, and recharge. The other important factor which will influence the length of the wedge is the vertical leakage through the adjacent aquifers. Usually the interface in the lower aquifer is relatively longer than the upper one (Figure 17.14).

The aquifer geometry and fresh–sea-water relationship at Digha on the eastern coast of India is illustrated in Figure 17.15. Electrical resistivity survey and drilling indicated two saline water zones separated by a freshwater aquifer due to clay layers within a depth of 170 m. The shallow saline water (20–30 m) appears to be of connate origin as indicated by hydrochemical and isotopic data (Singhal, 1963; Shivanna *et al.*, 1993).

As the freshwater and sea water are miscible fluids, the zone of contact between them is of a transitional type. It is characterized by the phenomena of molecular diffusion and especially hydrodynamic dispersion. The thickness of the interface depends on fluctuations of tides and groundwater level, aquifer permeability, climate and recharge and discharge from wells (Table 17.9). Sea-water intrusion in coastal areas can be monitored from groundwater level and water quality data (Figure 17.16).

Sea-water encroachment in coastal aquifers can be prevented and controlled by adopting certain measures, e.g. artificial recharge, subsurface barriers, modification of pumping pattern and suitable design of wells and other abstraction works (Todd, 1980; UNESCO, 1987). In oceanic islands, radial wells (Maui

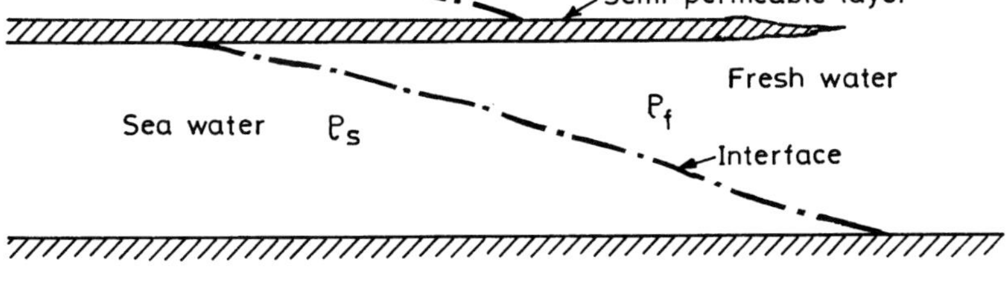

FIGURE 17.14 Presence of a semi-permeable layer leading to development of two fresh–sea-water interfaces in a coastal aquifer (after Reilly, 1990).

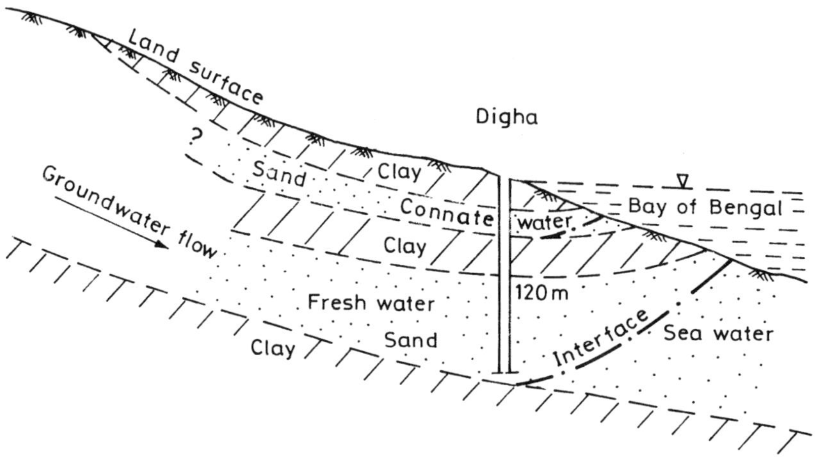

FIGURE 17.15 Schematic hydrogeological profile across the coastal town of Digha, West Bengal, India, showing fresh–sea-water relationship (simplified after Goswami, 1995).

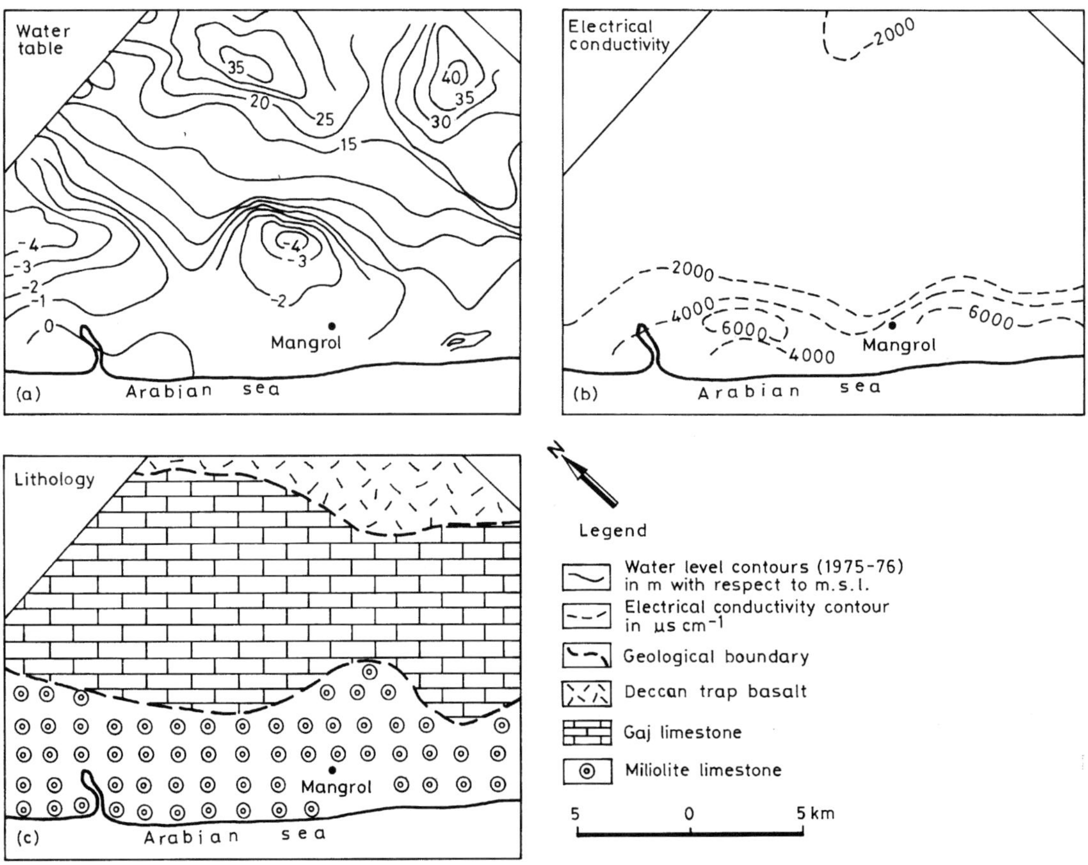

FIGURE 17.16 (a) Water-table, and (b) electrical conductivity (EC) contours of groundwater in the coastal part of Saurashtra, Western India (1975–76); (c) geological map of the area. Note that in the coastal belt, occupied by Miliolite limestone, the water-table gradients are inland, due to overpumping, resulting in an increase in EC caused by sea-water encroachment (redrawn after Desai *et al.*, 1979).

TABLE 17.9 Factors affecting thickness of fresh-water-sea water interface in coastal aquifers

Factors	*Nature of transition zone*	*Example*
Tidal activity		
Negligible	Narrow	Mediterranean area
Intense	Thick	Atlantic area
Hydraulic conductivity		
High	Thick	Hawaii Islands
Low	Narrow	
Climate		
Semi-arid	Thick	Israel
Humid	Narrow	Eastern coast of India

type), tunnels, galleries and shafts are more suitable depending on topography and rock type to avoid sea water intrusion (see section 12.5). An example of numerical modelling of fresh–sea-water relationship from a volcanic island is given in section 17.7.6.

17.7 Groundwater modelling techniques

17.7.1 Introduction

Groundwater modelling has been established as a useful tool for studying various problems related to quantity and quality of groundwater. This technique has been used mainly as an aid for: (i) prediction of the effects of probable stresses (e.g. groundwater withdrawal or injection of a contaminant) on the groundwater system, and (ii) management of these stresses for hazard-free exploitation of groundwater resources. Models also provide a useful input to the identification and estimation of aquifer parameters, designing of data collection network for the system quantification, economic and social management of groundwater, etc. Earlier, analogue methods like sand tank model and electric analogue models were commonly used for simulating a groundwater system. However, with the advent of powerful and versatile computers and development of efficient software, computer modelling has almost entirely replaced other modelling techniques. Presently, available computer codes for groundwater modelling have facilities not only for predicting but also visualizing the dynamic behaviour of groundwater flow and pollutant migration. Once a validated model is constructed, it can be used for selecting best management options by studying the aquifer response to a number of scenarios.

Before dealing with the subject further, an appreciation of the limitations of models may also be necessary. In spite of all the sophistication in hardware and software, several simplifying assumptions are inescapable in the estimation of various parameters characterizing a groundwater system, as well as during its conceptualization and modelling. Under these constraints, even the most carefully constructed model can, at best, be an approximate representation of the reality and not reality itself. Obviously, therefore, a model can never be perfectly deterministic for all purposes. But it can certainly be useful for a particular objective provided it captures most of the relevant features of a real system.

The salient work elements of groundwater modelling are:

(a) conceptualization of the system,
(b) mathematical description of the process to be simulated,
(c) solution of mathematical equations by different methods,
(d) model calibration and refinement, and
(e) prognosis for some probable scenarios.

A flow diagram of these activities is shown in Figure 17.17.

17.7.2 Conceptualization

Conceptualization of the system to be modelled is the most important and a critical activity in modelling. It consists of understanding the type

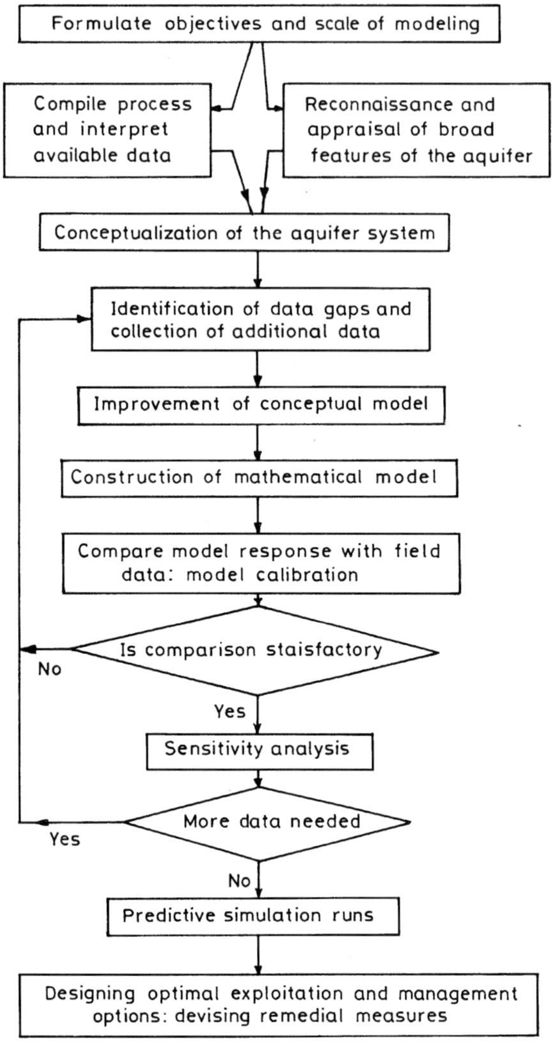

FIGURE 17.17 Flow diagram of various activities in groundwater modelling.

and nature of the aquifer, all its broad features and various processes in it, and quantification of the spatio-temporal distribution of the characteristic parameters. The required details of these features and parameters will, of course, depend upon the particular objective and scale of the model. For example, in a localized study such as a pumping test, one has to consider much more detailed features than is necessary in the case of a watershed modelling. Similarly, in the case of a regional study, small-scale inhomogeneities or anomalies need not be taken into consideration while conceptualizing a groundwater system.

Usually, all the data required for an accurate conceptualization may not be available at the beginning. However, one should start by using the existing data supported by a judicious use of geological insights and other available information. At the same time, one should also work out a well-designed pragmatic plan for the collection and compilation of additional data through a network of observation points in space and time to fill the gaps in the database. This is necessary for improving the conceptual model and making it more representative. The work plan will, naturally, depend upon the objective of modelling on the one hand and the resources of time, money and manpower on the other. Useful details for conceptualization of a groundwater system for modelling are given by Mercer and Faust (1981). Essentially, data on the physical framework of the groundwater system to be modelled, and also the stresses on it, constitute the basis of conceptualization.

The physical framework of an aquifer system is characterized by aquifer geometry, nature of the aquifer, and spatial distribution of hydrogeological parameters, viz. transmissivity and storativity, which control the movement and storage of groundwater. We have described earlier (Chapter Nine) the various field methods to estimate the hydraulic properties of rocks. It may be re-emphasized that one must keep in view the uncertainties that arise due to the assumptions made during the estimation of these parameters. Also, the parameters used in conceptualization should represent only that aquifer which is to be modelled. While conceptualizing a groundwater system, all the available geological data including lithologs, hydraulic gradients etc., should be utilized with discretion. For mass transport modelling, one is also required to know the spatial distribution of porosity, permeability and dispersivity values in the system. A few studies have been conducted, though with limited success, to develop suitable stochastic algorithms which help to estimate these parameters and data on hydraulic heads.

The upper and lower boundaries of a groundwater system can generally be defined quite reliably from drilling data; on the other hand, it is much more difficult to delineate lateral boundaries. It is quite possible that a certain boundary may not cut across the entire thickness of the aquifer. One should carefully analyse the hydraulic gradients and various hydrogeological parameters to ascertain the appropriate boundary conditions such as constant head, no flow or constant flow etc., which may exist at the boundaries of the aquifer under study. These hydraulic conditions should be estimated over at least a few selected points along the boundaries and then suitably interpolated at all the points.

Among the various data inputs for groundwater modelling, the hydraulic head is the most significant parameter and should be monitored at a large number of points exercising all the usual precautions (Chapter Three). The historical values of hydraulic heads are required for model calibration and validation.

The input into and output from the aquifer are termed as hydrological stresses. Recharge due to precipitation, infiltration from surface water bodies like rivers, lakes, tanks, unlined canals, irrigation return seepage, and lateral groundwater inflow etc., comprise the input to a groundwater system. Various prevalent techniques for the estimation of the natural recharge which is the main input to an aquifer system, have been described in section 17.1.2. For mass transport modelling, concentration of contaminants at the source points and relationship between fluid density and concentration of solutes in it are also required. Groundwater withdrawals, evapotranspiration, lateral groundwater outflow, effluence to surface water bodies etc., comprise the output from an aquifer.

The correctness of a conceptual model obviously depends upon the accuracy and adequacy of data, however, modelling must be considered as an iterative and dynamic process. A model is progressively modified as more data becomes available and the understanding of the system improves. Here the experience and imagination of the modeller plays an important role and that is why it is said that modelling is both an art and a science.

Fractured rocks

Conceptualization of fractured rocks is relatively much more difficult since one needs to know the orientation and density of fractures, apertures and smoothness of individual fractures, degree of connectivity, and extent of fractures. Various representations of fractured media include parallel plate model, discrete fracture model, double porosity model, equivalent porous medium model etc. (Chapter Seven). All these methods, relying on the knowledge of characteristic parameters of individual fractures, have been mostly academic in nature. This is particularly true for regional studies because of serious difficulties encountered in an accurate characterization of individual fractures. All the same, models dealing with individual fractures help enhance our basic understanding of the processes in fractures, even though their applicability for prognosis may be limited.

The details in the conceptual model depend on the purpose for which the model is being developed, i.e. whether it is for groundwater flow or solute transport studies. A comparatively simpler conceptual model can be used to predict volumetric flow of water, while for solute transport greater details about fracture characteristics are required, as the heterogeneity of the fracture system greatly influences the travel time and solute concentrations.

At the present level of the state-of-the-art of all the modes of representing a fractured formation, hydraulically equivalent continuum and double-porosity models appear to be more pragmatic approaches on a regional scale. For example, a continuum approach was used for predicting drawdown during the construction of a 420 m deep shaft at the URL in Manitoba, Canada, where the magnitude and temporal variations of drawdown could be predicted to within an order of magnitude. Besides, dual-porosity models have been used for parameter estimation as for the fractured volcanic tuff formation of Yucca Mountain, where a repository for radioactive waste disposal is planned, and also for simulating petroleum and geothermal reservoirs (Anon., 1996). The dual-porosity models are particularly useful for the study of solute transport where there is a possibility of substantial diffusive transfer of mass into the rock matrix blocks, as in fractured clays and dolomites. Further, hydrogeological modelling of fractured rocks which involve migration of contaminants, especially reactive waste and multiphase organic components, is more complex.

Data interpolation
Depending upon the complexity of a groundwater system, various parameters defining its physical framework are highly variable. But it is not possible to estimate them at all the points in an aquifer due to constraints of resources. It therefore becomes imperative to interpolate a parameter at unmeasured locations by using the values of a parameter at the measurement points. It amounts to an effective integration of information on a localized scale to evolve the best regional picture. One simple method in vogue has been linear triangular interpolation. Here the value of a parameter at a point within a triangle is estimated by linear interpolation of the values of that parameter at three vertices of the triangle. In doing so, one presumes that the value of the parameter varies linearly in space and its value at an unmeasured point depends only on three measured values.

An alternative approach for interpolation is to utilize the 'structure', i.e. spatial correlation of values over a large area. This method is called kriging which is based on the theory of regionalized variables. It is a method for optimizing the estimation of a variable which is distributed in space and is measured at a set of points. The value at a point is assumed to depend on all the measured values in the neighbourhood.

17.7.3 Mathematical formulation

Once an aquifer system is conceptualized, the process to be modelled should be expressed mathematically. The basic partial differential equation, which describes groundwater flow, utilizes the principle of mass conservation as well as momentum conservation, may be expressed as (Trescott and Larson, 1977)

$$\frac{\partial}{\partial x}\left(K_{xx}\frac{\partial h}{\partial x}\right)+\frac{\partial}{\partial y}\left(K_{yy}\frac{\partial h}{\partial y}\right)+\frac{\partial}{\partial z}\left(K_{zz}\frac{\partial h}{\partial z}\right)$$
$$\pm W = S_s\frac{\partial h}{\partial t} \qquad (17.17)$$

where h is the hydraulic head, K_{xx}, K_{yy} and K_{zz} are the principal components of the hydraulic conductivity tensor which are assumed to be colinear with the Cartesian co-ordinate system, W is the volume flux per unit volume of porous material (a positive sign for the inflow and negative sign for the outflow), S_s is the specific storage, and t is the time.

For movement of a non-reactive solute in groundwater regime, the condition of mass balance of the solute should also be satisfied. The mathematical equation describing this process may be expressed as

$$\frac{\partial}{\partial x}\left(D_{xx}\frac{\partial C}{\partial x}\right)+\frac{\partial}{\partial y}\left(D_{yy}\frac{\partial C}{\partial y}\right)+\frac{\partial}{\partial z}\left(D_{zz}\frac{\partial C}{\partial z}\right)$$
$$-\frac{\partial}{\partial x}(CV_x)+\frac{\partial}{\partial y}(CV_y)+\frac{\partial}{\partial z}(CV_z)$$
$$-\frac{C_o W}{\eta_e}=\frac{\partial C}{\partial t} \qquad (17.18)$$

where V_x, V_y and V_z are the components of velocity of groundwater along the three coordinate axes, C is the concentration of the solute at any point, D_{xx}, D_{yy}, D_{zz} are the principal components of the coefficient of hydrodynamic dispersion, η_e is the effective porosity, and C_o is the concentration of the solute in the source or sink.

The above mathematical formulation assumes that the medium is a porous continuum in which Darcy's law holds. The partial differential equations may get simplified under different situations. For example, in case the Dupuit approximation is valid, i.e. there is no vertical flow in the system, equation (17.17) is reduced to

$$T\left(\frac{\partial^2 h}{\partial x^2}+\frac{\partial^2 h}{\partial y^2}\right)\pm W = S\frac{\partial h}{\partial t} \qquad (17.19)$$

where T is the transmissivity and S is the storativity of a confined aquifer.

For a homogeneous isotropic medium, equation (17.19) is further simplified to

$$\frac{\partial}{\partial x}\left(T_{xx}\frac{\partial h}{\partial x}\right)+\frac{\partial}{\partial y}\left(T_{yy}\frac{\partial h}{\partial y}\right)\pm W = S\frac{\partial h}{\partial t} \qquad (17.20)$$

Similarly the equation for two-dimensional solute transport in groundwater gets simplified to

$$\frac{\partial}{\partial x}\left(D_{xx}\frac{\partial C}{\partial x}\right)+\frac{\partial}{\partial y}\left(D_{yy}\frac{\partial C}{\partial y}\right)-\frac{\partial}{\partial x}(CV_x)$$
$$-\frac{\partial}{\partial y}(CV_y)-\frac{C_o W}{\eta_e}=\frac{\partial C}{\partial t} \qquad (17.21)$$

17.7.4 Solution of mathematical equation

Sometimes in the literature, the main modelling process is identified with solving of the appropriate partial differential equation to obtain the spatiotemporal distribution of hydraulic heads in the case of groundwater flow modelling, or solute concentration in the case of mass transport modelling. Exact analytical solution of any of the partial differential equations is possible only for very simple situations. For most real life problems, numerical methods are used. In a numerical model, a partial differential equation is approximated by an algebraic equation which may be valid in a small sub-region for a particular time. A number of such equations can be written for all the sub-regions into which the entire domain of study may be suitably divided. Thus, one obtains a set of simultaneous algebraic equations relating all the unknown variables which are characteristic of these sub-regions.

Finite difference, integrated finite difference and finite elements are the main numerical methods for discretizing a partial differential equation. The difference between various methods lies in the process of sub-dividing the space into discrete blocks. These discrete blocks form a grid which may be either mesh-centered or block-centered. In the former, the representative nodes are located on the intersection of grid lines whereas in the latter they are centered between grid lines. The time period of interest is also divided into a number of intervals. The closeness of the spatial blocks and time intervals depends upon the gradient of parameter variation, desired accuracy, and availability of data. A sample of non-uniform dividing of the space is shown in Figure 17.18 using a conventional or a nested square approach.

In the finite difference method, the distribution of nodes is regular but the nodal spacing may either be uniform or non-uniform along an orthogonal co-ordinate system. The integrated finite difference method (IFDM) is more flexible than the finite difference method in that the nodes can be distributed in an irregular mesh. Further, the governing equations of IFDM are formulated in an integral form using the basic conservation laws. Despite greater flexibility in the mesh design, larger requirements of data limits the utility of IFDM. The finite element method (FEM) approximates the partial differ-

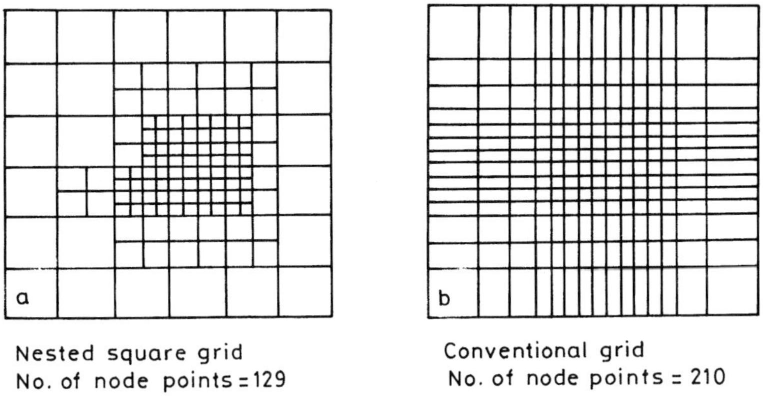

FIGURE 17.18 Comparison of non-uniform grid of node points: (a) nested square grid, and (b) conventional grid.

ential equation by an integral approach. In FEM, a region is divided into sub-regions called elements which can be of any shape. The flexibility in the shape of elements enables a more realistic simulation of the various internal and external boundaries. The value of a variable within an element is obtained by interpolating values of the variables at corner nodes. The integral representation of a partial differential equation in FEM is achieved by using either a weighted residual Galerkin scheme or the variational method. For further details on the use of FEM modelling, the reader is referred to Remson *et al.* (1971) and Pinder and Grey (1977). In many cases, it is seen that results of modelling by adopting different methods, do not differ significantly since any sophistication in modelling is largely cancelled out by the inadequacy of field data.

Solving matrix equation

The algebraic equations thus obtained from partial differential equations can be written in a matrix form which can then be solved by either direct or iterative methods. Gaussian elimination procedure or some of its variations have been used as a direct method for solving the matrix equation to find the unknown variables at subsequent time intervals. However, on account of large requirements of computer memory, round-off errors and the problem of numerical stability in the case of a sparse matrix, direct methods have not been popular in solving a matrix equation.

In contrast, an iterative approach avoids the need for storing large matrices, and thus saves on memory as well as time resources of a computer. An iterative method starts from a first approximation of the unknown value of the parameter to be computed, and its estimate is progressively improved until a sufficiently accurate solution is obtained. The iterative process can be considerably hastened by making a correct initial approximation and use of acceleration factors. The iteration is terminated when the value of the unknown parameter satisfies a pre-set error criterion. Some of the commonly used iterative methods are successive over-relaxation method, alternating direction implicit procedure, iterative alternative direction implicit procedure and strongly implicit procedure. Mercer and Faust (1981) and Javandel *et al.* (1984), among others, have reviewed the basic principles of these techniques. A number of versatile computer codes have been developed at the US Geological Survey which can easily be used by a hydrogeologist without going deeply into the theoretical aspects of modelling.

Some commercially available modelling software

A modular computer software (MODFLOW) for modelling the three-dimensional groundwater flow regime was developed by McDonald and Harbaugh (1983). Later, Pollock (1990) developed another software, MODPATH, to model mass transport in a groundwater system. This is a three-dimensional particle-tracking programme which is interfaced with MODFLOW to produce a plot of flow-path lines. Another three-dimensional mass transport modelling software MT3D (Zheng, 1990) was also interfaced with MODFLOW. These models gained wide acceptance because they are well documented, validated, and verified. All these modelling programmes, however, required the input data to be entered as numerical text through standard files which is quite a time-consuming process.

Recognizing the importance of visualization for improving the modelling process, Graphical User Interface (GUI) was incorporated in the computer software which enabled a substantially fast operation. Using this facility, a model grid can be displayed on the monitor screen for graphically inputting the model parameters using menu options and cursor controls. Some of the widely used modelling packages having a GUI are FLOWPATH (Franz and Guiger, 1990), Model CAD and Visual MODFLOW. This facility helped in getting rid of the strenuous task of manually creating formatted input data files.

With these software capabilities, users can now display a model in plan view (layer by layer) as well as in a vertical cross-section during any phase of the model development or prognosis. These software packages are operated onscreen by selecting various options using the mouse. The visualization of simulation results typically include features like transient or steady-state equipotential, drawdown contours, hydrographs, velocity vectors, path lines and iso-concentrations. Construction of a groundwater model, its refinement, and prognosis have now become much easier and faster. It is also possible to display an animated migration of contaminants in an aquifer system.

17.7.5 Model calibration

Before a deterministic model can be used for prediction purposes, it should be calibrated using the available historical data. The computed parameter, say, hydraulic head (in the case of a groundwater flow model) or solute concentration (in the case of a mass transport model) should be satisfactorily matched with the field values. The matching criterion will depend upon the objective of modelling and the required accuracy. In the case of a mismatch, various parameters should be gradually modified. Essentially, this process is based mostly on a trial-and-error approach and the experience and insight of the modeller plays a significant role in model calibration.

Even after calibration, it may be possible that the model is not a unique one. Perhaps, a number of combinations of the model parameters may produce the same degree of matching. The model calibration should, therefore, always be improved with the additional data. Sensitivity analysis of the model can provide a useful insight to plan for additional data collection.

17.7.6 Case studies

We give here a few groundwater case studies in different situations to illustrate the potential of the groundwater modelling technique. The first one relates to groundwater flow modelling in a mine environment developed to evolve an optimal groundwater pumping scheme from the mines. The other two provide examples for studying the problems of groundwater pollution caused by sea water upconing, and industrial waste respectively.

Simulation of regional hydrodynamics

A semi-confined aquifer in South India comprising sandstone, sands, and gravel/pebbles of Tertiary age occurs below an opencast mine. The aquifer has been very heavily pumped to reduce the upward hydrostatic pressure and thus avoid the flooding of the mine floor. This extensive pumping substantially lowered the hydraulic heads in a large area of about 2300 km^2. Besides stopping of artesian flows from wells in the area, the decline of hydraulic heads has threatened sea-water instrusion in areas which are close to the sea. A modelling study was undertaken to quantify future trends of the aquifer response due to the large scale pumping and thereby evolve an optimal groundwater pumping scheme so that the regional lowering of hydraulic heads may be minimal without creating the hazard of flooding of opencast mines (Gupta *et al.*, 1989).

The task of conceptualizing and quantifying the aquifer system was beset with a number of difficulties due to the complexity of the system and paucity of data. However, a careful analysis of the available data on the physical framework and the stresses on the groundwater flow regime helped develop a conceptual model of the system.

The aquifer shows an unconfined condition over an area of about 420 km^2 and is semi-confined in the remaining region of the study area. The input to the unconfined aquifer has mainly been the recharge due to rainfall, whereas in the remaining area there has been a downward leakage from the upper aquifer and an upward feed from the lower confined aquifer. The transverse flows to the aquifer under study were calculated on the basis of differences between the interpolated hydraulic heads in the consecutive aquifer layers, the inferred values of vertical hydraulic conductivity, and thickness of the intervening leaky layers. The available values of the transmissivity at some points, and data on the saturated thickness of the aquifer as obtained from drilling logs were used to work out spatial distribution of transmissivity.

Initially, only a regional model was prepared using a coarse grid of nodes. The entire area was divided into 136 nodes with a spacing of 4 km; the finite difference technique was used in the modelling. Several iterative runs were made to progressively refine the physical framework, as well as the input and output parameters, until a satisfactory match between the observed and computed hydraulic heads was obtained under both steady and dynamic states. A comparison of steady-state hydraulic heads (observed in the field and computed) is shown in Figure 17.19. Barring some localized mismatch, the general trend of computed and field contours agree. The mean annual input and output quantities for the initial steady state as obtained after a number of interactions are illustrated in Figures 17.20(a) and 17.20(b). The dynamic calibration of the model was carried out for a period of 30 years. A sample hydrograph in a well as observed in the field and that computed from the model are shown in Figure 17.21.

After obtaining a reasonably representative regional model through this process of calibration, a nested square fine grid was adopted in the mine area to evolve a scheme of optimal pumping from the mines. There were a number of constraints from the feasibility point of view on the location of pumping wells. These were followed in working out the location of wells and pumping rates. With this scheme, the desired lowering of hydraulic heads below the mines could be achieved while the regional decline of heads was minimized.

Groundwater regime in a small island

Potable groundwater in small coral islands occurs in the form of a thin floating lens over saline water. The salinity of groundwater increases with depth. Most of the drinking-water needs in these islands are met from groundwater. However, in view of the very small thickness of fresh groundwater lens, one has to be cautious not to resort to excessive pumping otherwise the upconing of sea water may cause increased salinity of shallow layers. In order to estimate the safe limit of groundwater pumping, modelling of a vertical cross-section of the aquifer may be very useful. Such a study was carried out in a coral island located in the Arabian sea (courtesy C.P. Gupta). The upper aquifer in this island consists of highly permeable coral sands. According to the available geological information, the sands mixed with corals extend up to a depth of 8 m. This layer is underlain by carbonate rocks up to a depth of about 1000 m above the basalt basement.

The rainfall occurs mainly during the months from June to September. Despite an average annual rainfall of about 1500 mm and a high rate of percolation due to high vertical hydraulic conductivity, the potential of fresh groundwater is quite meagre. This is attributable to a large subsurface outflow of freshwater to the sea leaving a small fraction of infiltrated water in the aquifer for utilization. As a result of excessive groundwater withdrawal, a progressive increase of the electrical conductivity (EC) of groundwater over a few years has been noted. The EC of groundwater ranges from 1000–2000 μS cm^{-1} in the central part of the island and is more than 2000 μS cm^{-1} in other peripheral areas.

Since the relative levels of water-table and sea water play a crucial role in the mixing of freshwater and sea water (section 17.6.2), they were regularly monitored during the study. It was found that the average water-table is only about 0.1–0.2 m above the mean sea level.

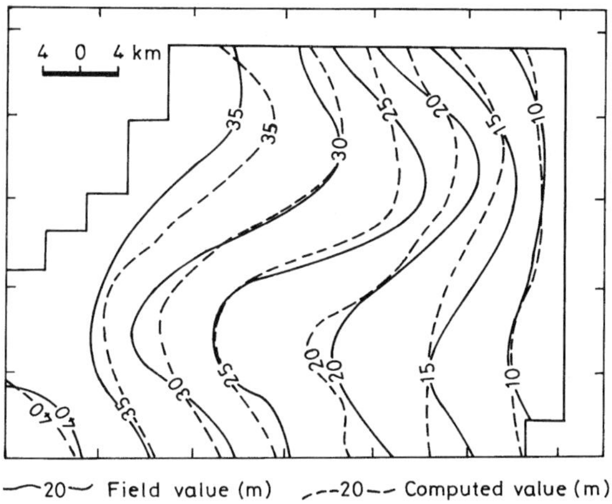

FIGURE 17.19 Regional groundwater modelling: contours of initial hydraulic heads.

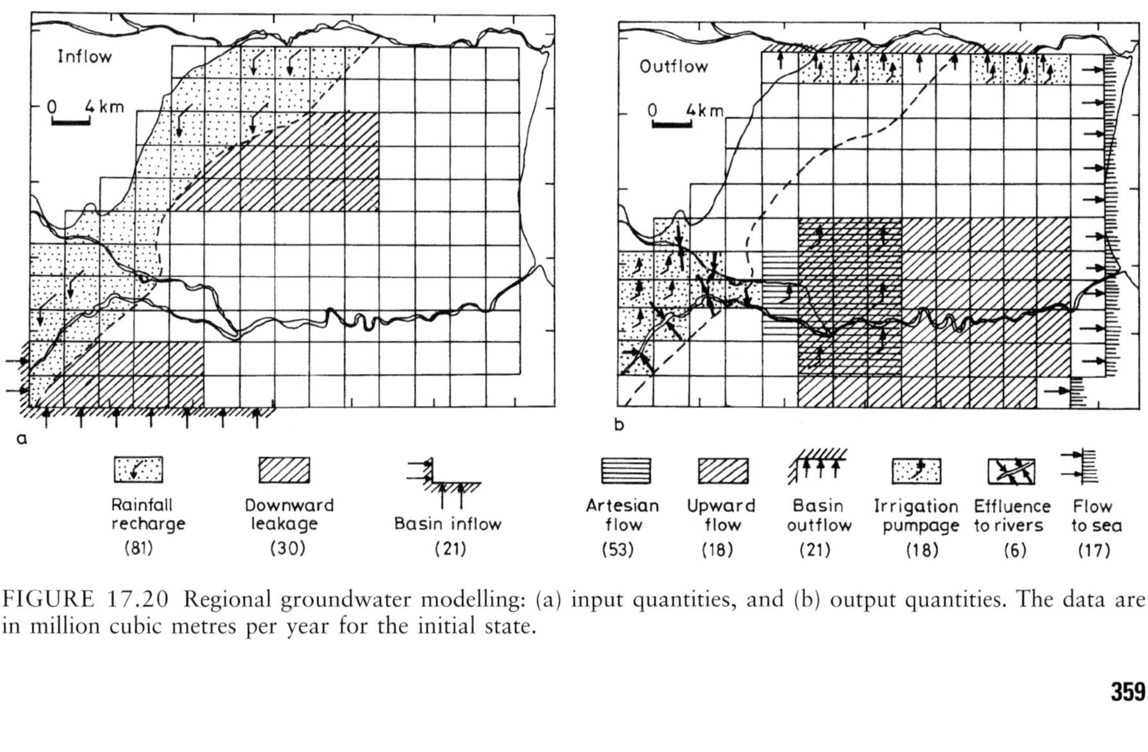

FIGURE 17.20 Regional groundwater modelling: (a) input quantities, and (b) output quantities. The data are in million cubic metres per year for the initial state.

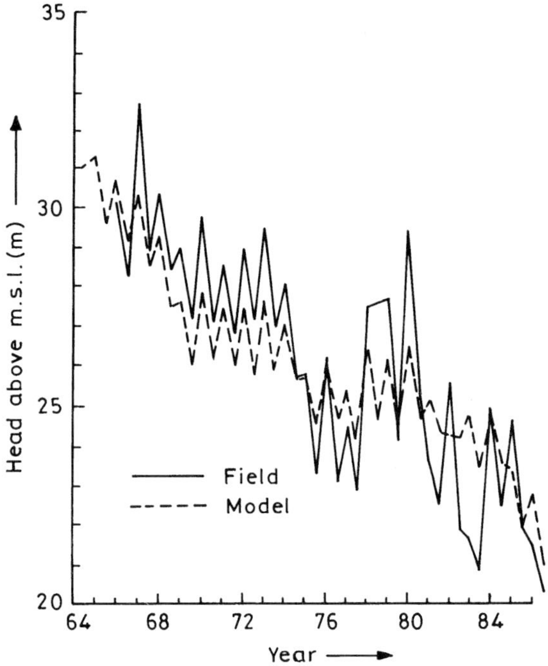

FIGURE 17.21 Regional groundwater modelling: comparison of field and computed model hydrographs for a well.

Hydrogeological parameters were estimated from short duration pumping tests data using a numerical method.

A computer code SUTRA (Voss, 1984) was used for modelling a vertical cross-section of the island. The code was suitably modified to take into account the effect of tides, non-uniform recharge which takes place only during the monsoon period (June to September), and water-table conditions.

Numerical values of the following characteristic parameters were adopted from a number of studies in other coral islands, e.g. Souza and Voss (1987); Griggs and Peterson (1993); Underwood *et al.* (1992), and Oberdorfer *et al.* (1990).

Aquifer compressibility	2.5×10^{-9} Pa^{-1}
Water compressibility	4.4×10^{-10} Pa^{-1}
Dynamic viscosity of fluid	1×10^{-3} kg m^{-1}s^{-1}
Freshwater density	1000 kg m^{-3}
Sea-water density	1025 kg m^{-3}
Solute mass concentration in sea water	0.0357 kg salt/kg sea water

Since the salinity variation with depth near the water-table is to be estimated accurately, a close nodal spacing was used in a layer of the upper 3 m and a sparse one for deeper zones. Nodes representing the island–ocean boundary were assigned oscillating pressure to simulate the tidal fluctuations having a periodicity of 24 h and an amplitude of 0.9 m. Solute concentration at all these nodes was assumed to be the same as that of sea water. Appropriate values of recharge and groundwater abstraction were assigned to top nodes. The criterion used for progressive modification of parameters to calibrate the model was a satisfactory match between the computed and observed values of (a) water-table above average sea level; (b) salinity at the water-table, and (c) tidal efficiency schedules were tested and the resultant increase in the salinity was calculated (Figure 17.22). As shown in Figure 17.23, the maximum amount of groundwater that can be pumped from a well at a point A should be less than 13 m^{3}d^{-1} if the salinity at the water-table is required to be less than 2.5% that of sea water. The rise in salinity due to pumping a well at A at this rate was predicted within a radius of 200 m.

The model was also used to examine the rise in the water-table and reduction in the outflow of fresh groundwater to the sea that may be achieved with the help of a subsurface dam. Such a dam may be constructed by drilling three rows of closely spaced 50 mm diameter boreholes and suitably grouting them. Assuming that this specially designed subsurface dam may reduce the hydraulic conductivity of the coastal elements from 250 md^{-1} to 40 md^{-1}, the water-table was found to rise by about 6 cm during a period of five years.

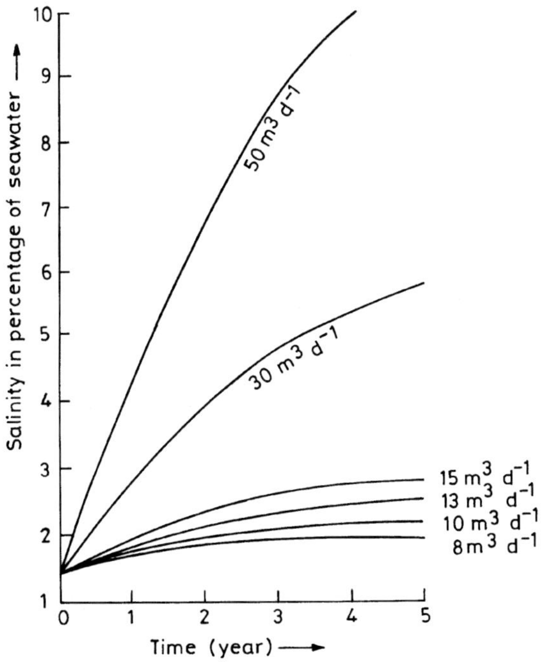

FIGURE 17.22 Modelling of groundwater system in a small island: increase in salinity at water-table for different rates of pumping from a water supply well.

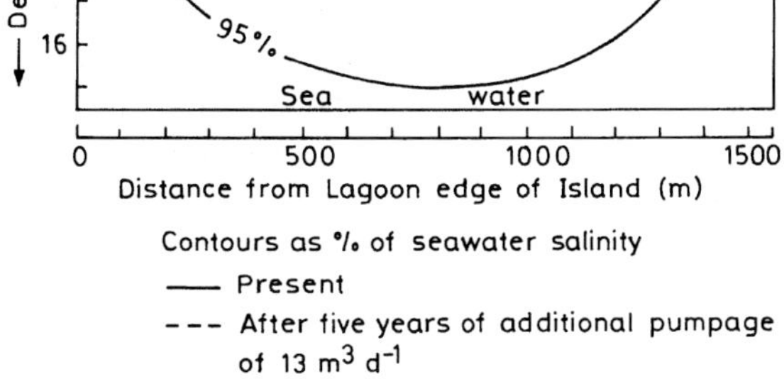

FIGURE 17.23 Change in salinity along a vertical section due to additional pumping of groundwater at a point.

Industrial pollution study

Groundwater pollution due to a tanning centre located in a river basin in South India has been investigated by Gupta *et al.*, (1996). About 650 tanneries have been operating in an area of about 1600 km^2. These tanneries discharge their effluent (mostly untreated) through unlined channels into the river, contaminating the groundwater regime. According to the available reports, the effluent from the tanneries contains 20 000–30 000 mg l^{-1} of total dissolved solids (TDS). At some locations, the TDS concentration in groundwater has been as high as 8000 mg l^{-1}. The situation is very precarious as most of the requirements for irrigation, industry and drinking purposes in the basin are being met from groundwater resources.

Main aquifers in the area are weathered and fractured crystalline rocks. However, along the river course, one encounters sand layers and pebbles having higher permeability and good groundwater potential. The seepage rate of polluting effluent through the river bed is quite high. Although hydrogeological parameters like transmissivity, specific yield, recharge to the aquifer, and groundwater pumpage in the study area were estimated at a few locations, porosity and dispersivity required for mass transport modelling were not known. The former were interpolated to evolve a spatial distribution but the latter were only guesstimated using the geological information and data from similar studies elsewhere. A uniform porosity of 0.2 was assumed for the entire area. The longitudinal dispersivity (α_L) was taken as 30 m and transverse dispersivity (α_T) as 10 m. Historical water-table data monitored every month at a number of wells for the last 12 years was used to calibrate the model.

The TDS concentration in groundwater obtained from chemical analysis of water samples from dug wells represented the cumulative value for the entire depth of the aquifer at a point. However, C_0 which is the source concentration, i.e. actual quantity of TDS reaching the water-table, was not known. It had to be, therefore, inferred only from the TDS content in the effluent discharged at tannery outlets. The TDS load reaching the water-table was assumed to be 12% of that contained in the effluent at the surface.

A finite difference approach coupled with the method of characteristics (Konikow and Bredehoeft, 1978) was used to construct groundwater flow and mass transport models. Initially, each node was assumed to be associated with a certain TDS concentration which moved with the groundwater velocity at that point. The numerical solution gave the values of head, h and concentration, C at all the points.

The water-levels obtained from the groundwater flow model were compared with those observed in the field. The transmissivity distribution was progressively modified until a satisfactory match was obtained at most of the points. The groundwater flow velocities (Figure 17.24) were computed using FLOWPATH computer code (Franz and Guiger, 1990). The solute transport model was calibrated using the historical data for 12 years.

It was seen that the path lines of pollutant

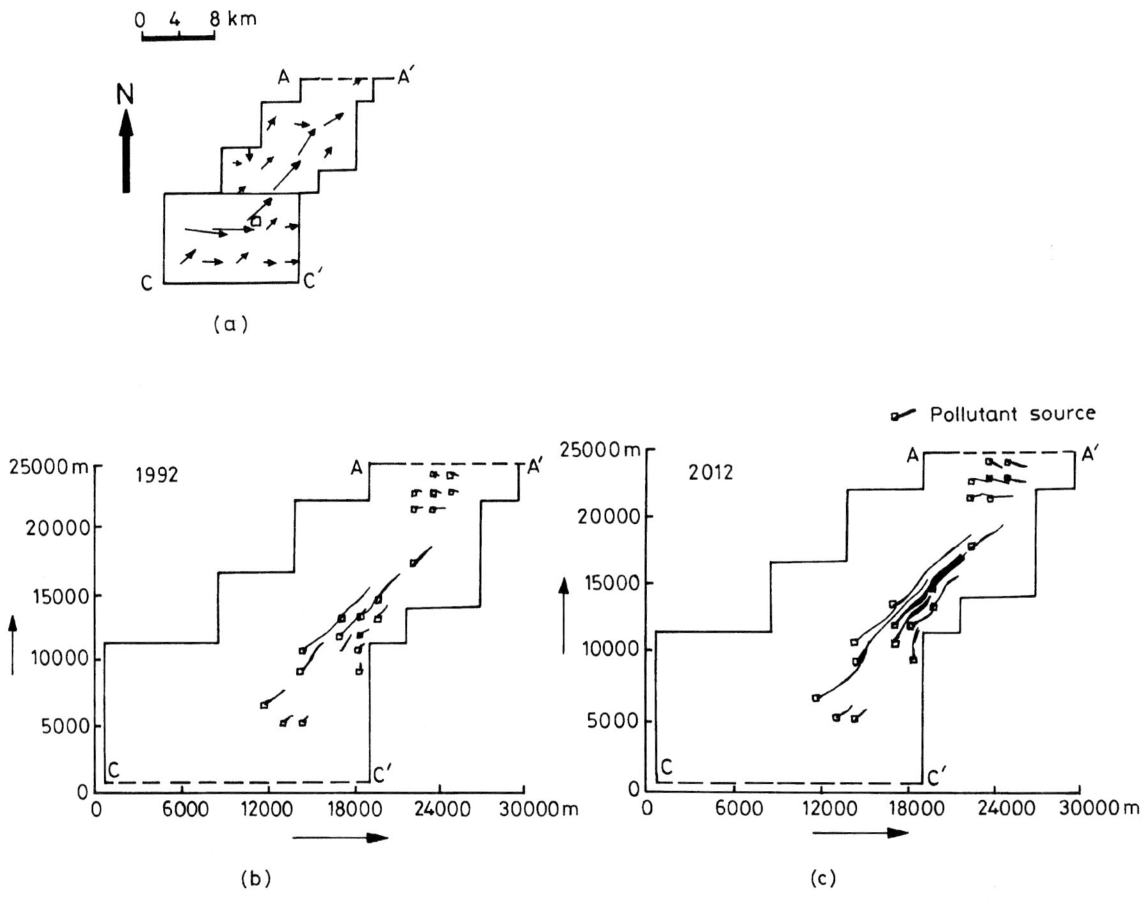

FIGURE 17.24 Modelling of groundwater pollution due to industrial waste: (a) a graphical view of velocity vectors; (b) modelled path lines of pollutant migration for the initial state (1992); (c) predicted trends after two decades.

migration after two decades were considerably extended (Figure 17.24). Actual concentration of TDS at many locations was also found to be higher. Despite lacunae in the data base for modelling of pollutant migration in the aquifer, it was found indisputably that if the tannery effluents continue to be discharged at the present level, both as regards the volume and TDS concentration, groundwater pollution will appreciably increase in the future. The sensitivity analysis helped to identify major data gaps and the required additional field investigations.

17.8 Groundwater management and conservation

17.8.1 Terminology

Groundwater resources should be utilized in a planned manner to avoid any undesirable effects, viz. ground subsidence, sea-water intrusion, uneconomic pumping lift etc. It is therefore necessary to define the desirable quantity of groundwater which can be withdrawn safely from a basin without causing any adverse effect.

Basin yield

One of the earliest concepts of basin yield was that of **safe yield** which was defined by Meinzer (1923) as the practicable rate of withdrawing water from an aquifer for human use without depleting the supply to the extent that withdrawal at this rate is no longer economically feasible. Withdrawal in excess of safe yield is regarded as **overdraft**. Lately, the concept of safe yield has been discarded as it considers a fixed quantity of available water without changing hydrological regime and the socio-economic condition. Therefore, alternative yield concepts as given below have been suggested (Todd, 1980; ASCE, 1987).

Perennial yield is the practicable rate at which groundwater can be withdrawn perennially under specified operating conditions without producing any adverse effects. The **maximum perennial yield** is the maximum quantity of water which can be made available if all possible methods of recharging the basin are adopted. **Mining yield** is the quantity of extractable water which exceeds the recharge. It is a nonrenewable resource which is like a mineral or petroleum deposit. This can be practised in areas where groundwater storage is otherwise of no value. This can be used to contribute to the economic development of a water scarcity area, e.g. in the Sahara desert. The degree of mining of groundwater in some countries is excessive, as illustrated in Figures 17.25 and 17.26 from Saudi Arabia and Qatar (Lloyd, 1991). Mining yield can be reduced by economic use of water and by finding alternative sources of water supply.

Dynamic and static resources

Groundwater resources can also be classified as static and dynamic. The **static resource** is the amount of groundwater available in the aquifer below the zone of water level fluctuation. The **dynamic resource** can be defined as the amount of groundwater available in the zone of water-level fluctuation. The useable groundwater resource is essentially a dynamic resource which is recharged annually or periodically. The static storage should be utilized during drought periods and the volume to be used from static storage is to be determined considering the need and economics of its exploitation. The groundwater resources of an area can be augmented by adopting artificial recharge, interbasin water transfer and water conservation measures.

17.8.2 Artificial recharge

Artificial recharge is the additional input of water into the ground besides natural infiltration. The main purpose of artificial recharge is to augment the groundwater resource. This also helps in checking pollution migration, sea-water instrusion in coastal aquifers and land subsidence. Artificial recharge is important in groundwater

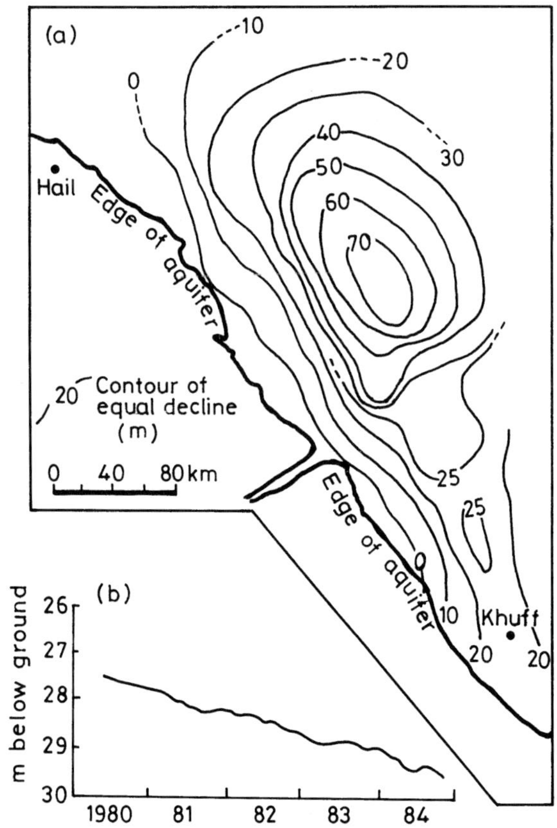

FIGURE 17.25 Contours of equal lowering of water-levels in North-Central Saudi Arabia due to groundwater mining (after Lloyd, 1991).

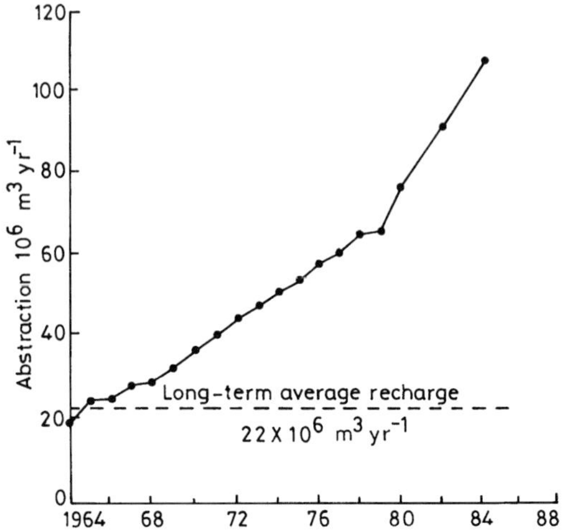

FIGURE 17.26 Over-abstraction of groundwater in comparison to long-term average recharge in Qatar (after Lloyd, 1991).

management as it provides natural storage for water available for use during dry periods.

The main source of water for artificial groundwater recharge is storm run-off and river water. Sewage and wastewater is also being increasingly used for artificial recharge after necessary treatment. The choice of a particular method of artificial recharge depends on the nature of aquifers, their hydraulic conductivity and storativity, and the quality of recharging water.

Surface spreading methods are used in unconfined aquifers where the strata are permeable down to the zone of saturation, and the water-table is deep to provide adequate storage. Flood plains, alluvial fans and certain glacial deposits like eskers and moraines are suitable for recharge by spreading methods.

Recharge basins are also constructed to recharge unconfined aquifer. These are made by excavation or by building dykes or levees, e.g. in the Rhine valley, Germany. A groundwater mound is formed beneath the recharge basin due to infiltration (Figure 17.27); the shape and size of the mound depends upon the basin size and shape, recharge rate and duration, depth of water-table and aquifer parameters. The infiltration rate is initially high but usually decreases with time due to accumulation of suspended material, and by the growth of algae and bacteria. Therefore, in order to maintain higher infiltration rates, the basin should be dried periodically to disrupt the growth of algae and bacteria. It also helps in maintaining higher rates of infiltration with the development of shrinkage cracks.

Percolation tanks

These are a type of infiltration basin built in the ephemeral river bed by providing a low

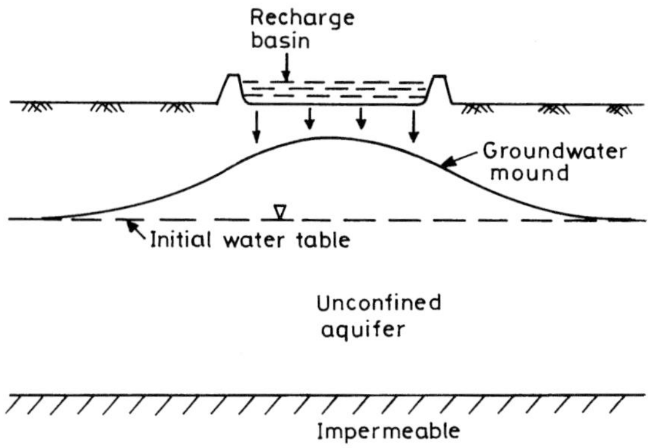

FIGURE 17.27 Development of a groundwater mound below a recharging basin.

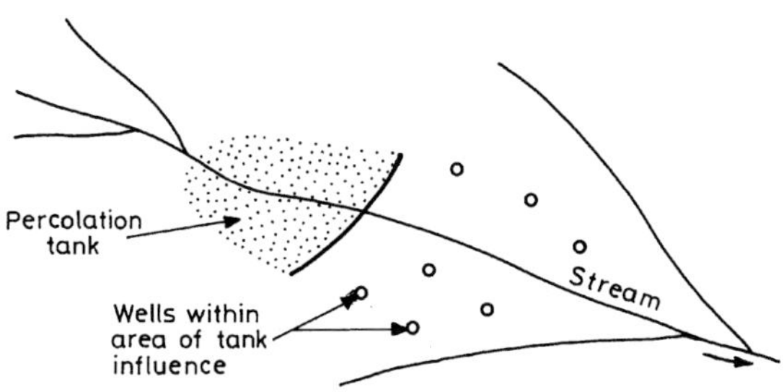

FIGURE 17.28 Schematic diagram showing location of wells in the area influenced by recharge due to a percolation tank.

earthen dam to store rain water (Figure 17.28). Recharge by percolation (infiltration) tanks is an ancient practice of water conservation in the hard rock formations the world over. The capacity of a percolation tank usually ranges between 0.14 and 3.0 million m^3. Studies in crystalline rocks of south India indicate that the rate of infiltration from percolation tanks varies from 0.5–1.5 md^{-1}; the initial rate of infiltration is high and decreases with time due to silting. In the case of non-perennial tanks, the recharge rates are comparatively greater. This is due to the removal of silt by farmers during the dry season which is used as fertilizer. The recharge estimates from percolation tanks in Deccan basalts of India range from 30% to 60% of impounded water depending on bedrock and climatic conditions (Maggirwar, 1992).

Recharge through wells

This is practised in areas where recharge due to surface spreading and basins is not feasible, e.g. in confined aquifers due to the presence of extensive clay layers overlying the aquifer, or in metropolitan areas where land values are too high. A cone of recharge is formed which is the reverse of cone of depression developed due to pumping (Figure 17.29). Under a steady-state condition, the equation for rate of recharge, Q_r in a confined aquifer can be written as

$$Q_r = \frac{2\pi Kb\,(h_w - h_o)}{\ln\,(r_o/r_w)} \qquad (17.22)$$

and for an unconfined aquifer

$$Q_r = \frac{\pi K\,(h_w^2 - h_o^2)}{\ln\,(r_o/r_w)} \qquad (17.23)$$

where Q_w is the rate of recharge, and the remaining terms are indicated in Fig. 17.29.

As in the case of surface spreading, the recharge rates through wells are initially high but decrease with time, unless well rehabilitation and well development methods like surging and pumping are adopted at intervals. Recharge rates from wells vary from 200 to about 2000 m^3d^{-1} and in exceptional cases, the recharge rates are as high as 51 000 m^3d^{-1}, especially in limestones (Todd, 1980). In the case of recharge through wells, it is essential that the recharge water is pre-treated for the removal of suspended material and undesirable chemical constituents to avoid clogging.

There are several excellent case histories of artificial recharge by wells. For example, in the coastal plain of Israel, seasonal storage is created by recharging through wells, with winter rainfall and the stored water being extracted in summer from the same wells; here imported water from the Jordan river is also used for

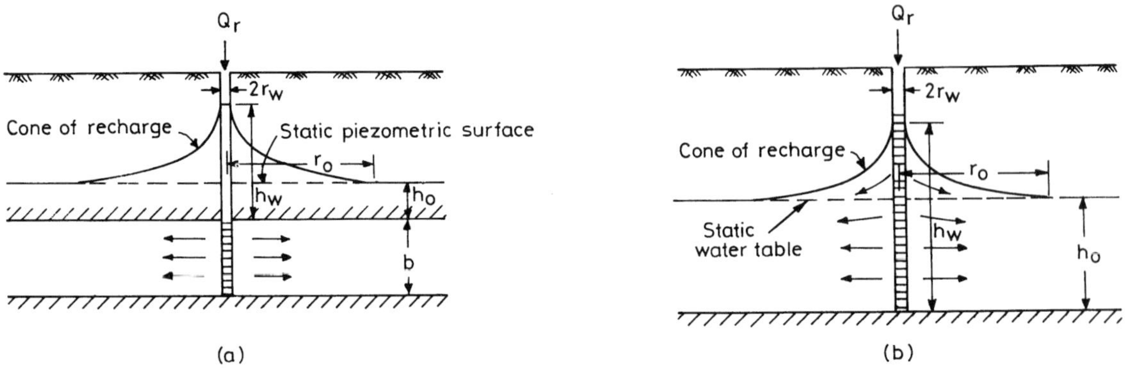

FIGURE 17.29 Cone of recharge developed in: (a) confined aquifer, (b) unconfined aquifer.

recharge purposes. Cooling water is used for recharging chalk aquifers through wells in England. Recharge wells are also used extensively to control sea-water intrusion by creating a freshwater ridge as in Los Angeles and Orange Counties in the USA (Todd, 1980).

Induced recharge

If a well is located adjacent to a stream or lake, the withdrawal of groundwater from the well causes flow from the surface water body to a groundwater reservoir (Figure 17.30). It is effective in cases where there is hydraulic continuity between the aquifer and river bed. Induced recharge also causes self-purification of water and regulates its temperature. Withdrawal of groundwater along the river banks will not only augment the groundwater supply but can also mitigate floods. Sometimes collector wells are constructed in the flood plains of river beds for inducing large flows of water from rivers, as along the Mississipi and Ohio rivers in the USA (section 15.1.4). In many developing countries also, large-scale groundwater withdrawal by induced recharge is planned, e.g. in Sudan from the river Nile (Lloyd, 1991) and in India from the river Yamuna.

Subsurface dams

The purpose of a subsurface dam is to check the outflow of groundwater from a sub-basin thereby raising the groundwater storage on the upstream side. Subsurface dams are feasible in narrow and gently sloping valleys where the bedrock occurs at a shallow depth, overlain by valley-fill deposit. A subsurface dam consists of an impervious wall built of impervious material like clay, bricks and concrete; tar felt, resin, polythene sheets and bitumen are also used. The structure usually extends to a depth of 1–3 m below the land surface. Such groundwater dams were also constructed in ancient times. There are examples of such structures built in Roman times from the islands of Sardinia in Italy, and from Tunisia. In recent years, they have gained popularity in many other countries e.g. India, southern and eastern Africa, Brazil and Japan (Sinha and Sharma, 1990).

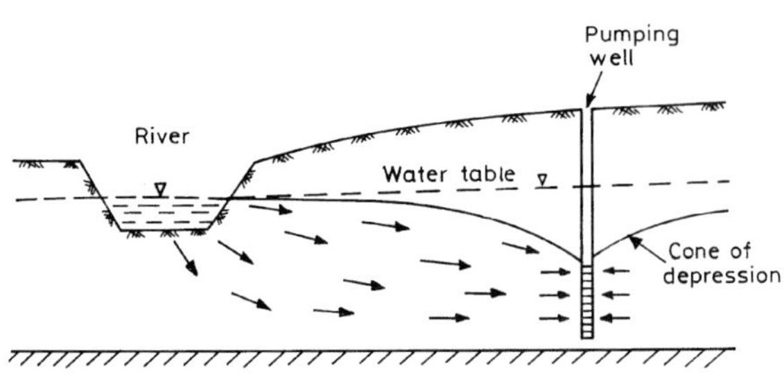

FIGURE 17.30 Induced recharge due to pumping near a stream.

17.8.3 Interbasin water transfer

A river basin is usually regarded as a basic unit for water resources planning and development. This is based on the concept that each basin has its own characteristics and that development in the upstream part of the basin will affect the downstream regime, both qualitatively and quantitatively. Therefore, there should be necessary co-ordination between water resources development in different parts of a basin. Further, for the proper utilization of water in a basin, and to meet the requirements in other adjacent basins within the country or in the adjoining countries, interbasin water transfer is advocated. It not only supplies additional surface water but also augments groundwater resources by recharge.

Interbasin water transfer is not a new concept. It has been practised in the past in several countries. For example, interbasin water transfer projects were implemented in the past in China, the former USSR, the USA, Mexico and India. A notable scheme of long-distance transfer of water in the former USSR is from the Irtysh–Karganda scheme in central Khazakstan where a 450 km long canal with a maximum capacity of 75 m^3s^{-1} was built. In China, interbasin water transfer programmes were implemented as far back as 200 BC. Singapore now buys most of its water from Malaysia but this arrangement may not last long due to the growing demand for water in Malaysia.

In India, interbasin water transfer has been in practice for more than the last 600 years when in the Moghul times, Western Yamuna Canal and the Agra Canal were built for transferring surplus waters from the Himalayan rivers to other parts in the Gangetic Plain. In recent years, several other such interbasin transfer programmes have been implemented such as Rajasthan canal, Beas-Sutlej Link Project and Telugu Ganga (IWRS, 1996). Proposals for long-distance transfer of waters are also under investigation. For example, the Ganga–Cauvery link envisages diversion of surplus waters (1680 m^3s^{-1}) of river Ganga in the north to water scarcity hard rock areas in the south over a distance of more than 2600 km and over a head of 550 m. About 17.5% of the total water proposed to be transferred through such link canals in granitic terrains is likely to be recharged to groundwater reservoirs (Arumugam *et al.*, 1992).

In China, several interbasin water transfer projects are being planned to transfer water from the south to the north in the North China Plain (Zhengzhou *et al.*, 1996). The aim is to solve the water shortage of the cities of Beijing and Tianjin and other adjoining areas by diverting about 20×10^9 to 30×10^9 m^3 of water from the south.

17.8.4 Other methods

Other methods of water conservation include afforestation, evaporation control, reduction in canal losses, and efficient methods of irrigation, viz. drip and sprinkler methods. Roof catchment and desalination of saline water and sea water can also solve to some extent the problems of drinking-water supply.

The various possible methods of water management should be planned in advance so that the feasibility of different alternatives can be assessed by pilot studies. Other aspects which should be considered in the management process include social aspects, environmental impact and economic limitations etc. Experience in developing countries indicate that people's participation in water management programmes leads to greater success in their implementation.

Appendix

ENGLISH-METRIC UNIT CONVERSION TABLE (from *Ground Water*, Vol. 33, No. 5, p. 873, reprinted by permission of the *Journal of Ground Water*). To convert A to B, multiply A by C, to convert B to A, divide B by C

A	B	C	A	B	C
Length			*Hydraulic conductivity/Permeability*		
inch	metre	2.540E-2	gal/day/ft^2	cm/sec	4.716E-5
feet	metre	.3048	gal/day/ ft^2	ft/day	0.1337
yard	metre	.9144	gal/day/ ft^2	metre/day	4.075E-2
mile	kilometre	1.609	gal (UK)/day/ ft^2	metre/day	4.893E-2
inch	centimetre	2.540			
Area					
sq inch	sq centimetre	6.452	darcy	ft/day	2.433
sq feet	sq metre	9.290E-2	darcy	metre/day	0.7416
sq yard	sq metre	.8361			
sq mile	sq kilometre	2.590	*Transmissivity*		
acre	sq kilometre	4.047E-3	gal/day/ft	sq metre/day	1.242E-2
			gal (UK)/day/ft	sq metre/day	1.492E-2
			sq ft/sec	sq metre/day	8.027E3
Volume			sq ft/day	sq metre/	9.290E-2
cu feet	cu metre	2.832E-2			
cu inch	cu centimetre	1.639E-1			
			Force and pressure		
			pound(f)	newton	4.448
gallon	litre	3.785	pounds/sq. in	pascal	6.895E3
gallon (UK)	litre	4.546	lb/sq./ft	pascal	4.788E1
barrel (petr)	litre	1.590E-2	atmosphere	pascal	1.013E5
acre-feet	cu metre	1.234E-3	psi	kg/cm^2	7.031E2
million gal	cu metre	3.785E-3	ft of H_20 (4°C)	psi	0.4335
gallon (UK)	gallon (US)	1.200			
Velocity and gradient					
feet/sec	metre/sec	0.3048	*Temperature*		
mile/hour	metre/sec	0.4470	Fahrenheit	Celsius	5 (F-32)/9
feet/mile	metre/km	0.1894	Celsius	Fahrenheit	1.8 (C) +32
Flow rate			Kelvin	Celsius	K-273.2
gal/min	litre/sec	6.309E-2			
gal/min	cu metre/day	5.300			
gal (UK)/min	litre/sec	7.577E-2			
10^6 gal/day	cu metre/day	3.785E-3			
cu ft/sec (cfs)	litre/sec	2.832E-1			
acre-feet/day	litre/sec	1.458E-1			

NOTE: (1) The (E) notation indicates exponentiation: 2.540E-2=2.540.10^{-2}. (2) Unless otherwise noted, all gallons are U.S. gallons. (3) The darcy is unit of permeability (k), not of hydraulic conductivity (K). (4) A newton (force) = kg.m/s^2; A pascal (pressure) = kg/m.s^2; each in a unit in Sl. (5) Under 'Temperature', entries are formulae, not multipliers.

References

To save space, some frequently used references are abbreviated:

AGU	: American Geophysical Union
ASCE	: American Society of Civil Engineers
AWRC	: Australian Water Resources Council
BGS	: British Geological Survey
CGWB	: Central Ground Water Board, Govt. of India
CSC	: Commonwealth Science Council
GSI	: Geological Survey of India
IAH	: International Association of Hydrogeologists
IAEA	: International Atomic Energy Agency
IAHS	: International Association of Hydrological Sciences
IGC	: International Geological Congress
IGW	: Proceedings International Workshop on Appropriate Methodlogies for Development and Management of Ground Water Resources in Developing Countries, NGRI, India.
NGRI	: National Geophysical Research Institute, Hyderabad, India.
UNESCO	: United Nations Educational, Scientific and Cultural Organisation
WHO	: World Health Organisation

Abelin, H. and Birgersson L. (1985) Migration experiments in the Stripa Mine, design and instrumentation, in *Design and Instrumentation of In-situ Experiments in Underground Laboratories for Radioactive Waste Disposal* (eds B. Come, P. Johnston and A. Muller), Balkema, Rotterdam, pp. 180–90.

Adyalkar, P.G. (1983) Groundwater survey and exploration techniques in karstic regions, in *Seminar on Assessment, Development and Management of Ground Water Res.*, CGWB, New Delhi, Spl. Papers, pp. 45–56.

Adyalkar, P.G. and Mani, V.V.S. (1972) An attempt at estimating the transmissibilities of trappean aquifers from specific capacity values. *J. Hydrol.* **17**, 237–41.

Adyalkar, P.G. and Mani, V.V.S. (1974) Application of groundwater hydraulics to a basaltic water-table aquifer. *J. Hydrol.* **21**, 211–18.

Adyalkar, P.G. and Rao, S.S. (1979) Hydrodynamic method of assessing groundwater recharge by precipitation in Deccan Trap terrain – a case study. *J. Geol. Soc. India*, **20**(3), 134–37.

Agterberg, F.P., Banham-Carter, G.F. and Wright, D.F. (1988) Statistical pattern integration for mineral exploration. *Proceedings Cogeodata Symposium on Computer Applications in Resources Exploration*, Espoo, Finland.

Ahmed, F., Andrawis, A.S. and Hagaz, Y.A. (1984) Landsat model for groundwater exploration in the Nuba Mountains, Sudan. *Adv. Space Res.*, **4**(11), 123–31.

Ahn, J.S. (1988) Environmental isotope-aided studies on river water and groundwater interaction in the region of Seoul and Taegu, in *Proceedings Seminar on Isotope Applications in Hydrology in Asia and the Pacific*, IAEA, Beijing, 104–40.

Albertson, P.E., Bourne, S.G. and Herrington, G.W. (1992) An integrated groundwater and landuse GIS for impact assessment. *Photogram. Eng. Remote Sens.*, **58**(8), 1203–1207.

Alfirevic, S. (1966) Hydrogeological investigations of submarine springs in the Adriatic. *IAH Mem.* **6**, pp. 225–64, Belgrade.

Alfoldi, L. (1980) Changes in the physical conditions of aquifers due to withdrawal of large volume of groundwater, in *Studies and Reports in Hydrology*, **28**, pp. 242–53, UNESCO, Paris.

Aller, L., Bennett, T., Lehr, J.H. and Petty, R.J. (1987) DRASTIC: a standardised system for evaluating groundwater pollution using hydrogeological settings. United States Environmental Protection Agency, Washington, Report 600/2–85/018.

Al-Saafin, A.K., Bader, T.A., Shehata, W. *et al.* (1990) Groundwater recharge in an arid karst area in Saudi Arabia, in *Selected Papers on Hydrogeology* (eds E.S. Simpson and J.M. Sharp, Jr.), Vol. 1, IAH/Verlag Heinz Heise, Hanover, pp. 29–41.

Alumbaugh, D.L. and Morrison, H.F. (1993) Crosshole electromagnetic tomography for groundwater monitoring. *AGU 1993 Fall Meeting*, San Francisco, Abstract Volume, Supplement to EOS, October 26, 1993, p. 306.

Alumbaugh, D.L. and Morrison, H.F. (1995) Theoretical and practical considerations for crosswell electromagnetic tomography using a cylindrical geometry, *Geophysics*, **60**, 846–70.

Angadi, K.S. (1986) Hydrogeological studies of the

lower parts of Ghataprabha Basin, Belgaum and Bijapur Districts of Karnataka, India, PhD thesis, University of Roorkee.
Angino, E.E. and Wilbur, C.K. (1990) The retention of Cd, Pb and Zn by fractured shale units: simulation of reactivity in the unconsolidated zone, in *Selected Papers on Hydrogeology*, Vol. 1 (eds E.S. Simpson and J.M. Sharp. Jr.), Verlag Heinz Heise, Hanover, pp. 169–77.
Anonymous (1995) Standard Guide for Design of Ground-water Monitoring Systems in Karst and Fractured-rock Aquifers, *Am. Soc. for Testing Materials*, D 5717–95, pp. 435–51.
Anonymous (1996) *Rock Fractures and Fluid Flow – Contemporary Understanding and Applications*, National Academy Press, Washington, D.C., 551 pp.
Attenucci, J.C., Brown K., Crosswell, P.L. and Kevany, M.J. (1991) *Geographic Information Systems: A guide to the Technology*, Von Nostrand Reinhold, New York.
Arihood, L.D. (1994) Hydrogeology and paths of flow in the carbonate bedrock aquifer, Northwestern Indiana. *Water Resources Bull.*, **30**(2), p. 205.
Aronoff, S. (1989) *Geographic Information Systems: A Management Perspective*. WDL Publications, Ottawa, Canada, 294 pp.
Arumugam, K., Sharda, A.K. and Hasan, Z. (1992) Augmented availability of water for artificial recharge due to interbasin transfer of waters, in *Proceedings Workshop on Artificial Recharge of Groundwater in Granitic Terrain*, CGWB, Bangalore, pp. 65–79.
ASCE (1987) *Groundwater Management*, 3rd edn, Manuals and Report on Engineering Practice, No. 40, Am. Soc of Civil Engrs. New York, p. 263.
Athavale, R.N. (1985) Nuclear tracer techniques for measurement of natural recharge in hard rock terrains, in *Proceedings International Workshop on Rural Hydrogeology and Hydraulics of Fissured Basement Zones* (ed. B.B.S. Singhal), University of Roorkee, pp. 71–80.
Athavale, R.N. and Rangarajan, R. (1988) Natural recharge measurements in the hard rock regions of semi-arid India using tritium injection – a review, in *Estimation of Natural Recharge of Groundwater* (ed. I. Simmers), NATO, Advanced Science Institute Series, Vol. 222, 175–94.
Athavale, R.N., Rangarajan, R. and Rao, S.M. (1992) Determination of the flow direction of geothermal waters at Manikaran using the borehole tracer technique, *J. Geol. Soc. India*, **39**(4), pp. 329–37.
Athavale, R.N., Singh, V.S. and Subrahmanyam, K. (1983), A constant discharge device for aquifer tests on large diameter wells. *Ground Water*, **21**, 752–55.
Atkinson, S.F. and Thomlinson, J.R. (1994) An examination of groundwater pollution potential through GIS modelling. *Technical Papers of the ASPRS/ACSM Annual Meeting*, Bethesda, Maryland, pp. 71–80.
Avci, C.B. (1992) Parameter estimation for step-drawdown tests. *Ground Water*, **30**(3), 338–42.
Ayers, R.S. (1975) Quality of water for irrigation. *Proceedings Irrig. Drain. Div. Speciality Conf.* ASCE, Logan, Utah, pp. 24–56.
Azeemuddin, M., Roegiers, J.C., Suri, P. *et al.* (1995) Stress-dependent permeability measurement of rocks in triaxial cell, in *Rock Mechanics*, (eds J.J.K. Daeman and R.A. Schultz), Balkema, Rotterdam, 645–50.
Back, W. (1961) Techniques for mapping hydrochemical facies. *US Geol. Surv. Prof. Paper* 424–D.
Back, W. (1966) Hydrochemical facies and groundwater flow patterns in northern part of Atlantic Coastal Plain, *US Geol. Surv. Prof. Paper* 498-A, p. 42.
Back, W. and Hanshaw, B. (1965) Chemical geohydrology, in *Advances in Hydroscience* (ed. V.T. Chow), Vol. 2, Academic Press, New York, pp. 49–109.
Baker, C.P. and Panciera, E.C. (1990), A geographic information system for groundwater system planning. *Journal Soil and Water Conservation*, **45**, 246–48.
Bakliwal, P.C. and Grover, A.K. (1988) Signatures and migration of the Saraswati river in Thar desert, Western India, *Records GSI*, **116**(3–8), 77–86.
Ball, D.F. and Herbert, R. (1992) The use and performance of collector wells within the regolith aquifer of Sri Lanka. *Ground Water*, **30**(5), 683–89.
Ball, T.K., Cameron, D.B., Colman, T.B. and Roberts, P.D., (1991) Behaviour of radon in the geological environment: a review, *Q. J. Eng. Geol*, **24**, 169–82.
Balmes, C.P. (1994) The geochemistry of the Mahanagdong sector, Tongonan geothermal field, Philippines. Report No. 2, Geothermal Training Programme, The United Nations University, Reykjavik, Iceland, pp. 31–52.
Banks, D., Solbjorg, M.L. and Rohr-Torp, E. (1992) Permeability of fracture zones in a Precambrian granite. *Q. J. Eng. Geol.*, **25**, 377–88.
Barenblatt, G.E., Zheltov, I.P. and Kochina, I.N. (1960) Basic concepts in the theory of homogeneous liquids in fisssured rocks. *J. Appl. Mathematics and Mech., USSR*, **24**(5), 1286–303.
Baria, R. (ed.) (1990) *Proceedings of the International Conference on Hot Dry Rock Geothermal Energy*, Camborne School of Mines, Redruth, Cornwall.
Barker, G.R. (1988) Remote sensing: the unheralded component of geographic information systems, *Photogram. Eng. Remote Sens.*, **55**(11), 1619–27.
Barker, J.A. (1991) On the discrete Kernel method for simulating pumping tests in large diameter wells, *J. Hydrol.*, **124**, 177–83.

Barker, R.D., White, C.C. and Houston, J.F.T. (1992) Borehole siting in an African accelerated drought relief, in *The Hydrogeology of Crystalline Basement Aquifers in Africa* (eds E.P. White and W.G. Burgess), Geol. Soc. Spl. Publ. 66, 183–201.

Barraclough, D., Gardner, C.M.K., Wellings, S.R. and Cooper, J.D. (1994), A tracer investigation into the importance of fissure flow in the unsaturated zone of the British Upper Chalk. *J. Hydrol.*, **156**, 459–69.

Barton, C.C., Larsen, E., Page, W.R. and Howard, T.M. (1987) Characterising fractured rock for fluid-flow, geochemical, and paleostress modelling: Methods and preliminary results from Yucca Mountain, Nevada. U.S. Gol. Survey Open-File Report, 87, 36 pp.

Barton, N.R. (1973) Review of new shear strength criterion for joints. *Engineering Geology*, **8**, 287–332.

Bear, J. (1979) *Hydraulics of Groundwater*. McGraw-Hill, New York, p. 567.

Bear, J. (1992) *Dynamics of Fluids in Porous Media*. Elsevier, New York, p. 764.

Bear, J. (1993) Modeling flow and contaminant transport in fractured rocks, in *Flow and Contaminant Transport in Fractured Rocks* (eds J. Bear, C.F. Tsang and G. de Marsily), Academic Press, San Diego, pp. 1–36.

Bear, J., Tsang, C.F. and Marsily, G. de (eds) (1993) *Flow and Contaminant Transport in Fractured Rock*. Academic Press, San Diego, p. 545.

Bedient, P.B., Rifai, H. and Nowell, C.J. (1993) *Groundwater Contamination, Transport and Remediation*, Prentice-Hall, London.

Beeson, S. and Jones C.R.C. (1988) The combined EM/VES geophysical method for siting boreholes. *Ground Water*, **26**(1), 54–63.

Bennett, G.D., Rehman, A., Sheikh, I.A. and Ali, S. (1967) Analysis of aquifer tests in the Punjab region of West Pakistan. *US Geol. Surv. Water Supply Paper 1608G*, Washington, DC, p. 56.

Bentall, R. (1963) Shortcuts and special problems in aquifer tests. *US Geol. Surv. Water Supply Paper 1545C*, p. 117.

Beres Jr., M. and Haeni, F.P. (1991) Application of ground-penetrating-radar methods in hydrogeologic studies. *Ground Water*, **29**, 375–86.

Bernard, R., Taconet, O., Vidjal-Madjar, D., Thony, J.L., Vaeulin, M., Chapotin, A., Watterlot, G. and Lebrun, A. (1984) Comparison of three in-situ surface soil moisture measurements and application to C-band scatterometer calibration. *IEEE Trans., Geoscience and Remote Sensing*, **22**, 388–94.

Berner, E.K. and Berner, R.A. (1987) *The Global Water Cycle*, Prentice-Hall, New Jersey, p. 383.

Bhar, A.K. (1996) Mathematical modelling of spring flow, PhD thesis, University of Roorkee, p. 177.

Bhattacharya, P.K. and Patra, H.P. (1968) *Direct Current Geoelectric Sounding Principles and Interpretation*, Elsevier, Amsterdam.

Billings, M.P. (1972) *Structural Geology*, (3rd ed), Prentice-Hall, New Delhi, 606 pp.

Black, J.H. (1987) Flow and flow mechanisms in crystalline rock, in *Fluid Flow in Sedimentary Basins and Aquifers*, (eds J.C. Goff and B.P.J. Williams). Geol. Soc. Spl. Publ. No. 34, pp. 185–200.

Black, J.H. (1993) Hydrogeology of fractured rocks – a question of uncertainty about geometry, in *Mem. 24th Congress of IAH,* Oslo, Part 2, pp. 783–96.

Black, J.H., Barker, J.A. and Holmes, D.C. (1983) Investigating fissured groundwater systems using slug tests and sinusoidal tests, in *Proceedings International Symposium on Methods of Instrumentation for the Investigation of Groundwater Systems,* IAH-UNESCO, 11, pp. 415–24.

Black, J.H. and Holmes, D.C. (1985) Hydraulic testing within the cross-hole investigation programme at Stripa, in *Design and Instrumentation of In situ Experiments in Underground Laboratories for Radioactive Waste Disposal* (eds B. Come, P. Johnston and A. Muller), Balkema, Rotterdam, pp. 116–27.

Bloom, A.L. (1978). *Geomorphology*, Prentice-Hall, Englewood Cliffs, New Jersey, p. 510.

Bobba, A.G., Bukata, R.P. and Jerome, J.H. (1992) Digitally processed satellite data as a tool in detecting potential groundwater flow systems. *J. Hydrol*, **131**, 25–62.

Boeckh, E. (1992) An exploration strategy for higher-yield boreholes in the West African Cystalline basement, in *The Hydrogeology of Crystalline Basement Aquifers in Africa* (eds E.P. Wright and W.G. Burgess), The Geological Society, London, pp. 87–100.

Boehmer, W.K. (1993) Secondary type aquifers and the value of pumping tests for the evaluation of groundwater potential, in *Mem. 24th Congress of IAH*, Oslo, pp. 159–68.

Boehmer, W.K. and Boonstra, J. (1987) Analysis of drawdown in the country rock of composite dike aquifers. *J. Hydrol.*, **94**, 199–214.

Bogli, A. (1980) *Karst Hydrology and Physical Speleology*. Springer-Verlag, Berlin, p. 284.

Bonacci, O. (1995) Ground water behaviour in karst: example of the Ombla Spring (Croatia), *J. Hydrol.*, **165**, 113–34.

Boni, C.F., Bano, P. and Kovalevsky, V.S. (1984) Evaluation of water resources, in *Hydrogeology of Karstic Terrains – Case Histories* (eds A. Burger and L. Dubertret), Vol. 1, pp. 9–17.

Boonstra, J. (1989) *SATEM – Selected Aquifer Test Evaluation Methods: A Microcomputer Program*, Publ. 48, Int. Inst. Land Reclamation and Improvement, Wageningen, 89 pp.

Boonstra, J. and Boehmer, W.K. (1986) Analysis of data from aquifer and well tests in intrusive dikes, *J. Hydrol.*, 88, 301–17.

Boonstra, J. and Boehmer, W.K., (1989), Analysis of data from well tests in dikes and fractures. ILRI Reprint No. 57, in *Groundwater Contamination – Use of Models in Decision Making*, Kluwer Academic, Dordrecht, pp. 171–80.

Bosak, P., Ford, D.C., Glazek, J. and Horacek, I. (eds) (1989) *Paleokarst: Development in Earth Surface Processes*, Elsevier, Amsterdam, p. 725.

Boulton, N.S. (1963) Analysis of data from nonequilibrium pumping tests allowing for delayed yield from storage. *Proceedings Inst. Civil Engrs.* **26**, 469–82.

Boulton, N.S. (1970) Analysis of data from pumping tests in unconfined anisotropic aquifers. *J. Hydrol.*, **10**, 369–78.

Boulton, N.S. and Streltsova, T.D. (1976) The drawdown near an abstraction well of large diameter under non-steady conditions in an unconfined aquifer. *J. Hydrol.*, 30, 29–46.

Boulton, N. S. and Streltsova, T.D. (1977) Unsteady flow to a pumped well in a fissured water bearing formation. *J. Hydrol.*, 35, 257–69.

Boulton, N.S. and Streltsova, T.D. (1978) Unsteady flow to a pumped well in an unconfined fissured aquifer. *J. Hydrol.*, 37, 349–63.

Bourdet, D. and Gringarten, A.C. (1980) Determination of fissure volume and block size in fractured reservoirs by type-curve analysis. Paper SPE 9293 presented at the 1980 SPE Annual Fall Tech. Conf. and Exhibition, Dallas.

Bouwer, H. (1975) Predicting reduction in water losses from open channels by phreatophyte control. *Water Resources Res.*, **11**, 96–101.

Bouwer, H. (1978), *Groundwater Hydrology*, McGraw-Hill, New York, p. 480.

Bouwer, H. and Rice, R.C. (1976) A slug test for determining hydraulic conductivity of unconfined aquifers with completely or partially penetrating wells. *Water Res. Research*, **12**(3), 423–28.

Brace, W.F. (1978) A note on the permeability changes in geologic material due to stress, in *Rock Friction and Earthquake Prediction* (eds J.D. Byerlee and M. Wyss), Birkhauser Verlag, Basel, pp. 627–33.

Brace, W.F. (1980) Permeability of argillaceous and crystalline rocks. *Int. J. Rock Mech. min. Sci. & Geomech. Abst*, 17, 241–51.

Brace, W.F. (1984) Permeability of crystalline rocks: new in-situ measurements. *J. Geophys. Res.*, **89** (B6), 4327–30.

Bredehoeft, J.D. and Papadopulos, S.S. (1980) A method for determining the hydraulic properties of tight formations. *Water Resources Res.*, **16**(1), 233–38.

Briechie, D. and Voigt, R. (1980) Deep open pit mining and groundwater problems in the Rhenish Lignite District, in *Studies and Reports in Hydrology* 28, pp. 274–90, UNESCO, Paris.

Bromley, J., Mannstrom, B., Nisca, D. and Jamtlid, A. (1994) Airborne geophysics: application to a groundwater study in Botswana. *Ground Water*, **32**(1), 79–90.

Brown, R.H. (1963) Estimating the transmissivity of an artesian aquifer from the specific capacity of a well, in *Methods of Determining Permeability, Transmissibility and Drawdown* (compiled by R. Bentall), US Geol. Survey Water Supply Paper, 1536-I, pp. 336–38.

Buckley, D.K. and Zeil, P. (1984) The character of fractured rock aquifers in eastern Botswana, in Challenges in African Hydrogeology and Water Resources. (eds D.E. Walling, S.S.D. Foster and P. Wurzel), Proceedings of the Harare Symposium, July 1984, *IAHS Publ.* 144, pp. 25–36.

Burdon, D.J. and Al-Sharhan, A. (1968) The problem of the palaeokarstic Damman limestone aquifer in Kuwait. *J. Hydrol.*, **6**, 385–404.

Burdon, D.J. and Mazloum, S. (1961) Some chemical types of groundwater from Syria, in *Salinity Problems in the Arid Zones, Arid Zone Res. XIV*, UNESCO, pp. 73–90.

Burger, A. and Dubertret, L. (eds) (1984) *Hydrogeology of Karstic Terrains*, IAH Vol. 1, Heise, Hanover, p. 264.

Burrough, P.A. (1981) *Principles of Geographic Information Systems for Land Resource Assessment*, Oxford University Press, New York, p. 193.

Busenberg, E. and Plummer, L.N. (1992) Use of chlorofluorocarbons (CCl_3F and CCl_2F_2) as hydrologic tracers and age-dating, *Water Resources Res.*, **28**(9), 2257–83.

Campbell, M.D. and Lehr. J.H. (1973) *Water Well Technology*, McGraw-Hill, New York, p. 681.

Canter, L.W. (1985) Methods for the assessment of groundwater pollution potential, in *Groundwater Quality* (eds C.H. Ward, W. Giger and P.L. McCarty), John Wiley, New York, pp. 270–306.

Canter, L.W., Knox, R.C. and Fairchild, D.M. (1987) *Ground Water Quality Protection*, Lewis, Chelsea, Michigan, p. 562.

Carlsson, A. and Olsson, T. (1977) Variations in hydraulic conductivity in some Swedish rock types. *Proceedings 1st International Symposium Rock Store*, Stockholm, pp. 301–307.

Carlsson, L. and Carlstedt, A. (1977) Estimation of transmissivity and permeability in Swedish bedrock. *Nordic Hydrology*, **8**, 103–16.

Carruthers, R.M., Greenbaum, D., Jackson, P.D. *et al.* (1993) Geological and geophysical characterisation of lineaments in Southeast Zimbabwe and implica-

tions for groundwater exploration. *B.G.S. Tech. Report WC/93/7*.
Carruthers, R.M. and Smith, I.F. (1992) The use of ground electrical survey methods for siting water supply boreholes in shallow crystalline basement terrains, in *The Hydrogeology of Crystalline Basement Aquifers in Africa* (eds E.P. Wright and W.G. Burgess), Geological Society of London, Special Publication, No. 66, London, pp. 203–220.
Carvalho, J.M. (1993) Mineral and thermal water resources development in the Portugese Hercynian Massif, in *Mem. 24th Congress of IAH*, Oslo, pp. 548–61.
Cassiani, G. and Medina, Jr. M.A. (1997) Incorporating auxiliary geophysical data into groundwater flow parameter estimation. *Ground Water*, **35**(1), 79–91.
Castany, G., Groba, E. and Romijn, E. (eds) (1985) Hydrogeological Mapping in Asia and the Pacific Region, in *IAH Proceedings Workshop*, Bandung, 1983, Heise, Hanover, p. 480.
CGWB (1980) Groundwater studies in the Vedavati river basin – parts of Karnataka and Andhra Pradesh (unpublished report), CGWB, p. 49.
CGWB (1989) *Hydrogeological Map of India*, 2nd edn.
CGWB (1995a) *Ground Water Resources of India*, Faridabad, 147 pp.
CGWB (1995b) *Hydrogeological Atlas of Orissa*, Bhubaneswar, p. 41.
CGWB (1995c) *Bhu-Jal News*, **10**(1), p. 46.
Chachadi, A.G., Mishra, G.C. and Singhal, B.B.S. (1991) Drawdown at a large-diameter observation well. *J. Hydrol.*, **127**, 219–33.
Chandrasekharam, D. (1996) Geothermal energy: the future renewable energy source of India, *Abstracts 30th IGC*, Beijing, Vol. 1, p. 379.
Charlie, W.A., Veyera, G.E., Durnford, D.S. and Doehring, D.O. (1994) Discussion on the paper 'Geology of the Chinese nuclear test site near Lop Nor, Xinjiang, Ugyur Autonomous Region, China', by J.R. Matzko. *Eng. Geol.*, **38**, 177–79.
Chebotarev, I.I. (1950) Genetic types of subterranean waters and their utilisation. *Water and Water Engineering*, October 1950, 146–50.
Chen, M. and Cai, Z. (1995) Groundwater resources and hydro-environmental problems in China. *Episodes*, **18** (1 and 2), pp. 66–68.
Cherry, J.A. (1989) Hydrogeologic contaminant behaviour in fractured and unfractured clayey deposits in Canada, in *Contaminant Transport in Groundwater* (eds H.E. Kobus and W.Kinzelbach), Balkema, Rotterdam, pp. 11–20.
Chiles, J.P. and Marsily, G. de. (1993) Stochastic models of fracture systems and their use in flow and transport modeling, in *Flow and Contaminant Transport in Fractured Rocks*, Academic Press, San Diego, pp. 169–231.
Chilton, P.J. and Foster, S.S.D. (1993) Hydrogeological characterisation and water supply potential of basement aquifers in tropical Africa, in *Mem 24th Congress of IAH*, Oslo, part 2, pp. 1083–110.
Chilton, P.J., Lawrence, A.R. and Stuart, M.E. (1995) The impact of tropical agriculture on groundwater quality, in *Groundwater Quality* (eds H. Nash and G.J.H. McCall), Chapman & Hall, London, pp. 113–22.
Clauser, C. (1992) Permeability of crystalline rocks *EOS, Trans. Am. Geophys. Union* 73, pp. 233–38.
Collie, M.J. (ed) (1982) *Geothermal Energy – Recent Developments*, Noyes Data Corp., New Jersey, USA, p. 444.
Colwell, R.N. (ed) (1960) *Manual of Photographic Interpretation*, Am. Soc. Photogramm, Falls Church.
Colwell, R.N. (ed) (1983) *Manual of Remote Sensing*, Vols I, II, 2nd edn, Am. Soc. Photogramm, Falls Church.
Come, B., Johnston, P. and Muller, A. (eds) (1985) *Design and Instrumentation of In-situ Experiments in Underground Laboratories for Radioactive Waste Disposal*, Balkema, Rotterdam, pp. 474.
Cook, J.C. (1975) Radar transparencies of mines and tunnel rocks. *Geophysics*, **40**, 865–85.
Cooper, H.H., Bredehoeft, J.D. and Papadopulos, I.S. (1967) Response of finite-diameter well to an instantaneous charge of water. *Water Resources Res.*, **3**, 263–69.
Couture, R.A., Seitz, M.G. and Steindler, M.J. (1983) Sampling of brines in cores of Precambrian granite from Northern Illinois. *J. Geophys, Res.*, **88**(B9), 7331–34.
Craig, H. (1963) The isotopic geochemistry of water and carbon in geothermal areas, in *Nuclear Geology in Geothermal Areas* (ed. E. Tongiorg), CNR, Rome, pp. 17–53.
Cruse, K. (1979) A review of water well drilling methods. *Q.J. Eng. Geol.*, **12**, 79–95.
Csallany, S.C. (1965) The hydraulic properties and yields of dolomite and limestone aquifers. *Proceedings Dubrovnik Symposium on Hydrogeology of Fractured Rocks*, Vol. I, 120–38.
Csallany, S.C. and Walton, W.C. (1963) Yield of shallow dolomite wells in Northern Illinois. *Illinois State Water Survey Report of Investigation*, 46, Urbana, Illinois, p. 43.
CSC (1984) Water resources of small islands. *Proceedings Regional Workshop on Water Resources of Small Islands*, Part 2, Commonwealth Science Council, London, 530.
Custodio, E. (1985) Low permeability volcanics in the Canary Islands (Spain), *Mem. IAH*, **17**(2), 562–73.
Daily, W. and Owen, E. (1991) Crosshole resistivity tomography. *Geophysics*, **56**, 1228–35.
Das, S., Kar, A. and Behera, S.C. (1996) Studies on

hydrogeology and groundwater development possibilities in alluvial and fractured basement aquifers in Keonjhar district Orissa. *Proceedings Hydro 96*, Indian Inst. of Tech., Kanpur.

Davis, S.N. (1969) Porosity and permeability of natural material, in *Flow Through Porous Media* (ed. R.J.M. DeWiest), Academic Press, New York, pp. 54–90.

Davis, S.N. (1974) Changes of porosity and permeability of basalt with geologic time, in *Proceedings Symposium on Hydrology of Volcanic Rocks*, (unpublished), Lanzarote, Spain.

Davis, S.N. and DeWiest, R.J.M. (1966) *Hydrogeology*, John Wiley, New York, 463 pp.

Davis, S.N. and Turk, L.J. (1964) Optimum depth of wells in crystalline rocks. *Ground Water*, **2**, 6–11.

Davis, S.N., Whittemore, D.O. and Fabryka-Martin, J. (1988) Uses of chloride/bromide ratios in studies of potable water. *Ground Water* **36**(2), 338–50.

Davison, C.C. (1985) Far-field hydrogeological monitoring at the site of Canada's Underground Research Laboratory, in *Design and Instrumentation of In-Situ Experiments in Underground Laboratories for Radioactive Waste Disposal* (eds B. Come, P. Johnston and A. Muller), Balkema, Rotterdam, pp. 142–56.

Deane, R.A. and Domville, A.R. (1973) Side looking radar systems and their potential application of earth-resources surveys. *ESRO Report*, CR-137B, Paris, Vol. 3.

Deike, R.G. (1990) Dolomite dissolution rates and possible Holocene dedolomitization of water-bearing units in the Edwards Aquifer, Southcentral Texas – a summary, in *Selected Papers on Hydrogeology* (eds E.S. Simpson and J.M. Sharp, Jr.), Vol. 1, IAH, Verlag Heinz Heise, Hanover, pp. 97–109.

Deolankar, S.B. (1980) The Deccan basalts of Maharashtra, India – their potential as aquifers. *Ground Water*, **18**(5), 434–37.

Desai, B.I., Gupta, S.K., Shah, M.V. and Sharma, S.C. (1979) Hydrochemical evidence of sea water intrusion along the Mangrol Chorwad coast of Saurashtra, Gujarat. *Hydrol. Sci. Bull.*, **24**(1), 71–82.

deStadelhofen, C.M. (1994) *Anwendung Geophysikalisher Methoden in der Hydrogeologie*, Springer Verlag, Berlin.

Detay, M. and Poyet, P. (1990) Influence of the development of saprolite reservoir and its state of saturation on the hydrodynamic characteristics of drilling in the crystalline basement, in *Selected Papers on Hydrogeology*, (eds E.S. Simpson and J.M. Sharp, Jr.) Vol. 1, IAH, Verlag Heinz Heise, Hanover, pp. 463–71.

DeWiest, R.J.M. (1969) *Flow Through Porous Media*, Academic Press, New York, p. 530.

Dietvorst, E.J.L., DeVries, J.J. and Gieske, A. (1991) Coincidence of well fields and tectonic basins in the Precambrian shield area of south east Botswana. *Ground Water*, **29**, 864–77.

Dietz, H.B. (1985) The in-situ test program for site characterization of basalt, in *Design and Instrumentation of In-situ Experiments in Underground Laboratories for Radioactive Waste Disposal* (eds B. Come, P. Johnston and A. Muller), Balkema, Rotterdam, pp. 82–92.

Dingman, S.L. (1994) *Physical Hydrology*, Macmillan, New York, p. 575.

Dissanayake, C.B.(1991) The fluoride problem in the groundwater of Sri Lanka – environmental management and health. *Int. J. Env. Studies*, **38**, 137–56.

Dobrin, M.B. (1976) *Introduction to Geophysical Prospecting*, McGraw-Hill, New York.

Domenico, P.A. (1972) *Concepts and Models in Groundwater Hydrology*, McGraw-Hill, New York, p. 405.

Downing, R.A., Allen, D.J., Barker, J.A. *et al.* (1984). Geothermal exploration at Southampton in the U.K., *Energy Exploration and Exploitation*, **2**(4), 327–42.

Dreier, R.B., Solomon, D.K., and Beaudoin, C.M. (1987) Fracture characterization in the unsaturated zone of a shallow land burial facility, in *Flow and Transport Through Unsaturated Fractured Rock* (eds D.D. Evans and T.J. Nicholson), AGU Monograph 42, 51–59.

Drever, J.I. (1988) *The Geochemistry of Natural Waters*, 2nd edn, Prentice Hall, New Jersey, p. 437.

Driscoll, F.G. (ed) (1986) *Groundwater and Wells*, Johnson Division, St. Paul, Minnesota, p. 1089.

Drury, S.A. (1987) *Image Interpretation in Geology*, Allen and Unwin, London, p. 243.

Duba, D. (1973) Guidelines for estimates of wells spacing in weathered zone of hard rocks in Mysore State. *Ground Water Studies* No. 122, Dept. of Mines and Geology, Govt. of Mysore, Bangalore, (unpublished), p. 9.

Duncan, R.A. and Pyle, D.G. (1988) Rapid eruption of the Deccan flood basalts, Western India in *Deccan Flood Basalts* (ed. K.V. Subbarao). *Mem. Geol. Soc. India*, 1–9.

Dutay, M., and Poyet P. (1990) Application of remote sensing in field hydro-engineering geology, the artificial intellligence approach. *Proceedings International Symposium Remote Sensing and Water Resources*, Enchede, IAH.

Dverstorp, B., Anderson, J. and Nordqvist, J. (1992) Discrete fracture network interpretation of field tracer migration in sparsely fractured rock. *Water Resources Res.*, **28**(9), 2327–43.

Eagon, H.B. and Johe, D.E. (1972) Practical solutions for pumping tests in carbonate rock aquifers. *Ground Water*, **10**, 6–13.

Earlougher, R.C. Jr. (1977) *Advances in Well Test Analysis*, Monograph Series, SPE, Richardson, Texas.

Edmunds, W.M. and Savage, D. (1991) Geochemical characteristics of groundwater in granites and related crystalline rocks, in *Applied Groundwater Hydrology* (eds R.A. Downing and W.B. Wilkinson), Oxford University Press, Oxford, pp. 266–82.

Edwards, K.B. and Jones, L.C. (1993) Modelling pumping tests in weathered glacial till. *J. Hydrol.*, **150**, 41–60.

Elachi, C. (1983) Microwave and infrared satellite remote sensing, (ed. R.N. Colwell), in *Manual of Remote Sensing*, 2nd edn, Vol. I, Am. Soc. Photogramm, Falls Church.

Elachi, C., Roth, L.E. and Schaber, G.G. (1984) Spaceborne radar subsurface imaging in hyperarid regions. *IEEE Trans* GE-22, 382–87.

Elder, J. (1981) *Geothermal Systems*, Academic Press, London, p. 508.

El-Kadi A.I., Oloufa, A.A., Eltahan, A.A. and Malik H.U. (1994) Use of a geographic information system in site-specific ground-water modelling. *Ground Water*, **32**(4) 617–25.

Ellyett, C.D. and Pratt, D.A. (1975) A review of the potential applications of remote sensing techniques to hydrogeological studies in Australia, *AWRC, Tech Paper No. 13*, 147 pp.

Erdelyi, M. and Galfi, G. (1988) *Surface and Subsurface Mapping in Hydrogeology*, John Wiley, New York, p. 384.

Evans, B.M. and Myers, W.L. (1990) A GIS-based approach to evaluating regional groundwater pollution potential with DRASTIC. *Journal Soil and Water Conservation*, **45**. 242–45.

Evans, D.D. and Nicholson, T.J. (eds) (1987) *Flow and Transport Through Unsaturated Fractured Rocks*, Monograph 42, AGU Washington, p. 187.

Fabbri, A.G., Fungi, K.B. and Kasrand, T. (1988) Issues in the integration of spatially distributed data ancillary to remotely sensed images, in *Image Analysis and Processing II*, (eds V. Contori, V. Di Gesu and S. Levialdi), Plenum, New York pp. 653–64.

Faillace, C. (1973) Location of groundwater and the determination of the optimum depth of wells in metamorphic rocks of Karamoja, Uganda, in *Proceedings 2nd International Convention on Groundwater*, Palermo, Italy.

Falkland, A.C. (1984) Development of groundwater resources on coral atolls; experiences from Tarawa and Christmas islands, Republic of Kiribati, in *Water Resources of Small Islands, Republic of Karibati. CSC Tech. Publ. Series No.154*, London, pp. 436–52.

Farvolden, R.N., Meneley, W.A., Le Breton, E.G. *et al.* (1963) *Early Contributions to the Groundwater Hydrology of Alberta.* Res. Council of Alberta, Bull. 12, p. 123.

Feddes, R.A., Kabat, P., Van Bakel, P.J.T., Bronswijk, J.J.B. and Halbertsma, J. (1988) Modelling soil water dynamics in the unsaturated zone – state of the art. *J. Hydrol.*, **100**, 69–111.

Fernandopulle, D., Vos, J., La Moneda, E. and Medina, L. (1974) Groundwater resources of the island of Gran Canaria, in *Symposium on Hydrogeology of Volcanic Rocks,* Lanzarote, Spain, UNESCO, (unpublished).

Ferris, J.G., Knowles, D.B., Brown, R.H. and Stallmen, R.W. (1962) *Theory of Aquifer Tests*, US Geol. Surv. Water Supply Paper 1536-E, p. 174.

Fetter, C.W. (1988) *Applied Hydrogeology*, 2nd edn, Merrill, Columbus, p. 592.

Fetter, C.W. (1993) *Contaminant Hydrogeology*, Macmillan, New York, 458 pp.

Florquist, B.A. (1973) Techniques for locating water wells in fractured crystalline rocks. *Ground Water*, **11**(3), 26–28.

Ford, D.C. and Williams, P.W. (1989) *Karst Geomorphology and Hydrology*. Unwin Hyman, London. p. 601.

Franz, T. and Guiger, N.F. (1990) *FLOWPATH: Steady State Two-Dimensional Horizontal Aquifer Simulation Model*, Waterloo Hydrogeologic Software, Ontario, Canada.

Freeze, R.A. and Cherry, J.A. (1979) *Groundwater*, Prentice Hall, New Jersey, p. 604.

Gale, J.E. (1982a) Fundamental hydraulic characteristics of fractures from field and laboratory investigations, in *Papers of the Groundwater in Fractured Rock Conference*, AWRC, Conference Series No.5, Canberra, pp. 79–94.

Gale, J.E. (1982b) Assessing the permeability characteristics of fractured rocks in *Recent Trends in Hydrogeology* (ed. T.N. Narasimhan) Geol. Soc. Am. Special Paper 189, 163–81.

Gale, J.E., Rouleau, A. and Witherspoon, P.A. (1982) Hydrogeologic characteristics of a fractured granite, in *Papers of the Groundwater in Fractured Rock Conference*, AWRC Conference Series No.5, Canberra, pp. 95–108.

Gardner, C.M.K., Bell, J.P., Cooper, J.D., Darling, W.G. and Reeve, C.E. (1991) Groundwater recharge and water movement in the unsaturated zone, in *Applied Groundwater Hydrology* (eds R.A. Downing and W.B. Wilkinson), Oxford University Press, Oxford, p. 54.

Garrels, R.M. and Christ, C.L. (1965) *Solutions, Minerals and Equilibria*, Harper and Row, New York, p. 450.

Gascoyne, M. and Kamineni, D.C. (1993) The hydrogeochemistry of fractured plutonic rocks in the Canadian shield, in *Mem. 24th Congress of IAH*, Oslo, 440–49.

Gascoyne, M., Ross, J.D. and Watson, R.L. (1996) Highly saline pore fluid in the rock matrix of a granitic batholith in the Canadian shield, *Abstracts*

30th IGC, Vol. 3, Beijing, p. 269.

Gerard, A., Baumgartner, J., Baria, R. *et al.* (1996) European Hot Dry Rock Research project at Soultz-sous-Forets. *30th IGC, Abstracts*, Vol. 1, Beijing, p. 385.

Germain, D. and Frind, E.O. (1989) Modelling of contaminant migration in fracture networks: effects of matrix diffusion, in *Contaminant Transport in Groundwater*, (eds H.E. Kobus and W. Kinzelbach), Balkema, Rotterdam, pp. 267–74.

Ghosh, A. (1990) Role of radial cracks, freeface and natural discontinuities of fragmentation from bench blasting, PhD thesis, University of Arizona.

Giusti, E.V. (1977) Hydrogeology and 'geoesthetics' applied to land use planning in the Puerto Rican Karst, in *Karst Hydrogeology* (eds J.S. Tolson and F.L. Doyle), Mem. 12th Congress of IAH, Alabama, pp. 149–68.

Goel, P.S., Datta P.S., Rama *et al.* (1975) Tritium tracer studies on groundwater recharge in the alluvial deposits of Indo-Gangetic Plains of Western U.P., Punjab and Haryana, in *Approaches and Methodologies for Development of Groundwater Resources* (eds R.N. Athavale and V.B. Srivastava) NRGI, Hyderabad, pp. 309–22.

Goswami, A.B. (1995) A Critical Study on Water Resources of West Bengal, PhD thesis (unpublished), Jadavpur University, p. 309.

Grant, M.A., Donaldson, I.G., and Bixley, P.F. (1982) *Geothermal Reservoir Engineering*, Academic Press, New York, p. 367.

Graziadei, W. and Zotl, J.G. (1984) Karst water gallery and hydroelectric power plant Muhau – the water supply of Innsbruck, Austria, in *Hydrogeology of Karstic Terrains* (eds A. Burger and L. Dubertret), IAH, Vol. 1, 113–16.

Greenbaum, D. (1992) Structural influences on the occurrence of groundwater in SE Zimbabwe, in *The Hydrogeology of Crystalline Basement Aquifers in Africa* (eds E.P. Wright and W.G. Burgess), Geol. Soc. Spl. Publ. No. 66, London, pp. 77–85.

Greene, E.A. and Rahn, P.H. (1995) Localized anisotropic transmissivity in a karst aquifer, *Ground Water*, **33**(5), 806–16.

Greenman, D.W., Swarzenski, W.V. and Bennett, G.D. (1967) *Groundwater Hydrology of the Punjab, West Pakistan with Emphasis on Problems Caused by Canal Irrigation. US Geol. Surv. Water Supply Paper*. 1608-H, Washington, p. 66.

Gregory, K.J. and Walling, D.E. (1973) *Drainage Basin Form and Process – a Geomorphic Approach*, Edward Arnold, London, p. 458.

Griffiths, D.H. and Turnbill J.H. (1985) A multi-electrode array for resistivity surveying. *First Break*, **3**, 16–20.

Griggs, J.E. and Peterson, F.L. (1993) Groundwater flow dynamics and development strategies at the atoll scale. *Ground Water*, **31**(2), 209–20.

Gringarten, A.C. (1982) Flow-test evaluation of fractured reservoirs, in *Recent Trends in Hydrogeology* (ed. T.N. Narasimhan), Geol. Soc. Am. Special Paper, 189, pp. 237–63.

Gringarten, A.C. and Witherspoon, P.A. (1972) A method of analyzing pump test data from fractured aquifers, in *Int. Soc. Rock Mech. and Int. Assoc. Eng. Geol. Proceedings Symp*, Rock Mech, Vol. 3B, 1–9, Stuttgart.

Grisak, G.E. and Pickens, J.F. (1980) Solute transport through fractured media 1. The effect of matrix diffusion. *Water Resources Res.*, **16**(4), 719–30.

Grisak, G.E., Pickens, J.F. and Cherry, J.A. (1980) Solute transport through fractured media 2. Column study of fractured till. *Water Resources Res.*, **16**(4), 731–39.

GSI (1969) *Geohydrological Map of India*, Geol. Surv. India, Calcutta.

GSI (1991) *Geothermal Atlas of India*, Geol. Surv. India, Spl. Publ. 19, p. 144.

Guillen, J.A. (1984) Hydrogeological facts about dike aquifers and underground water circulation in Tahiti, in *Proceedings Regional Workshop on Water Resources of Small Islands, Suva, Fiji*. CSC, Tech. Publ. Series, No. 154, London, 455–72.

Gupta, C.P., Thangarajan, M. Gurunadha Rao, V.V.S. *et al.* (1989) Simulation of regional hydrodynamics in the Neyveli aquifer. *IGW*, Vol. II, 607–18.

Gupta C.P., Thangarajan, M., Guranadha Rao, V.V.S. *et al.* (1996) Groundwater polution in the upper Palar Basin, Tamilnadu *Proceedings Workshop on Challenges in Groundwater Development*, Central Board of Irrigation and Power, New Delhi, 200–13.

Gupta, R.P. (1991) *Remote Sensing Geology*. Springer-Verlag, Berlin, p. 356.

Gupta, S.K. (1996) Evaluation of fracture characteristics for hydrogeological modelling. M. Tech. Dissertation, University of Roorkee, Roorkee.

Gustafsson, P. (1993) SPOT satellite data for exploration of fractured aquifers in a semi-arid area in southern Botswana. *Memo. of the 24th Congress of IAH*, Oslo, pp. 562–75.

Hadzisehovic, M., Dangic, A., Miljevic, N., Sipka, V. and Golobocanin, D. (1995) Geochemical-water characteristics of the Surdulica aquifer. *Ground Water*, **33**(1), 112–23.

Haeni, F.P., Lane Jr., J.W. and Lieblich, D.A. (1993) Use of surface-geophysical and borehole-radar methods to detect fractures in crystalline rocks, Mirror Lake Area, Grafton County, near Hampshire. *Memoires of the 24th Congress of IAH*, Oslo 1993, pp. 577–87.

Halliday, S.L. and Wolfe, M.L. (1991) Assessing

groundwater pollution potential from nitrogen fertilizer using a Geographic Information System. *Water Resources Bull.*, 27(2), 237–45.

Hamill, L. and Bell, F.G. (1986), *Groundwater Resource Development*, Butterworths, London, p. 344.

Handa, B.K. (1974) Environmental pollution – occurrence of high nitrate ion concentrations in groundwater from some parts of India. *IAH Mem. X*, pp. 80–85.

Handa, B.K.(1975) Natural waters, their chemistry pollution and treatment with a chapter on the use of saline water. *CGWB Tech. Report 2* (unpublished), p. 264.

Handa, B.K. (1988) Fluoride occurrence in natural waters in India and its significance. *Bhu-Jal News*, CGWB, 3(2), 31–37.

Handa, B.K. (1994) Ground water contamination in India, in *Proceedings Regional Workshop on Env. Aspects of Groundwater Dev.*, (ed. D.C. Singhal), Ind. Assoc. Hydrol., Kurukshetra, 1–33.

Hansmann, B.C., Meijerink, A.M.J. and Kodituwakku, K.A.N. (1992), An inductive approach for groundwater exploration. *ITC J.* 1992–93, 269–76.

Hantush, M.S. (1956) Analysis of data from pumping test in leaky aquifers. *Trans. Amer Geophys. Union*, 37, 702–14.

Hantush, M.S. (1964) Hydraulics of wells, in *Advances in Hydroscience*, (ed. V.T. Chow), Vol. 1, Academic Press, New York, pp. 281–32.

Hantush, M.S. and Jacob. C.E. (1955) Non-steady radial flow in an infinite leaky aquifer. *Trans. Amer Geophys. Union*, 36, 95–100.

Hasan, S.E. (1996*) Geology and Hazardous Waste Management.* Prentice Hall, Upper Sadle River, New Jersey, 387 pp.

Hatton, S.M. and Dincer, T. (1979) Using Landsat imagery to study the Okavango Swarup, Botswana, in *Satellite Hydrology 1979*, (eds M. Deutsch, B.R. Wiesnet and A. Rango), Proceedings American Water Resources Association.

Hazell J.R.T., Cratchley C.R. and Jones, C.R.C. (1992) The hydrogeology of crystalline aquifers in northern Nigeria and geophysical techniques used in their exploration in *The Hydrogeology of Crystalline Basement Aquifers in Africa* (eds E.P. Wright and W.G. Burgess) Geological Society of London, Special Publication, No. 66, London, pp. 155–82.

Hazen, A. (1893) Some physical properties of sands and gravels with special reference to their use in filtration. Massachusetts State Board of Health, *24th Ann. Report*, pp. 541–56, Boston.

Heath, M.J. (1985) Solute migration experiments in fractured granite, Southwest England, in *Design and Instrumentation of In-situ Experiments in Underground Laboratories for Radioactive Waste Disposal* (eds B. Come, P. Johnston and A. Muller), Balkema, Rotterdam, pp. 191–200.

Heath, R.C. (1976) Design of ground-water level observation-well programs. *Ground Water*, **14**(2), 71–77.

Hem, J.D. (1989) *Study and Interpretation of the Chemical Characteristics of Natural Water*, 3rd edn, *US Geol. Surv.* Water Supply Paper 2254, p. 263.

Hendry, M.J. (1982) Hydraulic conductivity of a glacial till in Alberta. *Ground Water*, **20**, 162–69.

Henriksen, H. (1995) Relation between topography and well yield in boreholes in crystalline rocks, Sognag Fjordane, Norway. *Ground Water*, **33**(4), 635–43.

Herbert, Jr, R.B. (1992) Evaluating the effectiveness of a mine tailing cover. *Nordic Hydrology*, **23**, 193–208.

Herbert, R. and Kitching, R. (1981) Determination of aquifer parameters from large-diameter dug well pumping tests. *Ground Water*, **19**(6), 593–99.

Heywood, I., Cornelius, S. and Carver, S. (1997) *Introduction to Geographical Information Systems*, Longman, London, 336 pp.

Hoek, E. and Bray, J. (1977) *Rock Slope Engineering*, 2nd edn. The Instituion of Mining and Metallurgy, London, p. 402.

Hollaway, J. (1996) Environmental problems in Zimbabwe from gold panning, in *Centre for Resources Studies Perspective*, No. 52, pp. 25–28, Ontario.

Hopkins, P.L., Eaton, R.R. and Sinnock, S. (1987) Effect of drift ventilation on repository hydrology and resulting solute transport implications, in *Flow and Transport Through Unsaturated Fractured Rock* (eds D.D. Evans and T.J. Nicholson), *Geophy. Monograph* 42, AGU, pp. 177–83.

Hord, R.M. (1982) *Digital Image Processing of Remotely Sensed Data.* Academic Press, New York, 256 pp.

Hord, R.M. (1986) *Remote Sensing – Methods and Applications.* Wiley, New York, 362 pp.

Horne, R.N. (1990) *Modern Well Test Analysis – A Computer Aided Approach.* Petroway Inc., Palto Alto, Clif., p. 185.

Houston, J. (1992) Rural water supplies: comparative case histories from Nigeria and Zimbabwe, in *The Hydrogeology of Crystalline Basement Aquifers in Africa* (eds E.P. Wright and W.G. Burgess), Geol. Soc. Spl. Publ. No. 66, London, pp. 243–57.

Huntley, D. (1986) Relations between permeability and electrical resistivity in granular aquifers. *Ground Water*, **24**, 466–75.

Huntley, D., Normenson, R. and Steffey, D. (1992) The use of specific capacity to assess transmissivity in fractured rock aquifers. *Ground Water*, **30**(3), 396–402.

Huntoon, P.W. (1992a) Hydrogeologic characteristics

and deforestation of the stone forest karst aquifers of South China. *Ground Water*, **30**(2), 167–76.

Huntoon, P.W. (1992b) Exploration and development of ground water from the Stone Forest karst aquifers of South China. *Ground Water*, **30**(3), 324–30.

Hyder, Z. and Butler, J.J. (1995) Slug test in unconfined formations: an assessment of the Bouwer and Rice technique. *Ground Water*, **33**(1), 16–22.

IAEA (1983) *Guidebook on Nuclear Techniques in Hydrology*, Tech. Reports Series No. 91.

IAHS–UNESCO (1965) *Hydrology of Fractured Rocks*, Vol. 1 and 2, Proceedings of the Dubrovnik Symposium, IAHS – UNESCO.

Inamdar, P.M. and Kumar, D. (1994) On the origin of bole beds in Deccan traps. *J. Geol. Soc. Ind.*, **44**(3), 331–34.

Ineson, J. and Downing, R.A. (1964) The groundwater component of river discharge and its relationship to hydrogeology. *J. Inst. Water Engrs.*, **18**, 519–41.

Irmay, S. (1954) On the hydraulic conductivity of unsaturated soils. *Trans. Am. Geophys. Union*, **35**, 463–67.

ISI (1983) *Characteristics for Drinking Water* IS:10500, Indian Standards Inst., New Delhi, pp. 6–11.

IWRS (1996) *Interbasin Transfer of Water for National Development – Problems and Prospects*. Indian Water Resources Society, Roorkee, p. 35.

Jacobson, G. (1982) Groundwater in fractured rock aquifers of the Australian Capital Territory – a valuable resource or a major problem, in *Papers of the Groundwater in Fractured Rock Conference* 1982, AWRS, Series No. 5, pp. 109–18, Canberra.

Jagannathan, V. (1993) Deep seated fractured aquifers in crystalline rocks. *J.Geol. Soc. Ind.*, **41**(3), 280–82.

Jalludin, M. and Razack, M. (1993) Groundwater occurrence in weathered and fractured basalts, Republic of Djibouti, in *Mem. 24th Congress of IAH*, Oslo, Part 1, pp. 248–58.

Javendal, I., Doughty, C. and Tsang, C.F. (1984) *Groundwater Transport: Handbook of Mathematical Models*. Water Resources Monograph Series 10, AGU, p. 228.

Jayasena, H.A.H., Singh, B.K. and Dissanayake, C.B. (1986) Groundwater occurrences in the hard rock terrains of Sri Lanka – a case study. *Aqua*, **4**, 214–19.

Jensen, J.R. (1986) *Introductory Digital Image Processing*. Prentice Hall, Englewood Cliffs, New Jersey, 379 pp.

Jones, L., Lemar, T. and Tsai, C. (1992) Results of two pumping tests in Wisconsin age weathered till in Iowa. *Ground Water*, **30**(4), 529–38.

Jorgensen, D.G. (1991) Estimating geohydrologic properties from borehole geophysical logs. *Groundwater Monitoring and Review*, Summer 1991, 123–29.

Jouanna, P. (1993), A summary of field test methods in fractured rocks, in *Flow and Contaminant Transport in Fractured Rock* (eds J. Bear, C.F. Tsang and G. de Marsily), Academic Press, San Diego, pp. 437–543.

Kaehler, C.A. and Hsieh, P.A. (1994) *Hydraulic Properties of a Fractured Rock Aquifer, Lee Valley, San Diego County, California*. US Geol. Surv. Water Supply Paper 2394, p. 64.

Kaila, K.L. (1988) Mapping the thickness of Deccan Trap flows in India from DSS studies and inferences about a hidden Mesozoic basin in the Narmada–Tapti region, in *Deccan Flood Basalts* (ed. K.V. Subbarao) *Geol. Soc. Ind.*, Mem. 10, 91–116.

Kakar, Y.P. (1988) Ground water quality monitoring in India. *Jalvigyan Sameeksha*, NIH, Roorkee, Vol. III, 52–63.

Kakar, Y.P. (1990) Groundwater pollution due to industrial effluents. *Bhu-Jal News*, CGWB, Vol. 2, 1–12.

Kalinsky, R.J., Kelly, W.E., Bogardi, I. *et al.* (1994) Correlation between DRASTIC vulnerabilities, and incidents of VOC contamination of municipal wells in Nebraska. *Ground Water*, 32(1), 31–34.

Kalkoff, S.J. (1993) Using Geographic Information Systems to determine the relation between stream quality and geology of the Roberts Creek watershed, Clayton County, Iowa. *Water Resources Bull.* (AWRA), **29**(6), 989–96.

Karanth, K.R. (1987) *Ground Water Assessment, Development and Management*, Tata McGraw-Hill, New Delhi, p. 720.

Karanth, K.R. (1989) *Hydrogeology*. Tata McGraw-Hill, New Delhi, p. 458.

Karanth, K.R. (1992) Pegmatites: A potential source of siting high yielding wells. *J. Geol. Soc. Ind.*, **39**, 77–81.

Karundu, A.J. (1993) Hydrogeology of fractured bedrock systems of Nyabisheki catchment in SW Uganda. *Mem. 24th Congress of IAH*, Oslo, 90–102.

Kazi, A. and Sen, Z. (1985) Volumetric RQD : an index of rock quality, in *International Symposium on Fundamentals of Rock Joints*, Bjorkliden, pp. 95–102.

Kehew, A.E., Straw, W.T., Steinmann, W.K. *et al.* (1996) Groundwater quality and flow in a shallow glacio-fluvial aquifer impacted by agricultural contamination. *Ground Water*, **34**(3), 491–500.

Kelly, W.E. (1977) Geoelectric sounding for estimating aquifer hydraulic conductivity. *Ground Water*, **15**, 420–25.

Keppler, A., Drost, W. and Wohnlich, S. (1996) Comparison of different radioactive tracer within a single fracture of granitic rock at the Grimsel Test Site, in *Hardrock Hydrogeology of the Bohemian Massif* (eds J. Krasny and J.Mls), Acta Universitatis Caroline, Geologica Vol. 40, 247–55, Univerzita Karlova, Czech Republic.

Kessels, W. and Kuck, J. (1995) Hydraulic communication in crystalline rock between the two boreholes of the Continental Deep Drilling Project in Germany. *Int. J. Rock Mech. Min. Sc. & Geomech. Abstr.*, **32**(1), 37–47.

Khair, K. and Haddad, F. (1993) The fractured carbonate rocks of Lebanon and their seasonal springs, in *Hydrogeology of Hard Rocks, Mem. 24th Congress of IAH,* Oslo, Part 2, pp. 1135–44.

Khan, A.A. and Sattar, G.S. (1994) Over exploitation of groundwater and its environmental impact in the Barind, Bangladesh, in *Regional Workshop on Env. Aspects of Groundwater Dev.*, (ed. D.C. Singhal), Roorkee, pp. IV–337–44.

Khan, S.H.A. and Raja, J. (1989) Management of groundwater resources in complex aquifer systems of Rajmahal traps of Sahibganj district, Bihar, in *IGW* (ed. C.P.Gupta), pp. 649–60.

Khare, M.C. (1981) Sedimentological and hydrogeological studies of parts of Durg and Raipur Districts, M.P., India. PhD thesis (unpublished), University of Roorkee, p. 351.

Kimblin, R.T. (1995) The chemistry and origin of groundwater in Triassic sandstone and Quaternary deposits: northwest England and some UK comparisons. *J. Hydrol.*, **172**, 293–311.

Kiraly, L. (1975) Report on the present knowledge on the physical characteristics of karstic rocks, in *Hydrogeology of Karstic Terrains,* (eds A. Burger and L. Dubertret), IAH, Paris, Ser. B. No.3, 53–67.

Kittu, N. (1990) High yielding bore wells in basaltic terrain at Yerangaon, Nagpur district, Maharashtra. *Bhu-Jal News*, **5**(4), Coverstory.

Kittu, N. and Mehta, M. (1990) Numerical technique for evaluation of aquifer parameters on large diameter wells in hard rocks, in *Proceedings All India Seminar on Modern Techniques of Rain Water Harvesting, Water Conservation and Artificial Recharge for Drinking Water*, Pune, pp. 259–62.

Klinck, B. and Crawford, M. (1995) Landfill – assessing the hazards. *Earthworks*, BGS, Nottingham, Issue 7, p. 8.

Klitten, K. (1991) Hydrofracturing of hard rock borewells. Lecture notes (unpublished), Regional Training Course on Groundwater Exploration and Assessment, University of Roorkee, Vol. 2, pp. 129–57.

Kloska, M., Ostrowski, L. Pusch, G. and Yerby, P. (1989) Gas pulse-test – a new method for low permeability formations, in *Rock at Great Depth,* Vol. 1 (eds V. Maury and D. Fourmaintraux), Balkema, Rotterdam, pp. 219–26.

Knopman, D.S. and Hollyday, E.F. (1993) Variation in specific capacity in fractured rocks, Pennsylvania. *Ground Water*, **31** (1), 135–45.

Koefoed, O. (1979) *Geosounding Principles, (1) Resistivity Sounding Measurements*, Elsevier, Amsterdam, 276 pp.

Komatina, M. (1975) Development conditions and regionalization of karst, in *Hydrogeology of Karstic Terrains* (eds A. Burger and L. Dubertret), IAH, Paris, pp. 21–29.

Konikow, L.F. and Bredehoeft, J.D. (1978) Computer model of two dimensional solute transport and dispersion in groundwater, in *Techniques of Water Resources Investigation*, US Geol. Surv., Chapter C-2, 90 pp.

Korkmaz, N. (1990) The estimation of groundwater recharge from spring hydrographs. *J. Hydrol. Sci.*, **35**(2), 209–17.

Krampe, K. (1983) Hydrogeological requirements for rural water supply in Central Java, Indonesia, in *Proceedings International Symposium on Ground Water in Water Resources Planning*, Vol. 1, UNESCO–IAH–IAHS, Kolbanz, Germany, 201–207.

Krasny, J. (1993) Hydrogeological map of the Czech Republic: a quantitative and standarized approach to representation of groundwater in hard rocks. *Mem. 24th Congress of IAH,* Oslo, pp. 1051–57.

Krasny, J. (1996) State of the art of hydrogeological investigations in hard rocks: the Czech Republic, in *Hardrock Hydrogeology* of the *Bohemian Massif* (eds J. Krasny and J. Mls), Acta Universitatis Carolinae, Geologica Vol. 40, pp. 89–101. Univerzita Karlova.

Krishnamoorthy, T.M., Nair, R.N. and Sarma, T.P. (1992) Migration of radionuclides from a granite repository. *Water Resources Res.*, **28**(7), 1927–34.

Krishnan, M.S.(1949) *Geology of India and Burma*, The Madras Law Journal Office, Madras, p. 544.

Kruger, P. and Otte, C. (eds) (1973) *Geothermal Energy*. Stanford University Press, Stanford, p. 360.

Krumbein, W.C. and Monk, G.D. (1942) Permeability as a function of the size parameters of unconsolidated sand. Am. Inst. *Mining Eng. Tech. Publ.*, Littleton, Colorado, pp. 153–63.

Kruseman, G.P. and de Ridder, N.A. (1970) *Analysis and Evaluation of Pumping Test Data*, Intl. Inst. for Land Reclamation and Improvement, Bulletin 11, Wageningen, p. 200.

Kruseman, G.P. and de Ridder, N.A. (1990) *Analysis and Evaluation of Pumping Test Data*, 2nd edn, Intl. Inst. for Land Reclamation and Improvement, Publication 47, Wageningen, p. 377.

Kukillaya, J.P., Kunhi, A.M. and Abdul Rahman, A.K. (1992) Basic and ultrabasic dykes as potential aquifers in hard rock terrains. *Bhu-Jal News*, 7(4), pp. 14–19.

Kulkarni, K.M., Rao, S.M., Singhal, B.B.S. *et al.* (1989) Origin of saline groundwaters of Haryana State, India, in *Regional Characterization of Water Quality* (ed. S. Ragone), IAHS Publ. No.182, pp. 125–32.

LaMoreaux, P.E. (1986) *Hydrology of Limestone Terranes*, IAH Intl. Contributions to Hydrogeology, Vol. 2, Verlag Heinz Heise, Hanover, p. 342.

LaMoreaux, P.E., LeGrand, H.E. and Stringfield, V.T. (1975) Progress of knowledge about hydrology of carbonate terrains, in *Hydrogeology of Karstic Terrains* (eds A. Burger and L. Dubertret), IAH, Paris, 41–52.

LaRiccia, M.P. and Rauch, H.W. (1977) Water well productivity related to photo-lineaments in carbonates of Fredrick Valley, Maryland, in *Hydrologic Problems in Karst Regions* (eds R.R. Dilamarter and S.C. Csallany), Western Kentucky University, Bowling Green, Kentucky, pp. 228–34.

Larsson, I. (1984) *Groundwater in Hardrocks, Studies and Reports in Hydrology,* Vol. 33, UNESCO, Paris.

Laslett, G.M. (1982) Censoring and edge effects in areal and line transect sampling of rock traces. *J. Int. Assoc. Math. Geol.*, **14**, 125–40.

Lattman, L.H. and Parizek, R.R. (1964) Relationship between fracture traces and the occurrence of groundwater in carbonate rocks. *J.Hydrol.*, **2**, 73–91.

Laubach, S. (1992) Fracture networks in selected Cretaceous sandstones of the Green River and San Juan Basins, Wyoming, New Mexico, and Colorado. Geological studies relevant to horizontal drilling: Examples from Western North America. (eds J.M. Schmoker, E.B. Coalson, and C.A. Brown). Rocky Mountain Assoc. of Geologists. pp. 61–73.

Lawrence, C.R. (1975) *Geology, Hydrodynamics and Hydrochemistry of the Southern Murray Basin,* Vol. 1, Geol. Surv. of Victoria, Mem. 30, Mines Dept., Victoria, p. 357.

Lee, C.F., Zhang, J.M. and Zhang, Y.X. (1996) Evaluation and origin of the ground fissures in Xian, China. *Eng. Geol.*, **43**(1), 45–55.

Lee, C.H., Chang, J.L. and Hsu, K.T. (1996) Investigation of hydraulic aperture at surface exposed rock fractures in-situ. *Geotechnique*, **46**(2), 343–49.

Lee, C.H., Deng, B.W. and Chang, J.L. (1995) A continuum approach for estimating permeability in naturally fractured rocks. *Eng. Geol.*, **39**, 71–85.

Lee, C.H. and Farmer, I. (1993) *Fluid Flow in Discontinuous Rocks*, Chapman & Hall, London, p. 169.

Lee, K. (1969) Infrared exploration for shoreline springs of Mono Lake, California test site. *Proceedings Sixth International Symposium, on Remote Sensing of Evnironment*, Ann Arbor, Michigan, 1275–87.

Lee, K.H. and Xie, G.O. (1993) A new approach to imaging with low-frequency electromagnetic fields. *Geophysics*, **58**, 780–96.

LeGrand, H.E. (1964) System for evaluation of contamination potential of some waste disposal sites. *J. Am. Water Work Assoc.*, **56**, 959–74.

LeGrand, H.E. (1967) Groundwater of the Piedmont and Blue Ridge Provinces of the Southeastern States. *US Geol. Surv. Circular*, 238, p. 11.

LeGrand, H.E. (1970) Movement of agricultural pollutants with groundwater, in *Agricultural Practices and Water Quality*, Iowa State University Press, Ames, Iowa, pp. 303–313.

Lerner, D.N. (1995) Chlorinated solvent pollution of an industrialized urban area: summary of findings and implications for groundwater protection and cleanup, in *Groundwater Quality* (eds H. Nash and G.J.H. McCall), Chapman & Hall, London, pp. 185–90.

Levens, R.L., Williams, R.E. and Rushton, D.R. (1994) Hydrogeologic role of geologic structures. Part 2: analytical models. *J. Hydrol.*, **156**, 245–63.

Lillesand, T.M. and Kiefer, R.W. (1987) *Remote Sensing and Image Interpretation*, 2nd edn, Wiley, New York, 721 pp.

Linsley, R.K., Kohler, M.A. and Paulhus, J.L.H. (1982) *Hydrology for Engineers*, 3rd edn, McGraw-Hill, New York.

Lloyd, J.W. (1991) Hydrogeology in developing countries, in *Geosciences in Development* (eds D.A.V. Stow and D.J.C. Laming), Balkema, Rotterdam, pp. 107–14.

Lloyd, J.W. and Heathcote, J.A. (1985) *Natural Inorganic Hydrochemistry in Relation to Ground Water – An Introduction*, Clarendon Press, Oxford, p. 296.

Loiselle, M. and Evans, D. (1995) Fracture density distributions and well yields in coastal Maine. *Ground Water*, **33**(2), 190–96.

Long, A. (1995) Lecture notes in training workshop on *Environmental Isotope Techniques in Hydrology*, Roorkee, India, p. 232.

Long, J.C.S., Remer, J.S., Wilson, C.R. and Witherspoon P.A. (1982) Porous media equivalents for networks of discontinuous fractures. *Water Resources Res.*, **18**(3), 645–58.

Long, J.C.S. and Witherspoon, P.A. (1985) The relationship of degree of interconnection to permeability in fracture networks. *J. Geophys. Res.*, **90**(B4), 3087–98.

Louis, C. (1974) Rock hydraulics, in *Rock Mechanics,* (ed. L. Muller), Springer-Verlag, Vienna, pp. 299–387.

Louis, C. (1977) Suggested methods for determining hydraulic parameters and characteristics of rock masses. *Int. Soc. of Rock Mechanics*, unpublished report, p. 137.

Louis, C. (1984) An introduction to rock hydraulics, lecture notes (unpublished), SIMESCOL, Paris, p. 16.

Lusczynski, N.J. (1961) Head and flow of ground water

of variable density. *J. Geophys. Res.*, **66**, 4247–56.
Mackie, C.D. (1982) Multi-rate testing in fractured rock aquifers, in *Proceedings AWRC Conference on Groundwater in Fractured Rocks*, Canberra, 139–50.
Madhure, V.B. and Sakare, S.V. (1990) Hydro-fracturing approach towards improving borewell yield. *Proceedings All India Seminar on Modern Techniques of Rain Water Harvesting, Water Conservation and Artificial Recharge*, Pune, 753–72.
Maggirwar, C.N. (1992) Techno-economic feasibility of groundwater recharge and water conservation, *Proceedings Workshop on Artificial Recharge of Groundwater*, CGWB, Raipur, India, 117–31.
Maguire, D.J., Goodchild, M.F. and Rhind, D.W. (eds) (1991) *Geographical Information Systems*, Vol. 1. Principles; Longman, London.
Mahoney, J.J. (1988) Deccan Traps, in *Continental Flood Basalts* (ed. J.D. Macdougall), Kluwer Academic, Dordrecht, The Netherlands, pp. 151–94.
Maillet, R. (1947) The fundamental equations of electrical prospecting. *Geophysics*, **12**, 529–56.
Maini, T. and Hocking, G. (1977) An examination of the feasibility of hydrologic isolation of a high level waste repository in crystalline rocks. Invited paper, Geologic Disposal of High Radioactive Waste Session, *Ann. Meet. Geol. Soc. Am.*, Seattle, Washington.
Mallik, S.B., Bhattacharya, D.C. and Nag, S.K. (1983) Behaviour of fractures in hard rocks – a study by surface geology and radial VES method. *Geophysics*, **21**, 181–89.
Mandel, S. and Shiftan, Z.L. (1981), *Groundwater Resources – Investigations and Development*. Academic Press, New York, p. 269.
Markussen, L.M. and Moller, H.M.F. (1995) Groundwater Quality under Copenhagen, in *Groundwater Quality* (eds H. Nash and G.J.H. McCall), Chapman & Hall, London, pp. 175–84.
Marsden, S.S. and Davis, S.N. (1967) Geological subsidence. *Scientific American*, **216**(6), 93–100.
Marsh, T.J. and Davies, P.A. (1983) The decline and partial recovery of groundwater levels below London. *Proceedings Inst. Civil Engrs.*, Part 1, 74, 263–76.
Marsily, G. de, (1986) *Quantitative Hydrogeology*, Academic Press, New York, p. 440.
Maslia, M.L. and Prowell, D.C. (1990) Effect of faults on fluid flow and chloride contamination in a carbonate aquifer system. *J. Hydrol.*, **115**(1), 1–49.
Mather, J.D. (1993) Pollution of shallow hard rock systems: an overview, in *Mem. 24th Congress of IAH, Olso.*
Mather, J.D. (1995) Preventing groundwater pollution from landfilled waste – is engineered containment an acceptable solution? in *Ground Water Quality* (eds H. Nash and G.J.H McCall), Chapman & Hall, London, pp. 191–95.
Mather, P.M. (1987). *Computer Processing of Remotely-sensed Images, an Introduction*. Wiley, New York, pp. 352.
Mathers, S. and Zalasiewicz, J. (1994) *A Guide to the Sedimentology of Unconsolidated Sedimentary Aquifers*, Tech. Report WC/93/32, BGS, p. 117.
Matthess, G. (1982) *The Properties of Groundwater*, John Wiley, New York, p. 406.
Matthews, R.A. (1985) High elevation aquifer-potential sources of water and electrical supply. *Mem. 17 of IAH*, Tucson, pp. 720–28.
Matzko, J.R. (1994) Geology of the Chinese nuclear test site near Lop Nor, Xinjiang Ugyur Autonomous Region, China. *Eng. Geol.*, **36**, 173–81.
Maxey, G.B. (1964) Geology: part I, Hydrogeology, in *Handbook of Applied Hydrology* (ed. V.T. Chow), McGraw-Hill, New York, pp. 4–38.
McCann, D.M., Baria, R., Jackson, P.D. and Green, A.S.P. (1986), Application of cross-hole seismic measurements in site investigation surveys. *Geophysics*, **51**(4), 914–29.
McCauley, J.F., Schaber, G.C., Breed, C.S., *et al.* (1982) Subsurface valleys and geoarcheology of the eastern Sahara revealed by shuttle radar. *Science*, **218**, 1004–19.
McConnell, C.L. (1993) Double porosity well testing in the fractured carbonate rocks of the Ozarks. *Ground Water*, **31**, 75–83.
McDonald M.G. and Harbaugh, A.W. (1983) A modular three dimensional finite difference groundwater flow model, *US Geol. Surv. Techniques of Water Resources Investigations*, Chapter A1, p. 586.
McDowell, P.W. (1979) Geophysical mapping of water filled fracture zones in rocks. *International Association of Engineering Geology Bulletin*, **19**, 258–64.
McEdwards, D.G. and Tsang, C.F. (1977) Variable rate multiple well testing analysis, in *Proceedings Int. Well-Testing Symposium*, Berkeley, pp. 92–99.
McFarlane, M.J., Chilton, P.J. and Lewis, M.A. (1992) Geomorphological controls on borehole yields; a statistical study in an area of basement rocks in central Malawi, in *The Hydrogeology of Crystalline Basement Aquifers in Africa* (eds E.P. Wright and W.G. Burgess), Geol. Soc. Spl. Publ. No. 66, 131–54.
Mehta, B.C. Das, S. and Srivastava, S.K. (1994) Environmental hazards of high iron in groundwater of Bhubaneswar, Orissa. *Proceedings Regional Workshop on Environmental Aspects of Groundwater Dev.*, (ed. D.C. Singhal), Kurukshetra University, I-50–I-57.
Mehta, N.S., Rajawat, A.S., Bahuguna, I.M. *et al.*, (1994) Remote sensing for mineral exploration: Geological potential of ERS-I SAR data covering

parts of Aravalli hills and Thar desert (Rajasthan, India), *Project Report*, SAC/RSA/RSAG/ERS-1/GEO/PR/01/94, p. 67.

Meijerink, A.M.J. (1974) *Photo Hydrological Reconnaissance Survey*. ITC, Enschede, The Netherlands, p. 371.

Meijerink, A.M.J. de Brouwer, H.A.M., Mannacrts, C.M., and Valenzuela, C.R. (1994) *Introduction to the Use of Geographic Information Systems for Practical Hydrology*. ITC Publication number 23, Enschede.

Meinzer, O.E. (1920) Quantitative methods of estimating ground water supplies. *Bull. Geol. Soc. Am.*, **31**, 329–38.

Meinzer, O.E. (1923) *Outline of Groundwater Hydrology, with Definitions*, US Geol. Surv. Water Supply Paper 494.

Mekel, J.F.M. (1970) *ITC Textbook of Photo-interpretation*, Chap VIII.1. The use of aerial photographs in geology. ITC, Enschede.

Mekel, J.F.M. (1978). ITC textbook of photo-interpretation. Chap VIII. The use of aerial photographs and other images in geological mapping. ITC, Enschede.

Mercer, J.W. and Faust, C.R. (1981) *Groundwater Modelling*, National Water Well Association, USA, p. 60.

Merin, I.S. (1992) Conceptual model of ground water flow in fractured siltstone based on analysis of rock cores, borehole geophysics, and thin sections. *GWHR*, Fall 1992, 118–25.

Meyer, P.D., Valocchi, A.J. and Eheart, J.W. (1994) Monitoring network design to provide initial detection of groundwater contamination. *Water Resources Res.*, **30**(9), 2647–59.

Meyer, R.R. (1963) A chart relating well diameter, specific capacity and the coefficients of transmissibility and storage, in *Methods of Determining Permeability, Transmissibility and Drawdown* (compiled by R. Bentall), US Geol. Surv. Water Supply Paper 1536–I, pp. 338–40.

Michalski, A. (1990) Hydrogeology of the Brunswick (Passaic) Formation and implications for groundwater monitoring practice. *GWMR*, Fall 1990, pp. 134–43.

Milanovic, P.T. (1981) *Karst Hydrology*, Water Resources Publication, Colorado.

Miller, D.W. (1985) Chemical contamination of ground water, in *Ground Water Quality* (eds C.H. Ward, W. Giger and P.L. McCarty), John Wiley, New York, pp. 39–52.

Miller, V.C. and Miller, C.F. (1961) *Photogeology*, McGraw-Hill, New York.

Minzi, S., Fagao, H. Ruifen, L., and Baoling, N. (1988), Application of oxygen and hydrogen isotopes of waters in Tengchong hydrothermal systems of China, in *Isotope Applications in Hydrology in Asia and the Pacific*, IAEA, Beijing, pp. 23–34.

Mishra, S.K., Singhal, B.B.S. and Singhal, D.C. (1989) Analysis of pumping test data from borewells in fractured rocks of Karnataka, in *IGW* (ed. C.P. Gupta), Vol. I, 15–25.

Mithal, R.S. (1969) A reappraisal of groundwater distribution and provinces of India, in *Proceedings Symposium Groundwater Studies in Arid and Semiarid Regions* (eds R.S. Mithal and B.B.S. Singhal), Roorkee, pp. 495–526.

Moench, A.F. (1984) Double-porosity models for a fissured ground water reservoir with fracture skin. *Water Resource Res.*, **20**(7), 831–46.

Moike, J.G. (1980) *Digital Processing of Remotely Sensed Images*, NASA SP-431, US Govt. Printing Office, Washington, DC.

Mojtabai, N., Cetintas, A., Farmer, I.W. and Savely, J. (1989) In-place and excavated block size distributions. *30th US Symposium Rock Mechanics*, West Virginia University, pp. 537–44.

Montgomery, C.W. (1988) *Physical Geology*, 2nd edn, William C. Brown, Dubuque, p. 555.

Moore, W.S. (1996) Large groundwater inputs to coastal waters revealed by ^{226}Ra enrichments. *Nature*, **380**, 612–14.

Morris, D.A. and Johnson, A.I. (1967) *Summary of Hydrologic and Physical Properties of Rock and Soil Materials, as Analyzed by the Hydrologic Laboratory of the U.S. Geological Survey 1948–1960*. US Geol. Surv. Water Supply Paper, 1839-D, p. 42.

Morrison, R.J., Prasad, R.A. and Brodie, J.E. (1984) Chemical hydrology of small tropical islands. *CSC Tech. Publ. Science*, **154**, 211–23.

Motz, L.H. (1982), Lower Hawthorn aquifer on Sanibel Island, Florida. *Ground Water,* **20**(2), 170–78.

Nace, R.L. (ed.) (1971) *Scientific Framework of World Water Balance*, UNESCO Tech. Papers Hydrol. 7, p. 27.

Nair, A.R., Sinha, U.K., Joseph, T.B., *et al.* (1993) Application of environmental radioisotopes for studying groundwater recharge process in arid areas, in *Proceedings 2nd National Symposium on Environment*, Jodhpur, India, 188–90.

Nair, V.N. (1981) Sedimentological and hydrogeological studies of Chorward. Madhavpur area, Junagadh District, Gujarat, India, PhD thesis, (unpublished) University of Roorkee, p. 328.

Navada, S.V., Nair, A.R., Rao, S.M., *et al.* (1993) Groundwater recharge studies in arid region of Jalore, Rajasthan using isotope techniques. *J. Arid Environments*, **24**, 125–33.

Neerdael, B., Wemaere, I., Thury, M. and Gautschi, A. (1996) Characterisation of argillaceous formations as potential host rock for geological disposal in Belgium and in Switzerland. *Abstract Vol 2, 30th IGC*, Beijing.

Nekut, A.G. (1995) Crosswell electromagnetic tomography in steel-cased wells. *Geophysics*, **60**, 912–20.

Neuman, S.P. (1972) Theory of flow in unconfined aquifers considering delayed response of the water table. *Water Resources Res.*, **8**, 1031–49.

Neuman, S.P. (1975) Analysis of pumping test data from anisotropic unconfined aquifers considering delayed gravity response. *Water Resources Res.*, **11**(2), 329–42.

Neuzil, C.E. (1982) On conducting the modified slug test in tight formations. *Water Resources Res.*, **18**(2), 439–41.

Neuzil, C.E. (1986) Groundwater flow in low permeability environments. *Water Resources Res.*, **22**(8), 1163–95.

Neuzil, C.E. and Tracy, J.V. (1981) Flow through fractures. *Water Resources Res.*, **17**(1), 191–99.

Newman, G.A. (1995) Crosswell electromagnetic inversion using integral and differential equations. *Geophysics*, **60**, 899–911.

Niedzielski, H. (1993) Productivity of the basalt aquifer of the eastern part of Chichinautzin Group, Mexico, in *Mem. 24th Congress of IAH*, Oslo, Part I, 281–89.

Nilsson, B. and Sidle, R.C. (1996) Flow and solute transport in a contaminated clay till in Denmark, *Abstracts 30th IGC*, Vol. 3, p. 270, Beijing.

Nishidai, T. (1993) Early results from 'Fuyo-1' Japan's Earth Resources Satellite (JERS-1). *Int. J. Remote Sensing*, **14**, 1825-33.

Oberdorfer, J.A., Hogan, P.J. and Buddemeier, R.W. (1990) Atoll island hydrogeology: flow and fresh water occurrence in a tidally dominated system. *J. Hydrol.*, **120**, 327–40.

Oda, M., Saitoo, T. and Kamemura, K. (1989) Permeability of rock masses at great depth, in *Proceedings Symposium Rock at Great Depth* (eds. V. Maury and D. Fourmaintraux), Balkema, Rotterdam, pp. 449–56.

O'Leary, D.W., Friedman, J.D. and Pohn, H.A. (1976) Lineament, linear and lineation: some proposed new standards for old terms. *Geol. Soc. Am. Bull.*, **87**, 1463–69.

Olivieri, A.W., Eisenberg, D.M. and Cooper, R.C. (1986) Groundwater contamination – site ranking methodology. *J. Env. Eng.*, ASCE, **112**, 757–69.

Ollier, C. (1969), *Volcanoes*, MIT Press, Cambridge, USA, p. 177.

Olsson, O., Falk, L., Forslund, O., Lundmark, L. and Sandberg, E. (1992) Borehole radar applied to the characterization of hydraulically conductive fracture zones in crystalline rock. *Geophysical Prospecting*, **40**, 109–42.

Otkun, G. (1977) More about Palaeocene karst aquifer in Saudi Arabia, in *Karst Hydrogeology* (eds J.S. Tolson and F.L. Doyle), *Mem. 12th Congress, IAH*, UAH Press, Alabama, 25–38.

Owen, M. (1981) The Thames groundwater scheme, in *Case Studies in Groundwater Resources Evaluation* (ed. J.W. Lloyd), Clarendon Press, Oxford, pp. 186–202.

Owoade, A. (1993) Some field observations on the groundwater potential in tropical hard rock catchments, in *Mem. 24th Congress of IAH*, Oslo, Part 2, pp. 1161–65.

Pahl, P.J. (1981) Estimating the mean length of discontinuity trace. *Int. J. Rock Mech. Min. Sci. & Geomech. Abstr.*, **18**, 221–28.

Palacky, G.J., Ritsema I.L. and deJong, S.J. (1981) Electromagnetic prospecting for groundwater in Precambrian terrains in the Republic of Upper Volta (Burkino Faso). *Geophysical Prospecting*, **29**, 932–55.

Paloc, H. and Back, W. (eds) (1992) *Hydrogeology of Selected Karst Regions*, IAH contributions to Hydrogeology, Vol. 13, Verlag-Heinz, Hanover, p. 514.

Panda, B.B. and Kulatilake, H.S.W. (1995) Study of the effect of joint geometry parameters on the permeability of jointed rocks, in *Rock Mechanics* (eds J.J.K. Daemen and R.A. Schultz), Balkema, Rotterdam, pp. 273–78.

Papadopulos, I.S. and Cooper, H.H. (1967) Drawdown in a well of large diameter. *Water Resources Res.*, **3**(1), 241–44.

Padadopulos, S.S., Bredehoeft, J.D. and Cooper, H.H. (1973) On the analysis of slug test data. *Water Resources Res.*, **9**, 1087–89.

Parasnis, D.S. (1986) *Principles of Applied Geophysics*, 4th edn, Chapman & Hall, London, 402 pp.

Parizek, R.R. (1976) Lineaments and groundwater, in *Interdisciplinary application and interpretations of EREP data within the Susquehanna River Basin*, (eds G.T. McMurthy and G.W. Peterson), Pennsylvania State Univ., pp. 459–86.

Patel, S.C. and Mishra, G.C. (1983) Analysis of flow to large-diameter well by discrete kernel approach. *Ground Water*, **21** (5), 573–76.

Payne, B.N. (1967) Isotope techniques in groundwater hydrology, in *Methods and Techniques of Groundwater Investigation and Development*, Water Resources Series, No. 33, UNESCO, pp. 107–13.

Pearce, B.R. (1982) Fractured rock aquifers in central Queensland, in *Papers of the Groundwater in Fractured Rock Conference 1982*, AWRC Conference Series No. 5, pp. 161–72.

Perumal, A. (1990) Hydromorphogeological investigations in parts of Athur Valley, Tamilnadu, India, using remote sensing data. PhD thesis (unpublished), University of Roorkee.

Peterson, F.L. (1972) Water development on tropic volcanic islands – type example: Hawaii. *Ground Water*, **10** (5), 18–23.

Peterson, F.L. (1984) Hydrogeology of high oceanic islands. *Proceedings Regional Workshop on Water Resources of Small Islands*, Part 2, CSC, London, 431–35.

Philip, G., Gupta, R.P. and Bhattacharya, A. (1989) Channel migration studies in middle Ganga basin, India, using remote sensing data. *Int. J. Remote Sens.*, **10**, 1141–49.

Philip, G., Gupta, R.P. and Bhattacharya, A. (1991) LANDSAT image enhancement for mapping fluvial palaeofeatures in parts of middle Ganga Basin, Bihar. *J. Geol. Soc. Ind.*, **37**, 63–74.

Philip, G. and Singhal, B.B.S. (1992) Importance of geomorphology for hydrogeological study in hard rock terrain, an example from Bihar Plateau through remote sensing. *Ind. J. Earth Sci.*, **19**(4), 177–88,

Pinder, G.F. and Grey, W.G. (1977) *Finite Element Simulation in Surface and Subsurface Hydrology*, Academic Press, New York, p. 295.

Piper, A.M. (1944) A graphic procedure in the geochemical interpretation of water analyses. *Trans. Amer. Geophys. Union*, **25**, 914–28.

Piper, A.M. (1953) A graphic procedure in the geochemical interpretation of water analyses. *U.S. Geol. Surv. Ground Water Notes*, **12**, 14 pp.

Pitale, U. and Padhi, R. (1996) Geothermal environment and energy potential of mobile crustal belt and intraplate tectonic zones in India, *Abstracts 30th IGC*, Vol. 1, Beijing, p. 386.

Plafker, G. (1969) Tectonics of the March 27, 1964, Alaska Earthquake, *US Geol. Surv. Prof. Paper* 543–I, 174 pp.

Pointet, T. (1989) Hydrogeological prospecting of hard rock aquifers in semi-arid climates for small scale irrigation. *Abstracts 28th IGC*, Vol. 2, p. 620, Washington.

Pollock, D.W. (1990) *Documentation of Computer Programs to Compute and Display Path Lines Using Results from the US Geol. Surv. Modular Three Dimensional Finite Difference Groundwater Flow Model.* US Geol. Surv. Open-File Report 89, pp. 381–88.

Powell, R.L. (1977) Joint patterns and solution channel evolution in Indiana, in *Karst Hydrogeology* (eds J.S. Tolson and F.L. Doyle), UAH Press, Huntsville, Alabama, pp. 255–69.

Price, N.J. and Cosgrove, J. (1990) *Analysis of Geological Structures*, Cambridge University Press, 502 pp.

Prickett, T.A. (1965) Type curve solution to aquifer tests under water-table conditions. *Ground Water*, 3(3), 5–14.

Priest, S.D. (1993) *Discontinuity Analysis for Rock Engineering*, Chapman & Hall, London, 473 pp.

Pruess, K. and Wang, J.S.Y. (1987) Numerical modelling of isothermal and nonisothermal flow in unsaturated fractured rock – a review. *Geophysical Monograph* 42, Am. Geophys. Union, pp. 11–21.

Put, M. and Ortiz, L. (1996), Large scale in-situ measurement of the hydraulic parameters of the Boom clay formation, *Abstract 30th IGC*, Vol. 3, p. 268, Beijing.

Radhakrishna, I. *et al.* (1979) Evaluation of Shallow Aquifers in Lower Maner Basin, unpublished report, NGRI, Hyderabad, 115 pp.

Radhakrishna, I., Athavale R.M., Singh, V.S. and Govil,P.K.(1976) Hydrogeological and hydrochemical studies in Lower Maner Basin. *Technical Report No. GH3-HG1*, Indo-German Collaboration Project on Exploration and Management of Groundwater Resources, NGRI, Hyderabad.

Rai, K.L. (1994) Environmental aspects of base-metal mining in India: selected case studies, in *Mineral Development and Environment* (eds K.L. Rai and G. Singh), Recent Researches in Geology, Vol. 15, pp. 325–40, Hindustan Publishing Corporation (India), New Delhi.

Raju, K.C.B. (1985) Artificial recharge through percolation tanks and subsurface dykes in *Proceedings Seminar Artificial Recharge of Groundwater*, CGWB, Ahmedabad, 12c–1 to 12c–21.

Raju, N.J., Reddy, T.V.K. and Nayudu, P.T. (1996) Electrical resistivity surveys for groundwater in the Upper Gunjanaeru Catchment, Cuddapah district, Andhra Pradesh. *J. Geol. Soc. Ind.*, **47**(6), 705–16.

Ramey, Jr., H.J. (1977) Petroleum engineering well test analysis – state of the art, in *Proceedings Invitational Well Testing Symposium*, Lawrence Berkeley Lab., California, 5–13.

Rana, R.S. and Vishwakarma, L.L. (1990) Occurrence of artesian conditions in the Sina river basin of drought prone Karjat Taluka, Ahmednagar district in Maharashtra. *Proceedings All India Seminar*, Pune, 207–13.

Randall-Roberts, J.A. (1993) Geophysical prospecting for water in Precambrian Gneiss, Oaxaca, Mexico, *Mem. of the 24th Congress of IAH*, Oslo, 1993, pp. 648–54.

Rao, S.H. (1975) Hydrogeology of parts of the Deccan Traps. PhD thesis (unpublished), Poona University, India, p. 190.

Rao, S.M., Jain, S.K., Navada, S.V., Nair, A.R. and Shivana, K. (1987), Isotopic studies of seawater intrusion and interrelations between water bodies: some field examples, in *Isotope Techniques in Water Resources Development*, IAEA, Vienna, pp. 403–25.

Rasmuson, A. and Neretnieks, I. (1981) Migration of radionuclides in fissured rock: the influence of micropore diffusion and longitudinal dispersion. *J. Geoph. Res.*, **86**(B5), 3749–58.

Raven, K.G. Novakowski, K.S. and Lapsevic, P.A.

(1988) Interpretation of field tracer tests of a single fracture using a transient solute storage model. *Water Resources Res.*, **24** (12), 2019–32.

Ray, R.G. (1965) *Aerial Photographs in Geologic Interpretation and Mapping*. USGS Prof. Paper 373.

Raymahashay, B.C. (1996) *Geochemistry for Hydrologists*. Allied Publishers, New Delhi, p. 190.

Razack, M. and Huntley, D. (1991) Assessing transmissivity from specific capacity in a large and heterogeneous alluvial aquifer. *Ground Water*, **29**(6), 856–61.

Read, M.D., Meredith, P.G. and Murrell, S.A.F. (1989) Permeability measurement techniques under hydrostatic and deviatoric stress conditions, in *Rock at Great Depth*, (eds V. Maury and D. Fourmaintraux) Vol. 1, Balkema, Rotterdam, pp. 211–17.

Read, R.E. (1982) Estimation of optimum drilling depth in fractured rock, in *Papers of the Groundwater in Fractured Rocks*, AWRC Conference, Series No. 5, Canberra, pp. 191–97.

Reader's Digest Association (1980) *Natural Wonders of the World*. Reader's Digest Association, Pleasantville, New York.

Reboucas, A. daC. (1993) Groundwater development in the Precambrian shield of South America and west side of Africa, in *Hydrogeology of Hard Rocks*. Mem. 24th Congress IAH, Oslo, Pt. 2, pp. 1101–15.

Rector, J.W. (ed.), (1995) Cross well methods. *Geophysics*, special issue, Vol. 60, No. 3.

Reddy, T.N. and Raj, P. (1997) Hydrogeological conditions and optimum well discharges in granitic terrain in parts of Nalgonda district, Andhra Pradesh, India. *J. Geol. Soc. Ind.*, **49**(1), 61–74.

Reilly, T.E. (1990) Simulation of dispersion in layered coastal aquifer systems. *J. Hydrol.*, **114**, 211–28.

Remson, I., Hornberger, G.M. and Molz, F.J. (1971) *Numerical Methods in Subsurface Hydrology*. Wiley, New York, p. 389.

Richards C.J., Roaza, H. and Roaza, R.M. (1993) Integrating Geographic Information Systems and Modflow for Groundwater Resources Assessment, *Water Resources Bulletin* (AWRA), Vol. 29, No. 5, pp. 847–53.

Rifai, H.S., Hendricks, L.A., Kilborn, K. and Bedient, P.B. (1993) A geographic information system (GIS) user interface for delineating wellhead protection areas. *Ground Water,* **31**, 480–88.

Roaza, H., Roaza, R.M. and Wagner, J.R. (1993) Integrating geographic information systems in groundwater applications using numerical modelling techniques. *Water Resources Bull.*, (AWRA), **29**(6), 981–88.

Rogbeer, G. (1984) Mauritius – water resources, in *Water Resources of Small Islands*, CSC Tech. Publ. Series No. 182, 114–39.

Rogers, R. B. and Kean, W.F. (1981) Monitoring groundwater contamination at a flash disposal site using surface electrical resistivity methods. *Ground* Water, **18**(5), 472–78.

Romani, S. and Singhal, B.B.S. (1970) A study of some of the thermal springs of Kulu district, H.P. *Indian Geohydrology*, **6**, 57–68.

Rovey, C.W.II and Cherkauer, D.S. (1995) Scale dependency of hydraulic conductivity measurements. *Ground Water*, **33**(5), 769–80.

Rowe, R.K., Booker, J.R. and Hammoud, A. (1989) The effect of multi-directional matrix diffusion on contaminant transport through fractured systems, in *Contaminant Transport in Groundwater*, (eds H.E. Kobus and W. Kinzelbach), Balkema, Rotterdam, 259–66.

Rubin, L.A. and Fowler, J.C. (1978) Ground-probing radar for delineation of rock fractures. *Engineering Geology*, **12**, 163–70.

Ruden, F. (1993) A revival of the qanat principle for rural water supply in a third world context, in *Mem. 24th Congress of IAH*, 1183–90, Oslo.

Ruhland, M. (1973) Methode d'etude de la fracturation naturelle des roches associe á divers modeles structureaux, *Bulletin des Sciences Geologiques*, **26**, 91–113.

Ruland, W.W., Cherry, J.A. and Feenstra, S. (1991) The depth of fractures and active groundwater flow in a clayey till plain in Southern Ontario. *Ground Water*, **29**(3), 405–17.

Ruprecht, J.K. and Schofield, N.J. (1991) Effects of partial deforestation on hydrology and salinity in high salt storage landscapes. *J. Hydrol.*, 12(9), 19–38.

Rushton, K.R. (1986) Vertical flow in heavily exploited hard rock and alluvial aquifers. *Ground Water*, **24**(5), 601–8.

Rushton, K.R. (1988) Numerical and conceptual models for recharge estimation in arid and semi-arid zones, in *Estimation of Natural Groundwater Recharge*, (ed. I. Simmers), Reidel, Dordrecht, pp. 223–38.

Rushton, K.R. and Holt, S.M. (1981) Estimating aquifer parameters for large-diameter wells. *Ground Water*, **19**, 505–9.

Rushton, K.R. and Phadtare, P.N. (1989) Artificial recharge pilot projects in Gujarat, India, in *Ground Water Management Quantity and Quality*, Proceedings IAHS Symposium, Benidorm, Spain, Publ. No. 188, pp. 533–45.

Rushton, K.R. and Raghava Rao, S.V. (1988) Groundwater flow through a Miliolite limestone aquifer. *J. Hydrol. Sciences*, **33**, 449–64.

Rushton, K.R. and Redshaw, S.C. (1979) *Seepage and Ground Water Flow*, John Wiley, New York, p. 339.

Rushton, K.R. and Singh, V.S. (1983) Drawdown in large-diameter wells due to decreasing abstraction rates. *Ground Water*, **21**(6) 670–77.

Rushton, K.R. and Singh, V.S. (1987) Pumping test analysis in large diameter wells with a seepage face by kernel function technique. *Ground Water*, **25**(1), 81–90.

Russell, C.E., Hess, J.W. and Tyler, S.W. (1987) Hydrogeologic investigations of flow in fractured tuffs, Rainier Mesa, Nevada Test Site (eds D.D. Evans and T.J. Nicholson), *AGU Monograph*, 42, 43–50.

Ryerson, R.A., Morain, S.A. and Budge, A.M. (eds) (1997) *The Manual of Remote Sensing*, 3rd edn, Earth Observing Platforms and Sensors (CD-ROM). American Society of Photogrammetric Engineering and Remote Sensing.

Sabins, F.F. (1983) Geologic interpretation of space shuttle radar images of Indonesia. *Am. Assoc. Petro. Geol. Bull.*, **67**, 2076–99.

Sabins, Jr., F.F. (1987) *Remote Sensing Principle and Interpretation*, 2nd edn, Freeman, San Francisco, 449 pp.

Sakthivadivel, R. and Rushton, K.R. (1989) Numerical analysis of large diameter wells with a seepage face. *J. Hydrol.*, **107**, 43–57.

Sami, K. (1996) Evaluation of the variations in borehole yield from a fractured Karoo aquifer, South Africa. *Ground Water*, **34**(1), 114–20.

Sanchez-Vila, X., Carrera, J. and Girardi, J.P. (1996) Scale effects in transmissivity. *J. Hydrol.*, **183**(1), 1–22.

Schoeller, H. (1959) *Arid Zone Hydrology: Recent Developments*, UNESCO, Paris, p. 125.

See, R.B., Naftz, D.L. and Qually, C.L. (1992), GIS-assisted regression analysis to identify sources of selenium in streams. *Water Resources Bull.*, **28**(2), 315–30.

Sekhar, M., Kumar, S.M. and Sridharan, K. (1993) A leaky aquifer model for hard rock aquifers, in *Mem. 24th Congress of IAH*, Oslo, Part 1, 338–48.

Shahin, M. (1985) Hydrology of the Nile Basin, in *Development in Water Science* 21, Elsevier, Amsterdam, p. 575.

Sharma, M.L. and Hughes M.W. (1985) Groundwater recharge estimation using chloride, deuterium and oxygen-18 profiles in the deep coastal sands of Western Australia, *J. Hydrol.*, **81**, 93–109.

Sharp Jr., J.M. (1993) Fractured aquifers/reservoirs: approaches, problems and opportunities, *Mem. 24th Congress of IAH*, Oslo, Part 1, pp. 23–38.

Sharp Jr., J.M., Boulton, R.C.S. and Fuller, C.M. (1993) Permeability and fracture patterns in a weathered and variably welded tuff: implications for flow and transport. *Mem. of 24th Congress of IAH*, Oslo, Part 1, 103–14.

Shata, A.A. (1982) Hydrogeology of the Great Nubian Sandstone basin, Egypt. *Q. J. Eng. Geol.*, **15**, 127–33.

Sheila, B.B. and Banks, D. (eds) (1993) Hydrogeology of Hard Rocks, *Mem. 24th Congress of IAH*, Oslo, Norway.

Shelton, M.L. (1982) Ground water management in basalts. *Ground Water*, **20**, 86–93.

Shiklomanov, I.A. (1990) Global water resources. *Nature and Resources*, **26**(3), 34–43, UNESCO, Paris.

Shivanna, K. *et al.* (1993) Isotope and geochemical evidence of past seawater salinity in Midnapore groundwaters, in *International Symposium on Application of Isotope Techniques in Studying Past and Current Env. Changes in the Hydrosphere* (unpublished) IAEA/UNESCO, Vienna.

Siddiqui, S.H. and Parizek, R.R. (1971) Geohydrological factors influencing well yields in folded and faulted rocks in Central Pennsylvania. *Water Resources Res.*, **7**(5), 1295–312.

Sidle, W.C. and Lee, P.Y. (1995) Estimating local groundwater flow conditions in a granitoid: preliminary assessments in the Waldoboro Pluton Complex, Maine. *Ground Water*. **33**(2), 291–303.

Siegal, B.S. and Abrams, M.J. (1976). Geologic mapping using Landsat data. *Photogramm. Eng. Remote Sensing*, **42**, 325–37.

Silar, J. (1996) Groundwater residence time in crystalline rocks, in *Hardrock Hydrogeology of the Bohemian Massif* (eds J. Krasny and J. Mls), Acta Universitatis Carolinae Geologica, Univerzita Karlova, Czech Republic, Vol. 40, 279–88.

Simmons, G.R. (1985) In situ experiments in granite in underground laboratories – a review, in *Design and Instrumentation of In-situ Experiments in Underground Laboratories for Radioactive Waste Disposal* (eds B. Come, P. Johnston and A. Muller), Balkema, Rotterdam, pp. 56–81.

Singh, G. (1988) Impact of coal mining on mine water quality. *Int. J. Mine Waters*, **7**(3), 49–59.

Singh, V.S. and Gupta, C.P. (1986) Hydrogeological parameter estimation from pump test on a large diameter well. *J. Hydrol.*, **87**, 223–32.

Singhal, B.B.S. (1963) Occurrence and geochemistry of ground water in the coastal region of Midnapur, West Bengal, India. *Econ. Geol.*, **58**, 419–33.

Singhal, B.B.S. (1973) Some observations on the occurrence, utilisation and management of groundwater in the Deccan Trap areas of Central Maharashtra, India, *Proceedings International Symposium on Development of Ground Water Resources*, Vol. 3, Madras, India, pp. V 75–81.

Singhal, B.B.S. and Gupta, B.L. (1966) Analysis of pumping test data from a well in the Indo-Gangetic alluvium of India and its bearing on the aquifer characteristics, *J. Hydrol.*, **4**, 121–40.

Singhal, B.B.S., Singhal, D.C. and Awasthi, A.K. (1987) Study of borewells and dugwells in the state of

Karnataka, India. *Tech. report prepared for NABARD*, Bombay (unpublished), 223 pp.
Singhal, B.S.S and Singhal, D.C. (1990) Evaluation of aquifer parameters and well characteristics in fractured rock formations of Karnataka, India, in *Selected Papers on Hydrogeology* (eds E.S. Simpson and J.M. Sharp, Jr.), IAH, Vol. 1, 951–62.
Singhal, D.C. and Sri Niwas (1983) Estimation of aquifer transmissivity from surface geoelectrical measurements, in *Proceedings Symposium Methods and Instrumentation for the Investigation of Ground Water Systems*, UNESCO-IAH, Noordwijkerhout, The Netherlands, pp. 405–14.
Sinha, B.P.C. and Kakar, Y.P. (1974) Geochemistry of fluoride in groundwater of north-Central Rajasthan, India, *IAH Mem.* X, pp. 130–32.
Sinha, B.P.C. and Sharma, S.K. (1990) Groundwater dams – concepts and case histories. *Bhu-Jal News*, 5, 3–13.
Skinner, A.Ch. (1985) Groundwater protection in fissured rocks, in *International Contributions to Hydrogeology* 6, IAH, Hanover, pp. 123–43.
Skjernaa, L. and Jorgensen, N.O. (1993), Detection of local fracture systems by azimuthal resistivity survey: examples from south Norway. *Memoires of the 24th Congress of IAH*, Oslo, 1993, 662–71.
Slater, P.N. (1980). *Remote Sensing – Optics and Optical Systems*, Addison Wesley, Reading, 575 pp.
Smedley, P.L. (1992) Relationship between trace elements in water and health, with special reference to developing countries. *Interim Report 1992*. BGS Tech. Report WD/92/39, p. 33.
Smeldley, P. and West, J. (1995) Arsenic in water – implications for health. *Earthworks*, **7**, BGS, Nottingham, p. 3.
Smellie, J.A.T., Laaksoharju, M. and Wikberg, P. (1995) Aspo, SE Sweden, a natural groundwater flow model derived from hydrogeochemical observations. *J. Hydrol.*, **172**, 147–69.
Smith, L. and Schwartz, F.W. (1993) Solute transport through fracture network, in *Flow and Contaminant Transport in Fractured Rock*, (eds J. Bear, C.F. Tsang and G. de Marsily), Academic Press, San Diego, pp. 129–66.
Snow, D.T. (1968a) Rock fracture spacings, openings and porosities. *J. Soil Mech. Found. Div.*, ASCE, 94 (SMI), 73–91.
Snow, D.T. (1968b), Hydraulic character of fractured metamorphic rocks of the front-range and implications to the Rocky Mountain Arsenal Well. *Colorado School of Mines Q.*, Denver, **63**(1), 167–99.
Sokolov, A.A. and Chapman, T.G. (eds) (1974) Methods of water balance computations: an international guide for research and practice. *Studies and Reports in Hydrology*, No.7, UNESCO Press, Paris, p. 127.
Souza, W.R. and Voss, C.I. (1987) Analysis of an anisotropic coastal aquifer system using variable density flow and solute transport simulation. *J. Hydrol.*, **92**, 17–41.
Spies B.R. and Habashy, T.M. (1995) Sensitivity analysis of crosswell electromagnetics. *Geophysics*, **60**, 834–45.
Sridharan, K., Mohan Kumar, M.S. and Sekhar, M. (1992) Groundwater flow and storage in hard rock aquifers, *Proceedings Workshop on Artificial Recharge of Groundwater in Granite Terrain*, CGWB, 86–107.
Sridharan, K., Sekhar, M. and Mohan Kumar, M.S. (1990) Analysis of aquifer-water table-aquitard system. *J. Hydrol.*, **114**, 175–89.
Srikantha, N.P., Narsimha Rao, B.S.R. and Devee, S. (1985) A realistic approach for hydrological parameters determination in areas adjoining Brahmaputra river bank. *Proceedings All India Seminar on Ground Water Development*, Vol. I, Lucknow, T-II/1–8.
Sri Niwas and Singhal, D.C. (1981) Estimation of aquifer transmissivity from Dar-Zarrouk parameters in porous media. *J. Hydrol*, **50**, 393–99.
Sri Niwas and Singhal, D.C. (1985) Aquifer transmissivity of porous media from resistivity data, *J. Hydrol.*, **82**, 143–53.
Srivastava, N. (1993) Technique for processing and interpreting groundwater data using GIS. M.Tech. thesis, (unpublished). Department of Earth Sciences, University of Roorkee, Roorkee, India.
Stafford, D.B. (ed) (1991) *Civil Engineering Applications of Remote Sensing and Geographic Information System*. The Amer. Soc. Civil Engineering, New York.
Star, J. and Estes, J. (1990) *Geographic Information Systems: An Introduction*. Prentice Hall, Englewood Cliffs, New Jersey.
Stasko, S. and Tarka, R. (1996) Hydraulic parameters of hard rocks based on long-term field experiment in the Polish Sudetes, in *Acta Universitis Carolinae, Geologica*, (eds J. Krasny and J. Mls), **40**(2), 167–78.
Stearns, H.T. (1942) Hydrology of volcanic terrains, in *Hydrology* (ed. O.E. Meinzer), Dover Publ., New York, pp. 678–703.
Stevenson, D.R. Kozak, E.T., Gascoyne, M., Davison, C.C. and Brown, A. (1996) Hydrogeologic characteristics of massive domains of sparsely fractured rock in the granitic Lac du Bonnet Batholith, southeastern Manitoba, Canada. *Abstract 38th IGC*, Vol.3, p. 269, Beijing.
Strahler, A.N. (1964) Geology: Part II, Quantitative geomorphology of drainage basins and channel networks, in *Handbook of Applied Hydrology* (ed.

V.T. Chow), McGraw-Hill, New York, 4–39 to 4–70.

Streltsova, T.D. (1976a) Hydrodynamics of groundwater flow in fractured formation. *Water Res. Research*, **12**(3), 405–14.

Streltsova, T.D. (1976b) Advances and uncertainties in the study of groundwater flow in fissured rocks, in *Advances in Groundwater Hydrology* (ed. Z.A. Saleem). *Am. Water Resources Assoc.*, 48–56.

Streltsova, T.D. (1977) Storage properties of fractured formations, in *Hydrologic Problems in Karst Regions*, (eds R.R. Dilamarter and S.C. Csallany), Kentucky University, pp. 188–92.

Streltsova–Adams, T.D. (1978) Well hydraulics in heterogeneous aquifer formations, in *Advances in Hydroscience* (ed. V.T. Chow), Vol. 11, Academic Press, New York, pp. 357–423.

Struckmeier, W.F. (1993) Hydrogeological mapping in hard rock terrains, *Mem. of the 24th Congress of IAH*, Oslo.

Subramanian, P. (1992) Hydrogeology and its variations in the granites and asociated rock formations in India. *Proceedings Workshop on Artificial Recharge of Groundwater in Granitic Terrains*, CGWB, Bangalore, pp. 1–13.

Subramanian, P.R. (1993) High yielding borewell at Khairi village in basaltic terrain in Maharashtra. *Bhu-Jal News*, **8** (3 & 4), 35–36.

Sudicky, E.A. and Frind, E.O. (1982) Contaminant transport in fractured porous media: Analytical solutions for a system of parallel fractures. *Water Resources Res.*, 18(6), 1632–42.

Sukhija, B.S., Nagabhushanam, P. and Reddy, D.V. (1996) Groundwater recharge in semi-arid regions of India: an overview of results obtained using tracers. *Hydrogeology J.*, **4**(3), 50–71.

Sukhija, B.S. and Rao, A.A. (1983) Environmental tritium and radiocarbon studies in the Vedavati river basin, Karnataka and Andhra pradesh, India. *J. Hydrol.*, **60**, 185–96.

Sukhija, B.S., Reddy, D.V., Nagabhushanam, P. and Chand, R. (1988) Validity of the environmental chloride method for recharge evaluation of coastal aquifers, India. *J. Hydrol.*, **99**, 349–66.

Talbot, J.C., Buckley, D.K. and Herbert, R. (1993) *Hydraulic Fracturing: Further Investigations on the use of Low Yielding Boreholes in the Basement Rocks of Zimbabwe*, Tech. Report WD/93/16, BGS, Keyworth, UK.

Tanwar, B.S. (1983) Groundwater studies of the alluvial belt of Yamuna and Ghaggar Doab in Haryana. PhD thesis (unpublished), Kurukshetra University, Kurukshetra, India, p. 205.

Taylor Jr., G.C. (1959), Groundwater provinces of India, *Econ. Geol.*, **54**. 683–97.

Taylor, R.W. and Fleming, A.H. (1988) Characterizing jointed systems by azimuthal resistivity surveys. *Ground Water*, **26**, 464–74.

Telford, W.M., Geldart, L.P., Sheriff, R.E. and Keys, D.A. (1988) *Applied Geophysics*, Oxford & IBH Pub., New Delhi, 860 pp.

Theis, C.V. (1935) Relation between the lowering of the piezometric surface and the rate and duration of discharge of a well using ground-water storage. *Trans. Am. Geoph. Union* Pt. 2, 519–24.

Thornthwaite, C.W. (1948) An approach toward a rational classification of climate. *Geographical Review*, **38**, 55–94.

Thorstenson, D.C., Fisher, D.W. and Croft, M.G. (1979) The geochemistry of Fox Hills–Basal Hill Creek aquifer in southwestern North Dakota and northwestern South Dakota. *Water Resources Res.*, **15**(6), 1479–98.

Thrailkill, J. (1985) Flow in a limestone aquifer as determined from water tracing and water levels in wells. *J. Hydrol.*, **78**, 123–36.

Tickell, S.J. (1994) *Dryland Salinity Hazard Map – The Northern Territory*. Northern Territory of Australia, Darwin, Power and Water Authority, Report 54/94D p. 43.

Todd, D.K. (1980) *Groundwater Hydrology*, 2nd edn. John Wiley, New York, p. 535.

Toth, J. (1966) Groundwater geology, movement, chemistry, and resources near Olds, Alberta. *Res. Council of Alberta, Canada, Geol. Div.*, Bull. 17.

Trescott, P.C. and Larson, S.P. (1977) Solution of three-dimensional groundwater flow equations using the strongly implicit procedure, *J. Hydrol.*, **35**, 49–60.

Trevett, J.W. (1986) *Imaging Radar for Resources Surveys*, Chapman & Hall, London, p. 313.

Trimmer, D., Bonner, B., Heard, H.C. and Duba, A. (1980) Effect of pressure and stress on water transport in intact and fractured gabbro and granite. *J. Geophys. Res.*, **85**(B12), 7059–71.

Tsang, Y.W. and Tsang, C.F. (1987) Channel flow through fractured media. *Water Resources Res.*, **23**, 467–79.

Tsang, Y.W. and Witherspoon, P.A. (1985) Effects of fracture roughness on fluid flow through a single deformable fracture, in *Mem. 17 of IAH*, Tucson, pp. 683–94.

Tyagi, N.K. (1994) Prevention and amelioration of waterlogging and salinity in India: an overview, in *Proceedings Regional Workshop on Environmental Aspects of Groundwater Development*, (ed. D.C. Singhal), Kurukshetra, III–1 to III–10.

Uhl, V.W. (1979) Occurrence of groundwater in the Satpura Hills region of Central India. *J. Hydrol.*, **41**, 124–41.

Uhl, V.W., Joshi, V.G., Alpheus, A. and Sharma, G. (1975) The application of step-drawdown pumping tests for determining well losses in consolidated rock aquifers, in *Hydrology and Water Resources in Arizona and the Southwest*, Vol. 5, pp. 133–46.

Uma, K.O., Egboka, B.C.E. and Onuoha, K.M. (1989) New statistical grain size method for estimating the hydraulic conductivity of sandy aquifers, *J. Hydrol.*, **108**, 343–66.

Underwood, M.R., Peterson, F.L. and Voss, C.I. (1992), Groundwater lens dynamics of atoll islands. *Water Resources Res.*, **28**(11), 2889–902.

UNESCO/IAHS (1967) *Proceedings Dubrovnik Symposium on Hydrogeology of Fractured Rocks* Vol. I and II, UNESCO, Louvain, Belgium.

UNESCO (1972) *Ground Water Studies* (eds R.H. Brown, A.A. Konoplyantsev, J. Ineson and V.S. Kovalevsky), UNESCO, Paris.

UNESCO (1975) Analytical and investigational techniques for fissured and fractured rocks, *in Ground Water Studies* (eds R.H. Brown *et al.*) Studies and Reports in Hydrology (Chapter 14), Supplement 2, UNESCO, Paris.

UNESCO (1977) Planning and design of groundwater level networks, in *Groundwater Studies*, Chapter 7.6, Studies and Reports in Hydrology 7, UNESCO, Paris.

UNESCO (1979) *Water in Crystalline Rocks*, SC-77/WS/71, UNESCO, Paris, p. 344.

UNESCO (1980) *Aquifer Contamination and Protection*, Studies and Reports in Hydrology 30, p. 440.

UNESCO (1983) *International Legend for Hydrogeological Maps*, Tech. Document SC-84/WS/7, UNESCO, Paris, p. 51.

UNESCO (1984a) *Groundwater in Hard Rocks*, Studies and Reports in Hydrology 33, UNESCO, Paris.

UNESCO (1984b) *Guide to the Hydrology of Carbonate Rocks*, Studies and Reports in Hydrology, 41, p. 345.

UNESCO (1987) *Groundwater Problems in Coastal Areas. Studies and Reports in Hydrology* 45, UNESCO, Paris, p. 596.

UNESCO/UNEP (1989) *Geology and the Environment*, Vol. II (ed. G.S. Vartanyan), UNESCO, Paris, 201 pp.

Vacher, H.L. and Ayers, J.F. (1980) Hydrology of small oceanic islands – utility of an estimate of recharge inferred from the chloride concentration of the fresh water lenses. *J. Hydrol.*, **45**, 21–37.

van Golf-Racht, T.D. (1982) *Fundamentals of Fractured Reservoir Engineering*, Elsevier, Amsterdam, p. 710.

Veeger, A., Davis, S., Long, A. and Cuslodio, E. (1989) Groundwater chemistry related to sources of recharge on the island of La Palma, Canary Islands, Spain, *Abstracts Vol 3, 28th IGC*, Washington, pp. 288–89.

Velasquez, E. (1983) Underground water in volcanic rocks. *Proceedings Int. Symposium on Ground Water in Water Resources Planning,*, Vol. I, UNESCO-IAHS, Koblenz, Germany, 201–207.

Vengosh, A. and Rosenthal, E. (1994) Saline groundwater in Israel: its bearing on the water crisis in the country. *J. Hydrol.*, **156**, 389–430.

Vengosh, A., Starinsky, A., Kolodny, Y. and Chivas, A.R. (1994), Boron isotope geochemistry of thermal springs from the northern rift valley, Israel. *J. Hydrol.*, **162**, 155–69.

Venkateswarlu, J., Sen, A.K., Dubey, J.C. *et al.* (1990) *Water 2000 AD: The Scenario for Arid Rajasthan*, Central Arid Zone Res. Inst., Jodhpur, India. p. 49.

Verma, M.N. and Jolly, P.B. (1992) *Hydrogeology of Helen Springs – Explanatory Notes for 1:250 000 Scale Map*, Power and Water Authority, Report 50, Darwin.

Verma, M.N. and Qureshi, H. (1982) Groundwater occurrence in fractured rocks in Darwin Rural Area, N.T., in *AWRC Conference on Groundwater in Fractured Rocks*, Series No. 5, 229–39.

Versey, H.R. and Singh, B.K. (1982) Groundwater in Deccan basalts of the Betwa Basin, India. *J. Hydrol.*, **58**, 279–306.

Verstappen, H. Th. (1983) *Applied Geomorphology*, Elsevier, Amsterdam, p. 437.

Vogel, P. and Giesel, W. (1989) Propagation of dissolved substances in rocks: theoretical consideration of the relationship between systems of parallel fractures and homogeneous aquifers, in *Contaminant Transport in Groundwater* (eds H.E. Kobus and W. Kinzelbach), Balkema, Rotterdam, pp. 275–80.

Voss, C.I. (1984) *A Finite Element Simulation Model for Saturated and Unsaturated Fluid Density Dependent Groundwater Flow with Energy Transport of Chemically Reactive Single Species Solute Transport.* US Geol. Surv. Water Resources Investigation Report 84–4369, p. 410.

Vrba, J. (1988) Impact of nitrogenous fertilizers on groundwater quality in some European countries, in *Geology and the Environment*, Vol. 1 (ed. K.I. Sytchev), UNESCO, Paris, pp. 148–56.

Walder, J. and Nur, A. (1984) Porosity reduction and crustal pore pressure development. *J. Geophy. Res.*, **89** (B13), 11539–48.

Walker, G.P.L. (1973) Length of lava flows. *Philosophical Trans. of the Royal Soc. of London,* 274(B), 107–18.

Walker, W.H. and Walton, W.C. (1961) Groundwater development in three areas of Central Illinois. *Report of Investigation* 41, Illinois State Water Surv., Urbana, p. 43.

Waltham, A.C. (1989) *Ground Subsidence*, Blackie, Glasgow & London, p. 202.

Walton, W.C. (1962) *Selected Analytical Methods for*

Well and Aquifer Evaluation, Bull. 49, Illinois State Water Survey, Urbana, p. 81.

Walton, W.C. (1970) *Groundwater Resource Evaluation*, McGraw-Hill, New York, p. 664.

Walton, W.C. (1979) Review of leaky artesian aquifer test evaluation methods. *Ground Water*, **17**(3), 270–83.

Walton, W.C. and Csallany, S. (1962) *Yields of Deep Sandstone Wells in Northern Illinois*. Illinois State Water Survey, Report of Inv. 43, 47 pp.

Walton, W.C. and Stewart, J.W. (1961) Aquifer tests in the Snake river basalts. *Am. Soc. Civil Engrs., Trans.*, **126**, part III, 612–32.

Wang, J.S.Y. and Narasimhan, T.N. (1993) Unsaturated flow in fractured porous media, in *Flow and Contaminant Transport in Fractured Rock* (eds J. Bear, C.F. Tsang and G. de Marsily), Academic Press, San Diego, pp. 325–95.

Ward, C.H., Giger, W. and McCarty, P.L. (eds) (1985), *Ground Water Quality*, John Wiley, New York, p. 547.

Warren, M.A. (1963) Graphic shortcuts in applying the nonequilibrium formula to ground water problems, in *Shortcuts and Special Problems in Aquifer Tests* (eds W.C. Walton *et al.*), US Geol. Surv. Water Supply Paper 1545-C, 14–32.

Warwick, D., Hartopp, P.G. and Viljoen, R.P. (1979) Application of the thermal infrared linescanning technique to engineering mapping in South Africa. *Q. J. Engeg Geol.*, **12**, 159–79.

Water Res. Board (1972) *The Hydrogeology of the London Basin*, Water Res. Board, Reading, UK, p. 139.

Waters, P., Greenbaum, D., Smart, P.L. and Osmaston, H. (1990) Applications of remote sensing to groundwater hydrology. *Remote Sensing Reviews*, **4**(2), 223–64.

Watson, I. and Burnett, A.D. (1993) *Hydrology – An Environmental Approach*, Buchanan Books, Cambridge, p. 702.

Weeks, E.P. (1977) Aquifer tests – the state of art in hydrology, in *Proceedings Invitational Well-Testing Symposium*, Lawrence Berkeley Lab., Berkeley, California, pp. 4–26.

Weeks, E.P. (1987) Effect of topography on gas flow in unsaturated fractured rock: concepts and observations. *Geophysical Monograph 42, Am. Geophys. Union*, 165–70.

Wels, C. and Smith, L. (1994) Retardation of sorbing solutes in fractured media. *Water Resources Res.*, **30**(9), 2547–63.

Wenzel, L.K. (1942), *Methods for Determining Permeability of Water Bearing Materials*. US Geol. Surv. Water Supply Paper 887, p. 192.

Wesslen, A., Gustafson, G. and Maripuu, P. (1977) Groundwater and storage in rock caverns, pumping tests as an investigation method, *Proceedings Ist Int. Symposium Rock Store*, Stockholm, pp. 359–66.

White, D.E. (1957) Magmatic, connate, and metamorphic waters. *Bull. Geol. Soc. Am.*, **68**, 659–82.

White, D.E., Hem, J.D. and Waring, G.A. (1963) Chemical composition of sub-surface waters, in *Data of Geochemistry*, 6th edn (ed. M. Fleischer), US Geol. Surv. Prof. Paper 440F, 67 pp.

White, R.B. and Gainer, R.B. (1985) Control of groundwater contamination at an active uranium mill. *Ground Water Monitoring Review*, **5**(2), 75-82.

WHO (1984) *Guidelines for Drinking Water Quality*, Vols 1, 2 and 3, WHO, Geneva.

Wilcox, L.V. (1955) *Classification and Use of Irrigation Waters*. US Dept. Agri. Circ. 969, Washington, D.C., p. 19.

Wilhelm, H., Rabbel, W., Luschen, E., Li, Y.D. and Bried, M. (1994) Hydrological aspects of geophysical measurements in crystalline rocks of the Black Forest, *J. Hydrol.*, **157**, 325–47.

Wilkins, A., Subbarao, K.V., Ingram, G. and Walsh, J.N., (1994) Weathering regimes within the Deccan Basalts, in *Volcanism* (ed. K.V. Subbarao), Wiley, New Delhi, 217–31.

Wilkinson, W.B. and Brassington, F.C. (1991) Rising groundwater levels – an international problem, in *Applied Groundwater Hydrology* (eds R.A. Downing and W.B. Wilkinson), Oxford University Press, Oxford, 35–53.

Williamson, K.H., Fernandez, J.C., Aunzo, Z.P., Sussman, D. and Prabowo, H.T. (1996) Large scale geothermal development in Asia – UNDCAL's experience. *Abstracts, 30th IGC*, Vol 1. p. 384, Beijing.

Williamson, W.H. (1982) The use of hydraulic techniques to improve the yield of bores in fractured rocks, *Paper of the Groundwater in Fractured Rock Conference,* AWRC conference series No.5, Canberra, pp. 273–83.

Wilson, C.R. and Witherspoon, P.A. (1985) Steady state flow in rigid networks of fractures. *Water Resources Res.*, **10**(2), 328–35.

Wilt, M.J. *et al.* (1995), Crosswell electromagnetic tomography: system design considerations and field results. *Geophysics,* **60**, 871–85.

Witherspoon, P.A., Wang, J.S.Y., Iwai, K. and Gale, G.E. (1980) Validity of cubic law for fluid flow in deformable rock fracture. *Water Resources Res.*,**16**(6), 1016–24.

Wolf, P.R. (1983) *Elements of Photogrammetry*, 2nd edn, McGraw-Hill, New York.

Woolley, D.R. (1982) Depth-yield relationships of bores in fractured rocks in New South Wales, in *AWRC Conference on Groundwater in Fractured Rock Conference Ser.* No. 5, Canberra, pp. 283–92.

Worthington, M.H. (1984) An introduction to geophysical tomography. *First Break*, **2**, 20–26.

Wright, E.P. (1984) Drilling for groundwater in the Pacific region, in *Water Resources of Small Islands*, Tech. Proceedings (Part 2) of the Regional Workshop on Water Resources of Small Islands, CSC, London, pp. 525–29.

Wright, E.P. (1985) Groundwater in the Third World, in *Hydrogeology in the Service of Man*, Mem. IAH, Cambridge, UK, 53–64.

Wright, E.P. (1992) The hydrogeology of crystalline basement aquifers in Africa, in *The Hydrogeology of Crystalline Basement Aquifers in Africa* (eds E.P. Wright and W.G. Burgess), Geol. Soc. Special Publ. No. 66, The Geological Society, London, pp. 1–27.

Wright, E.P. and Burgess, W.G. (eds) (1992) *The Hydrogeology of Crystalline Basement Aquifers in Africa,* Geol. Soc. Special Publ. No. 66, The Geological Society, London, p. 264.

Xuanjiang, H. (1994) The mutual relationship between urban development, surface settlement and its counter measure. *Proceedings – Int. Conference on Geoscience in Urban Development (LANDPLAN IV),* Beijing, China, pp. 271–76. China Ocean Press, Beijing.

Yashpal, Sahai, B., Sood, R.K. *et al.* (1980) Remote sensing of the lost Saraswati river, *Proceedings Ind. Acad. Sc. (Earth Planet Sc.)*, Vol. 89, 317–31.

Yin, Zhi-Yong and Brook, G.A. (1992) The topographic approach to locating high-yielding wells in crystalline rocks: does it work? *Ground Water*, **30**, 96–102.

Yin, Z., Zhenmin, H., Xu, H. and Xu, J. (1996) *Field Trip Guide* T.337, 30th IGC, Beijing p. 30.

Yuan, D. (ed) (1996) *Karst Processes and the Carbon Cycle.* Newsletter, Project 379, Institute of Karst Geology, Guilin, China, p. 188.

Yuan, D., Drogue, C., Dai, A. *et al.* (1990) Hydrology of the karst aquifer at the experimental site of Guilin in Southern China. *J. Hydrol.*, **115**(2), 285–96.

Yuan, D., Li, B. and Liu, Z. (1996) Karst in China. *Episodes*, **18**(1&2), 62–65.

Zaporozec, A. (1972) Graphical interpretation of water quality data. *Ground Water*, **10**(2), 32–43.

Zeil, P., Volk, P. and Saradeth, S. (1991) Geophysical methods for lineament studies in groundwater exploration. A case history from SE Botswana. *Geoexploration*, **27**, 165–77.

Zeizel, A.J., Walton, W.C. Prickett, T.A. and Sasman, R.T. (1962) *Groundwater Resources of Dupage County*, Illinois, Illinois State Water Survey and Geol. Surv. Coop. Ground Water Report No.2.

Zektser, I.S. and Dzhamalov, R.G. (1988) *Role of Groundwater in the Hydrological Cycle and in Continental Water Balance.* IHP III Project, 2.3, UNESCO, Paris, p. 133.

Zektser, I.S. and Loaiciga, H.A. (1993) Groundwater fluxes in the global hydrological cycle: past, present and future. *J. Hydrol.*, **144**(2), 405–27.

Zhao, J. and Brown, E.T. (1992) Thermal cracking induced by water flow through joints in heated granite. *Int. J. Rock Mech. Min. Soc. and Geomech. Abstr.* **29**(1), 77–82.

Zheng, C. (1990) *A Modular Three-Dimensional Transport Model for Simulation of Advection, Dispersion and Chemical Reactions of Contaminants in Groundwater Systems.*, Environmental Protection Agency, Ada, Oklahoma, USA, R.S. Ken Environmental Research Laboratory.

Zhengzhou, Y., Zhenmin, H., Shimin, X. and Junxiang, X. (1996) *Field Trip Guide, T337, Experiments, Exploitation and Utilization of Groundwater in the North China Plain.* 30th IGC, Geol. Publ. House, Beijing, p. 29.

Zhiming, W., Shokang, H., Feng, X., Sen, L. and Jiahe, Z. (1996) Isotope-geochemical studies of geothermal water in Chenrian uranium mine, in *Abstracts of Uranium Geology Theses*, Bureau of Geology, China National Nuclear Corporation, Beijing.

Zhou, Q., Becker, A. and Morrison H.F. (1993) Audiofrequency electromagnetic tomography in 2-D. *Geophysics*, **58**, 378–90.

Zhou, W. and Apted, M.J. (1996) Flow and transport simulation in fractured rock with low permeability strata. *Abstracts 30th IGC,* Vol.3, p. 267. Beijing.

Zohdy, A.A.R., Eaton, G.P. and Mabey, D.R. (1974) *Application of Surface Geophysics to Groundwater Investigations*, US Geological Survey, Chapter 1, 116 pp.

Index